WORLD SCIENTIFIC HANDBOOK OF EXPERIMENTAL RESULTS ON HIGH SPEED PENETRATION INTO METALS, CONCRETE AND SOILS

WORLD SCIENTIFIC HANDBOOK OF EXPERIMENTAL RESULTS ON HIGH SPEED PENETRATION INTO METALS, CONCRETE AND SOILS

Gabi Ben-Dor
Anatoly Dubinsky
Tov Elperin

Ben-Gurion University of the Negev, Israel

NEW JERSEY · LONDON · SINGAPORE · BEIJING · SHANGHAI · HONG KONG · TAIPEI · CHENNAI · TOKYO

Published by

World Scientific Publishing Co. Pte. Ltd.

5 Toh Tuck Link, Singapore 596224

USA office: 27 Warren Street, Suite 401-402, Hackensack, NJ 07601

UK office: 57 Shelton Street, Covent Garden, London WC2H 9HE

Library of Congress Cataloging-in-Publication Data

Names: Ben-Dor, Gabi, 1950–   author. | Dubinsky, Anatoly, author. |
    Elperin, Tov, author. | World Scientific (Firm)
Title: World Scientific handbook of experimental results on high speed
    penetration into metals, concrete and soils / [compiled by] Gabi Ben-Dor
    (University of the Negev, Israel), Anatoly Dubinsky (University of the
    Negev, Israel) & Tov Elperin (University of the Negev, Israel).
Other titles: Handbook of experimental results on high speed penetration into
    metals, concrete and soils
Description: New Jersey : World Scientific, 2016.
Identifiers: LCCN 2016014236 | ISBN 9789813109346 (hc : alk. paper)
Subjects: LCSH: Penetration mechanics--Research. | Materials--Experiments. |
    Materials--Testing. | Projectiles--Speed.
Classification: LCC TA354.5 .W67 2016 | DDC 620.1/126--dc23
LC record available at http://lccn.loc.gov/2016014236

British Library Cataloguing-in-Publication Data
A catalogue record for this book is available from the British Library.

Desk Editor: V. Vishnu Mohan

Typeset by Stallion Press
Email: enquiries@stallionpress.com

Printed in Singapore

Preface

This handbook contains a vast compendium of systematized data regarding experimental investigations of high speed penetration. The handbook can be considered as a natural addition to the previous monographs of the authors ([Ben-Dor *et al.*, 2006], [Ben-Dor *et al.*, 2013]) devoted to engineering models of penetration since it provides the researches in the field with extensive data for the analysis of existing models and developing new penetration models.

The handbook includes data on about 14,000 experiments on penetration into metals, concrete and reinforced concrete, and geological media which were published in the open literature (journal papers, reports, conference proceedings) during the last 70 years. The bibliography contains 468 items. We express our gratitude to the authors of studies containing the results of their experiments which are referenced in this handbook.

The handbook contains Introduction, Table of Contents, Six Chapters, List of References and Author Index. Chapters 1–3 play an auxiliary role while Chapters 4-6 contain information about the experiments.

Chapter 1 introduces the general principles of selecting and presenting the material. This chapter is in fact a continuation of the present preface and we strongly suggest to the readers to familiarize themselves with this chapters before turning their attentions to the experiments.

Chapter 2 describes the notations used throughout this handbook. In addition to a traditional description of the parameters, we present the notations for describing the structure of multilayer shields and typical structure of reinforcements using meshes. If certain notation in the main text of the handbook is not clear (including abbreviations) we recommend referring to this chapter.

Chapter 3 includes drawings of strikers that are referred to in Chapters 4–6; these drawings are simplified and secondary details are not shown. Some typical drawings are supplemented with tables containing geometrical dimensions. It must be noted that the handbook also includes experimental results obtained with the strikers that are not described in Chapter 3.

The structure of Chapters 4–6 is as follows. The first section of each chapter includes information helping to navigate through the experiments described in this chapter while all the rest sections contain descriptions of experiments. The titles of the sections are ordered by the last name of the author(s) of corresponding publication.

The authors intend this handbook to serve as a compass in a large sea of experimental studies on high-speed penetration into metals, concrete and geological media. They hope that this handbook will be indispensable for experimentalists in penetration mechanics; developers of new analytical and numerical penetration models (for estimating the adequacy of the models); developers of new types of armor and weapons. The handbook can be also useful for graduate students in engineering mechanics for broadening their scientific horizons.

The authors are indebted to Mrs. V. Orlova for her help in technical editing of the manuscript and preparation of the camera-ready version.

G. Ben-Dor
A. Dubinsky
T. Elperin

Contents

Chapter 1

Introduction

In this handbook we collected experiments on high-velocity penetration into various types of shields. High-speed penetration is accompanied mainly by local interaction of a striker with a shield and corresponds approximately to the range of impact velocities from one hundred fifty up to one thousand and five hundred meters per second. However, we analyzed also a certain number of experiments with relatively small impact velocities when a striker interacts with a whole plate and hypervelocity penetration when penetration proceeds under conditions of a hydrodynamic flow regime. The systematic information on experiments in hydrodynamic penetration regime which were performed before 1992 can be found in [Anderson *et al.*, 1992]

The main goal of this handbook is to systemize experimental data on basic integral parameters which characterize the initial and the final states of the striker and of the shield, without analyzing penetration process. Consequently, in the handbook we do not present all information about the experiment that was available in the source. In addition to the parameters which characterize mechanical and geometric properties of the striker and the shield we present also (if available) the following parameters: striking (impact) and residual velocities of projectile or depth of penetration; changes of mass and size of projectile; angles that determine the initial and residual position of the projectile (including a case of ricochet); ballistic limit velocity; basic characteristics of plug and deformation of the shield. We do not include detailed information on (i) deformation of the striker and the shield; (ii) chemical composition of alloys or concrete; (iii) material processing (e.g. annealing etc.). Data arrangement in the handbook is almost always

1

different from the source because of data unification requirements and editorial considerations.

In order to avoid cumbersome notations we adopted the following system of numbering figures, formulas and tables. For figures and formulas we used a triple positional numeration (chapter number, section number, formula/figure number separated by a dot). For tables a triple positional numeration is used only in Chapters 2-3 while in the rest of the chapters tables are labeled by their ordinal numbers in the section.

Chapters 4-6 are organized as follows. The first subsection of each chapter includes indices to facilitate search for experimental data that satisfy certain requirements (shape of striker, thickness of shield, etc.). Other subsections include information about the experiments whereby each subsection is devoted to one original publication. Short description of this publication appears in the title of the subsection (as a rule family names of authors and year of publication). All subsections starting from the second appear in alphabetical order by their description in the References.

In order to provide maximum (in reasonable limits) concentration of the material on the pages of the book and to meet a number of editorial requirements, information about strikers and shields is presented in a textual form rather than in tables. Tables are designed to be read from left-to-right in each line.

The numerical values of all parameters are given in SI units except for the hardness that is given in a traditional form. Symbol # (commonly after a number of a table) indicates that data in the source were presented in a graphical form. In reference to the table the symbol # (if it is present) is ignored.

Data and experimental results are presented in a reasonably unified form (see Chapters 2 and 3) so that most notations have the same meaning through the whole handbook. Only a few notations are introduced and valid inside the section. In case when a figure from Chapter 3, that illustrates the shape of a striker, is referred in the section, some special notations used in this section can be found in the figure.

Material characteristics are specified as comprehensively as in the source (with a few exceptions). A reader may attempt to recover the missing data from other sources using the specification of the material and taking a risk of possible error.

In case when the same experimental data appear in different sources, they are presented only in a section dedicated to one of these sources with references to other sources.

Sketches of the strikers in Chapter 3 are not their exact drawings and do not include all fine details of their shapes. In particular, the sketches do not include internal cavities when their effect is not a goal of the investigation. Clearly, classification of the shape of the strikers is a matter of convention, e.g. more bullets can be categorized as having ogive shape while neglecting peculiarities of their structure.

In the handbook we present only information about the name and the shape of the striker which appears in the source. Reference to the particular class of striker's shape is given only when there are good reasons to believe that the authors of the source mean exactly this particular shape (probably with small deviations). In other cases we refer only to the specification of a striker (this is related, mainly, with bullets).

When caliber of a projectile (usually in its label) in the handbook is given in inches zero is added in the integral part and notation of inch (double prime) in some cases, e.g. 0.50 or 0.50″ instead of .50.

When the striking angle in oblique impact of projectile, θ_s, is not indicated, the impact is considered to be normal ($\theta_s = 0°$). Small deviations of the angle θ_s are ignored.

When a source presents a numerical value of a parameter, we usually retained the same number of digits after the decimal point as in the source and did not round off the parameter. In the case when the experimental results in the source are presented in graphical form, one must take into account the uncertainties related with plotting a symbol that denotes a certain experimental value. It is not always clear which part of the symbol corresponds to the measured parameter, especially in the older studies where experimental points were plotted by hand. The problem remains even for high precision retrieval of data points from the plot that can be easily achieved.

Short description of original publications and references to published sources are given in a traditional form. In case of a single author the last name of the author and the year of publication are indicated; in case of two authors their last names and the year of publication are indicated; in case of three or more authors only the last name of the first author with adding et al. and the year of publication are indicated. When the source does not include the names of the authors, abbreviated title and the year of publication are indicated.

Author index includes names of all authors and editors presented in References. In the text of the handbook, publications are cites as they are indicated in square brackets in References. Consequently, if the coauthor does not appear at the first or at the second position in the author's list, the reference to his/her publication is presented in the Author index for this coauthor but the last name of this coauthor cannot be seen at the corresponding pages of Chapters 1-6.

Chapter 2

Nomenclature and notations

2.1 Parameters

Unified compendium of notations for parameters that is valid through the whole book is presented in Table 2.1.1. Only in very few instances when notations from Table 2.1.1 have different meaning they are explained separately.

Table 2.1.1. Basic nomenclature.

Designation	Description
a	Coefficient in Eq. (2.1.1)
A	Areal density
A_{red}	Reduction in area
AA	Aluminum Alloy
AISI	American Iron and Steel Institute
AP	Armor Piercing
B	Bulk modulus
BLV	Ballistic Limit Velocity
c_l	Length of rectangular shield
$\tilde{c}_l$	Free span length of rectangular shield
c_w	Width of rectangular shield
$\tilde{c}_w$	Free span width of rectangular shield
$c_{l,w}$	Length and width of rectangular shield if they are the same
$\tilde{c}_{l,w}$	Free span length and width of rectangular shield if they are the same
Charpy	Charpy V-notched impact test
CRH	Caliber Radius Head of ogive-nose projectile (Fig. 3.3.1)
d^{fib}	Diameter of fiber
$d^{\bullet\bullet\bullet}_{\bullet\bullet\bullet}$	Diameter of rebar (superscript $\neq fib$), for details, see Section 2.4
D	(Maximum) diameter of projectile
D^{core}	(Maximum) diameter of the core of projectile

Designation	Description
$\hat{D}$	Diameter of predrilled hole in shield
D_{fr}^{pl}	Plug diameter on the front size of shield
D_{re}^{pl}	Plug diameter on the rear size of shield
$D_{\min}^{pl}$	Minimum diameter of plug
$D_{\max}^{pl}$	Maximum diameter of plug
D_r	Residual maximum diameter of projectile
D^{sh}	Diameter of shield
$\widetilde{D}^{sh}$	Free span diameter of shield
DOP	Depth of Penetration
E	Young's modulus (tensile modulus, elastic modulus)
ECC	Engineered Cementitious Composite
f_c'	Unconfined compressive strength
f_r	Modulus of rupture (flexural strength)
f_{cr}	Cracking strength
f_{sl}	Slump
FFV	Försvarets Fabriksverk
FMJ	Full Metal Jacket
FRC	Fiber Reinforced Concrete
FSP	Fragment simulating projectile
FT	Fracture Toughness
$g_{\dots}^{\dots}$	Distance between adjacent rebars; for details see Section 2.4
G	Shear modulus (modulus of rigidity)
h^{def}	Maximum deflection (bulging) of shield
h^{lip}	Average lip height at exterior surface of shield
HB	Brinell hardness number
HM	Meyer hardness
HRA	Rockwell A Scale Hardness
HRB	Rockwell B Scale Hardness
HRC	Rockwell C Scale Hardness
HRH	Rockwell H Scale Hardness
HS	Horizontal Shot
HT	Heat Treating (heat treated, heat treatment)
HV	Vickers Hardness
K_{Ic}	Fracture toughness
k_S	Penetrability of shield, or "S-number"*
L	Initial length of projectile
L^{core}	Initial length of the core of projectile

Designation	Description
L_{cyl}	Initial length of cylindrical part of projectile ($L_{cyl} = L$ for cylindrical projectiles)
L_{nose}	Initial length of projectile nose
L^{fib}	Length of fiber in FRC
L^{pl}	Length of plug
L_r	Residual length of projectile
MRC	Mesh Reinforced Concrete
m	Striking (initial) mass of projectile
m^{core}	Striking (initial) mass of projectile core
m_r	Residual mass of projectile
m^{fr}	Mass of fragments (debris)
m^{pl}	Mass of plug
m^{sh}	Initial mass of shield
N	Number of petals
N_{perf}	Number of perforated layers
p	Coefficient in Eq. (2.1.1)
P	"Perpendicular" depth of penetration: maximum penetration measured from the plane of the original shield surface
$\widetilde{P}$	Depth of penetration measured along the channel path to the location of projectile (length of trajectory)
P^{pl}	Plug displacement
P^{cr}	Front crater depth in concrete shield
$\widetilde{P}_x$	x-coordinate (Fig. 2.2.1a-b) of projectile tip corresponding to DOP
$\widetilde{P}'_x$	x-coordinate (Fig. 2.2.1a-b) of projectile tail corresponding to DOP
$\widetilde{P}_y$	y-coordinate (Fig. 2.2.1a-b) of projectile tip corresponding to DOP
$\widetilde{P}'_y$	y-coordinate (Fig. 2.2.1a-b) of projectile tail corresponding to DOP
PC	Plain concrete
PE	Polyethylene
q^{agg}	Size of coarse aggregate in concrete
q^{agg}_{max}	Maximum size of coarse aggregate in concrete
q^{grain}	Size of sand grains
RQD	Rock Quality Designation
RC	Reinforced Concrete
RFHA	Rolled Face Hardened Armor
RHA	Rolled Homogeneous Armor
SAE	Society of Automotive Engineers/Materials Classification System

Designation	Description
SDT	Shield Damage Type
STS	Special Treatment Steel
$t^{\bullet\bullet\bullet}$	Distance between mesh and front or rear surface of shield; for details see Section 2.4
T	Thickness of monolithic shield (default) or of layer (when implied in text)
T_i	Thickness of ith (in the direction of penetration) layer
TA	Titanium Alloy
T_{lin}^{F}	Liner thickness at front face of shield; for details see Section 2.4
T_{lin}^{R}	Liner thickness at rear face of shield; for details see Section 2.4
T^{pl}	Plug thickness
T_{room}	Room temperature
T^{sh}	Initial temperature of shield
T_{sum}	Sum of thicknesses of plates in shield
v_{bl}	Ballistic Limit Velocity (BLV)
v^{pl}	Residual velocity of plug
v_r	Residual velocity of projectile
$\bar{v}_r$	$= v_r / v_{bl}$
v_s	Striking (initial, impact) velocity of projectile
$\bar{v}_s$	$= v_s / v_{bl}$
$\tilde{v}_s$	$= v_s / v^{sound}$
v^{sound}	Velocity of sound
VAR	Vacuum-arc-remelted
VS	Vertical shot
w_{perf}	Critical perforation energy
w_s	Striking (initial, impact) energy of projectile
w_r	Residual energy of projectile
β	Semi-vertex angle of cone-nose projectile
γ	Density or bulk density of shield (if not indicated otherwise)
δ	Elongation
δ_v	$= (v_s - v_r)/v_s$
ΔD	Change of projectile diameter after impact
ΔL	Change of projectile length after impact
Δm	Mass loss of projectile after impact
Δm^{sh}	Mass loss of shield
Δv	$= v_s - v_r$ (projectile velocity drop)

Designation	Description
ε_f	Fracture strain
ε_u	Tensile strain
ε_y	Yield strain
θ_r	Residual angle of oblique projectile; for details see Section 2.2
θ_{ric}	Minimum striking angle of oblique projectile when ricochet occurs; for details see Section 2.2
θ_s	Striking (initial, impact) angle of oblique projectile; for details see Section 2.2
ζ^{hole}	Ratio of diameter of residual hole at exterior surface of shield to diameter of projectile
η_{por}	Porosity
η_{wat}^{sh}	Water content
η_{FM}	Fineness modulus
μ	Mass of reinforcement (MRC or FRC) per $1\,\mathrm{m}^3$ of mixture**
ν	Poisson's ratio
ρ	Radius of circle of ogive
$\rho_{\cdots}^{\cdots}$	Reinforcement quantity (subscript $\neq$m and $\neq$V); for details see Section 2.4
ρ_m	Ratio of mass of reinforcement (MRC or FRC) to mass of mixture**
$\widetilde{\rho}_m$	Ratio of mass of reinforcement (MRC or FRC) to mass of cement**
ρ_V	Ratio of volume of reinforcement (MRC or FRC) to volume of mixture
$\widetilde{\rho}_V$	Ratio of volume of reinforcement (MRC or FRC) to volume of cement**
σ_f	Fracture strength (breaking strength)
σ_p	Proportional limit
σ_{ss}	Ultimate shear strength
σ_{tsp}	Splitting tensile strength
σ_u	Ultimate tensile strength (tensile strength\stress, ultimate strength)
σ_y	Yield strength
$\widetilde{\sigma}_y$	0.2% offset yield strength
ψ	Caliber radius head of ogive-nose projectile
ω	Deflection angle of shield
ω^{pl}	Plug angular velocity
Subscripts	
r	Residual
s	Striking
Superscripts	

Designation	Description
pl	Plug
sh	Shield
Symbols	
#	Indicates that data were taken from figure

*For details, see [Young, 1969], [Young, 1972], [Young, 1997].

**Type of reinforcement is described in the text of corresponding section.

Lambert and Jonas [1976] suggested the following relationship between the impact, the residual and the ballistic limit velocities for processing the results of ballistic tests:

$$v_r = \begin{cases} 0 & \text{if } 0 \le v_s \le v_{bl}, \\ a[(v_s)^p - (v_{bl})^p]^{1/p} & \text{if } v_s > v_{bl}, \end{cases} \qquad (2.1.1)$$

where $0 \le a \le 1$, $p > 1$, v_{bl} are constants determined from the experiments (in practical applications the restrictions on the values of the parameters are ignored).

Energy approach [Recht and Ipson, 1963] yields $p = 2$, and in a case of constant mass of a shield (absence of plug) $a = 1$.

Additional information regarding this approximation can be found in [Ben-Dor *et al.*, 2001], [Ben-Dor *et al.*, 2002], [Ben-Dor *et al.*, 2006], [Ben-Dor *et al.*, 2013].

In the present handbook, Eq. (2.1.1) is considered as an approximate formula and the values of coefficients are indicated if they are presented in the original study.

2.2 Oblique impact

Striking (initial, impact) θ_s and residual θ_r angles of oblique impact of projectile are defined as angles between the direction of projectile motion at corresponding stage of penetration and the negative direction of *y*-axis (Fig. 2.2.1a-b). It is assumed that *x*-axis, *y*-axis, vector of impact and residual velocities are located in a plane that is perpendicular to the non-deformed initial front surface of the shield.

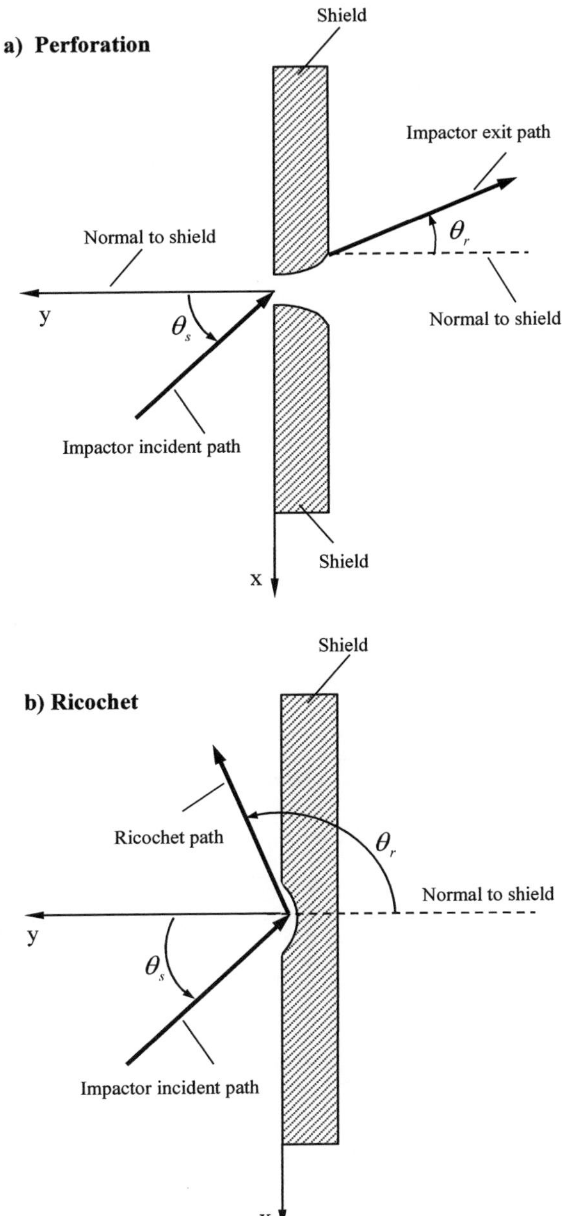

Fig. 2.2.1a-b. Explanations to notations related to oblique impact.

Consequently, normal impact, perforation and ricochet are associated with $\theta_s = 0$, $|\theta_r| \leq 90°$ and $\theta_r > 90°$, respectively. In the case of incomplete penetration, arbitrary value of θ_r is possible.

2.3 Layered shields

In this handbook we employ a generalized version of unified notations for describing the structure of multi-layer shields which have been suggested in [Ben-Dor *et al.*, 2012]. In the general case, a shield consists of plates made from different materials and/or having different thicknesses, and the adjacent plates are either in contact or are separated by an air gap. The notation for the structure of a shield in the general case reads:

$$[(M_1)\,T_1]\Theta_1 \cdots [(M_i)\,T_i]\Theta_i \cdots [(M_n)\,T_n], \tag{2.3.1}$$

where i is ordinal number of a plate in a shield in the direction of penetration; M_i identifies material of the ith plate; T_i is thickness of ith plate in mm;

$$\Theta_i = \begin{cases} + & \text{if } \ i\text{th and}\,(i+1)\text{th plates are in contact,} \\ |A_i| & \text{if } \ \text{there is an air gap between } i\text{th and}\,(i+1)\text{th plates} \end{cases} \tag{2.3.2}$$

and A_i is the width of air gap between ith and $(i+1)$th plates in mm.

If air gaps are present between all plates and they have the same width A, the unifies notation for a shield reads:

$$\{[(M_1)\,T_1] + \cdots + [(M_i)\,T_i] + \cdots + [(M_n)\,T_n]\} \setminus A . \tag{2.3.3}$$

If all plates are made from the same material M then the following two notations

$$(M)\{[T_1]\Theta_1 \cdots [T_i]\Theta_i \cdots [T_n]\}, \tag{2.3.4}$$

$$(M)\{[T_1] + \cdots + [T_i] + \cdots + [T_n]\} \setminus A \tag{2.3.5}$$

are equivalent and correspond to the notations given by Eqs. (2.3.1) and (2.3.3). In the latter case the label of material can be omitted if material is described in the text.

Notation for a shield structure can be simplified when a shield consists of the same sub-structures, e.g. identical plates. In the following this simplification is illustrated by several examples of equivalent notations:

$$(M)\{[T]+[T]+[T]\} \sim 3\times(M)[T], \qquad (2.3.6)$$

$$[T_a]+[T_a]\backslash A\backslash[T_b] \sim 2\times[T_a]\backslash A\backslash[T_b]. \qquad (2.3.7)$$

Representative situations are also illustrated in Fig. 2.3.1.

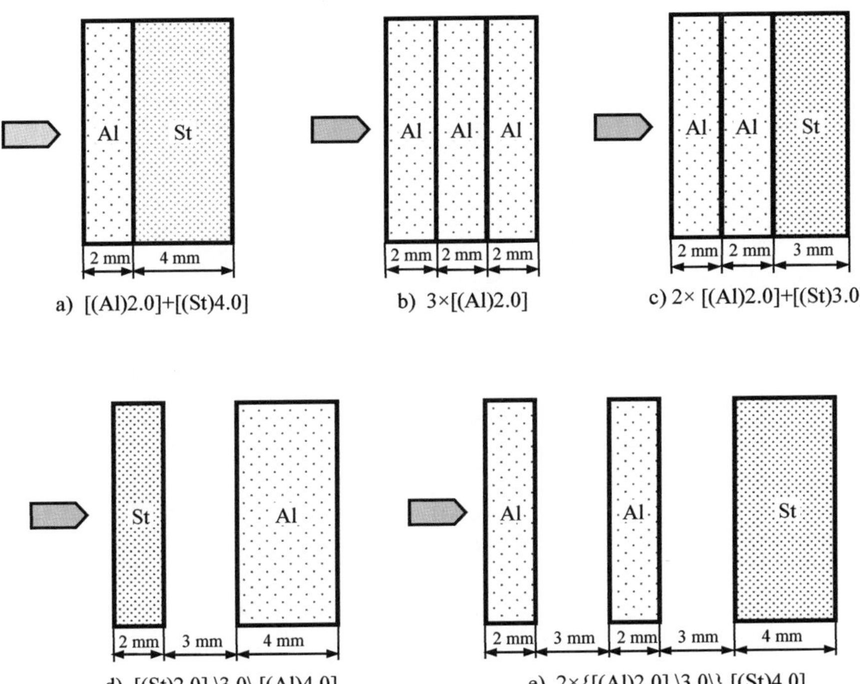

a) [(Al)2.0]+[(St)4.0]

b) 3×[(Al)2.0]

c) 2× [(Al)2.0]+[(St)3.0]

d) [(St)2.0] \3.0\ [(Al)4.0]

e) 2×{[(Al)2.0] \3.0\} [(St)4.0]

Fig. 2.3.1. Examples of notations for shield structure.

2.4 Mesh reinforcement of concrete

Standard description of mesh reinforced concrete shields based on [Barr, 1990] is presented in Fig. 2.4.1.

Monolithic shield is a concrete rectangular parallelepiped (T is its thickness) with one or two built-in meshes made from metal or other material. In some cases the frontal and/or the rear faces of a parallelepiped are covered with steel liners.

Reinforcing meshes are located near the frontal and/or the rear faces of a shield. Parameters associated with the frontal and/or the rear faces are denoted using superscripts F and R, correspondingly. In particular, t^F and t^R are the distances from a mesh to the nearest face of a shield; T_{lin}^F and T_{lin}^R are the thicknesses of a corresponding liners.

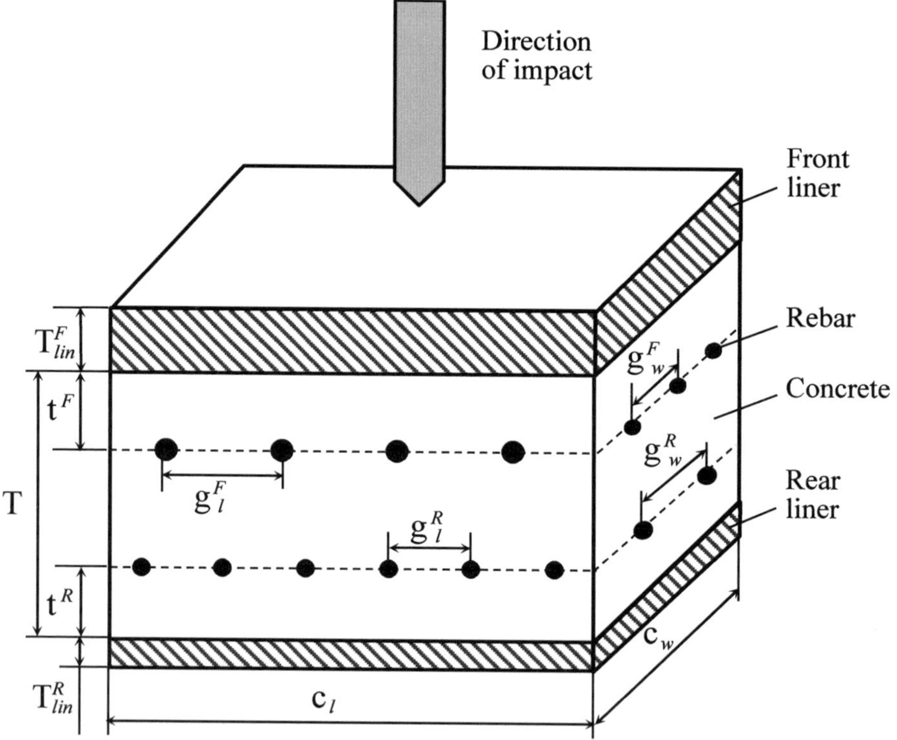

Fig. 2.4.1. Mesh reinforced concrete shield.

Rectangular mesh is made from rebars (reinforcing bars). Mesh structure is described by projections of rebars on two side faces of a parallelepiped which are called *l-way* and *w-way* (see Fig. 2.4.1); their widths are c_l and c_w, correspondingly, where the lower index indicates the selected face of a parallelepiped.

Apart from the parameters indicated above, a shield is described by the following characteristics: $d_l^F, d_l^R, d_w^F, d_w^R$ — rebars diameters (if they have a circular cross-section); $s_l^F, s_l^R, s_w^F, s_w^R$ — cross-sectional areas of rebars; $g_l^F, g_l^R, g_w^F, g_w^R$ — distances between rebars; $n_l^F, n_l^R, n_w^F, n_w^R$ — number of rebars. Parameters $\rho_l^F, \rho_l^R, \rho_w^F, \rho_w^R$ (in %) [Barr, 1990)] are used for characterizing the reinforcement (at *each way and each face*):

$$\rho_\lambda^\Theta = \frac{n_\lambda^\Theta s_\lambda^\Theta}{c_\lambda T} \cdot 100 \approx \frac{s_\lambda^\Theta}{g_\lambda T} \cdot 100 ; \quad \Theta = F, R ; \quad \lambda = l, w. \tag{2.4.1}$$

If a certain parameter of a mesh is the same in both directions, then lower indices used in notation for this parameter, l and w, are comma-separated (e.g. $c_{l,w} = 1.5m$ instead of $c_l = c_w = 1.5m$). Similar simplification is used for parameters indicating a face of a shield in the upper index or for parameters characterized by combination of lower and upper indices (e.g. $d_{l,w}^{F,R} = 3mm$ instead of $d_l^F = d_w^F = d_l^R = d_w^R = 3mm$).

In cases when reinforcing cannot be described by the above scheme, a less formal description is used. When mesh is located between the frontal and rear meshes or mesh location is not indicated or is determined separately, notations for mesh parameters do not include upper indices.

Along with mesh reinforcing increasingly often fiber reinforcing is employed when concrete contains fibers made from various materials and having different shapes.

Chapter 3

Some shapes of projectiles used in experiments

3.1 Cylindrical projectiles

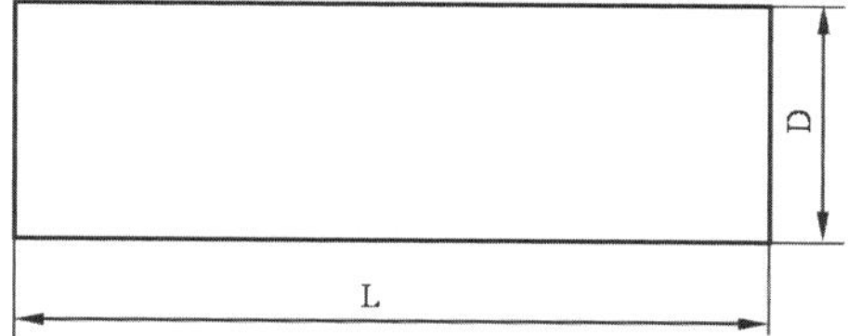

Fig. 3.1.1. Cylindrical projectile.

a)

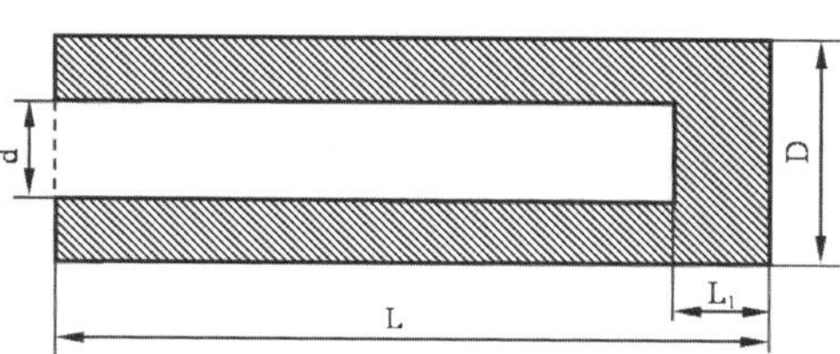

b)

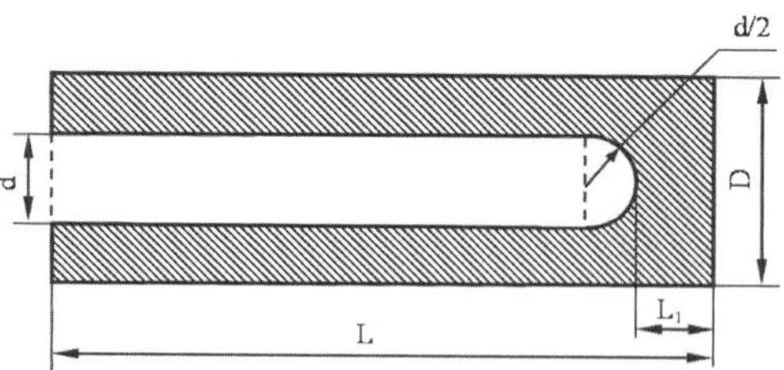

c)

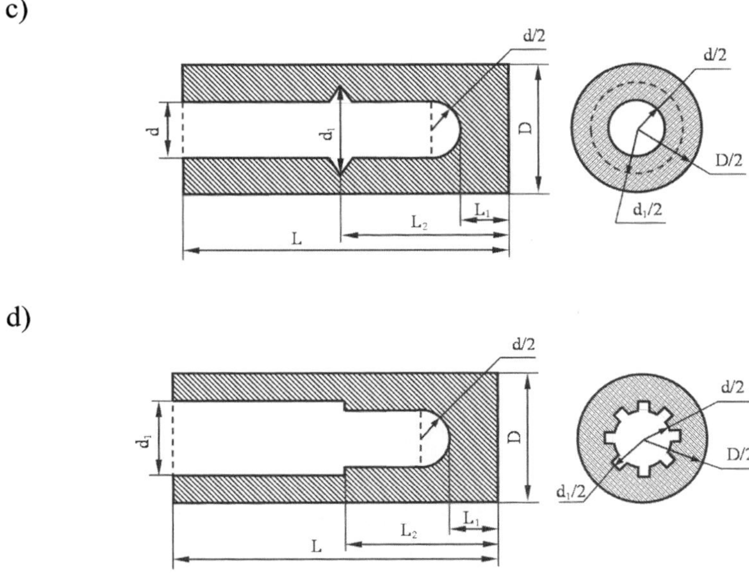

Fig. 3.1.2a-d. Cylindrical projectiles with inner cavity.

3.2 Cone-nose projectiles

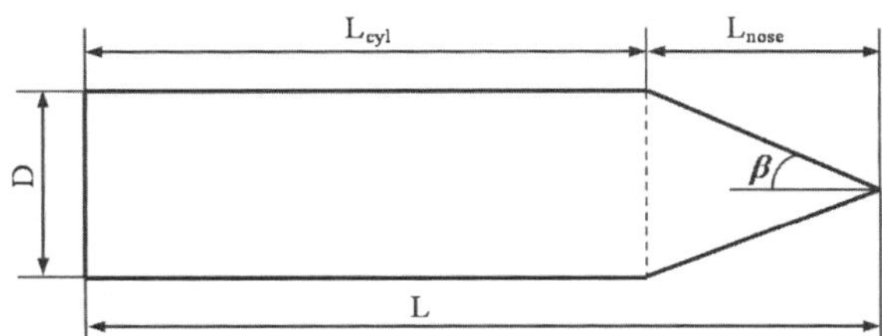

Fig. 3.2.1. Cone-nose projectile.

a)

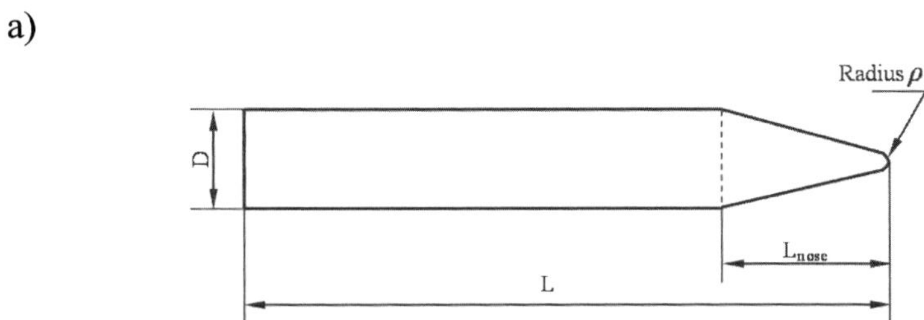

b)

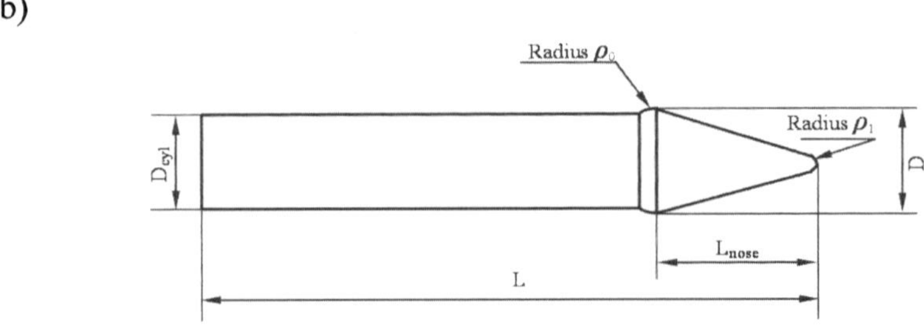

c)

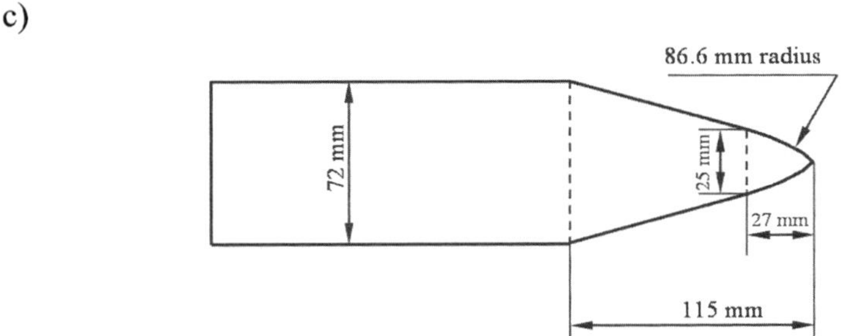

Fig. 3.2.2a-c. Cone-nose projectiles of different types.

3.3 Ogive-nose projectiles

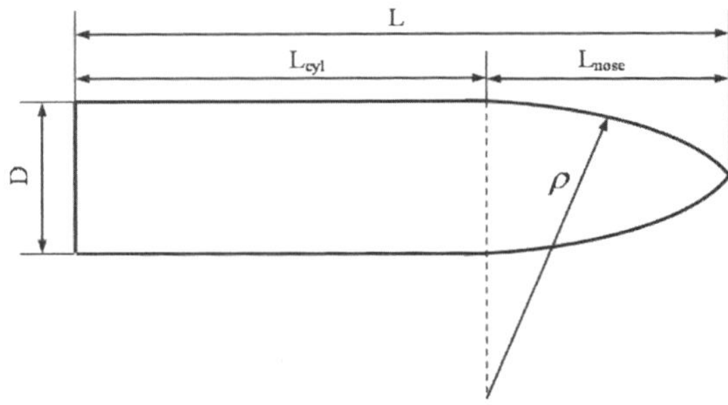

Fig. 3.3.1. Ogive-nose projectile, $\psi = CHR = \dfrac{\rho}{D}$.

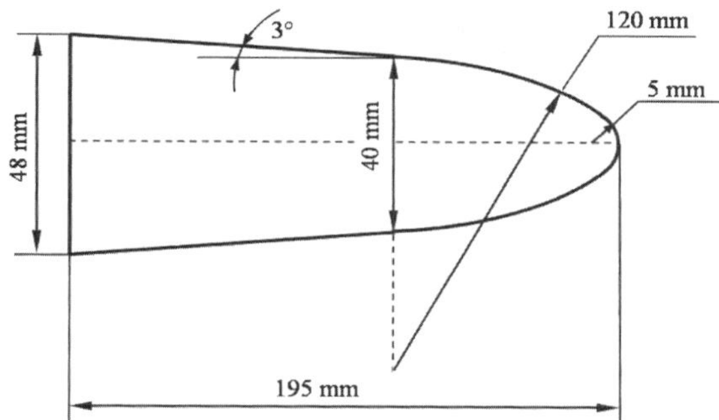

Fig. 3.3.2. Projectile having blunted ogive nose.

3.4 Hemispherical-nose projectiles

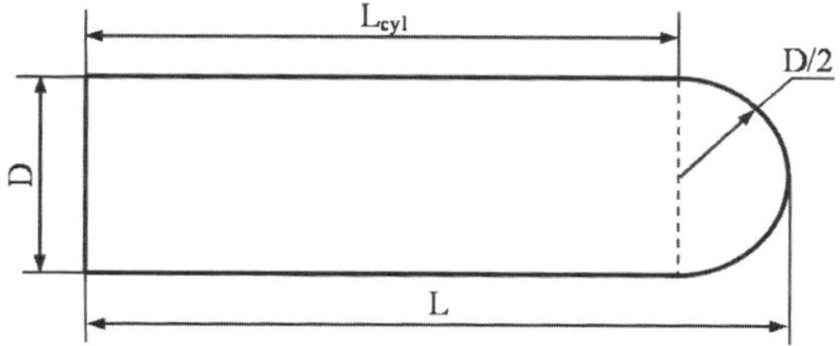

Fig. 3.4.1. Hemispherical-nose projectile.

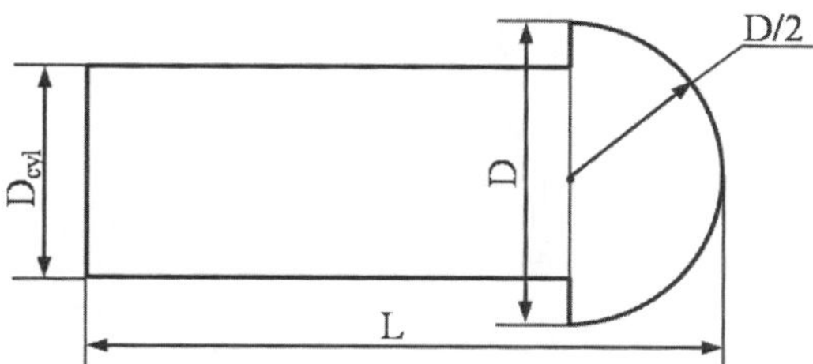

Fig. 3.4.2. Mushroom-shaped projectile.

3.5 Spherical projectiles (balls)

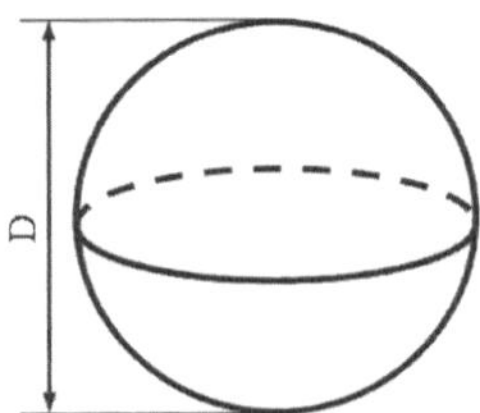

Fig. 3.5.1. Spherical projectile.

3.6 Flat nose projectiles

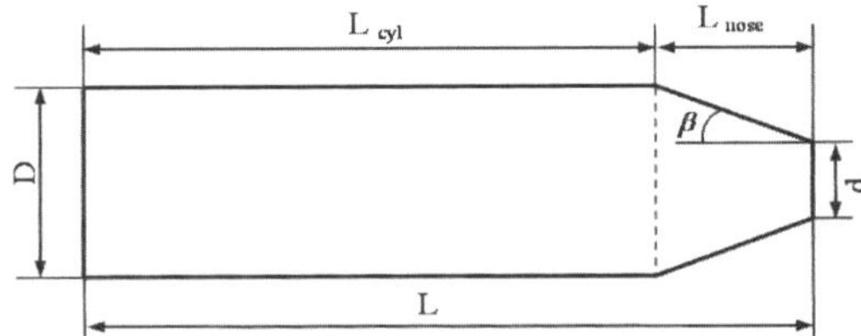

Fig. 3.6.1. Truncated cone-nose projectile.

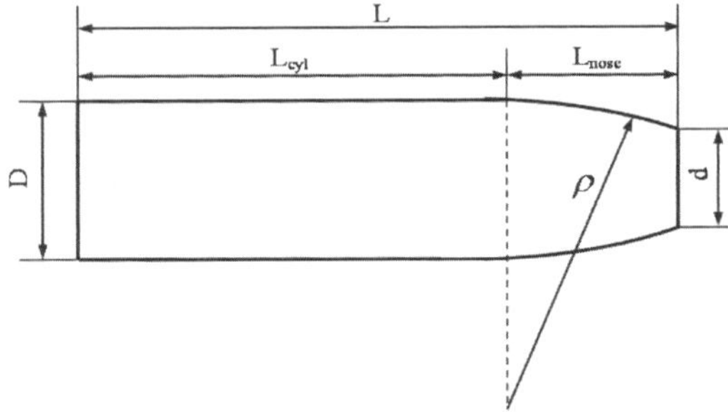

Fig. 3.6.2. Truncated ogive-nose projectile.

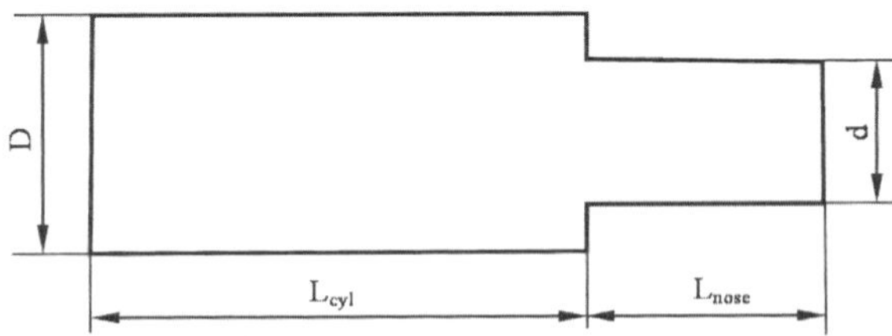

Fig. 3.6.3. Cylindrical-nose projectile.

a)

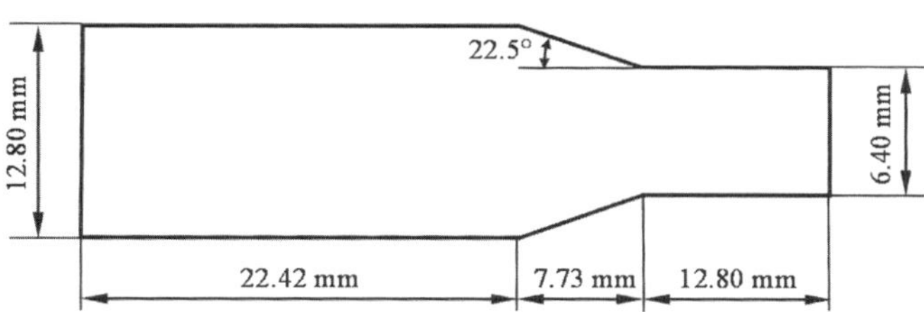

b)

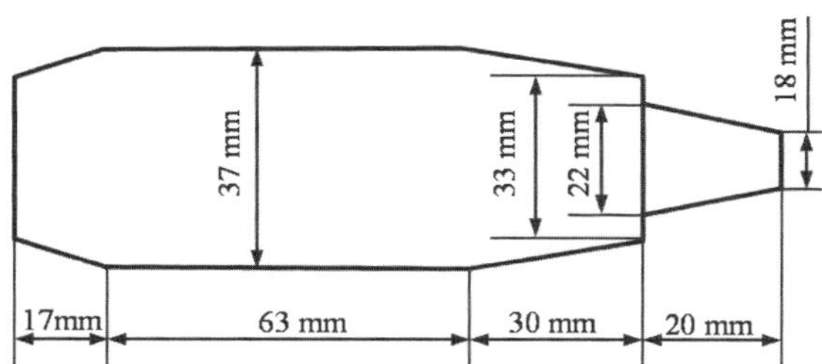

Fig. 3.6.4a-b. Cylindrical-conical nose projectile.

3.7 Conical pointed projectiles with a compound nose

a)

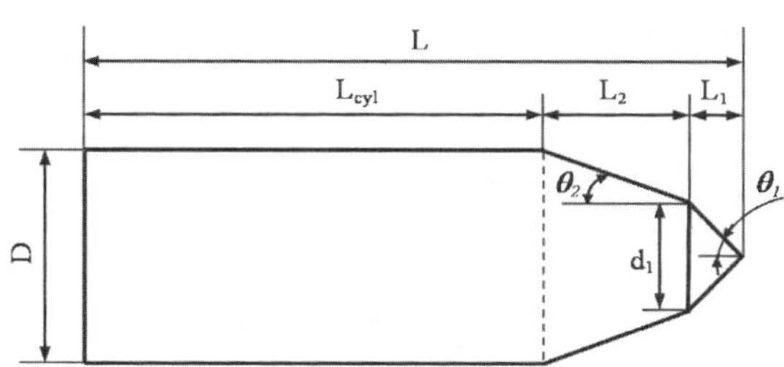

b)

c)

d)

e)

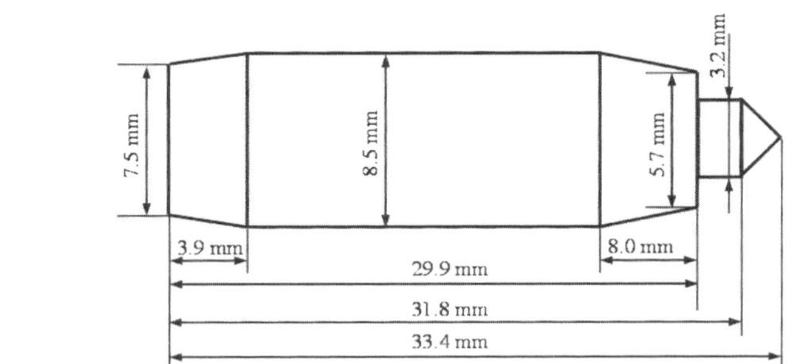

Fig. 3.7.1a-e. Conical pointed projectiles with a compound nose.

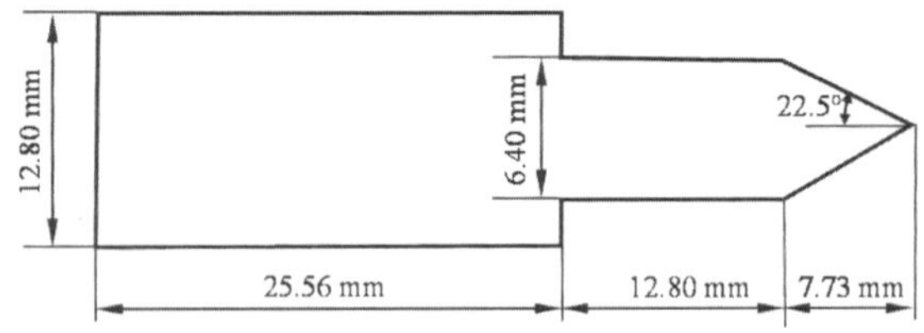

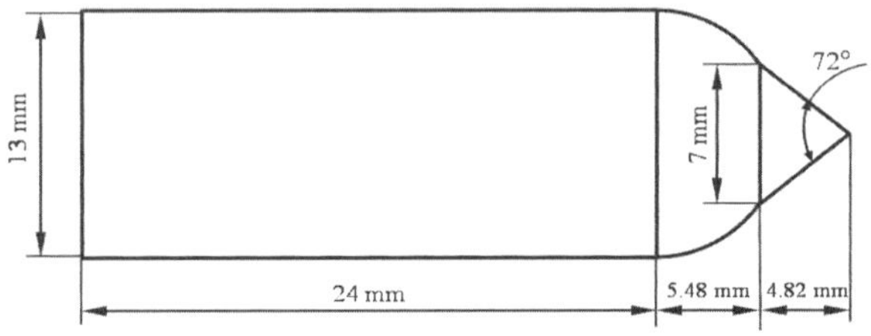

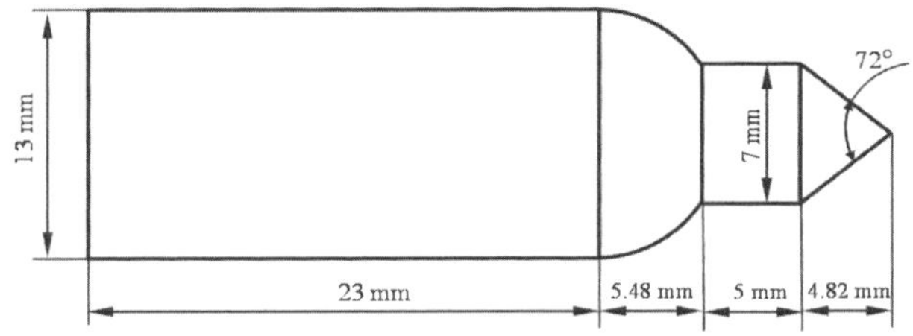

3.8 Spherical-segment nose projectiles

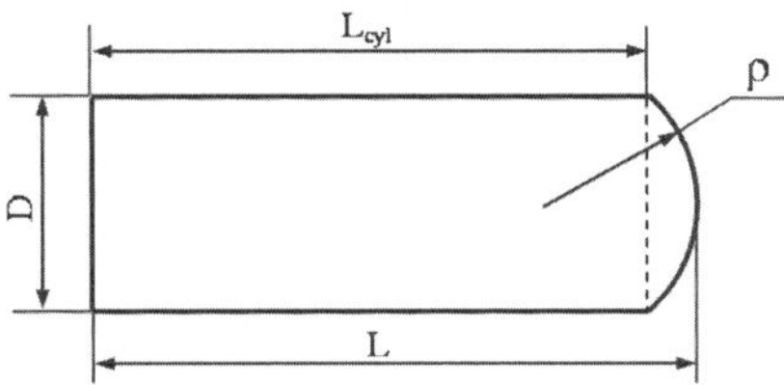

Fig. 3.8.1. Spherical-segment nose projectiles.

3.9 Hemispherical blunted projectiles with a compound nose

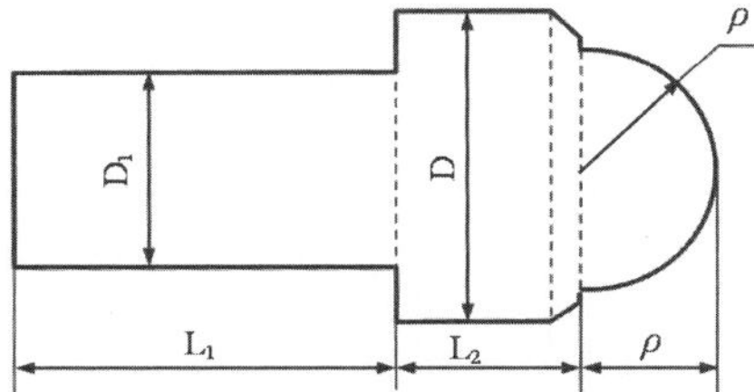

Fig. 3.9.1. Hemispherical blunted projectiles with a compound nose.

3.10 Fragment simulating projectiles (FSPs)

3.10.1 *Standard FSPs [MIL-DTL-46593B, 2006]*

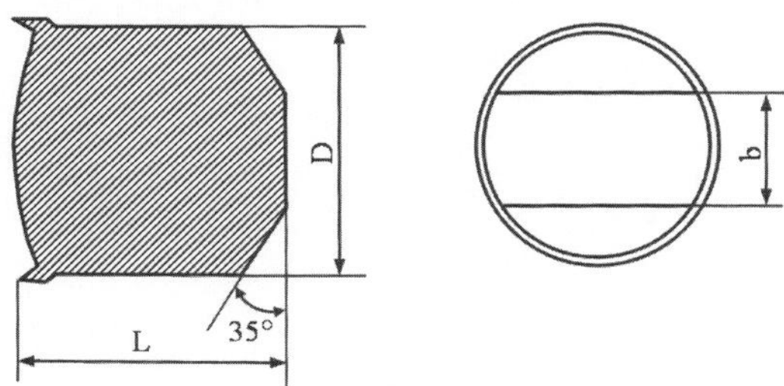

Fig. 3.10.1. Standard FSPs.

Table 3.10.1. Characteristics of standard FSPs.

FSP caliber	D (mm)	b (mm)	L (mm)	m (g)	HRC
0.22" (type 1*)	5.5+0.025	2.5–0.254	6.4	1.1±0.032	30±1
0.22" (type 2**)	5.5+0.025	2.5–0.254	6.4	1.1± .032	27±3
0.30"	7.5–0.025	3.5–0.254	8.6	2.9±0.032	30±1
0.50"	12.6±0.025	5.7–0.381	14.7	13.4±0.130	30±1
20 mm	19.9–0.051	9.3–0.305	22.9	53.8±0.259	30±1

*Armor shield, ** Body armor.

3.10.2 *First additional version of FSP*

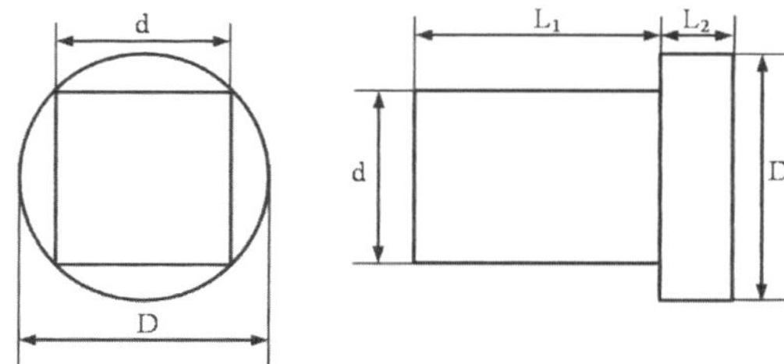

Fig. 3.10.2. First additional version of FSPs.

3.10.3 *Second additional version of FSP*

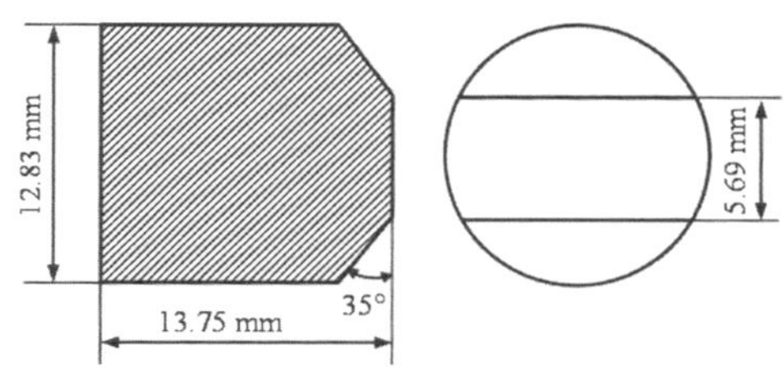

Fig. 3.10.3. Second additional version of FSP.

3.11   7.62 mm (0.30") caliber projectiles

3.11.1   *7.62 mm caliber AP M2 projectile*

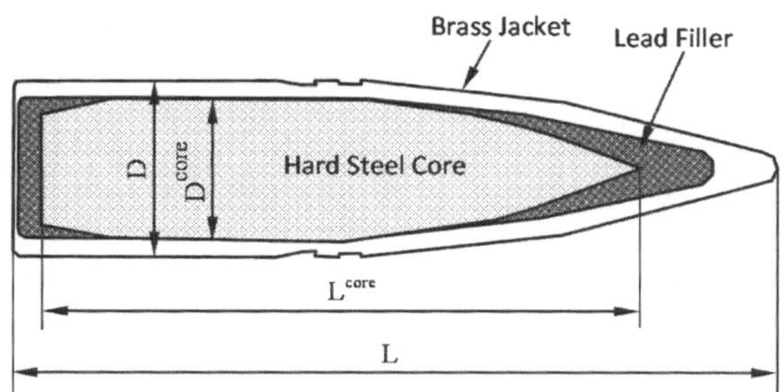

Fig. 3.11.1. 7.62 mm caliber AP M2 projectile.

Table 3.11.1. Typical characteristics of 7.62 mm caliber AP M2 projectile.

Projectile			Core		
L (mm)	D (mm)	m (g)	L^{core} (mm)	D^{core} (mm)	m^{core} (g)
35.3	7.85	10.8	27.4	6.2	5.3

3.11.2   *7.62 mm steel penetrator having ogive shape encased in copper jacket*

a) General view

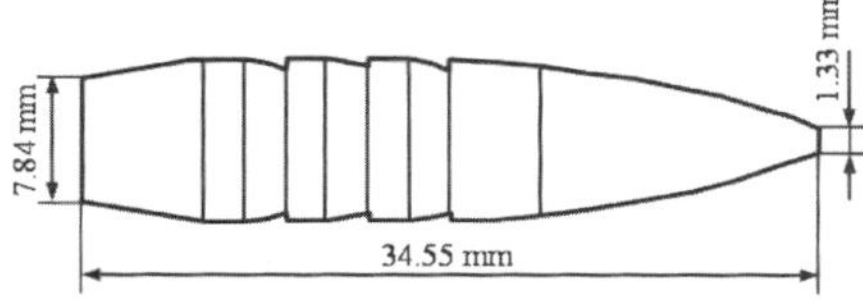

b) Jacket

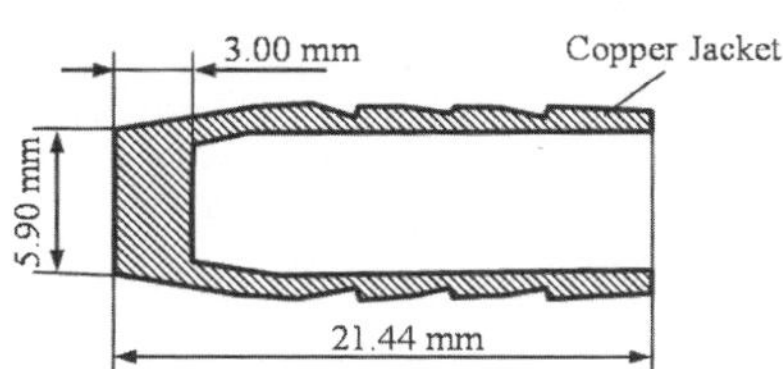

c) Core

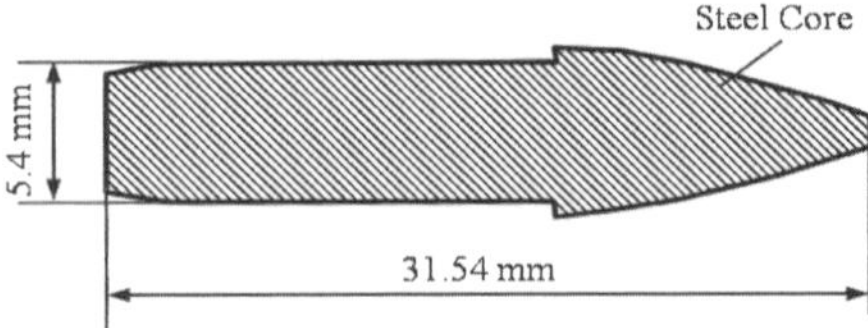

Fig. 3.11.2a-c. 7.62 mm steel penetrator having ogive shape encased in copper jacket.

3.11.3 *7.62 mm tungsten carbide penetrator having conical shape encased in brass sabot*

a) General view

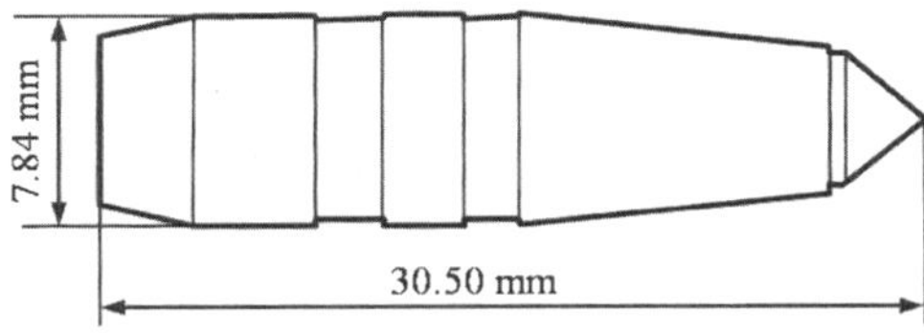

b) Sabot

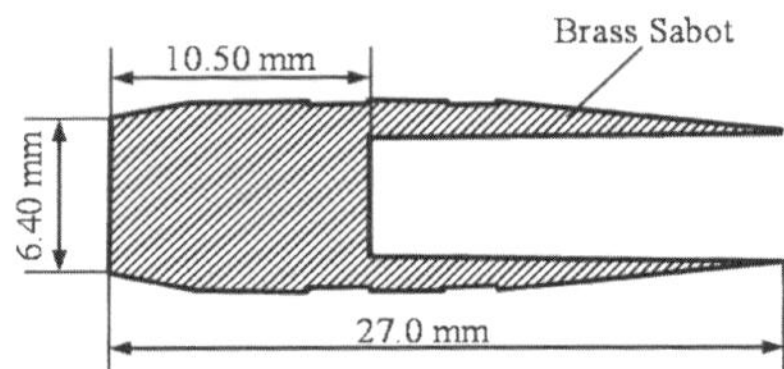

c) Core

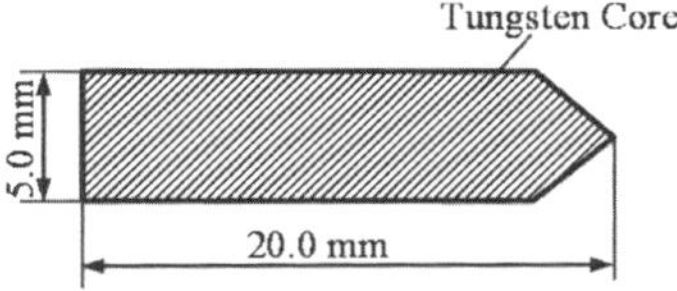

Fig. 3.11.3a-c. 7.62 mm tungsten carbide penetrator having conical shape encased in brass sabot.

3.11.4 *7.62 mm ball projectile*

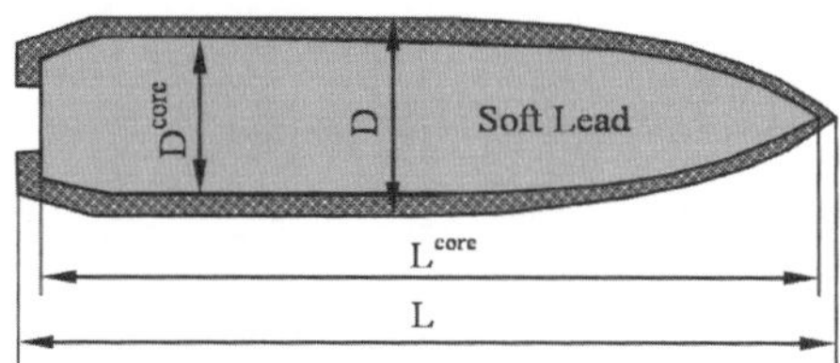

Fig. 3.11.4. 7.62 mm ball projectile.

Table 3.11.2. Typical characteristics of 7.62 mm ball projectile.

Projectile			Core		
L (mm)	D (mm)	m (g)	L^{core} (mm)	D^{core} (mm)	m^{core} (g)
28.6	7.83	9.5-10.0	27.35	6.62	4.5

3.11.5 *7.62 mm Kalashnikov steel-core ball round*

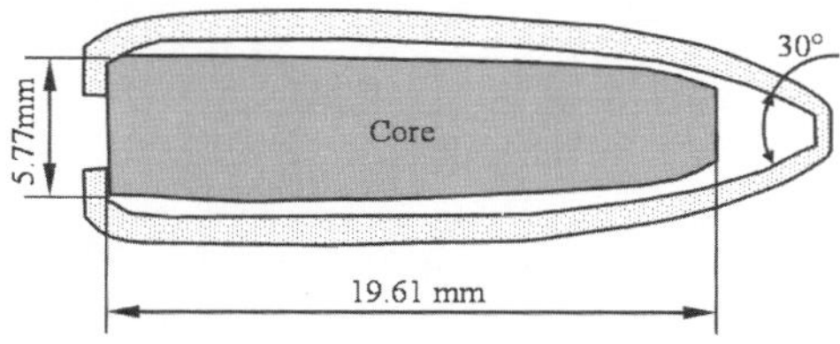

Fig. 3.11.5. 7.62 mm Kalashnikov steel-core ball round.

3.11.6 *7.62 mm FFV tungsten carbide AP round*

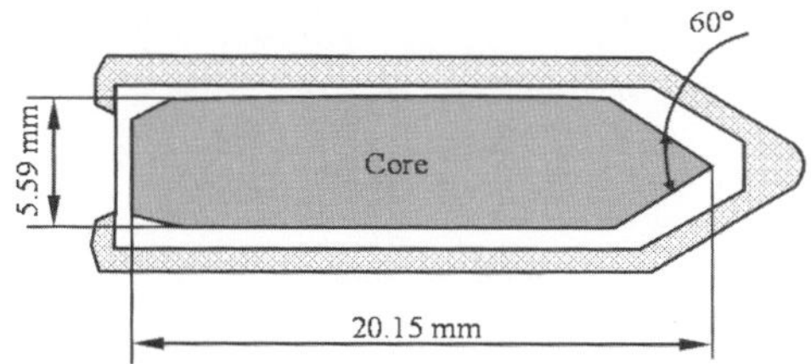

Fig. 3.11.6. 7.62 mm FFV tungsten carbide AP round.

3.12 9 mm caliber round nose full metal jacket bullet

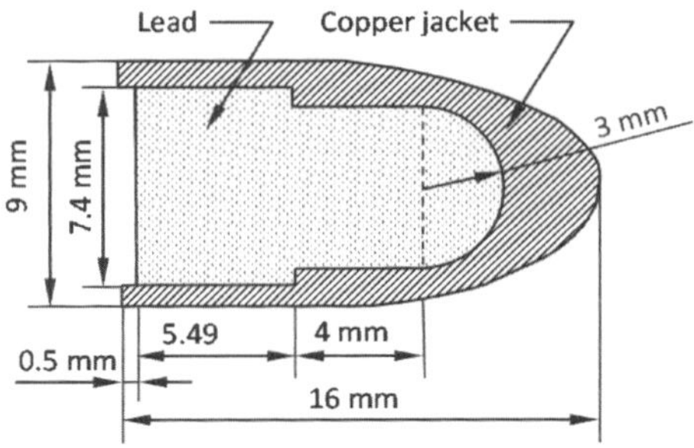

Fig. 3.12.1. 9 mm caliber round nose full metal jacket bullet.

3.13 12.7 mm (0.50") caliber AP M2 projectile

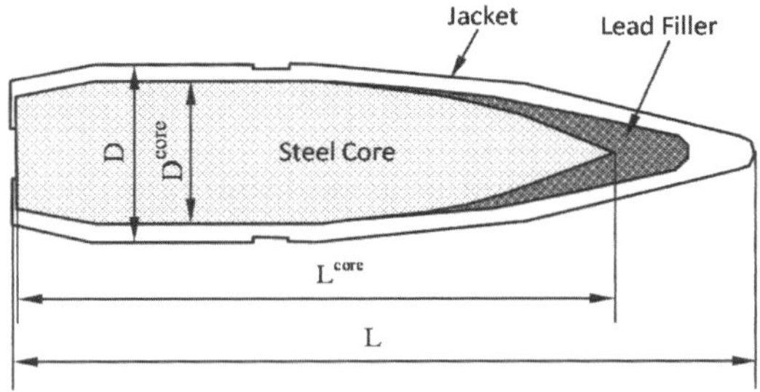

Fig. 3.13.1. 12.7 mm caliber AP M2 projectile.

Table 3.13.1. Typical characteristics of 12.7 mm caliber AP M2 projectile.

Projectile			Core		
L (mm)	D (mm)	m (g)	L^{core} (mm)	D^{core} (mm)	m^{core} (g)
58.7	12.98	45.9	47.5	10.9	25.9

3.14 14.5 mm caliber projectiles

3.14.1 *14.5 mm BS41 projectile*

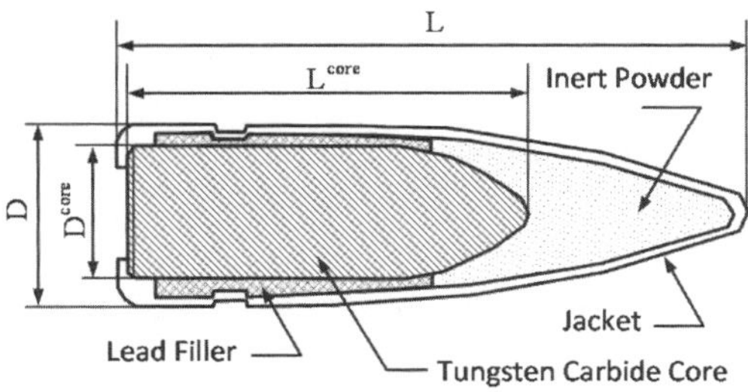

Fig. 3.14.1. 14.5 mm BS41 projectile.

Table 3.14.1. Typical characteristics of the 14.5 mm BS41 projectile.

Projectile			Core		
L (mm)	D (mm)	m (g)	L^{core} (mm)	D^{core} (mm)	m^{core} (g)
52.6	14.94	63.2	32.3	10.9	37.9

3.14.2 *14.5 mm B32 projectile*

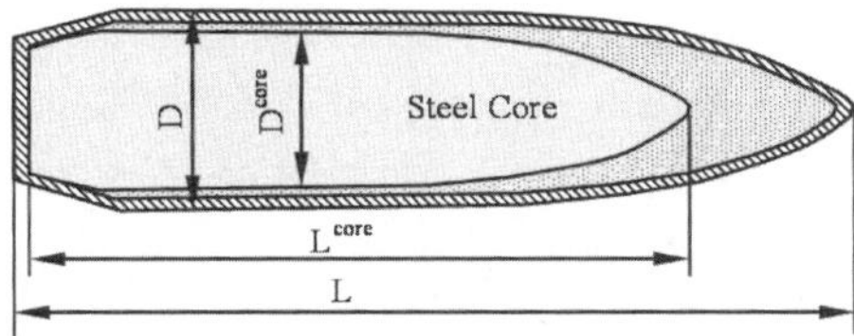

Fig. 3.14.2. 14.5 mm B32 projectile.

Table 3.14.2. Typical characteristics of the 14.5 mm B32 projectile.

Projectile			Core		
L (mm)	D (mm)	m (g)	L^{core} (mm)	D^{core} (mm)	m^{core} (g)
66.3	14.9	64.1	53.1	12.4	41.0

Chapter 4

Penetration into metals and alloys

4.1 Cumulative index of experiments

Classification of experiments with respect to penetrator type is presented in Tables 1–9. Notations M, L, S in the column Shield structure indicate monolithic, layers in contact and spaced (air gaps) shields, respectively. In column T we present a thickness of a shield in a case of a monolithic shield or the sum of the thicknesses of layers in the case of layers in contact or spaced shields; $T = \infty$ indicates a semi-infinite shield [Backman and Goldsmith, 1978].

In Table 9 for bullets and shells the values of caliber and diameter are usually slightly different. In cases when geometry of penetrator is unambiguously indicated in the original study, in Table 9 we present diameter of a striker and provide a reference to the corresponding figure. In the opposite case there is no reference to a figure, and diameter is selected either from the data in Chapter 3 for a striker having the same name or a caliber of a penetrator is adopted as its diameter.

Table 1. Experiments on penetration of cylindrical projectiles (Fig. 3.1.1).

Section	D (mm)	Shield structure	T (mm)	θ_s	T/D
4.4	5.8	M	20.6	0	3.55
4.10	4.0	L	2.0	0	0.50
4.14	20	M	12.1	0	0.61
4.15	20	M	12	0	0.60
4.16	20	M	6-20	0	0.30-1.00
4.19	20	M	20	0	1.00
4.22	8.86	M	∞	0	∞
4.35	12.5	M	1.3-6.4	0	0.10-0.53
4.36	12.5	M, L	1.18-6.4	0	0.09-0.51
4.37	12.7	M	1.59-6.48	0	0.13-0.51

Section	D (mm)	Shield structure	T (mm)	θ_s	T/D
4.43	12.62	L	12.0	0	0.95
4.45	12.7	M, L	6.0	0	0.47
4.48	20	M	11.9-12.2	0	0.60-0.61
4.49	20	M, L, S	6.0-12.0	0	0.30-0.60
4.50	20	L	12.0	0	0.60
4.63	5.71	M, L	1.25-3.45	0	0.22-0.60
4.66	30	M	10.5	0	0.35
	30.8	M	5.3-10.5	0	0.17-0.34
4.75	5.54	M	0.25-6.35	0	0.05-1.15
	7.59	M	1.27-2.29	0	0.17-0.30
4.78	12.7	M	1.27-3.81	0	0.10-0.30
4.79	7.62	M, L, S	6.7	0	0.88
4.81	12.3	M	3.175-19.05	0	0.26-1.55
4.82	12.3	M	3.175-25.4	≥ 0	0.26-2.07
4.85	1.8-25.9	M	6.3-152.4	≥ 0	4.0-30
4.90	19	M	1.0	0	0.05
4.91	19	M	0.5-3.0	0	0.03-0.16
4.92	19	L	1.0-3.0	0	0.05-0.16
4.94	30	M	10.5	0	0.35
4.96	7.94	M	9.52	0	1.20
4.99	12.7	M	3.2-9.5	≥ 0	0.25-0.75
4.102	19	M	1.0	0	0.05
4.106	12.7	M	3.2-6.4	0	0.25-0.50
4.111	12.8	M	0.81-1.91	≥ 0	0.06-0.15
4.112	12.8	M	0.82	0	0.06
4.115	13	M	1.0	0	0.08
4.121	12.5	M	3.2-12.75	0	0.26-1.02
	12.3	M	3.2-12.7	0	0.25-1.03
4.122	7.82	M	6.35	0	0.81
4.123	12.5	M	1.15-6.42	0	0.09-0.51
	25.0	M	1.15-2.98	0	1.05-0.12
4.133	13	L	25.40-28.58	0	1.95-2.20
4.135	32	M	1.0	0	0.031
	40	M	1.0	0	0.025
	43	M	6.0	0	0.14
	63	M	15-25	0	0.24-0.40
	85	M	6.0-12	0	0.07-0.14
4.138	76.2	M	18.5-34.5	≥ 0	0.24-0.45
4.139	7.62	M, S	15.2-54.0	≥ 0	1.99-7.09
4.141	12.6	M	0.5	0	0.04
4.142	12.6	M, L, S	0.5-1.0	0	0.04-0.08
4.148	12.7	M	1.0-2.0	0	0.08-0.16
4.149	12.7	M	2.3-12.8	>0	0.18-1.01
	19.1	M	13.1	>0	0.69
4.153	5.9-17.4	M	0.5-50.8	≥ 0	0.03-4.18
4.154	12.57	M, L	1.6-6.4	0	0.13-0.51

Section	D (mm)	Shield structure	T (mm)	θ_s	T/D
4.155	15	M	1	>0	0.07
4.156		M		≥0	$0.15 \leq T/L \leq 1.45$
4.158	13	M	4	0	0.31
4.162	10	M	5.0-15.0	0	0.50-1.50
4.165	25.4	M	3.175-12.7	0	0.13-0.50
4.169	12.5	M	1.3-6.5	0	0.10-0.52
4.176	14.5	M	30-60	0	2.07-4.14
4.189	6.35	M, L	9.53	0	1.50
4.190	6.23	M	1.55-12.7	0	0.25-2.04
4.193	4.8	M	∞	0	∞
4.197	9.52	M, L	1.2-9.5	0	0.13-1.00
4.198	7.62	M	4.76-7.62	0	0.62-1.00
4.199	20.0	M	12	0	0.60
4.200	12.7	M, L, S	2.0-6.0	0	0.16-0.47
4.201	12.68	M, L	0.4-0.5	≥0	0.03-0.04
4.203	7.87	M	25.4-76.2	0	3.23-9.68

Table 2. Experiments on penetration of cone-nose projectiles (Fig. 3.2.1).

Section	D (mm)	Shield structure	T (mm)	θ_s	T/D	Note
4.5	12.8	M	0.81-1.52	≥0	0.06-0.12	
4.23	10.16	M	∞	0	∞	
4.26	6.35-12.7	M	1.27	0	0.20-1.00	
4.54	6.1	M	20	0	3.28	
4.57	6.1	M	10	≥0	1.64	
4.68	9.53	M	25.4	0	2.67	*
	10.2	M	25.4	0	2.49	**
4.69	7.1	M	∞	0	∞	
4.70	8.31	M	12.7-76.2	0	1.53-9.17	
4.81	12.3	M	3.175-6.35	0	0.26-0.52	
	12.5	M	3.175	0	0.25	
4.82	12.3	M	3.175-6.35	≥0	0.26-0.52	
4.83	20	M	70	0	3.50	
4.99	12.7	M	3.2-9.6	≥0	0.25-0.76	
4.112	12.8	M	0.82	0	0.06	
4.115	13	M, L	1.0-4.0	0	0.08-0.31	
4.116	13	M	1.0	0	0.08	
4.120	12.45	M	3.175	0	0.26	*
	12.32	M	3.175	0	0.26	*
	12.5	M	3.175	0	0.25	*
4.137	6.5	M	3.2-4.9	0	0.49-0.75	
4.141	12.6	M	0.5	0	0.04	
4.142	12.6	M, L, S	0.5-1.0	0	0.04-0.08	
4.144	66-160	M	7-38	0	0.08-0.45	
4.154	12.57	M, L, S	1.6-6.4	0	0.13-0.51	

Section	D (mm)	Shield structure	T (mm)	θ_s	T/D	Note
4.155	15	M	1	>0	0.07	
4.157	13	M	0.5-1.0	0	0.04-0.08	
4.158	13	M	4	0	0.31	
4.160	7.1	M	25.4	0	3.58	*
	7.1	M	25.4	0	3.58	**
4.186	12.7	M	1.27-1.59	≥0	0.10-0.13	
4.189	6.35	M, L	9.53	0	1.50	
4.191	4.76	M	1.6-12.7	0	0.34-2.67	
	6.35	M	4.78-12.7	0	0.75-2.00	***
4.192	4.76	M	3.1-12.8	0	0.65-2.69	***
4.193	4.8	M	∞	0	∞	*
4.195	36.2	M	4.76	0	0.13	

*   Slightly modified shape (Fig. 3.2.2a).

**   Slightly modified shape (Fig. 3.2.2b).

***   Slightly modified shape (Fig. 3.6.1).

Table 3. Experiments on penetration of ogive-nose projectiles (Fig. 3.3.1).

Section	D (mm)	Shield structure	T (mm)	θ_s	T/D
4.19	20	M	20	0	1.00
4.22	8.86	M	∞	0	∞
4.42	12.62	M, L, S	2; 6	0	0.16-0.48
4.43	12.62	L	12.0	0	0.95
4.44	12.62	M, L	6.0	0	0.48
4.48	20	M	11.9-12.3	0	0.60-0.62
4.49	20	M, L, S	6.0-12.0	0	0.30-0.60
4.50	20	L	12.0	0	0.60
4.51	20	M	20	0	1.00
4.52	20	M	20-80	0	1.00-4.00
4.53	20	M	20-80	0	1.00-4.00
4.55	20	M	10	>0	0.50
4.56	20	M	20	≥0	1.00
4.58	20	M	20	0	1.00
4.69	7.1	M	∞	0	∞
4.72	7.11	M	∞	0	∞
4.77	17.0	M	55	0	3.24
	19.9	M	55	0	2.76
4.89	10	M	0.74-1.0	0	0.07-0.10
	15	M	0.5-2.0	0	0.03-0.13
4.91	19	M	0.5-3.0	0	0.03-0.16
4.92	19	L	1.0-3.0	0	0.05-0.16
4.100	15	M	1.0	0	0.07
4.102	19	M	1.0	0	0.05
4.113	15.0	M	0.5-1.5	0	0.03-0.10

Section	D (mm)	Shield structure	T (mm)	θ_s	T/D
4.114	6.7	M	25	0	3.73
4.151	12.9	M	26.3	≥ 0	2.04
4.152	7.11	M	127-470	0	17.86-66.10
4.155	15	M	1	>0	0.07
4.159	11.3	L, S	85.0-220.0	≥ 0	7.52-19.47
4.163	6.76	M	76.2	0	11.27
4.164	6.76	M	∞	0	∞
4.181	6.2	M	20	0	3.23
4.187	7.11	M	∞	>0	∞

Table 4. Experiments on penetration of hemispherical-nose projectiles (Fig. 3.4.1).

Section	D (mm)	Shield structure	T (mm)	θ_s	T/D
4.15	20	M	12	0	0.60
4.28	14.9	M	1.36	0	0.09
4.32	6.0	M, L, S	152.4	0	25.40
4.33	5.7	M, L, S	150-175	0	26.32-30.70
4.34	12.7	M	1.0-10.0	0	0.08-0.79
4.41	12.62	M, L, S	2; 3; 6	0	0.16-0.48
4.67	7.11	M	121-297	0	17.02-41.77
4.69	7.1	M	∞	0	∞
	7.1*	M	∞	0	∞
4.71	7.11	M	82-216	0	11.53-30.38
	5.08	M	90-203	0	17.72-39.96
4.83	20	M	70	0	3.50
4.90	19	M	1.0	0	0.05
4.91	19	M	0.5-3.0	0	0.03-0.16
4.92	19	L	1.0-3.0	0	0.05-0.16
4.115	13	M	1.0	0	0.08
4.116	13	M	1.0	0	0.08
4.118	8.1	M	19.1-50.8	≥ 0	2.36-6.27
	9.7	M	25.3	≥ 0	2.61
	10.2	M	19.1-51.0	≥ 0	1.87-5.00
	13.0	M	19.1-50.8	≥ 0	1.47-3.91
4.136	8.7	M, L, S	3.0	0	0.34
4.141	12.6	M	0.5	0	0.04
4.155	15	M	1	>0	0.07
4.157	13	M	0.5-1.0	0	0.04-0.08
4.158	13	M	4	0	0.31
4.163	6.76	M	76.2	0	11.27
4.164	6.76	M	∞	0	∞
4.167	6.81	M	50.8	>0	7.46
4.186	12.7	M	1.27-4.76	≥ 0	0.10-0.37
4.188	17.0	L	43	0	2.53
4.194	12.7	M	1	0	0.08

Section	D (mm)	Shield structure	T (mm)	θ_s	T/D
4.203	7.87	M	25.4-76.2	0	3.23-9.68

* Mushroom-shaped projectiles (Fig. 3.4.2).

Table 5. Experiments on penetration of spherical projectiles (Fig. 3.5.1).

Section	D (mm)	Shield structure	T (mm)	θ_s	T/D
4.11	6.35	M	9.53-1.27	≥ 0	1.50-0.20
4.12	4.0	M	∞	0	∞
4.25	12.7	M	1.587-6.35	0	0.12-0.50
4.26	6.35-12.7	M	1.27	0	0.20-1.00
4.30	4.0	M	∞	0	∞
	6.1	L	31.75-43.2	0	5.20-7.08
	6.35	M, L	∞	0	∞
	6.35	M, L	1.27-31.75	0	0.20-5.00
	6.4	M	∞	0	∞
4.31	4.0	M	17.3-27.9	0	4.33-6.98
	4.4	M	25.4	0	5.77
	6.4	M	17.3-59.2	0	2.70-9.25
	12.7	M	24.9-27.2	0	1.96-2.14
4.39	8	M	0.4	0	0.05
4.47	6	M	4.0-8.0	0	0.67-1.33
	6.3	M	4.0-10.0	0	0.63-1.59
	16	M	6.0-12.0	0	0.38-0.75
4.64	5.55	M	2	0	0.36
4.80	6.35	M	1.27-6.35	0	0.20-1.00
	9.53	M	1.27-6.35	0	0.13-0.67
4.98	9.5	M	∞	≥ 0	∞
4.107	4.76	M	∞	0	∞
4.119	10.3	M	0.9-19.0	0	0.09-1.84
	6.37	M	1.9-19.3	0	0.30-3.03
4.124	12.7	M	1.6-6.4	0	0.13-0.50
4.129	6.4	M	51	0	7.97
4.130	6.37	M	1.00-4.00	0	0.16-0.63
	10.3	M	1.50-3.37	0	0.15-0.33
4.131	6.35	M	1.8-9.9	0	0.28-1.56
	10.3	M	4.85-6.35	0	0.47-0.62
4.134	6.37	M	2.0-6.0	0	0.31-0.94
	10.3	M	2.0-7.4	0	0.19-0.72
	16.0	M	2.0-12.7	0	0.13-0.79
4.175	12.7	M	2	0	0.16
4.177	6.35	M	∞	0	∞
	9.52	M	∞	0	∞
4.179	11.4	M	0.58-1.60	0	0.05-0.14
4.180	4.7	S	40.0	≥ 0	8.51

Section	D (mm)	Shield structure	T (mm)	θ_s	T/D
4.182	8.0	M	3.0	0	0.38
4.183	3.175	M	∞	0	∞
4.196	6.0	M	0.87-1.63	0	0.15-0.27
4.202	6.0	L	5.0	0	0.83
4.204	6.35	M	6.35-50.8	≥ 0	1.00-8.00
	11.91	M	6.35-76.2	≥ 0	0.53-6.40

Table 6. Experiments on penetration of fragment simulating projectiles (FSPs).

Section	Projectile shape	D (mm)	Shield structure	T (mm)	θ_s	T/D
4.13	Fig. 3.10.1	12.98	M	14	0	1.08
4.27	Fig. 3.10.1	19.9	M	32.3-40.9	0	1.62-2.06
4.38	Fig. 3.10.1	20	M	3.8-4.7	0	0.19-0.24
4.59	Fig. 3.10.1	20	M	∞	0	∞
4.74	Fig. 3.10.1	20	M	40	0	2.0
4.76	Fig. 3.10.1	12.6	M	18.36-25.45	0	1.46-2.02
	Fig. 3.10.1	19.9	M	25.35-37.97	0	1.27-1.91
4.78	Fig. 3.10.3	12.7	M	1.27-3.81	0	0.10-0.30
4.84	Fig. 3.10.1	19.9	M	25.2-57.2	0	1.27-2.87
4.108	Fig. 3.10.1	12.6	M	40.4	0	3.21
		19.9	M	59.7	0	3.00
4.109	Fig. 3.10.1	19.9	M	76.5	0	3.84
		5.5	M	7.6-7.7	0	1.38-1.40
		12.6	M	31.5	0	2.50
4.110	Fig. 3.10.1	12.6	M	25.0-38.7	0	1.98-3.07
		19.9	M	38.7-63.1	0	1.94-3.17
4.126	Fig. 3.10.3	12.83	M	6	0	0.47
4.143	Fig. 3.10.1	12.6	M	3.81-12.7	0	0.30-1.01
4.153	Fig. 3.10.2	5.9-14.9	M	0.5-50.8	≥ 0	0.03-4.18
4.170	Fig. 3.10.1	19.9	M	25.1-50.9	0	1.26-2.56
		12.6	M	18.8-26.3	0	1.49-2.09
4.179	Fig. 3.10.1	5.5	M	0.58-1.60	0	0.11-0.29
4.184	Fig. 3.10.1	2.65-12.9	*	11.0-15.8	0	*

* Witness packs with plates separated by polystyrene foam (25.4 mm).

Table 7. Experiments on normal penetration of truncated cone-nosed projectiles into monolithic shields.

Section	D (mm)	T (mm)	T/D
4.17	20	15-30	0.75-1.50
4.48	20	12.0-12.4	0.60-0.62
4.185	15.6	2.5	0.16
4.195	36.2	4.76	0.13

Table 8. Experiments on penetration of monolithic shields with projectiles having complex shapes.

Section	Projectile shape	D (mm)	T (mm)	θ_s	T/D
4.23	Fig. 3.7.1a	10.16	∞	0	∞
4.35	Fig. 3.8.1	12.5	1.3	0	0.10
4.97	Fig. 3.2.2c	72	50	0	0.69
4.101	Fig. 3.7.1b	12.80	0.82-1.82	0	0.06-0.14
	Fig. 3.6.3	12.80	0.82-1.82	0	0.06-0.14
	Fig. 3.6.4	12.80	0.82-1.82	0	0.06-0.14
4.116	Fig. 3.7.1c	13	1.0	0	0.08
	Fig. 3.7.1d	13	1.0	0	0.08
4.140	Table 2	76.2	18.5-82.6	≥ 0	0.24-1.08
4.174	Fig. 3.7.1e	8.5	28.3-29.2	0	3.33-0.044

Table 9. Experiments on penetration of bullets and shells.

Section	Caliber or D (mm)	Figure	Shield structure	T (mm)	θ_s	T/D
4.2	6.24	–	M,L	3.0	0	0.48
4.3	7.62	–	M, L, S	8.0-16.0	0	1.05-2.10
4.6	7.62	–	M	6-19	0	0.79-2.49
	5.6	–	M	1-6	0	0.18-1.07
	9.0	–	M	4-6.35	0	0.44-0.71
4.7	5.56	–	M	2-6	≥ 0	0.36-1.08
4.8	5.56	–	M, L, S	0.8-8.1	≥ 0	0.14-1.46
4.9	5.56	–	M	0.8-6	≥ 0	0.14-1.08
4.13	7.85	Fig. 3.11.1	M	14	0	1.78
4.18	7.83	Fig. 3.11.4	M	6	0	0.77
	7.85	Fig. 3.11.1	M,L,S	12	0	1.53
	6.2	Fig. 3.11.1(core)	L	12	0	1.94
4.20	7.85	Fig. 3.11.1	M,L	20-60	0	2.55-7.64
	6.2	Fig. 3.11.1(core)	M,L	20-60	0	3.23-9.68
4.21	7.83	Fig. 3.11.4	M	20	0	2.55
	7.85	Fig. 3.11.1	M	20	0	2.55
4.24	7.85	–	L	6-6.5	0	0.76-0.83
4.27	7.85	Fig. 3.11.1	M	3.23-40.9	0	0.41-5.21
	12.98	Fig. 3.13.1	M	39.0-53.8	0	3.00-4.14
4.29	7.85	Fig. 3.11.1	M	∞	0	∞
	7.83	Fig. 3.11.4	M	∞	0	∞
4.40	6.0	–	M	6.5-9.5	0	1.08-1.58
	12.58	–	M	6.5-24.8	0	0.52-1.97
	37	–	M	15.1-49.1	0	0.41-1.33
4.46	12.58	–	M	7.2-14.4	0	0.57-1.14
4.60	9	–	M, L	1-2	0	0.11-0.22
4.61	7.62	–	M	5	0	0.66

Section	Caliber or D (mm)	Figure	Shield structure	T (mm)	θ_s	T/D
	7.62	-	M	6	0	0.79
4.62	7.62	Fig. 3.11.5	M	5	0	0.66
	7.62	Fig. 3.11.6	M	10	0	1.31
4.65	7.58	-	M	9.5-32	$\geq$0	1.25-4.22
4.73	7.85	Fig. 3.11.1	M, L	20.0-40.0	0	2.55-5.10
	6.2	Fig. 3.11.1(core)	M, L	20.0-40.0	0	3.23-6.45
4.76	7.85	Fig. 3.11.1	M	12.52-52.76	$\geq$0	1.59-6.72
	12.98	Fig. 3.13.1	M	37.95-77.47	0	2.92-5.97
	14.94	Fig. 3.14.1	M	63.35-74.00	0	4.24-4.95
4.84	7.85	Fig. 3.11.1	M	25.1-57.4	0	3.20-7.29
4.86	7.85	-	M	3.7-4.9	0	0.47-0.62
	12.98	-	M	5.4-12.4	0	0.42-0.96
4.87	7.85	-	M	13.3-20.2	0	1.69-2.57
	7.83	-	M	13.1-44.9	$\geq$0	1.67-5.73
	12.98	-	M	19.5-44.9	>0	1.50-3.46
	7.85	-	M	13.1-44.9	$\geq$0	1.67-5.72
	12.98	-	M	19.5-44.9	>0	1.50-3.46
4.88	6.2	-	M, L, S	4.7-40.0	$\geq$0	0.76-6.45
4.93	7.62	-	M	6.35	0	0.83
	5.56	-	M	6.35	0	1.14
	5.56	-	M	6.35	0	1.14
4.95	7.85	Fig. 3.11.1	M	20	0	2.55
4.103	7.62	-	M	9.5-19.1	0	1.25-2.51
4.104	7.62	-	M	25	0	3.28
4.105	6.1	-	M	15	$\geq$0	2.46
4.108	7.85	Fig. 3.11.1	M	38.1-76.2	00	4.85-9.71
4.109	7.85	Fig. 3.11.1	M	31.5-76.5	0	4.01-9.75
	12.98	Fig. 3.13.1	M	76.5	0	5.89
4.110	7.85	Fig. 3.11.1	M	38.8-63.5	0	4.94-8.09
	12.98	Fig. 3.13.1	M	76.5-102.0	0	5.89-7.86
4.117	20	-	M	12.7-38.1	$\geq$0	0.64-1.91
	20	-	M	12.7-25.4	$\geq$0	0.64-1.27
	12.98	-	M	3.17-38.1	$\geq$0	0.24-2.94
4.125	7.85	Fig. 3.11.1	M	12.0	0	1.53
4.126	9	Fig. 3.12.1	M	6-8	0	0.67-0.9
4.127	7.84	Fig. 3.11.2a-c	M	25-101.6	0	3.19-12.96
4.128	5.56	-	M, L, S	1.0-10.0	0	0.18-1.80
4.132	5.56	-	L	16	0	2.88
4.138	76.2	-	M	18.5-34.5	$\geq$0	0.24-0.45
4.139	76.2	-	M, S	15.2-105.5	$\geq$0	0.20-1.38
4.140	76.2	-	M	18.5-81.0	$\geq$0	0.24-1.06
4.145	9	-	M	6.35-8.00	0	0.71-0.89
4.146	9	-	M	∞	0	∞
4.147	6.2	Fig. 3.11.1(core)	M	6.6	0	1.06

Section	Caliber or D (mm)	Figure	Shield structure	T (mm)	θ_s	T/D
	6.2	Fig. 3.11.1(core)	M	∞	0	∞
4.150	20	–	M	9.52-25.4	>0	0.48-1.27
4.161	7.62	Fig. 3.11.1	M	20-50	0	2.62-6.56
4.166	14.9	Fig. 3.14.2	L	23-35.4	0	1.54-2.38
	14.9	Fig. 3.14.2	M	∞	0	∞
4.167	14.9	Fig. 3.14.2	M	1.75	≥ 0	0.12
4.168	7.62	–	M	∞	≥ 0	∞
4.170	7.85	Fig. 3.11.1	M	12.4-51.4	≥ 0	1.58-6.55
	12.98	Fig. 3.13.1	M	51.0-76.6	0	3.93-5.90
4.171	7.85	Fig. 3.11.1	M	4.52-8.36	>0	0.58-1.06
	12.98	Fig. 3.13.1	M	8.33-12.25	>0	0.64-0.94
4.172	7.85	Fig. 3.11.1	M	5.1-7.7	>0	0.65-0.98
	12.98	Fig. 3.13.1	M	3.1-7.7	>0	0.24-0.59
	14.94	Fig. 3.14.1	M	24.5	>0	1.64
	14.9	Fig. 3.14.2	M	15.4-18.8	>0	1.03-1.26
4.173	5.56	–	M	7	0	1.26
4.174	14.94	–	M	21.9-25.7	>0	1.47-1.72
	20	–	M	21.9-40.6	≥ 0	1.10-2.03
4.178	12.98	–	M	9.52-25.4	≥ 0	0.73-1.96
	37	–	M	25.4	≥ 0	0.69

4.2   [Abdel-Wahed *et al.*, 2010]

4.2.1   *General data on shields and projectiles*

Shields. Monolithic and layers-in-contact steel shields; $c_{l,w} = 200\,mm$, $T_{sum} = 3\,mm$. Material properties: $\gamma = 7810\,kg/m^3$, $E = 201\,GPa$, $G = 80\,GPa$, $\sigma_y = 305\,MPa$.

Projectiles. 7.62×39 mm AP projectiles consist of a penetrator and a jacket. Penetrator having a shape close to ogive-nose penetrator is made of high strength steel (for details see Table 1).

Table 1. Characteristics of penetrator.

D (mm)	L (mm)	L_{nose} (mm)	E (GPa)	γ (kg/m^3)	G (GPa)	σ_y (MPa)
6.24	23.8	9	220	7870	80	1500

4.2.2   *Test results*

Table 2# Test results for shields having different structure.

Shield structure	v_s (m/s)	v_r (m/s)	v_s (m/s)	v_r (m/s)	v_s (m/s)	v_r (m/s)
1×[3.0]	341	252	347	260	450	380
	552	499				
3×[1.0]	349	271	351	273	450	380
	552	488				
3×[1.0]\5.0	334	284	437	381	552	499
3×[1.0]\10.0	335	279	350	291	450	403
	554	507				
2×[1.5]	335	267	385	319	447	393
	489	434	560	504		
2×[1.5]\5.0	335	257	449	377	553	502

4.3   [Almohandes *et al.*, 1996]

See also [Eleiche *et al.*, 1996].

4.3.1   *General data on shields and projectiles*

Shields. Monolithic and layered (plates-in-contact and with air gaps) mild steel shields.

Table 1. Material properties of layers.

Label	σ_y (MPa)	σ_u (MPa)	ε_f (%)	HB
S1	210-247	316-346	29-32	96-105
S2	311	448	28	134

Projectiles. 7.62 mm bullets, $L/D = 4.2$; $m = 11.75\,\text{g}$.

4.3.2   *Test results*

Table 2. Test results for S1 shields with $T_{sum} = 8.0\,mm$.

v_s (m/s) →	706	755	775	805	826
Shield structure ↓					
[2.0]+[6.0]	477	514	550	609	658

v_s (m/s) $\rightarrow$	706	755	775	805	826
Shield structure $\downarrow$					
[6.0]+[2.0]	471	522	522	616	654
2×[4.0]	486	533	567	643	667
{[2.0]+[6.0] }\6.0	474	522	545	612	654
{[6.0]+[2.0] }\6.0	477	523	550	615	665
2×[4.0]\6.0	488	535	562	644	669
{2×[1.0]+[6.0]}\6.0	484	538	566	644	667
{[6.0]+2×[1.0]}\6.0	476	510	536	638	664
{2×[2.0]+[4.0]}\6.0	493	539	580	655	680
{[4.0]+2×[2.0]}\6.0	495	542	579	644	674
{4×[1.0]+[4.0]}\6.0	527	573	609	654	684
{[4.0]+4×[1.0]}\6.0	519	553	613	661	675
1×[(S2) 8.0]	446	489	528	611	647

Table 3. Test results for S1 shields and $v_s = 826 m/s$.

Shield structure	T_{sum} (mm)	v_r (m/s)
2×[4.0]\24.0	8.0	638
{2×[2.0]+[4.0]}\12.0	8.0	651
{[2.0]+[4.0]+[2.0]}\12.0	8.0	668
{[2.0]+[6.0]}\8.0	8.0	635
{[2.0]+3×[4.0]+[2.0]}\4.5	12.0	*
{[4.0]+[6.0]+[4.0]}\9.0	14.0	268
{2×[2.0]+2×[3.0]+[4.0]}\4.5	14.0	273
{[2.0]+[3.0]+[4.0]+[3.0]+[2.0]}\4.5	14.0	285
{[4.0]+[8.0]+[4.0]}\8.0	16.0	*
{[2.0]+[4.0]+[2.0]+ 2×[4.0]\4.5	16.0	*

* Incomplete penetration

Table 4. Test results for S2 shields and $v_s = 826 m/s$.

Shield structure	T_{sum} (mm)	v_r (m/s)
2×[4.0]\24.0	8.0	631
{[2.0]+[6.0]}\24.0	8.0	625
{2×[2.0]+[4.0]}\12.0	8.0	645
3×[4.0]\10.0	12.0	355
{[2.0]+[4.0]+[6.0]}\10.0	12.0	348
{4×[2.0]+[4.0]}\5.0	12.0	360
{[4.0]+[5.0]+[4.0]}\9.5	13.0	*

Shield structure	T_{sum} (mm)	v_r (m/s)
{[2.0]+[5.0]+[6.0]}\9.5	13.0	*
{2×[2.0]+2×[2.5]+[[4.0]}\4.75	13.0	*
[(S1) 6.0]+[(S2) 8.0]	14.0	261
2×[8.0]	16.0	*

* Incomplete penetration.

4.4 [Anderson *et al.*, 1999]

4.4.1 *General data on shields and projectiles*

Shields. Monolithic shields manufactured from two types of RHA steel, $T = 20.6\,mm$.

Projectiles. Cylindrical projectiles (Fig. 3.1.1) manufactured of different steels and D17 tungsten alloy; $D = 5.8\,mm$, $L = 58.0\,mm$.

Table 1. Material properties.

Material	γ (kg/m^3)	HV	σ_u (MPa)	δ (%)
Steel 1	7850	212±10	565±10	16.2±0.7
Steel 1	7850	425±15	1440±10	28.3
Steel 2	7850	465±15	1575±10	13.0±0.5
Steel 2	7850	750±15	2555±20	7.
Steel 3	7850	527±10	–	–
Steel 3	7850	562±10	–	–
Steel 3	7850	600±10	–	–
Steel 3	7850	640±10	–	–
Steel 4	8200	260±10	855±10	19.5±1.5
Steel 4	8200	280±10	–	–
Steel 4	8200	340±10	–	–
Steel 4	8200	440±10	–	–
Steel 4	8200	525±10	–	–
Steel 4	8200	550±10	1800±10	11±1
Steel 5	8830	227±10	–	–
Steel 5	8830	286±10	–	–
D17 tungsten alloy	17000	300±10	807±8	11
RHA 1	7850	486±10	1430±20	8±1
RHA 2	7850	550±10	1620±25	6±1

4.4.2 *Test results*

Table 2. Test results for RHA 1 shields.

Projectile material	v_{bl} (m/s)	Projectile material	v_{bl} (m/s)	Projectile material	v_{bl} (m/s)
Steel 1	1415±20	Steel 1	1395±15	Steel 2	1405±15
Steel 2	1280±20	Steel 3	1380±10	Steel 3	1338±5
Steel 3	1327±14	Steel 3	1320±5	Steel 4	1410±10
Steel 4	1390±5	Steel 4	1385±5	Steel 4	1385±5
Steel 4	1380±5	Steel 4	1357±10	Steel 5	1382±10
Steel 5	1352±10	D17	940±15		

Table 3. Test results for RHA 2 shields.

Impactor material	v_{bl} (m/s)	Impactor material	v_{bl} (m/s)	Impactor material	v_{bl} (m/s)
Steel 1	1395±15	Steel 1	1405±15	Steel 2	1400±20
Steel 2	1280±15	D17	940±15		

4.5 [Ansari *et al.*, 2010]

4.5.1 *General data on shields and projectiles*

Shields. Monolithic circular plates manufactured from commercially pure aluminum, $D^{sh} = 255\,mm$.

Projectiles. Hardened steel (HRC 56-57) cone-nose projectiles (Fig. 3.2.1); $D = 12.8\,mm$, $L = 40.8\,mm$, $\beta = 30°$, $m = 31.44\,g$.

4.5.2 *Test results*

Table 1. Test results for normal impact ($\theta_s = 0°$).

T (mm)	v_s (m/s)	v_r (m/s)	v_s (m/s)	v_r (m/s)	v_s (m/s)	v_r (m/s)
0.81	25	*	34.7	18.3	44.6	30.3
	54.3	44.8	62.5	61	78.3	71.5
	83.3	74.2	104.2	98.7		
1.52	40	5.6	49	18.5	62.5	42.5
	69.4	58	78.1	60	83.3	73.4
	96.2	82				

* Nonperforation

Table 2#. Test results for shield with $T = 0.81\,mm$ impacted at $\theta_s = 30°$.

v_s (m/s)	v_r (m/s)	v_s (m/s)	v_r (m/s)	v_s (m/s)	v_r (m/s)
34.7	12.7	36.8	18.9	56.6	38.2
69.4	56.3	77.9	70.1	89.1	78.6

Table 3. Test results for oblique impact ($\theta_s = 45°$).

T (mm)	v_s (m/s)	v_r (m/s)	v_s (m/s)	v_r (m/s)	v_s (m/s)	v_r (m/s)
0.8	36	5.7	44.6	34.2	62.8	53.6
	78.1	74.9	96.2	84.7		
1.5	53.7	*	65.8	48.6	78.1	53.4
	89.3	80.0	104.1	83.7		

* Ricochet

Table 4#. Test results for shield with $T = 0.81\,mm$ impacted at $\theta_s = 60°$.

v_s (m/s)	v_r (m/s)	v_s (m/s)	v_r (m/s)	v_s (m/s)	v_r (m/s)	v_s (m/s)	v_r (m/s)
44.8	29.1	56.7	36.2	65.7	49.9	96.0	81.4

4.6 [Awerbuch and Bodner, 1974]

4.6.1 *General data on shields and projectiles*

Shields. Monolithic metallic shields, $c_{l,w} = 250\,mm$.

Table 1. Mechanical properties of shield materials.

Label	Material	σ_y (MPa)	σ_u (MPa)	σ_{ss} (MPa)	HRC/HB
St-A	Steel Alloy	960	1078	637	35/–
St-B	Steel Alloy	1176	1470	784	45/–
St-C	Steel Alloy	1029	1294	686	42–
St-D	Steel Alloy	882	1176	294	38/–
St-E	Mild Steel	274	441	176	–/109
Al-A	1100-H14 Al	118	124	75	–/32
Al-B	6061-T6 Al	255	289	186	–/90

Label	Material	σ_y (MPa)	σ_u (MPa)	σ_{ss} (MPa)	HRC/HB
Al-C	6061-T6 Al	–	314	206	–/90
Al-D	6061-T6 Al	274	314	206	–/90

Table 2. Projectile description.

Label	Rifle type	Projectile type	D (mm)	m (g)
Projectile-1	Standard	Regular-lead	7.62	9.30
Projectile-2	Standard	AP	7.62	9.81*
Projectile-3	0.22 caliber	Regular-lead	5.6	2.63
Projectile-4	9 mm caliber Automatic	Regular-lead	9.0	7.40

* Core mass is 3.7 g

4.6.2 *Test results*

Table 3. Test results for Projectile-1.

Shield label	T (mm)	L^{pl} (mm)	v_s (m/s)	v_r (m/s)
St-A	6.0	5.0	850	500-600*
St-A	8.0	6.9	855	460
St-B	6.35	5.3	854	350-550
St-C	8.0	7.0	855	450-470
St-D	9.0	7.7	850	220-290
St-D	10.0	8.6	845	155-185
St-E	10.0	5.8	855	400
Al-C	9.6	2.5	845	748
Al-D	13.0	7.5	845	712
Al-D	19.0	12.0	836	568-585

* Fragmentation occurred

Table 4. Test results for Projectile-2.

Shield label	T (mm)	L^{pl} (mm)	v_s (m/s)	v_r (m/s)
St-A	6.0	5.0	835	550-650*
St-B	6.35	5.0	848	500-600*
St-D	12.0	9.0	855	300-390

* Fragmentation occurred.

Table 5. Test results for Projectile-3 and Al-A shield.

T (mm)	L^{pl} (mm)	v_s (m/s)	v_r (m/s)	T (mm)	L^{pl} (mm)	v_s (m/s)	v_r (m/s)
1.0	0.7	387	369	2.0	1.4	385	346
3.0	2.2	397	332	4.0	2.8	393	292
5.0	3.4	410	275	6.0	4.2	387	186

Table 6. Test results for Projectile-3 and Al-B shield.

T (mm)	L^{pl} (mm)	v_s (m/s)	v_r (m/s)	T (mm)	L^{pl} (mm)	v_s (m/s)	v_r (m/s)
1.0	0.65	379	357	2.0	1.5	399	341
3.0	2.0	397	309	4.0	2.5	389	268
5.0	3.5	401	174				

Table 7. Test results for Projectile-4.

Shield designation	T (mm)	L^{pl} (mm)	v_s (m/s)	v_r (m/s)
Al-A	4.0	2.80	416	355
Al-A	5.0	3.5	422	339
Al-A	6.0	4.2	428	330
Al-B	3.0	1.6	422	349
Al-B	5.0	3.6	416	291
Al-B	6.35	5.0	412	234

4.7   [Awerbuch and Bodner, 1977]

4.7.1   *General data on shields and projectiles*

Shields. Monolithic shields ($c_{l,w} = 250\,mm$) manufactured from commercially pure aluminum (*Al-1*), 6061-T6 aluminum (*Al-2*) and 1100-H14 aluminum (*Al-3*).

Projectiles. *0.22″* caliber lead bullet, $D = 5.56\,mm$, $m = 2.6\,g$.

Impact velocity. $v_s = 380 - 400\,m/s$.

Additional notation. $\delta_v = (v_s - v_r)/v_s$.

4.7.2 *Test results on dimensionless impactor velocity drop for Al-1 shields*

Table 1#. Test results for shields with *T*=2.0 *mm*.

θ_s (deg)	δ_v	θ_s (deg)	δ_v	θ_s (deg)	δ_v	θ_s (deg)	δ_v
0	0.09	0	0.11	10	0.11	20	0.09
30	0.12	30	0.11	40	0.14	49	0.16
60	0.21	70	0.35	74	1.00	76	1.00

Table 2#. Test results for shields with *T*=3.0 *mm*.

θ_s (deg)	δ_v	θ_s (deg)	δ_v	θ_s (deg)	δ_v	θ_s (deg)	δ_v
0	0.15	0	0.17	0	0.17	10	0.16
20	0.17	29	0.21	40	0.24	40	0.26
50	0.26	60	0.25	65	0.59	70	1.00

Table 3#. Test results for shields with *T*=4.0 *mm*.

θ_s (deg)	δ_v	θ_s (deg)	δ_v	θ_s (deg)	δ_v	θ_s (deg)	δ_v
0	0.22	0	0.25	0	0.27	10	0.20
20	0.26	30	0.31	40	0.35	50	0.43
60	0.64	61	0.68	61	0.69	62	0.76
62	1.00						

Table 4#. Test results for shields with *T*=5.0 *mm*.

θ_s (deg)	δ_v	θ_s (deg)	δ_v	θ_s (deg)	δ_v	θ_s (deg)	δ_v
0	0.30	0	0.32	0	0.35	1	0.38
10	0.30	10	0.31	10	0.35	20	0.34
20	0.36	20	0.38	30	0.38	30	0.42
30	0.44	30	0.52	35	0.46	40	0.51
40	0.53	40	0.55	40	0.66	45	0.66
44	0.72	45	0.74	46	0.70	47	0.71
48	0.75	45	0.99	47	0.99	49	1.00
51	0.99						

Table 5#. Test results for shields with T=6.0 *mm*.

θ_s (deg)	δ_v	θ_s (deg)	δ_v	θ_s (deg)	δ_v	θ_s (deg)	δ_v
0	0.45	0	0.47	1	0.51	0	0.54
10	0.51	10	0.52	10	0.53	20	0.52
20	0.54	20	0.61	19	0.67	29	0.65
30	0.68	30	0.71	30	0.73	30	0.74
35	0.59	35	0.72	35	0.79	39	0.99

4.7.3 *Test results on dimensionless impactor velocity drop for Al-2 shields*

Table 6#. Test results for shields with T=2.0 *mm*.

θ_s (deg)	δ_v	θ_s (deg)	δ_v	θ_s (deg)	δ_v	θ_s (deg)	δ_v
0	0.15	10	0.15	20	0.16	30	0.15
35	0.20	40	0.23	40	0.22	45	0.30
50	0.39	55	0.57	60	1.00		

Table 7#. Test results for shields with T=3.0 *mm*.

θ_s (deg)	δ_v	θ_s (deg)	δ_v	θ_s (deg)	δ_v	θ_s (deg)	δ_v
0	0.17	0	0.19	0	0.20	10	0.20
10	0.24	20	0.23	20	0.26	20	0.27
30	0.25	30	0.29	30	0.30	35	0.26
35	0.30	35	0.32	35	0.35	40	0.48
40	0.53	40	0.55	46	1.00		

Table 8#. Test results for shields with T=3.5 *mm*.

θ_s (deg)	δ_v	θ_s (deg)	δ_v	θ_s (deg)	δ_v	θ_s (deg)	δ_v
0	0.28	0	0.31	10	0.32	20	0.33
30	0.36	35	0.41	40	0.45	40	0.53
42	0.44	42	0.49	42	0.57	42	0.80
44	1.00						

Table 9#. Test results for shields with T=5.0 *mm*.

θ_s (deg)	δ_v	θ_s (deg)	δ_v	θ_s (deg)	δ_v	θ_s (deg)	δ_v
0	0.56	0	0.59	0	0.60	0	0.62
10	0.50	10	0.62	15	0.67	20	0.66
25	0.72	25	1.00				

4.7.4 *Test results for residual velocity and plug characteristics*

Note. Angular velocity of plug, ω^{pl}, in $\dfrac{rad}{s} \times 10^4$ units.

Table 10. Test results for Al-2 shields with T=3.0 *mm*.

θ_s (deg)	v_s (m/s)	v_r (m/s)	v^{pl} (m/s)	ω^{pl}
0	384	310	310	0
10	388	308	303	1.15
20	385	284	214	1.55
30	375	270	154	4.50
35	383	262	96	3.30
40	382	178	0	0

Table 11. Test results for Al-3 shields with T=5.0 *mm*.

θ_s (deg)	v_s (m/s)	v_r (m/s)	v^{pl} (m/s)	ω^{pl}
0	394	271	271	0
10	391	268	268	0
20	388	247	214	1.26
30	383	218	156	1.83
35	393	208	91	1.15
40	382	170	0	0.12
45	389	0	0	0

Table 12. Test results for Al-3 shields with T=6.0 *mm*.

θ_s (deg)	v_s (m/s)	v_r (m/s)	v^{pl} (m/s)	ω^{pl}
0	388	206	206	0
10	378	184	179	0.27
20	384	176	165	0.95
30	386	132	52	132
35	381	161	56	0

4.8 [Awerbuch, 1969]

See also [Awerbuch, 1970].

4.8.1 *General data on shields and projectiles*

Shields. Monolithic and layered (plates-in-contact and spaced plates), $c_{l.w} = 250\,mm$.
Projectiles. $0.22''$ caliber lead bullet, $D = 5.56\,mm$, $L = 12\,mm$, $m = 2.63\,g$.
Impact velocity. $v_s \approx 390\,m/s$.

Table 1. Mechanical properties of shield materials.

Label	Material	σ_y (MPa)	σ_u (MPa)	σ_{ss} (MPa)	γ (kg/m^3)
Al-1	1100-H14 AA	137	145	82	2695
Al-2	2024-T3 AA	345	482	282	2695
Al-3	Commercially pure aluminum	–	–	–	–
Al-4	Duralumin	–	–	–	–
St-1	Mild steel	–	686	–	7840

Note. The average values of some results obtained for the same experimental conditions are presented in [Awerbuch, 1970] in tables and figures. As a result, and also because of unavoidable errors in plotting experimental points, some indicators obtained for the same experimental conditions in the tables below may differ.

4.8.2 *Test results for monolithic shields*

Table 2#. Test results on velocity drop for normal impact.

Shields material	T (mm)	Δv (m/s)	T (mm)	Δv (m/s)	T (mm)	Δv (m/s)	T (mm)	Δv (m/s)
Al-1	1.0	18	2.0	38	2.0	40	2.0	42
	3.0	60	3.0	63	3.0	65	4.0	99
	4.0	101	5.0	135	5.0	139		
Al-2	1.0	22	2.0	58	3.0	87	3.0	89
	3.0	90	3.5	112	5.0	226	5.0	234

Shields material	T (mm)	Δv (m/s)	T (mm)	Δv (m/s)	T (mm)	Δv (m/s)	T (mm)	Δv (m/s)
St-1	0.8	44	0.8	46	0.9	56	0.9	60
	1.0	68	1.0	70	1.5	141	1.5	143
	2.2	391						

Table 3. Test results on residual velocity for normal impact.

Shields material	T (mm)	v_s (m/s)	v_r (m/s)	T (mm)	v_s (m/s)	v_r (m/s)	T (mm)	v_s (m/s)	v_r (m/s)
Al-1	1.0	387	369	2.0	385	347	3.0	397	333
	4.0	393	292	5.0	410	275	6.0	387	186
Al-2	1.0	379	357	2.0	399	341	3.0	397	310
	3.5	389	268	5.0	401	175			
St-1	0.9	382	322	1.5	388	260			

Table 4#. Test results on velocity drop for oblique impact against Al-2 shields.

T (mm)	θ_s (deg)	Δv (m/s)	θ_s (deg)	Δv (m/s)	θ_s (deg)	Δv (m/s)	θ_s (deg)	Δv (m/s)
1.0	0	20	10	19	20	23	30	23
	45	29	50	30	60	34	70	43
	80	62						
2.0	0	57	10	61	20	64	30	63
	35	80	40	87	40	92	45	114
	50	148	55	222	60	382		
3.0	0	87	0	91	10	97	20	94
	30	114	35	140	40	157	42	190
	45	386						
3.5	0	121	10	123	20	128	30	141
	35	163	40	182	40	205	42	196
	42	225	42	308				
5.0	0	226	0	230	0	235	10	243
	15	264	20	262	25	285	30	389

Table 5#. Test results on velocity drop for oblique impact against Al-3 shields.

T (mm)	θ_s (deg)	Δv (m/s)	θ_s (deg)	Δv (m/s)	θ_s (deg)	Δv (m/s)	θ_s (deg)	Δv (m/s)
1.0	0	19	10	19	20	21	40	23
	50	25	60	30	70	40		
2.0	0	36	10	42	20	40	30	45
	40	54	50	60	60	79	70	135
3.0	0	65	10	85	20	66	30	80
	40	90	50	100	60	139		

T (mm)	θ_s (deg)	Δv (m/s)	θ_s (deg)	Δv (m/s)	θ_s (deg)	Δv (m/s)	θ_s (deg)	Δv (m/s)
4.0	0	100	30	116	40	138	50	172
	60	243	70	390				
5.0	10	144	20	154	30	195	40	253
	50	390						
6.0	0	200	10	210	20	234	30	275
	40	390						

Table 6#. Test results on velocity drop for oblique impact against St-1 shields.

T (mm)	θ_s (deg)	Δv (m/s)	θ_s (deg)	Δv (m/s)	θ_s (deg)	Δv (m/s)	θ_s (deg)	Δv (m/s)
0.9	0	58	0	60	10	62	20	60
	30	70	30	72	40	79	40	81
	50	83	50	95	55	120	60	114
	60	130	70	173	80	383		
1.5	0	125	0	128	10	122	20	122
	30	130	40	171	50	262	52	282
	52	291	53	396				

Table 7. Test results for θ_s at which a bullet is stopped.

Shields material	T (mm)	θ_s (deg)	T (mm)	θ_s (deg)	T (mm)	θ_s (deg)	T (mm)	θ_s (deg)
Al-1	1.0	89	2.0	74	3.0	67	4.0	62
	5.0	47	6.0	38				
Al-2	2.0	60	3.0	45	3.5	45	5.0	22

Table 8. Test results for Al-3 shields: residual angle of projectile.

T (mm)	θ_s (deg)	θ_r (deg)	θ_s (deg)	θ_r (deg)	θ_s (deg)	θ_r (deg)	θ_s (deg)	θ_r (deg)
2.0	0	0	20	25	30	31	40	29
	50	41	60	63				
3.0	0	0	20	21	30	23	40	30
	50	32	60	47	65	40		
4.0	0	0	20	16	30	17	50	26
	60	33						
5.0	0	0	20	17	30	16	40	24
	45	26	50	27				
6.0	0	0	20	10	30	16	32	17

Table 9. Test results for Al-4 shields: residual angle of projectile.

T (mm)	θ_s (deg)	θ_r (deg)	θ_s (deg)	θ_r (deg)	θ_s (deg)	θ_r (deg)	θ_s (deg)	θ_r (deg)
2.0	0	0	10	8	20	−1	30	22
	40	30	50	46				
3.0	0	0	10	0	30	2	40	5
	45	5						
3.5	0	0	20	4	30	20	40	10
	42	10						
5.0	0	0	10	7	20	1	25	4

4.8.3 *Test results for layered shields*

Table 10#. Test results for Al-1 shields.

T_{sum} (mm)	Layered shield		Monolithic shields	T_{sum} (mm)	Layered shield		Monolithic shields
	Structure	Δv (m/s)	Δv (m/s)		Structure	Δv (m/s)	Δv (m/s)
0.9	–	–	12	1.0	–	–	18
1.8	2×[0,9]	31	33	2.0	2×[1.0]	34	40
2.0	2×[1.0]	44	40	2.7	3×[0.9]	52	56
3.0	3×[1.0]	71	63	3.6	4×[0.9]	88	85
4.0	4×[1.0]	101	100	4.0	2×[2.0]	106	100
4.5	5×[0.9]	107	117	5.0	5×[1.0]	148	135
5.4	6×[0.9]	166	159	6.0	6×[1.0]	218	200
6.0	3×[2.0]	189	200	6.0	3×[2.0]	192	200
6.0	3×[2.0]	222	200	6.3	7×[0.9]	194	–
6.3	7×[0.9]	205	–	6.3	7×[0.9]	268	–
7.0	7×[1.0]	230	–	7.0	7×[1.0]	392	–
7.2	8×[0.9]	231	–	7.2	8×[0.9]	242	–
7.2	8×[0.9]	254	–	8.1	9×[0.9]	290	–
8.1	9×[0.9]	393	–				

Table 11#. Test results for shield consisting of 0.9 mm St-1 plate and Al-3 plates having different thicknesses.

T_{sum} (mm)	Structure of shield	Δv (m/s)
1.9	[(St-1) 0.9]+[(Al-3) 1.0]	98

T_{sum} (mm)	Structure of shield	Δv (m/s)
1.9	{[(St-1) 0.9]+[(Al-3) 1.0]}\20	86
1.9	[(Al-3) 1.0]+[(St-1) 0.9]	73
1.9	{[(Al-3) 1.0]+[(St-1) 0.9]}\20	68
1.9	{[(Al-3) 1.0]+[(St-1) 0.9]}\20	76
2.9	[(St-1) 0.9]+[(Al-3) 2.0]	163
2.9	{[(St-1) 0.9]+[(Al-3) 2.0]}\20	117
2.9	[(Al-3) 2.0]+[(St-1) 0.9]	104
2.9	{[(Al-3) 2.0]+[(St-1) 0.9]}\20	119
3.9	[(St-1) 0.9]+[(Al-3) 3.0]	214
3.9	{[(St-1) 0.9]+[(Al-3) 3.0]}\20	164
3.9	[(Al-3) 3.0]+[(St-1) 0.9]	136
3.9	{[(Al-3) 3.0]+[(St-1) 0.9]}\20	173
3.9	{[(Al-3) 3.0]+[(St-1) 0.9]}\20	176
3.9	{[(Al-3) 3.0]+[(St-1) 0.9]}\20	190
4.9	[(St-1) 0.9]+[(Al-3) 4.0]	398
4.9	[(Al-3) 4.0]+[(St-1) 0.9]	217
4.9	{[(Al-3) 4.0]+[(St-1) 0.9]}\20	394
4.9	[(Al-3) 4.0]+[(St-1) 0.9]	222
4.9	{[(Al-3) 4.0]+[(St-1) 0.9]}\20	298
4.9	[(Al-3) 4.0]+[(St-1) 0.9]	227
4.9	{[(Al-3) 4.0]+[(St-1) 0.9]}\20	302
5.9	[(Al-3) 5.0]+[(St-1) 0.9]	396
5.9	{[(St-1) 0.9]+[(Al-3) 5.0]}\20	391

Table 12#. Test results for shield consisting of 1.5 mm St-1 plate and Al-3 plates having different thicknesses.

T_{sum} (mm)	Structure of shield	Δv (m/s)
2.5	[(St-1) 1.5]+[(Al-3) 1.0]	169
2.5	{[(St-1) 1.5]+[(Al-3) 1.0]}\20	138
2.5	[(Al-3) 1.0]+[(St-1) 1.5]	113
2.5	{[(Al-3) 1.0]+[(St-1) 1.5]}\20	147

T_{sum} (mm)	Structure of shield	Δv (m/s)
2.5	[(Al-3) 1.0]+[(St-1) 1.5]	134
2.5	[(Al-3) 1.0]+[(St-1) 1.5]	160
3.5	[(St-1) 1.5]+[(Al-3) 2.0]	274
3.5	{[(St-1) 1.5]+[(Al-3) 2.0]}\20	230
3.5	[(Al-3) 2.0]+[(St-1) 1.5]	165
3.5	{[(Al-3) 2.0]+[(St-1) 1.5]}\20	232
3.5	[(St-1) 1.5]+[(Al-3) 2.0]	314
3.5	[(Al-3) 2.0]+[(St-1) 1.5]	172
3.5	[(St-1) 1.5]+[(Al-3) 2.0]	406
3.5	[(Al-3) 2.0]+[(St-1) 1.5]	207
4.5	[(St-1) 1.5]+[(Al-3) 3.0]	406
4.5	{[(St-1) 1.5]+[(Al-3) 3.0]}\20	315
4.5	[(Al-3) 3.0]+[(St-1) 1.5]	207
4.5	{[(Al-3) 3.0]+[(St-1) 1.5]}\20	304
4.5	[(Al-3) 3.0]+[(St-1) 1.5]	234
4.5	[(Al-3) 3.0]+[(St-1) 1.5]	247
5.5	{[(St-1) 1.5]+[(Al-3) 4.0]}\20	405
5.5	[(Al-3) 4.0]+[(St-1) 1.5]	382
5.5	{[(Al-3) 4.0]+[(St-1) 1.5]}\20	402
5.5	[(Al-3) 4.0]+[(St-1) 1.5]	392

4.9 [Awerbuch, 1970]

4.9.1 *General data on shields and projectiles*

Shields. Monolithic shields manufactured of 1100-H14 aluminum (*Al-1*), 2024-T3 aluminum alloy (*Al-2*), commercially pure aluminum (*Al-3*) and mild steel (*St-1*).

Projectiles. *0.22″* caliber lead bullet, $D = 5.56 mm$, $m = 2.6 \, \text{g}$.

Impact velocity. $v_s = 380-400 m/s$ [Awerbuch and Bodner, 1977].

4.9.2 *Test results for normal impact*

Table 1#. Test results for Al-1 shields.

T (mm)	Δv (m/s)	T (mm)	Δv (m/s)	T (mm)	Δv (m/s)	T (mm)	Δv (m/s)
1.0	19	2.0	38	3.0	65	4.0	100
5.0	135	6.0	199				

Table 2#. Test results for Al-2 shields.

T (mm)	Δv (m/s)	T (mm)	Δv (m/s)	T (mm)	Δv (m/s)	T (mm)	Δv (m/s)
1.0	23	2.0	58	3.0	88	3.5	121
5.0	235						

Table 3#. Test results for St-1 shields.

T (mm)	Δv (m/s)	T (mm)	Δv (m/s)	T (mm)	Δv (m/s)	T (mm)	Δv (m/s)
0.8	48	0.9	60	1.5	128	2.0	394

4.9.3 *Test results for oblique impact against Al-3 shields*

Table 4#. Test results for shields with T=1.0 *mm*.

θ_s (deg)	Δv (m/s)	θ_s (deg)	Δv (m/s)	θ_s (deg)	Δv (m/s)	θ_s (deg)	Δv (m/s)
0	18	10	18	20	19	30	19
40	23	50	24	60	29	70	41
80	63	85	65	88	72		

Table 5#. Test results for shields with T=2.0 *mm*.

θ_s (deg)	Δv (m/s)	θ_s (deg)	Δv (m/s)	θ_s (deg)	Δv (m/s)	θ_s (deg)	Δv (m/s)
0	39	0	44	10	45	10	42
20	37	30	46	30	48	40	54
50	60	60	80	60	100	69	138
75	375						

Table 6#. Test results for shields with $T = 3.0$ *mm*.

θ_s (deg)	Δv (m/s)	θ_s (deg)	Δv (m/s)	θ_s (deg)	Δv (m/s)	θ_s (deg)	Δv (m/s)
0	61	0	65	0	68	10	64
20	65	30	79	40	88	40	97
50	99	65	230	70	384		

Table 7#. Test results for shields with $T = 4.0$ *mm*.

θ_s (deg)	Δv (m/s)	θ_s (deg)	Δv (m/s)	θ_s (deg)	Δv (m/s)	θ_s (deg)	Δv (m/s)
0	89	0	99	0	102	10	85
20	100	30	117	40	146	50	161
60	244	61	268	61	272	61	275
62	251	62	295	63	362	63	384

Table 8#. Test results for shields with $T = 5.0$ *mm*.

θ_s (deg)	Δv (m/s)	θ_s (deg)	Δv (m/s)	θ_s (deg)	Δv (m/s)	θ_s (deg)	Δv (m/s)
0	132	0	141	0	154	10	144
20	156	30	196	40	254	45	261
45	276	45	280	46	273	48	282
48	299	48	394				

Table 9#. Test results for shields with $T = 6.0$ *mm*.

θ_s (deg)	Δv (m/s)	θ_s (deg)	Δv (m/s)	θ_s (deg)	Δv (m/s)	θ_s (deg)	Δv (m/s)
0	200	0	205	10	209	20	231
20	234	20	259	30	258	30	275

4.10 [Babaei *et al.*, 2011]

4.10.1 *General data on shields and projectiles*

Shields. Two-layer (plates-in-contact) shields.

Table 1. Properties of shields layers.

Notation	Layers materials	E (GPa)	v	γ (kg/m^3)	σ_y (MPa)	T (mm)
St	Steel	845	0.33	7850	845	1
Al	Aluminum	103	0.33	2700	103	1

Table 2. Cylindrical projectiles (Fig. 3.1.1).

D (mm)	L (mm)	Material	E (GPa)	ν	γ (kg/m^3)	σ_y (MPa)	m (g)
4	32	Steel	200	0.33	7850	1800	3.16

Notations for penetration modes. N – no perforation, Y – yaw angle, S – projectile shatters.

4.10.2 *Test results*

Table 3. Test results for 2×[(Al) 1.0] shields.

v_s (m/s)	v_r (m/s)	ΔL (mm)	v_s (m/s)	v_r (m/s)	ΔL (mm)	v_s (m/s)	v_r (m/s)	ΔL (mm)
81.4	0	0.0	89.1	0	0.1	93.6	24.7	0.2
101.1	54.4	0.0	122.9	91.7	0.0	128.7	Y	–
143.5	119.2	0.0	175.2	155.5	0.3	188.5	S	–
191.1	174.8	0.3						

$v_{bl} = 91 m/s$.

Table 4. Test results for [(Al) 1.0]+[(St) 1.0] shields.

v_s (m/s)	v_r (m/s)	ΔL (mm)	v_s (m/s)	v_r (m/s)	ΔL (mm)	v_s (m/s)	v_r (m/s)	ΔL (mm)
122.1	0	0.3	143.4	0	0.4	146.9	36.0	0.4
150.2	45.3	0.4	169.5	99.1	0.6	187.7	135.3	0.5
216.3	171.7	0.5	283.6	Y, N	–	301.6	260.6	0.5
345.8	Y, N	–						

$v_{bl} = 145 m/s$.

Table 5. Test results for [(St) 1.0]+[(Al) 1.0] shields.

v_s (m/s)	v_r (m/s)	ΔL (mm)	v_s (m/s)	v_r (m/s)	ΔL (mm)	v_s (m/s)	v_r (m/s)	ΔL (mm)
134.6	0	0.4	141.8	0	0.6	151.5	0	0.5
156.4	18.4	0.5	170.9	89.0	0.6	183.4	S	–
202.2	151.2	0.4	234.0	Y	–	243.6	191.5	0.5
307.1	259.4	0.5						

$v_{bl} = 156 m/s$.

Table 6. Test results for 2×[(St) 1.0] shields.

v_s (m/s)	v_r (m/s)	ΔL (mm)	v_s (m/s)	v_r (m/s)	ΔL (mm)	v_s (m/s)	v_r (m/s)	ΔL (mm)
171.4	0	0.7	175.2	38.7	0.7	181.3	62.3	0.7
186.6	78.5	0.7	207.8	114.6	0.8	215.6	Y, N	–
239.2	167.1	0.9	268.5	S	4.3	271.5	212.0	0.9
289.7	S	5.4	330.2	280.7	1.2			

$v_{bl} = 173\,m/s$.

4.11 [Backman and Finnegan, 1976]

4.11.1 *General data on shields and projectiles*

Shields. Monolithic AA and steel shields.

Projectiles. SAE 52100 alloy steel spherical projectiles (Fig. 3.5.1); $D = 6.35\,mm$.

Notations for penetration modes (PMs). P – perforation, E – embedment, R – ricochet.

4.11.2 *Test results for 6061-T6 AA shields with T = 9.53 mm*

Table 1#. Results for $\theta_s = 0°$ (perforation).

v_s (m/s)	v_r (m/s)	v_s (m/s)	v_r (m/s)	v_s (m/s)	v_r (m/s)	v_s (m/s)	v_r (m/s)
734	7	788	222	890	430	1014	628
1193	825	1219	877	1332	984	1546	1195

Table 2#. Results for $\theta_s = 20°$.

v_s (m/s)	v_r (m/s)	θ_r (deg)	PM	v_s (m/s)	v_r (m/s)	θ_r (deg)	PM
211	52	151	R	319	44	178	R
466	21	180	R	507	21	180	R
794	213	11	P	822	251	10	P
1007	581	16	P	1113	687	16	P

v_s (m/s)	v_r (m/s)	θ_r (deg)	PM	v_s (m/s)	v_r (m/s)	θ_r (deg)	PM
1227	838	19	P	1313	918	20	P
1376	1004	21	P				

Table 3#. Results for $\theta_s = 45°$.

v_s (m/s)	v_r (m/s)	θ_r (deg)	PM	v_s (m/s)	v_r (m/s)	θ_r (deg)	PM
275	119	119	R	358	153	120	R
463	138	133	R	556	106	154	R
565	90	163	R	924	–	–	E
950	184	17	P	1102	490	37	P
1254	717	39	P	1436	861	43	P

Table 4#. Results for $\theta_s = 60°$.

v_s (m/s)	v_r (m/s)	θ_r (deg)	PM	v_s (m/s)	v_r (m/s)	θ_r (deg)	PM
262	165	111	R	477	276	116	R
736	252	130	R	874	182	138	R
1020	0	180	R	1114	–	–	E
1235	258	32	P	1366	484	51	P
1525	655	57	P				

4.11.3 *Test results for 2024-T3 AA shields*

Table 5#. Results for $T = 1.27\ mm$, $\theta_s = 0°$ (perforation).

v_s (m/s)	v_r (m/s)	v_s (m/s)	v_r (m/s)	v_s (m/s)	v_r (m/s)
224	157	260	199	295	242
328	292	854	783	1510	1391

Table 6#. Results for $T = 1.27\ mm$, $\theta_s = 60°$.

v_s (m/s)	v_r (m/s)	θ_r (deg)	PM	v_s (m/s)	v_r (m/s)	θ_r (deg)	PM
254	--	111	R	258	–	117	R
265	116	103	R	275	15	0	P
277	82	112	R	285	–	105	R
287	–	116	R	291	118	34	P
294	82	40	P	299	41	142	R

v_s (m/s)	v_r (m/s)	θ_r (deg)	PM	v_s (m/s)	v_r (m/s)	θ_r (deg)	PM
300	21	–	P	322	144	42	P
331	161	–	P	378	220	50	P
573	472	62	P	608	513	56	P
708	618	64	P	866	762	63	P

Table 7#. Results for $T = 3.18\ mm$, $\theta_s = 0°$ (perforation).

v_s (m/s)	v_r (m/s)	v_s (m/s)	v_r (m/s)	v_s (m/s)	v_r (m/s)	v_s (m/s)	v_r (m/s)
424	214	831	673	868	747	1085	945
1316	1186	1640	1480	1714	1549		

Table 8#. Results for $T = 3.18\ mm$, $\theta_s = 30°$.

v_s (m/s)	v_r (m/s)	θ_r (deg)	PM	v_s (m/s)	v_r (m/s)	θ_r (deg)	PM
200	29	125	R	280	–	142	R
327	10	180	R	390	11	180	R
407	11	134	R	448	158	0	P
526	292	17	P	762	526	26	P
851	689	30	P	1020	909	29	P
1104	976	30	P	1248	1102	30	P
1317	–	30	P				

Table 9#. Results for $T = 3.18\ mm$, $\theta_s = 45°$.

v_s (m/s)	v_r (m/s)	θ_r (deg)	PM	v_s (m/s)	v_r (m/s)	θ_r (deg)	PM
282	83	115	R	362	88	123	R
370	65	126	R	395	62	129	R
478	39	1	P	573	289	30	P
757	530	41	P	797	545	41	P
971	780	44	P	1130	991	45	P
1400	1187	45	P				

Table 10#. Results for $T = 3.18\ mm$, $\theta_s = 60°$.

v_s (m/s)	v_r (m/s)	θ_r (deg)	PM	v_s (m/s)	v_r (m/s)	θ_r (deg)	PM
328	204	108	R	446	237	108	R
473	229	113	R	479	256	108	R
505	230	112	R	536	–	104	R
592	–	109	R	635	21	180	R
644	56	150	R	669	77	–8	P

v_s (m/s)	v_r (m/s)	θ_r (deg)	PM	v_s (m/s)	v_r (m/s)	θ_r (deg)	PM
698	–	32	P	718	194	32	P
737	227	31	P	899	568	54	P
1048	719	59	P	1099	818	59	P
1243	920	60	P				

Table 11#. Results for $T = 6.35\ mm$, $\theta_s = 0°$ (perforation).

v_s (m/s)	v_r (m/s)	v_s (m/s)	v_r (m/s)	v_s (m/s)	v_r (m/s)	v_s (m/s)	v_r (m/s)
551	0	605	173	787	355	856	450
863	469	878	510	882	476	1145	816
1290	977	1346	993				

Table 12#. Results for $T = 6.35\ mm$, $\theta_s = 30°$.

v_s (m/s)	v_r (m/s)	θ_r (deg)	PM	v_s (m/s)	v_r (m/s)	θ_r (deg)	PM
316	78	131	R	367	82	149	R
399	70	133	R	652	–	–	E
729	143	3	P	850	394	20	P
872	425	25	P	1092	733	30	P
1297	952	30	P	1431	1093	31	P

Table 13#. Results for $T = 6.35\ mm$, $\theta_s = 45°$.

v_s (m/s)	v_r (m/s)	θ_r (deg)	PM	v_s (m/s)	v_r (m/s)	θ_r (deg)	PM
239	–	108	R	410	150	127	R
471	166	123	R	580	117	155	R
774	–	–	E	788	36	–	P
860	216	11	P	906	330	24	P
1048	528	36	P	1136	675	38	P
1281	837	44	P	1592	1069	45	P

Table 14#. Results for $T = 6.35\ mm$, $\theta_s = 60°$.

v_s (m/s)	v_r (m/s)	θ_r (deg)	PM	v_s (m/s)	v_r (m/s)	θ_r (deg)	PM
342	217	–	R	371	266	107	R
524	307	113	R	728	310	121	R
750	308	120	R	898	199	140	R
967	187	145	R	1032	–	–	E

v_s (m/s)	v_r (m/s)	θ_r (deg)	PM	v_s (m/s)	v_r (m/s)	θ_r (deg)	PM
1060	–	–	E	1128	203	32	P
1159	187	18	P	1210	384	42	P
1296	577	53	P	1388	641	59	P
1513	831	57	P	1551	850	60	P

4.11.4 *Test results for mild steel (SAE 1010) shields with* $T = 1.47$ *mm*

Table 15#. Results for $\theta_s = 0°$ (perforation).

v_s (m/s)	v_r (m/s)	v_s (m/s)	v_r (m/s)	v_s (m/s)	v_r (m/s)	v_s (m/s)	v_r (m/s)
307	0	445	229	627	426	720	529
870	679	1027	805	1118	854	1319	995

Table 16#. Results for $\theta_s = 30°$.

v_s (m/s)	v_r (m/s)	θ_r (deg)	PM	v_s (m/s)	v_r (m/s)	θ_r (deg)	PM
270	12	180	R	297	6	180	R
327	2	180	R	348	–	–	E
369	–	–	E	389	134	8	P
456	228	18	P	506	309	26	P
533	332	–	P	608	412	30	P
754	548	30	P	881	681	–	P
888	717	30	P	996	741	30	P

Table 17#. Results for $\theta_s = 45°$.

v_s (m/s)	v_r (m/s)	θ_r (deg)	PM	v_s (m/s)	v_r (m/s)	θ_r (deg)	PM
249	14	–	R	264	39	117	R
276	23	122	R	306	18	124	R
334	–	–	E	356	–	–	E
378	–	–	E	436	102	–	P
443	124	21	P	446	150	10	P
530	280	36	P	537	306	–	P
559	286	–	P	568	331	39	P
589	348	45	P	708	480	45	P

v_s (m/s)	v_r (m/s)	θ_r (deg)	PM	v_s (m/s)	v_r (m/s)	θ_r (deg)	PM
843	637	45	P	885	642	45	P
942	714	45	P	1105	820	45	P

Table 18#. Results for $\theta_s = 60°$.

v_s (m/s)	v_r (m/s)	θ_r (deg)	PM	v_s (m/s)	v_r (m/s)	θ_r (deg)	PM
200	126	99	R	277	122	106	R
363	147	120	R	380	60	116	R
395	85	122	R	404	109	118	R
447	12	159	R	475	–	–	E
544	34	0	P	580	134	30	P
615	240	50	P	663	276	54	P
794	378	60	P	933	536	62	P
496	692	60	P				

4.12 [Baranov *et al.*, 1991]

Note. Experimental data by I. Kh. Mingazov.

4.12.1 *General data on shields and projectiles*

Shields. Monolithic mild steel shields; $\gamma = 7850\ kg/m^3$, HD 1860 MPa.
Projectiles. Spherical projectiles (Fig. 3.5.1); $D = 4\,mm$,
$\gamma = 7850\ kg/m^3$, $m = 0.263$ g.

Note. Dynamical hardness, HD, is an indicator used by scientists from the former USSR; for details see [Vitman and Ioffe, 1948] and [Vitman and Stepanov, 1959].

4.12.2 *Test results*

Table 1. Test results.

v_s (m/s)	P (mm)	v_s (m/s)	P (mm)	v_s (m/s)	P (mm)	v_s (m/s)	P (mm)
406	1.30	533	1.73	610	2.05	766	2.50
811	2.40						

4.13 [Bartus, 2009]

4.13.1 *General data on shields and projectiles*

Shield. Monolithic Ti-5AL-5Mo-5V-3Cr (Ti-5553) plates after different heat treatment, $c_{l,w} = 305\,mm$.

Projectiles. 0.30 Caliber AP M2 projectiles (Fig. 3.11.1, Table 3.11.1) and 0.50 Caliber FSPs (Fig. 3.10.1, Table 3.10.1).

4.13.2 *Test results*

Table 1. Test results for BLV.

Heat treatment	Projectile type	T (mm)	v_{bl} (m/s)
STA	0.30 Caliber AP M2	13.9	729
BASCA	0.30 Caliber AP M2	14.0	698
STA	0.50 Caliber FSP	13.9	916
BASCA	0.50 Caliber FSP	13.8	720

Note. Processing schedules subjected to the STA (solution treated and aged) and BASCA (beta annealed, slow cooled and aged) heat treatment are illustrated in the original paper.

4.14 [Børvik *et al.*, 1999]

4.14.1 *General data on shields and projectiles*

Shields. Monolithic Weldox 460 E steel alloy shields; $T = 12.1\,mm$ mm; $\widetilde{D}^{sh} = 500\,mm$.

Projectiles. Arne oil hardened steel (HRC 53) cylindrical projectiles (Fig. 3.1.1); $D = 20\,mm$, $L = 80\,mm$.

4.14.2 *Test results*

Table 1. Test results on v_r, v^{pl} and m^{pl}.

v_s (m/s)	m (g)	v_r (m/s)	v^{pl} (m/s)	m^{pl} (g)
174	196.1	0	0	–
177	197.0	0	0	–

v_s (m/s)	m (g)	v_r (m/s)	v^{pl} (m/s)	m^{pl} (g)
179	196.8	0	0	–
182	196.7	0	0	–
184	196.8	31	45	27.8
185	196.9	0	0	–
189	196.9	43	67	27.8
189	196.7	40	67	28.0
190	196.9	44	72	27.9
190	196.7	42	64	27.7
199	196.9	67	104	27.8
200	197.0	71	104	27.8
225	196.9	114	169	27.3
244	196.9	133	188	28.1
285	197.3	181	225	27.6
304	196.6	200	242	27.6

$v_{bl} = 184.5\,m/s$; $a = 0.76$, $p = 2.36$ or $a = 0.85$, $p = 2.00$.

4.15   [Børvik *et al.*, 2002a]

4.15.1   *General data on shields and projectiles*

Shields. Monolithic Weldox 460 E steel alloy shields; $T = 12\,mm$; $\widetilde{D}^{sh} = 500\,mm$.

Projectiles. Cylindrical (Fig. 3.1.1, $L = 80\,mm$), hemispherical-nose (Fig. 3.4.1; $L_{cyl} = 74\,mm$) and truncated cone-nose (Fig. 3.6.1; $L_{cyl} = 68\,mm$, $L_{nose} = 30\,mm$, $d = 2\,mm$) projectiles with $D = 20\,mm$ and $m = 197g$ manufactured of Arne oil hardened steel (HRC 53).

4.15.2   *Test results*

Table 1. Test results for cylindrical projectiles.

v_s (m/s)	v_r (m/s)	v^{pl} (m/s)	m^{pl} (g)	ΔD (mm)	ΔL (mm)
182	0	0	–	0.11	0.24
184	31	45	27.8	0.17	0.20
185	0	0	–	0.17	0.22
190	42	64	27.7	0.19	0.22
200	71	104	27.8	0.29	0.28

v_s (m/s)	v_r (m/s)	v^{pl} (m/s)	m^{pl} (g)	ΔD (mm)	ΔL (mm)
225	114	169	27.3	0.49	0.47
244	133	188	28.1	0.50	0.53
285	181	225	27.6	0.77	0.72
304	200	242	27.6	1.01	0.90
400	291	–	36.0	–	–

$v_{bl} = 184.5\,m/s$; $a = 0.79$, $p = 2.24$.

Table 2. Test results for hemispherical-nose projectiles.

v_s (m/s)	v_r (m/s)	v^{pl} (m/s)	m^{pl} (g)	ΔD (mm)	ΔL (mm)
279	0	–	–	–	–
292	0	–	18.58	–	–
300	97	160	20.30	0.01	0.18
327	155	189	18.08	0.01	0.33
363	220	265	16.72	0.02	0.35
421	284	302	19.80	0.22	2.38
452	325	348	15.12	0.40	0.50

$v_{bl} = 292.1\,m/s$; $a = 0.81$, $p = 2.71$.

Table 3. Test results for truncated cone-nose projectiles (plug is not formed).

v_s (m/s)	v_r (m/s)	ΔD (mm)	ΔL (mm)	v_s (m/s)	v_r (m/s)	ΔD (mm)	ΔL (mm)
207	0	0.01	0.01	249	0	0.06	0.02
281	0	0.02	0.04	300	110	0.03	0.02
318	156	0.02	0.01	356	232	0.02	0.05
406	312	0.01	0.19				

$v_{bl} = 290.6\,m/s$; $a = 0.95$, $p = 2.52$.

4.16 [Børvik *et al.*, 2003]

See also [Børvik *et al.*, 1998].

4.16.1 *General data on shields and projectiles*

Table 1. Monolithic shields.

Material	E (GPa)	σ_y (MPa)	σ_u (MPa)	ε_f (%)	$\widetilde{D}^{sh}$ (mm)
Weldox 460 E steel	200	490	580	35	500

Table 2. Cylindrical projectiles (Fig. 3.1.1).

D (mm)	L (mm)	Material	E (GPa)	σ_y (MPa)	σ_u (MPa)	ε_f (%)	HRC
20	80	Arne oil hardened steel	204	~1900	2080	2.4	53

4.16.2 *Test results*

Table 3. Test results for shields with $T = 6\,mm$.

v_s (m/s)	m (g)	v_r (m/s)	v^{pl} (m/s)	m^{pl} (g)	T^{pl} (mm)	D_{fr}^{pl} (mm)	D_{max}^{pl} (mm)	ΔD (mm)	ΔL (mm)
77	196.9	0	0	0	–	–	–	0	0
97	196.8	0	0	0	–	–	–	0.02	0.02
98	195.9	0	0	0	–	–	–	0	0.02
111	196.6	0	0	0	–	–	–	0.01	0.04
124	196.3	0	0	0	–	–	–	0.03	0.03
137	196.7	0	0	0	–	–	–	0.06	0.02
145	196.7	0	0	0	–	–	–	0.02	0.02
146	196.6	38	50	12.0	5.9	20.3	20.4	0.10	0.06
157	196.5	64	87	12.3	6.0	20.0	20.2	0.08	0.02
165	196.5	96	120	12.8	6.1	19.8	20.5	0.05	0.03
165	196.5	98	112	12.5	5.9	20.2	20.3	0.09	0.08
185	196.5	132	169	13.6	6.0	20.2	20.6	0.15	0.14
201	196.6	158	176	13.8	6.0	20.3	20.6	0.20	0.15
234	197.0	201	241	14.0	6.0	20.4	20.8	0.33	0.19
296	196.8	260	-	14.3	6.1	20.6	21.1	0.50	0.32

$v_{bl} = 145.5 \pm 0.2\,m/s$; $a = 0.94$, $p = 2.71$.

Table 4. Test results for shields with $T = 8\,mm$.

v_s (m/s)	m (g)	v_r (m/s)	v^{pl} (m/s)	m^{pl} (g)	T^{pl} (mm)	D_{fr}^{pl} (mm)	D_{max}^{pl} (mm)	ΔD (mm)	ΔL (mm)
137	196.9	137	0	0	–	–	–	0.01	0.02
153	196.3	153	0	0	–	–	–	0.07	0.05
156	196.9	156	86	18.3	8.0	20.4	21.1	0.06	0.12
159	196.6	159	84	18.2	8.1	20.3	20.9	0.02	0.05
160	197.0	160	99	18.6	8.2	20.7	20.8	0.05	0.07
160	196.8	160	83	18.2	8.0	20.3	20.4	0.09	0.15
161	196.4	161	103	18.2	8.2	20.3	20.5	0.11	0.09
161	196.6	161	95	18.5	8.1	20.4	20.4	0.11	0.09
165	196.6	165	99	18.5	8.2	20.5	20.6	0.07	0.12
174	196.3	174	131	18.5	8.0	20.3	20.8	0.08	0.10
182	196.8	182	143	18.1	7.9	20.3	20.5	0.12	0.14

v_s (m/s)	m (g)	v_r (m/s)	v^{pl} (m/s)	m^{pl} (g)	T^{pl} (mm)	D_{fr}^{pl} (mm)	D_{max}^{pl} (mm)	ΔD (mm)	ΔL (mm)
191	196.5	191	168	18.6	8.1	20.4	20.8	0.15	0.10
251	196.8	251	223	18.8	8.0	20.2	20.8	0.38	0.38
298	196.8	298	261	18.7	7.9	20.7	21.3	0.58	0.47

$v_{bl} = 154.3 \pm 1.8\,m/s$; $a = 0.84$, $p = 3.30$.

Table 5. Test results for shields with $T = 9.7\,mm$.

v_s (m/s)	m (g)	v_r (m/s)	v^{pl} (m/s)	m^{pl} (g)	T^{pl} (mm)	D_{fr}^{pl} (mm)	D_{max}^{pl} (mm)	ΔD (mm)	ΔL (mm)
131	197.1	0	0	0	–	–	–	0.02	0.03
161	196.5	0	0	0	–	–	–	0.06	0.05
169	197.0	66	93	22.2	9.52	20.3	20.4	0.07	0.16
172	195.8	66	95	22.1	9.51	20.3	20.6	0.08	0.16
173	196.9	67	96	21.9	9.47	20.3	20.4	0.11	0.18
185	196.9	95	126	21.8	9.41	20.3	20.6	0.13	0.22
193	196.5	109	145	22.2	9.42	20.4	20.5	0.15	0.22
204	197.3	124	162	22.4	9.48	20.4	20.8	0.23	0.25
242	196.4	166	204	22.1	9.42	20.3	21.2	0.41	0.43
278	197.2	198	235	22.2	9.29	20.6	21.5	0.55	0.51
296	196.7	218	251	22.1	9.28	20.8	21.5	0.73	0.70

$v_{bl} = 165.3 \pm 4.1\,m/s$; $a = 0.76$, $p = 3.23$.

Table 6. Test results for shields with $T = 12\,mm$.

v_s (m/s)	m (g)	v_r (m/s)	v^{pl} (m/s)	m^{pl} (g)	T^{pl} (mm)	D_{fr}^{pl} (mm)	D_{max}^{pl} (mm)	ΔD (mm)	ΔL (mm)
174	196.1	0	0	0	–	–	–	0.14	0.22
177	197.0	0	0	0	–	–	–	0.15	0.13
179	196.8	0	0	0	–	–	–	0.16	0.13
182	196.7	0	0	0	–	–	–	0.11	0.24
184	196.8	31	45	27.8	11.7	20.5	21.3	0.17	0.20
189	196.9	43	67	27.8	11.7	20.6	21.3	0.19	0.27
189	196.7	40	67	28.0	11.7	20.4	21.4	0.21	0.34
190	196.9	44	72	27.9	11.8	20.4	21.3	0.19	0.28
190	196.7	42	64	27.7	11.7	20.4	21.2	0.17	0.22
199	196.9	67	104	27.8	11.6	20.4	21.3	0.17	0.22
200	197.0	71	104	27.8	11.6	20.4	21.4	0.29	0.28
225	196.9	114	169	27.3	11.5	20.5	21.2	0.49	0.47
244	196.9	133	188	28.1	11.6	20.7	21.7	0.50	0.53
285	197.3	181	225	27.6	11.4	20.7	21.7	0.77	0.70
304	196.6	200	242	27.6	11.4	21.0	21.9	1.01	0.90

$v_{bl} = 184.5 \pm 0.3\,m/s$; $a = 0.74$, $p = 2.42$.

Table 7. Test results for shields with $T = 16\,mm$.

v_s (m/s)	m (g)	v_r (m/s)	v^{pl} (m/s)	m^{pl} (g)	T^{pl} (mm)	D_{fr}^{pl} (mm)	D_{max}^{pl} (mm)	ΔD (mm)	ΔL (mm)
234	196.1	0	0	0	–	–	–	0.28	0.22
240	196.5	32	84	36.1	15.2	20.5	21.9	0.57	0.81
242	196.9	56	94	36.6	15.1	20.7	21.7	0.62	0.81
243	196.4	52	94	35.8	15.2	20.4	21.4	0.49	0.83
249	196.8	57	100	35.5	15.2	20.4	21.8	0.64	0.89
251	196.6	64	–	36.2	15.1	20.7	22.1	0.64	0.88
279	197.5	107	142	36.2	14.9	21.5	22.1	0.81	1.13
312	196.7	140	184	35.9	14.8	20.7	22.2	1.16	1.57
356	196.1	189	237	35.1	14.8	21.0	22.3	1.93	1.59

$v_{bl} = 236.9 \pm 2.6\,m/s$; $a = 0.63$, $p = 2.40$.

Table 8. Experimental data for $T = 20\,mm$.

v_s (m/s)	m (g)	v_r (m/s)	v^{pl} (m/s)	m^{pl} (g)	T^{pl} (mm)	D_{fr}^{pl} (mm)	D_{max}^{pl} (mm)	ΔD (mm)	ΔL (mm)
185	196.8	0	0	0	–	–	–	0.30	0.56
293	196.9	0	0	0	–	–	–	0.75	0.68
295	197.1	26	67	44.8	18.0	20.9	22.2	0.93	1.32
304	196.8	*	–	47.8	18.3	21.1	23.5	0.89	0.81
307	196.5	*	36	46.5	17.9	20.4	22.6	0.92	0.84
307	196.7	37	83	45.4	17.9	21.0	22.4	0.65	1.74
318	196.2	44	–	44.3	18.0	21.0	22.5	1.07	0.99
319	196.9	58	111	44.4	18.0	20.9	22.5	0.71	1.72
338	196.1	91	132	44.4	17.8	21.3	23.0	1.42	2.13
352	196.9	94	145	47.1	17.6	21.7	23.4	1.70	2.53
360	196.0	117	152	44.4	17.7	21.3	22.7	1.63	2.31
371	196.4	106	160	46.0	17.8	21.5	22.9	1.89	2.79
436	196.4	110	144	45.6	17.7	21.3	23.1	1.84	2.83
466	196.3	0	0	0	–	–	–	–	–

* Fragmentation of projectile occurs or it remains embedded in the shield

$v_{bl} = 293.9 \pm 1.4\,m/s$; $a = 0.57$, $p = 1.78$, $v_s < 400\,m/s$.

4.17 [Børvik *et al.*, 2004]

4.17.1 *General data on shields and projectiles*

Shields. Monolithic 5083-H116 AA shields ($\tilde{D}^{sh} = 500\,mm$); $\sigma_u = 347\,MPa$, $\tilde{\sigma}_y = 261\,MPa$, $\varepsilon_f = 17\%$.

Table 1. Truncated cone-nose projectiles (Fig. 3.6.1).

D (mm)	d (mm)	L_{cyl} (mm)	L_{nose} (mm)	Material	HRC	m (g)
20	2	68	30	Arne oil hardened steel	53	196±0.6

4.17.2 *Test results*

Table 2. Test results for shields with $T = 15\,mm$.

v_s (m/s)	v_r (m/s)	v_s (m/s)	v_r (m/s)	v_s (m/s)	v_r (m/s)	v_s (m/s)	v_r (m/s)
215	0	219	17	221	40	249	132
302	215						

$v_{bl} = 216.8 \pm 2.1\,m/s$.

Table 3. Test results for shields with $T = 20\,mm$.

v_s (m/s)	v_r (m/s)	v_s (m/s)	v_r (m/s)	v_s (m/s)	v_r (m/s)	v_s (m/s)	v_r (m/s)
230	0	246	0	252	34	261	74
303	176	365	276				

$v_{bl} = 249.0 \pm 3.0\,m/s$.

Table 4. Test results for shields with $T = 25\,mm$.

v_s (m/s)	v_r (m/s)	v_s (m/s)	v_r (m/s)	v_s (m/s)	v_r (m/s)	v_s (m/s)	v_r (m/s)
250	0	264	53	286	126	303	161
357	251						

$v_{bl} = 256.6 \pm 7.0\,m/s$.

Table 5. Test results for shields with $T = 30\,mm$.

v_s (m/s)	v_r (m/s)	v_s (m/s)	v_r (m/s)	v_s (m/s)	v_r (m/s)	v_s (m/s)	v_r (m/s)
310	0	319	73	331	144	352	167
373	191						

$v_{bl} = 309.7 \pm 4.7\,m/s$.

4.18   [Børvik *et al.*, 2009]

4.18.1   *General data on shields and projectiles*

Shields. Monolithic and layered (plates-in-contact and spaced plates) shields; $c_{l,w} = 300\,mm$.

Table 1. Properties of shields' materials.

Material	$\widetilde{\sigma}_y$ (MPa)	Material	$\widetilde{\sigma}_y$ (MPa)	Material	$\widetilde{\sigma}_y$ (MPa)
Weldox 500E	605	Weldox 700E	819	Hardox 400	1148
Domex Protect 500	1592	Armox 560T	1711		

Projectiles.   0.30 Caliber Ball   (Fig. 3.11.4,   Table 3.11.4),   and 0.30 Caliber AP M2   and   0.30 Caliber AP M2 (core)   (Fig. 3.11.1, Table 3.11.1).

4.18.2   *Test results for 0.30 Caliber Ball projectiles*

Table 2#. Test results for 1×[(Weldox 500E) 6.0] shields.

v_s (m/s)	v_r (m/s)	v_s (m/s)	v_r (m/s)	v_s (m/s)	v_r (m/s)	v_s (m/s)	v_r (m/s)
470	0	546	0	572	0	616	12
629	222	731	442	766	490	822	574
838	598	838	652				

$v_{bl} = 596\,m/s$; $a = 1.0$, $p = 2.3$.

Table 3#. Test results for 1×[(Weldox 700E) 6.0] shields.

v_s (m/s)	v_r (m/s)	v_s (m/s)	v_r (m/s)	v_s (m/s)	v_r (m/s)	v_s (m/s)	v_r (m/s)
533	0	590	0	635	0	662	0
667	99	682	436	697	442	700	427
778	532						

$v_{bl} = 666\,m/s$; $a = 1.0$, $p = 2.6$.

Table 4#. Test results for 1×[(Hardox 400) 6.0] shields.

v_s (m/s)	v_r (m/s)	v_s (m/s)	v_r (m/s)	v_s (m/s)	v_r (m/s)	v_s (m/s)	v_r (m/s)
759	0	764	234	769	231	774	336
778	330	797	385	789	553	816	616

$v_{bl} = 762\,m/s$; $a = 1.0$, $p = 3.0$.

Table 5#. Test results for 1×[(Domex Protect 500) 6.0] shields.

v_s (m/s)	v_r (m/s)	v_s (m/s)	v_r (m/s)	v_s (m/s)	v_r (m/s)	v_s (m/s)	v_r (m/s)
835	0	851	0	863	0	885	0
888	132	891	216	929	640	953	685
1011	730						

$v_{bl} = 886\,m/s$; $a = 1.0$, $p = 3.4$.

Table 6#. Test results for 1×[(Armox 560T) 6.0] shields.

v_s (m/s)	v_r (m/s)	v_s (m/s)	v_r (m/s)	v_s (m/s)	v_r (m/s)	v_s (m/s)	v_r (m/s)
874	0	904	0	913	0	920	267
923	273	926	363	931	357	932	333
965	631						

$v_{bl} = 918\,m/s$; $a = 1.0$, $p = 3.5$.

4.18.3 *Test results for 0.30 Caliber AP M2 projectiles*

Table 7#. Test results for 2×[(Weldox 500E) 6.0] shields.

v_s (m/s)	v_r (m/s)	v_s (m/s)	v_r (m/s)	v_s (m/s)	v_r (m/s)	v_s (m/s)	v_r (m/s)
515	0	568	0	578	0	612	0
637	156	640	180	663	237	716	385

$v_{bl} = 624\,m/s$; $a = 1.0$, $p = 2.2$.

Table 8#. Test results for 2×[(Weldox 700E) 6.0] shields.

v_s (m/s)	v_r (m/s)	v_s (m/s)	v_r (m/s)	v_s (m/s)	v_r (m/s)	v_s (m/s)	v_r (m/s)
553	0	650	0	658	0	677	115
680	168	686	218	724	450	745	448
750	374	799	287	833	564	916	694

$v_{bl} = 675\,m/s$; $a = 1.0$, $p = 2.4$.

Table 9#. Test results for 2×[(Hardox 400) 6.0] shields.

v_s (m/s)	v_r (m/s)	v_s (m/s)	v_r (m/s)	v_s (m/s)	v_r (m/s)	v_s (m/s)	v_r (m/s)
729	0	734	0	748	117	765	95
766	169	768	72	800	301	807	331
830	373						

$v_{bl} = 741\,m/s$; $a = 1.0$, $p = 2.0$.

Table 10#. Test results for 2×[(Domex Protect 500) 6.0] shields.

v_s (m/s)	v_r (m/s)	v_s (m/s)	v_r (m/s)	v_s (m/s)	v_r (m/s)	v_s (m/s)	v_r (m/s)
821	0	834	0	843	69	843	195
859	192	862	156	865	258	878	126
930	451	958	475				

$v_{bl} = 837\,m/s$; $a = 1.0$, $p = 2.1$.

Table 11. Test results for 2×[(Armox 560T) 6.0] shields.

v_s (m/s)	v_r (m/s)	v_s (m/s)	v_r (m/s)	v_s (m/s)	v_r (m/s)	v_s (m/s)	v_r (m/s)
609	0	853	0	864	0	876	0
874	237	895	147	912	75	926	195

$v_{bl} = 871\,m/s$; $a = 1.0$, $p = 1.5$.

Table 12#. Test results for 1×[(Weldox 700E) 12.0] shields.

v_s (m/s)	v_r (m/s)	v_s (m/s)	v_r (m/s)	v_s (m/s)	v_r (m/s)	v_s (m/s)	v_r (m/s)
644	0	683	96	692	0	713	282
745	128	747	313	754	320	783	387
829	561	843	467	880	526	922	602

Table 13#. Test results for 2×[(Weldox 700E) 6.0]\30.0 shields.

v_s (m/s)	v_r (m/s)	v_s (m/s)	v_r (m/s)	v_s (m/s)	v_r (m/s)	v_s (m/s)	v_r (m/s)
606	0	651	0	657	0	678	115
681	166	688	217	724	445	745	447
750	371	799	283	834	565	916	691

4.18.4 *Test results for 0.30 Caliber AP M2 projectiles (core only)*

Table 14#. Test results for 2×[(Weldox 700E) 6.0] shields.

v_s (m/s)	v_r (m/s)	v_s (m/s)	v_r (m/s)	v_s (m/s)	v_r (m/s)	v_s (m/s)	v_r (m/s)
630	0	638	0	649	138	658	229
663	161	665	0	673	244	687	243
725	393	807	535				

Table 15#. Test results for 2×[(Hardox 400) 6.0] shields.

v_s (m/s)	v_r (m/s)	v_s (m/s)	v_r (m/s)	v_s (m/s)	v_r (m/s)	v_s (m/s)	v_r (m/s)
658	0	701	0	722	156	725	135
735	241	741	137	755	166	762	232
807	407						

4.19 [Børvik *et al.*, 2010a]

4.19.1 *General data on shields and projectiles*

Shields. Monolithic 7075-T651 AA shields; $c_{l,w} = 500\,mm$, $T = 20\,mm$.

Projectiles. Hardened steel (HRC 52) projectiles with $D = 20\,mm$ and $m = 197\,g$, namely, cylindrical (Fig. 3.1.1, $L_{cyl} = 80\,mm$) and ogive-nose (Fig. 3.3.1; $L = 62\,mm$, $L_{nose} = 33\,mm$, $\psi = 3$) projectiles.

4.19.2 *Test results*

Table 1#. Test results for cylindrical projectiles.

v_s (m/s)	v_r (m/s)	v_s (m/s)	v_r (m/s)	v_s (m/s)	v_r (m/s)	v_s (m/s)	v_r (m/s)
184	0	189	71	200	61	209	90
257	161	321	247				

$v_{bl} = 183.8\,m/s$; $a = 0.89$, $p = 2.15$.

Table 2#. Test results for ogive-nose projectiles.

v_s (m/s)	v_r (m/s)	v_s (m/s)	v_r (m/s)	v_s (m/s)	v_r (m/s)	v_s (m/s)	v_r (m/s)
209	0	219	80	220	87	225	91
278	185	337	256				

$v_{bl} = 208.7\,m/s$; $a = 0.87$, $p = 2.54$.

4.20 [Børvik *et al.*, 2010b]

4.20.1 *General data on shields and projectiles*

Shields. Monolithic and layered (plates-in-contact) 5083-H116 Al shields; $\sigma_u = 317\text{–}325\,MPa$, $\tilde{\sigma}_y = 238\text{-}246\,MPa$, $\varepsilon_f = 16\%$.

Projectiles. Hardened steel (HRC 53) ogive-nose (Fig. 3.3.1; $D = 20\,mm$, $L_{cyl} = 62\,mm$, $L_{nose} = 33\,mm$, $\psi = 3$, $m = 197\,g$), and   0.30 Caliber AP M2 and 0.30 Caliber AP M2 (core) (Fig. 3.11.1, Table 3.11.1) projectiles.

4.20.2   *Test results for ogive-nose projectiles*

Table 1. Test results for 1× [20.0] shield.

v_s (m/s)	v_r (m/s)	v_s (m/s)	v_r (m/s)	v_s (m/s)	v_r (m/s)	v_s (m/s)	v_r (m/s)
242	0	248	52	248	51	271	133
293	174	360	281				

$v_{bl} = 244\,m/s$; $a = 1.00$, $p = 2.14$.

4.20.3   *Test results for 0.30 Caliber AP M2 projectiles*

Table 2. Test results for 1× [20.0] shield.

v_s (m/s)	v_r (m/s)	v_s (m/s)	v_r (m/s)	v_s (m/s)	v_r (m/s)	v_s (m/s)	v_r (m/s)
480	0	492	0	498	85	498	34
525	201	562	301	615	394	715	552
822	694						

$v_{bl} = 492\,m/s$; $a = 1.00$, $p = 2.14$.

Table 3. Test results for 2× [20.0] shield.

v_s (m/s)	v_r (m/s)	v_s (m/s)	v_r (m/s)	v_s (m/s)	v_r (m/s)	v_s (m/s)	v_r (m/s)
670	0	701	0	716	0	728	76
742	129	748	212	772	311	802	379
866	486	915	604				

$v_{bl} = 722\,m/s$; $a = 1.00$, $p = 2.08$.

Table 4. Test results for 3× [20.0] shield.

v_s (m/s)	v_r (m/s)	v_s (m/s)	v_r (m/s)	v_s (m/s)	v_r (m/s)	v_s (m/s)	v_r (m/s)
906	0	913	85	915	0	925	78
929	150	934	167	945	209	956	255

$v_{bl} = 912\,m/s$; $a = 1.00$, $p = 1.85$.

4.20.4 *Test results for 0.30 Caliber AP M2 (core only) projectiles*

Table 5. Test results for 1× [20.0] shield.

v_s (m/s)	v_r (m/s)	v_s (m/s)	v_r (m/s)	v_s (m/s)	v_r (m/s)	v_s (m/s)	v_r (m/s)
501	0	510	0	518	45	532	170
546	189	552	222	614	351	668	429
741	532	854	700				

$v_{bl} = 513\,m/s$; $a = 1.00$, $p = 2.05$.

Table 6. Test results for 2× [20.0] shield.

v_s (m/s)	v_r (m/s)	v_s (m/s)	v_r (m/s)	v_s (m/s)	v_r (m/s)	v_s (m/s)	v_r (m/s)
741	0	751	0	762	0	764	0
781	0	784	202	787	200	800	162
803	219	868	402	901	462	995	641

$v_{bl} = 767\,m/s$; $a = 1.00$, $p = 2.02$.

Table 7. Test results for 3× [20.0] shield.

v_s (m/s)	v_r (m/s)	v_s (m/s)	v_r (m/s)	v_s (m/s)	v_r (m/s)	v_s (m/s)	v_r (m/s)
651	0	1008	0	1015	0	1024	0
1028	376	1029	351	1033	377	1044	403
1059	400	1079	488	1093	522		

$v_{bl} = 1025\,m/s$; $a = 1.00$, $p = 2.60$.

4.21 [Børvik *et al.*, 2011]

4.21.1 *General data on shields and projectiles*

Shields. Monolithic AA6083-T4 shields; $c_{l.w} = 300\,mm$, $T = 20\,mm$.
Projectiles. 0.30 Caliber Ball (Fig. 3.11.4, Table 3.11.4) and 0.30 Caliber AP M2 projectiles (Fig. 3.11.1, Table 3.11.1).
Impact velocity $v_s \approx 830\,m/s$.

4.21.2 *Test results*

Table 1#. Test results for 0.30 Caliber Ball projectile.

θ_s (deg)	v_r (m/s)	θ_s (deg)	v_r (m/s)	θ_s (deg)	v_r (m/s)	θ_s (deg)	v_r (m/s)
0	597	0	617	0	637	0	644
0	688	6	119	15	562	15	572
15	621	30	466	30	503	30	522
45	449	45	435	45	306	45	128
45	0	60	0				

Table 2#. Test results for 0.30 Caliber AP M2 projectile.

θ_s (deg)	v_r (m/s)	θ_s (deg)	v_r (m/s)	θ_s (deg)	v_r (m/s)	θ_s (deg)	v_r (m/s)
0	712	0	729	6	124	15	687
15	713	15	726	15	734	30	656
30	687	30	701	30	723	30	733
45	274	45	330	44	528	45	608
45	629	45	638	60	0		

4.22   [Brooks and Erickson, 1971]

Data from: [Anderson *et al.*, 1992].

4.22.1 *General data on shields and projectiles*

Shields. Monolithic 4340 steel shields; $\gamma = 7850\ kg/m^3$, HB 294.

Projectiles.   Ogive-nose   projectiles   (Fig. 3.3.1; $D = 8.86mm$, $L = 43.56mm$, see also Table 1) and cylindrical projectiles (Fig. 3.1.1; see Table 2).

Table 1. Ogive-nose projectiles characteristics.

Label	m (g)	Material	γ (kg/m^3)	HB	δ (%)
Projectile-1	30.9	K-94 WC	14720	427	0
Projectile-2	36.4	W alloy	17360	227	2
Projectile-3	39.2	U-2Mo	18700	344	23
Projectile-4	39.2	U-2Mo	18700	298	0
Projectile-5	39.2	U-2Mo	18700	298	12
Projectile-6	39.2	U-2Mo	18700	290	22

Label	m (g)	Material	γ (kg/m^3)	HB	δ (%)
Projectile-7	38.8	U–1Mo	18500	354	24
Projectile-8	38.8	U–1Mo	18500	502	24
Projectile-9	38.8	U–1Mo	18500	358	26
Projectile-10	16.4	Steel	7800	684	5

Table 2. Characteristics of cylindrical projectiles.

D (mm)	L (mm)	Material	γ (kg/m^3)	HB	δ (%)	m (g)
8.86	34.00	W alloy	17360	227	2	36.4

4.22.2 *Test results for ogive-nose projectiles*

Table 3. Test results for Projectile-1.

v_s (m/s)	P (mm)	v_s (m/s)	P (mm)	v_s (m/s)	P (mm)	v_s (m/s)	P (mm)
434	24.1	568	34.3	691	44.0	805	52.0
920	64.0	1027	75.5	1138	87.0	1343	112.0

Table 4. Test results for Projectile-2.

v_s (m/s)	P (mm)	v_s (m/s)	P (mm)	v_s (m/s)	P (mm)	v_s (m/s)	P (mm)
561	20.3	793	35.6	1013	52.3	1227	57.9
1408	49.0	1311	42.5				

Table 5. Test results for Projectile-3.

v_s (m/s)	P (mm)	v_s (m/s)	P (mm)	v_s (m/s)	P (mm)	v_s (m/s)	P (mm)
411	3.0	561	6.0	667	12.0	777	36.1
891	42.4	1008	48.8	1121	35.0	1232	40.0
1306	46.0	1386	49.9				

Table 6. Test results for Projectile-4.

v_s (m/s)	P (mm)	v_s (m/s)	P (mm)	v_s (m/s)	P (mm)	v_s (m/s)	P (mm)
427	4.0	647	12.4	669	13.4	778	18.9
899	26.0	1013	45.0	1121	37.0	1208	41.0
1320	47.0	1863	47.0				

Table 7. Test results for Projectile-5.

v_s (m/s)	P (mm)	v_s (m/s)	P (mm)	v_s (m/s)	P (mm)	v_s (m/s)	P (mm)
661	22.5	781	33.5	898	46.0	1121	41.5
1306	45.5	1377	50.5				

Table 8. Test results for Projectile-6.

v_s (m/s)	P (mm)	v_s (m/s)	P (mm)	v_s (m/s)	P (mm)	v_s (m/s)	P (mm)
470	3.10	668	13.6	749	13.6	812	18.0
896	28.5	966	42.0	1058	38.0	1128	39.5
1311	46.5						

Table 9. Test results for Projectile-7.

v_s (m/s)	P (mm)	v_s (m/s)	P (mm)	v_s (m/s)	P (mm)	v_s (m/s)	P (mm)
400	1.5	532	4.5	644	25.0	765	37.0
884	45.0	988	54.0	1101	41.5	1189	45.0
1284	45.5	1373	50.0				

Table 10. Test results for Projectile-8.

v_s (m/s)	P (mm)	v_s (m/s)	P (mm)	v_s (m/s)	P (mm)	v_s (m/s)	P (mm)
386	2.5	529	5.5	648	12.0	763	24.0
885	48.5	987	47.0	1093	46.5	1193	46.0
1284	46.5	1367	49.0				

Table 11. Test results for Projectile-9.

v_s (m/s)	P (mm)	v_s (m/s)	P (mm)	v_s (m/s)	P (mm)	v_s (m/s)	P (mm)
412	2.5	540	7.0	654	17.0	764	33.5
884	43.0	1121	59.0	1191	45.0	1284	46.5
1364	50.0						

Table 12. Test results for Projectile-10.

v_s (m/s)	P (mm)	v_s (m/s)	P (mm)	v_s (m/s)	P (mm)	v_s (m/s)	P (mm)
499	21.5	591	22.5	828	35.5	931	41.0

4.22.3 *Test results for cylindrical projectiles*

Table 13. Test results.

v_s (m/s)	P (mm)	v_s (m/s)	P (mm)	v_s (m/s)	P (mm)	v_s (m/s)	P (mm)
373	0.5	655	5.5	884	14.5	1113	27.5
1311	36.5						

4.23 [Brooks, 1973]

See also [Anderson *et al.*, 1992].

4.23.1 *General data on shields and projectiles*

Shields. Monolithic metallic shields.

Table 1. Shield material properties.

Material	$\widetilde{\sigma}_y$ (MPa)	σ_u (MPa)	δ (%)	A_{red} (%)	HB
65-S Al	46	122	35	51	38
1020 Steel	320	420	28	49	125
4340 Steel	909	1033	17.5	57	302

Projectiles. Non-deforming steel and tungsten projectiles having different shapes with $D = 10.16\,mm$.

4.23.2 *Test results for cone-nose projectiles (Fig. 3.2.1)*

Table 2. Test results for 65-S Al plate and steel projectiles with $m = 25.8\,g$.

β (deg)	v_s (m/s)	P (mm)	β (deg)	v_s (m/s)	P (mm)
10^	650	105.2	10^^	762	> 137.2
15	644	94.7	15	761	127.0
20	661	92.7	20	760	118.6
25	654	85.3	25	768	111.5
30	670	84.6	30	793	113.8
40	657	76.2	40	771	99.6
50	646	71.4	50	752	90.9

β (deg)	v_s (m/s)	P (mm)	β (deg)	v_s (m/s)	P (mm)
60	632	67.1	60	774	92.2
90	622	64.5	90	758	86.4

$\wedge$ Projectile was deformed during penetration.

$\wedge\wedge$ Data are unreliable; measured from the original surface to the shoulder of the projectile.

Table 3. Test results for 1020 steel plate and tungsten projectiles with $m = 37.78\,g$, $\beta = 30°$.

v_s (m/s)	P (mm)	v_s (m/s)	P (mm)	v_s (m/s)	P (mm)	v_s (m/s)	P (mm)
504	18.3	588	24.4	737	36.6	815	44.2
910	52.8						

Table 4. Test results for 4340 steel plate and tungsten projectiles with $m = 37.78\,g$, $\beta = 30°$.

v_s (m/s)	P (mm)	v_s (m/s)	P (mm)	v_s (m/s)	P (mm)	v_s (m/s)	P (mm)
582	15.7	715	23.6	911	36.1	1042	46.2
1188	43.7$\wedge$						

$\wedge$ Projectile shattered during penetration.

4.23.3   *Test results for double-tapered projectiles (Fig. 3.7.1a)*

Table 5. Test results for 65-S Al plate and steel projectiles with $m = 25.8\,g$, $\theta_1 = 60°, \theta_2 = 10°$.

$\dfrac{d_1}{D}$	v_s (m/s)	P (mm)	$\dfrac{d_1}{D}$	v_s (m/s)	P (mm)	$\dfrac{d_1}{D}$	v_s (m/s)	P (mm)
0.5	611	82.6	0.5	756	123.2	0.6	614	78.7
0.6	754	115.6	0.7	671	88.4	0.7	785	116.3
0.8	631	74.2	0.8	862	128.5	0.9	615	67.8
0.9	785	101.1	1	613	64.0	1	766	91.7

4.24   [Buchar *et al.*, 2002a]

4.24.1   *General data on shields and projectiles*

Shields. Layered (plates-in-contact) shields.

Table 1. Layers of the shields.

Designation	Material	Comments
TS	TENAX Steel	First layer
AI	AREMA Iron	Interlayer
2P	2P Armor steel	Second layer

Projectiles. 7.62 mm AP projectiles.

4.24.2 *Test results*

Table 2. Test results on BLV.

Shield structure	T_{sum} (mm)	v_{bl} (m/s)
[(TS) 2.0]+[(2P) 4.0]	6.0	700
[(TS) 3.0]+[(2P) 3.0]	6.0	740
[(TS) 4.0]+[(2P) 2.0]	6.0	770
[(TS) 5.0]+[(2P) 1.0]	6.0	750
[(TS) 2.0]+[(AI) 0.5]+[(2P) 4.0]	6.5	705
[(TS) 3.0]+[(AI) 0.5]+[(2P) 3.0]	6.5	750
[(TS) 4.0]+[(AI) 0.5]+[(2P) 2.0]	6.5	800
[(TS) 5.0]+[(AI) 0.5]+[(2P) 1.0]	6.5	790

4.25 [Buyuk *et al.*, 2009]

4.25.1 *General data on shields and projectiles*

Shields. Monolithic 2024-T3/T351 Al shields; $\tilde{c}_{l.w} = 254\,mm$.
Projectiles. Chrome 52100 alloy steel spherical projectiles (Fig. 3.5.1); $D = 12.7\,mm$.

4.25.2 *Test results*

Table 1#. Test results for shields with $T = 1.587\,mm$.

v_s (m/s)	v_r (m/s)	v_s (m/s)	v_r (m/s)	v_s (m/s)	v_r (m/s)	v_s (m/s)	v_r (m/s)
87	0	91	0	101	0	108	0
119	0	119	8	120	19	122	49

v_s (m/s)	v_r (m/s)	v_s (m/s)	v_r (m/s)	v_s (m/s)	v_r (m/s)	v_s (m/s)	v_r (m/s)
122	29	127	39	130	56	132	37
141	84	151	96	160	113	164	119
175	140	181	144	188	151	199	161
205	169	225	192	267	232	271	238
275	245	316	285				

Table 2#. Test results for shields with $T = 3.175\,mm$.

v_s (m/s)	v_r (m/s)	v_s (m/s)	v_r (m/s)	v_s (m/s)	v_r (m/s)	v_s (m/s)	v_r (m/s)
160	0	191	0	196	0	202	0
211	0	215	0	218	0	219	27
220	24	220	45	221	39	225	61
226	65	227	68	231	80	236	100
242	108	245	112	247	120	252	128
257	134	258	139	260	143	262	146
269	156	271	161	348	252		

Table 3#. Test results for shields with $T = 6.350\,mm$.

v_s (m/s)	v_r (m/s)	v_s (m/s)	v_r (m/s)	v_s (m/s)	v_r (m/s)	v_s (m/s)	v_r (m/s)
311	0	319	0	404	0	405	73
409	117	409	0	410	60	412	133
413	0	418	133	420	136	422	139
426	148	434	177	437	171	442	179
451	188	460	219	470	232	482	263
501	251	510	286	538	331	558	345
572	376						

4.26 [Calder and Goldsmith, 1971]

4.26.1 *General data on shields and projectiles*

Shields. Monolithic 2024-0 Al shields; $\widetilde{D}^{sh} = 362\,mm$, $T = 1.27\,mm$.

Projectiles. Spherical projectiles (Fig. 3.5.1) and cone-nose projectiles (Fig. 3.2.1) with $\beta = 30°$ and $L = 19.05\,mm$.

4.26.2 *Test results for spherical projectiles*

Table 1. Test results for projectiles with $D = 12.7\,mm$, $m = 8.37\,g$.

v_s (m/s)	v_r (m/s)	m^{pl} (g)	v_s (m/s)	v_r (m/s)	m^{pl} (g)	v_s (m/s)	v_r (m/s)	m^{pl} (g)
151	96	0.135	208	179	0.187	284	263	0.269

Test results for projectiles with $D = 6.35\,mm$, $m = 1.05\,g$:
$v_s = 393\,m/s$, $v_r = 310\,m/s$.

4.26.3 *Test results for cone-nose projectiles*

Table 2. Test results for projectiles with $D = 12.7\,mm$, $L = 19.1\,mm$, $m = 11.66\,g$.

v_s (m/s)	v_r (m/s)	N	v_s (m/s)	v_r (m/s)	N	v_s (m/s)	v_r (m/s)	N
92	59	6	119	92	4	151	136	4
174	159	4	256	245	4			

Test results for projectiles with $D = 6.35\,mm$, $L = 15.9\,mm$, $m = 3.05\,g$:
$v_s = 262.4\,m/s$, $v_r = 245.7\,m/s$, $N = 4$.

4.27 [Cheeseman *et al.* 2008]

4.27.1 *General data on shields and projectiles*

Shields. Monolithic 2139-T8 Al shields.
Projectiles. 0.30 Caliber AP M2 (Fig. 3.11.1, Table 3.11.1), 0.50 Caliber AP M2 (Fig. 3.13.1, Table 3.13.1) projectiles and 20 mm FSPs (Fig. 3.10.1, Table 3.10.1).

4.27.2 *Test results*

Table 1. Test results for 0.30 caliber AP M2 projectiles.

T (mm)	HB	v_{bl} (m/s)	T (mm)	HB	v_{bl} (m/s)	T (mm)	HB	v_{bl} (m/s)
32.3	156	783	39.0	159	860	40.9	159	892

Table 2. Test results for 20 mm FSPs.

T (mm)	HB	v_{bl} (m/s)	T (mm)	HB	v_{bl} (m/s)	T (mm)	HB	v_{bl} (m/s)
32.3	156	710	39.0	159	996	40.9	159	1027
52.1	159	1425	52.1	166	1400			

Table 3. Test results for 0.50 caliber AP M2 projectiles.

T (mm)	HB	v_{bl} (m/s)	T (mm)	HB	v_{bl} (m/s)	T (mm)	HB	v_{bl} (m/s)
39.0	159	657	39.9	159	668	40.9	159	677
52.1	159	785	52.1	166	790	53.8	153	796
57.2	156	819	64.1	156	873			

4.28   [Chen *et al.*, 2014]

4.28.1   *General data on shields and projectiles*

Shields. Monolithic Q235 mild steel ($E = 210\,GPa$, $\gamma = 7800\,kg/m^3$, $v = 0.30$, $\sigma_y = 235\,MPa$, $\sigma_u = 400-490\,MPa$, $\delta = 35\%$) shields; $c_{l,w} = 350\,mm$, $T = 1.36\,mm$.

Projectiles. Quenched 45 steel ($\sigma_y = 355\,MPa$, $\sigma_u = 450-685\,MPa$) hemispherical-nose projectiles (Fig. 3.4.1); $D = 14.9\,mm$, $L = 21.4\,mm$, $m = 25.7\,g$.

4.28.2   *Test results*

Table 1. Test results on v_r.

v_s (m/s)	v_r (m/s)	v_s (m/s)	v_r (m/s)	v_s (m/s)	v_r (m/s)
352.1	314.2	317.1	288.7	277.0	245.7
259.5	232.7	187.7	152.6		

4.29   [Chocron *et al.*, 1999]

4.29.1   *General data on shields and projectiles*

Shields. Monolithic 6061-T6 Al shields.

Projectiles. 0.30 Caliber AP M2 projectile (Fig. 3.11.1, Table 3.11.1, $m = 10.8$ g, *Standard Projectile*) and 0.30 Caliber AP M2 projectile in which the lead nose was removed ($m = 9.6$ g , *Modified Projectile*).

4.29.2 *Test results for Standard Projectiles*

Table 1#. Results on depth of penetration.

v_s (m/s)	P (mm)	v_s (m/s)	P (mm)	v_s (m/s)	P (mm)	v_s (m/s)	P (mm)
608	30	618	31	620	31	708	39
710	38	771	43	773	43	795	45
800	45	827	49	857	53	860	52
867	53						

Table 2#. Results on residual velocity for shields with $T = 6.35$ mm.

v_s (m/s)	v_r (m/s)	v_s (m/s)	v_r (m/s)	v_s (m/s)	v_r (m/s)	v_s (m/s)	v_r (m/s)
281	0	292	0	315	91	320	114
379	261	408	279	437	329	515	371
843	780	847	786				

4.29.3 *Test results for Modified Projectile*

Table 3#. Results on depth of penetration.

v_s (m/s)	P (mm)	v_s (m/s)	P (mm)	v_s (m/s)	P (mm)	v_s (m/s)	P (mm)
666	32	675	33	746	40	752	39
827	47	852	48	852	49		

4.30 [Contiliano *et al.*, 1977]

4.30.1 *General data on shields and projectiles*

Shields. Monolithic and layered (plates-in-contact) shields made from different materials. For most of the shields $D^{sh} = 152\,mm$ and $T = 25.4\,mm$.

Projectiles. Spherical projectiles (Fig. 3.5.1) manufactured of different materials.

Note. Cases "projectile was broken" are marked with an asterisk.

4.30.2 *Test results on DOP into monolithic shields of non-deformed projectiles*

Table 1#. DOP of tungsten carbide projectiles into aluminum (1100-F) shields.

D (mm)	v_s (m/s)	P (mm)	D (mm)	v_s (m/s)	P (mm)	D (mm)	v_s (m/s)	P (mm)
6.4	9	0.1	6.4	278	5.4	6.4	371	8.3
4.0	61	0.6	4.0	139	1.4			

Table 2#. DOP of steel projectiles into aluminum (1100-F) shields.

D (mm)	v_s (m/s)	P (mm)	D (mm)	v_s (m/s)	P (mm)
4.0	49	0.3	4.0	113	0.7
4.0	181	1.3	12.7	8	0.2

Table 3#. DOP of tungsten carbide projectiles into RHA shields.

D (mm)	v_s (m/s)	P (mm)	D (mm)	v_s (m/s)	P (mm)	D (mm)	v_s (m/s)	P (mm)
4.0	60	0.2	4.0	138	0.4	6.4	186	0.8
6.4	229	1.0	6.4	397	2.1	6.4	501	2.7*
6.4	637	3.6*						

Table 4#. DOP of tungsten carbide projectiles (D = 6.35 mm) into titanium shields.

v_s (m/s)	P (mm)	v_s (m/s)	P (mm)	v_s (m/s)	P (mm)
318	1.9	238	1.3	463	2.9
818	6.3*	596	4.0	924	7.5*

4.30.3 *Test results on DOP into monolithic shields of deformed projectiles (D = 6.35 mm)*

Table 5#. DOP of Lead projectiles into Steel shields.

v_s (m/s)	P (mm)	v_s (m/s)	P (mm)	v_s (m/s)	P (mm)
650	1.8	914	2.9	1195	4.5
1442	5.9	1844	6.9		

Table 6#. DOP of lead projectiles into aluminum (1100-F) shields.

v_s (m/s)	P (mm)	v_s (m/s)	P (mm)	v_s (m/s)	P (mm)	v_s (m/s)	P (mm)
118	0.4	225	1.8	339	3.5	422	5.0
527	6.7	552	7.2	757	10.7	859	11.2
1182	15.3	1669	23.5	1532	19.4	1872	22.7

Table 7#. DOP of aluminum projectiles into aluminum (1100-F plate) shields.

v_s (m/s)	P (mm)	v_s (m/s)	P (mm)	v_s (m/s)	P (mm)	v_s (m/s)	P (mm)
386	1.2	461	1.7	631	2.5	698	2.8
947	4.3	1239	5.4	1693	5.9	2048	7.5

Table 8#. DOP of aluminum projectiles into steel (1020) shields.

v_s (m/s)	P (mm)	v_s (m/s)	P (mm)	v_s (m/s)	P (mm)
483	0.1	865	0.4	1062	0.7
1463	1.3	2028	2.1		

4.30.4 *Test results on DOP into two-layered shields consisting of aluminum (Al) and steel (St) plates*

Table 9#. DOP of tungsten carbide projectiles ($D = 6.35$ *mm*) into [(Al) 6.35]+ [(St) 25.4] shield.

v_s (m/s)	P (mm)	v_s (m/s)	P (mm)	v_s (m/s)	P (mm)
465	8.1	607	9.7	893	13.2
1143	15.6	1329	17.3		

Table 10#. DOP of tungsten carbide projectiles (D=6.35 *mm*) into [(St) 6.35]+ [(Al) 25.4] shield.

v_s (m/s)	P (mm)	v_s (m/s)	P (mm)	v_s (m/s)	P (mm)
611	8.6	721	10.4	793	12.9
922	17.2	1052	21.2*		

Table 11#. DOP of lead projectiles (D=6.10 *mm*) into [(Al) 6.35]+ [(St) 25.4] shield.

v_s (m/s)	P (mm)	v_s (m/s)	P (mm)	v_s (m/s)	P (mm)
551	5.3	870	6.2	1194	7.1
1529	8.3	1802	8.9		

Table 12#. DOP of lead projectiles (D=6.10 *mm*) into [(St) 17.8] + [(Al) 25.4] shield.

v_s (m/s)	P (mm)	v_s (m/s)	P (mm)	v_s (m/s)	P (mm)
883	9.7	977	10.4	1350	14.3
1636	15.4	1863	19.2		

4.30.5 *Test results on DOP into 8×{[(Al) 1.27]+[(St) 1.78]} shields consisting of aluminum (Al) and steel (St) plates (D = 6.35 mm)*

Table 13#. DOP of tungsten carbide projectiles.

v_s (m/s)	P (mm)	v_s (m/s)	P (mm)	v_s (m/s)	P (mm)
379	5.2	440	6.2	562	8.9
640	10.4	825	15.9		

Table 14#. DOP of lead projectiles.

v_s (m/s)	P (mm)	v_s (m/s)	P (mm)	v_s (m/s)	P (mm)	v_s (m/s)	P (mm)
237	0.1	283	0.1	609	0.2	910	0.3
1292	0.4	1475	0.5	1860	0.7		

4.30.6 *Test results on v_r vs. v_s for tungsten carbide projectiles (D = 6.35 mm)*

Table 15#. Test results for Aluminum shields.

T (mm)	v_s (m/s)	v_r (m/s)	T (mm)	v_s (m/s)	v_r (m/s)	T (mm)	v_s (m/s)	v_r (m/s)
1.27	214	170	1.27	397	362	1.27	746	727
1.27	1284	1245	6.35	436	267	6.35	555	432
6.35	778	674	6.35	1239	1131	12.7	626	371

T (mm)	v_s (m/s)	v_r (m/s)	T (mm)	v_s (m/s)	v_r (m/s)	T (mm)	v_s (m/s)	v_r (m/s)
12.7	771	563	12.7	1089	884*	12.7	1411	1212*
25.4	828	411	25.4	1064	699*	25.4	1416	1018*

Table 16#. Test results for steel shields with T=1.83 mm.

v_s (m/s)	v_r (m/s)	v_s (m/s)	v_r (m/s)	v_s (m/s)	v_r (m/s)
351	196	507	377	781	642
1076	943				

Table 17#. Test results for two-layered steel (St)–aluminum (Al) shields.

Shield structure	v_s (m/s)	v_r (m/s)	v_s (m/s)	v_r (m/s)
[(St) 1.83]+[(Al) 1.27]	363	168	915	776
	915	600*		
[(St) 1.83]+[(Al) 6.35]	609	349	928	686*
	1251	977*	1276	1151*
[(St) 1.83]+[(Al) 25.4]	952	354*		

4.31 [Contiliano *et al.*, 1978]

4.31.1 *General data on shields and projectiles*

Shields. Monolithic shields.

Table 1. Shield characteristics.

Material	Description	γ (kg/m^3)	D^{sh} (mm)	c_l (mm)	c_w (mm)
Aluminum-1	1100-F plate	2730	156	–	–
Aluminum-2	5083, HRB 75	2740	–	99	143
Cadmium	99.9% Pure, Open Cast	8810	152	–	–
Copper	Hot Rolled Electrolytic Though Pitch Plate	9010	156	–	–
Iron	Class 40, Gray	7130	159	–	–
Lead	99.9% Pure, Open Cast	1130	152	–	–
Steel-1	1020, Hot Rolled Plate	7860	156	–	–
Steel-2	RHA	7830	–	114	99
Titanium	Ti-6Al-4V	4480	–	111	111
Zinc	99.9% Pure, Open Cast	7190	149	–	–

Projectiles. Spherical projectiles (Fig. 3.5.1).

Table 2. Classification of test results.

Label	Description
A	Projectile remained embedded in the shield
B	Projectile was broken
C	DOP is larger than half of the shield thickness

4.31.2 *Test results for tungsten carbide projectiles*

Table 3. Test results for Aluminum-1 shields.

T (mm)	m (g)	D (mm)	v_s (m/s)	P (mm)	Test details
27.2	2.01	6.4	518	13.0	A, C
26.9	2.01	6.4	372	8.4	–
27.2	2.01	6.4	275	5.4	–
27.2	2.01	6.4	492	12.0	A
27.2	0.485	4.0	138	1.3	–
27.2	0.485	4.0	60	0.6	–
27.2	2.01	6.4	9	0.1	–

Table 4. Test results for Aluminum-2 shields.

T (mm)	m (g)	D (mm)	v_s (m/s)	P (mm)	Test details
25.9	2.01	6.4	255	2.2	–
25.9	2.01	6.4	441	4.6	–
25.9	2.01	6.4	550	6.4	A
25.9	0.485	4.0	60	0.3	–
25.9	0.485	4.0	139	0.7	–
25.9	2.01	6.4	340	3.2	–
25.9	2.01	6.4	709	8.4	A

Table 5. Test results for Cadmium shields.

T (mm)	m (g)	D (mm)	v_s (m/s)	P (mm)	Test details
25.9	2.01	6.4	319	6.5	A

T (mm)	m (g)	D (mm)	v_s (m/s)	P (mm)	Test details
28.4	2.01	6.4	331	5.9	–
26.9	2.01	6.4	213	3.2	–
26.9	2.01	6.4	454	9.3	A
27.9	0.485	4.0	61	0.5	–
26.9	0.485	4.0	139	1.1	–
27.7	2.01	6.4	271	4.5	–
24.9	2.01	6.4	9	0.1	–

Table 6. Test results for copper shields.

T (mm)	m (g)	D (mm)	v_s (m/s)	P (mm)	Test details
25.9	2.01	6.4	319	6.5	A
25.9	2.01	6.4	198	2.6	–
25.9	2.01	6.4	329	4.4	–
25.9	2.01	6.4	556	9.3	–
25.4	2.01	6.4	757	14.6	A, C
26.2	2.01	6.4	475	7.6	–
26.2	0.485	4.0	62	0.5	–
26.2	0.485	4.0	139	1.1	–
25.9	2.01	6.4	9	0.1	–

Table 7. Test results for iron shields.

T (mm)	m (g)	D (mm)	v_s (m/s)	P (mm)	Test details
26.2	2.01	6.4	536	4.7	–
26.7	2.01	6.4	881	10.3	A, B
26.2	2.01	6.4	340	2.3	–
26.7	2.01	6.4	228	1.6	–
26.7	2.01	6.4	688	5.7	A
26.9	0.485	4.0	62	0.3	–
26.7	2.01	6.4	231	1.7	–
26.4	2.01	6.4	392	3.2	–
26.7	2.01	6.4	9	0.04	–

Table 8. Test results for lead shields.

T (mm)	m (g)	D (mm)	v_s (m/s)	P (mm)	Test details
26.7	2.01	6.4	307	17.9	A, C
26.7	2.01	6.4	194	9.2	A

T (mm)	m (g)	D (mm)	v_s (m/s)	P (mm)	Test details
27.2	2.01	6.4	279	15.4	A, C
26.9	2.01	6.4	222	11.3	A
26.9	0.485	4.0	61	1.1	–
27.2	0.485	4.0	137	3.3	–
27.2	2.01	6.4	9	0.3	–
55.1	2.01	6.4	306	17.3	A
57.4	2.01	6.4	277	15.4	A
54.4	2.01	6.4	233	12.5	–
59.2	2.01	6.4	358	21.5	A
54.4	2.01	6.4	190	9.1	–
54.4	2.01	6.4	318	18.3	A

Table 9. Test results for Steel-1 shields.

T (mm)	m (g)	D (mm)	v_s (m/s)	P (mm)	Test details
25.7	2.01	6.4	691	5.6	B
25.4	2.01	6.4	1032	8.4	B
25.7	2.01	6.4	394	2.6	–
25.7	2.01	6.4	1322	10.1	B
25.7	2.01	6.4	245	1.5	–
25.7	0.485	4.0	61	0.2	–
25.7	0.485	4.0	137	0.4	–
25.4	2.01	6.4	9	0.0	–

Table 10. Test results for Steel-2 shields.

T (mm)	m (g)	D (mm)	v_s (m/s)	P (mm)	Test details
25.4	2.01	6.4	229	1.0	–
25.4	2.01	6.4	402	2.1	–
25.4	2.01	6.4	635	3.7	B
25.4	0.485	4.0	59	0.2	–
25.4	0.485	4.0	139	0.4	–
25.4	2.01	6.4	185	0.8	–
25.7	2.01	6.4	495	2.7	B

Table 11. Test results for titanium shields.

T (mm)	m (g)	D (mm)	v_s (m/s)	P (mm)	Test details
17.3	2.01	6.4	319	1.9	–

T (mm)	m (g)	D (mm)	v_s (m/s)	P (mm)	Test details
17.3	2.01	6.4	593	4.1	–
17.3	2.01	6.4	919	7.5	B
17.3	2.01	6.4	816	6.4	B
17.3	2.01	6.4	463	2.9	–
17.3	2.01	6.4	242	1.3	–
17.3	0.485	4.0	94	0.3	–
17.3	0.485	4.0	79	0.3	–
17.3	0.485	4.0	87	0.3	–
17.3	0.485	4.0	144	0.3	–

Table 12. Test results for zinc shields.

T (mm)	m (g)	D (mm)	v_s (m/s)	P (mm)	Test details
26.4	2.01	6.4	421	5.4	–
27.2	2.01	6.4	230	2.6	–
27.4	2.01	6.4	633	9.8	A
25.1	2.01	6.4	308	3.8	–
24.6	2.01	6.4	817	15.6	A, C
26.2	0.485	4.0	140	0.9	–
27.7	0.485	4.0	60	0.4	–
25.7	2.01	6.4	9	0.1	–

4.31.3 *Test results for chrome steel projectiles*

Table 13. Test results for Aluminum-1 shields.

T (mm)	m (g)	D (mm)	v_s (m/s)	P (mm)
27.2	0.25	4.0	111	0.7
27.2	0.25	4.0	49	0.3
25.9	0.25	4.0	181	1.3
27.2	8.23	12.7	9	0.2

Table 14. Test details for cadmium shields.

T (mm)	m (g)	D (mm)	v_s (m/s)	P (mm)
28.2	0.25	4.0	111	0.6
25.9	0.25	4.0	182	1.0

T	m	D	v_s	P
(mm)	(g)	(mm)	(m/s)	(mm)
24.9	0.25	4.0	48	0.3
24.9	8.23	12.7	9	0.2

Table 15. Test results for copper shields.

T	m	D	v_s	P
(mm)	(g)	(mm)	(m/s)	(mm)
26.2	0.25	4.0	107	0.5
25.9	0.25	4.0	49	0.2
26.2	0.25	4.0	160	0.8
25.9	8.23	12.7	9	0.1

Table 16. Test results for iron shields.

T	m	D	v_s	P
(mm)	(g)	(mm)	(m/s)	(mm)
26.4	0.25	4.0	51	0.1
26.2	0.25	4.0	114	0.3
26.7	0.25	4.0	185	0.6
26.7	8.23	12.7	9	0.1

Table 17. Test results for lead shields.

T	m	D	v_s	P
(mm)	(g)	(mm)	(m/s)	(mm)
27.2	0.25	4.0	108	1.6
27.2	0.25	4.0	49	0.7
26.9	0.25	4.0	150	2.4
27.2	0.25	4.0	127	1.9
27.2	8.23	12.7	9	0.4

Table 18. Test results for Steel-1 shields.

T	m	D	v_s	P
(mm)	(g)	(mm)	(m/s)	(mm)
25.4	0.345	4.4	223	0.6
25.4	0.345	4.4	90	0.3
25.7	0.25	4.0	50	0.1
25.4	8.23	12.7	9	0.0

Table 19. Test results for titanium shields.

T (mm)	m (g)	D (mm)	v_s (m/s)	P (mm)
17.3	0.25	4.0	132	0.3
17.3	0.25	4.0	123	0.2
17.3	0.25	4.0	124	0.2
17.3	0.25	4.0	188	0.3

Table 20. Test results for zinc shields.

T (mm)	m (g)	D (mm)	v_s (m/s)	P (mm)
26.2	0.25	4.0	108	0.5
25.7	0.25	4.0	50	0.2
27.4	0.25	4.0	168	0.7
25.7	8.23	12.7	9	0.17

4.32 [Copland and Scheffler, 2003]

4.32.1 *General data on shields and projectiles*

Shields. Monolithic and layered (plates-in-contact and with air gaps) RHA (HB 255) shields; $c_{l,w} = T_{sum} = 152.4\,mm$.

Projectiles. Uranium-3/4 titanium hemispherical-nose projectiles (Fig. 3.4.1); $D = 6\,mm$, $L = 120\,mm$, $m = 65\,g$.

4.32.2 *Test results*

Table 1. Test results on DOP.

Shield structure	v_s (m/s)	P (mm)	Shield structure	v_s (m/s)	P (mm)
1×[152.4]	1656	130	1×[152.4]	1599	129
1×[152.4]	1614	124	6×[25.4]	1607	140
6×[25.4]	1609	138	6×[25.4]	1611	139
6×[25.4]\1.55	1595	143	6×[25.4]\1.55	1607	148
6×[25.4]\1.55	1606	144	6×[25.4]\3.0	1606	143
6×[25.4]\3.0	1605	144	6×[25.4]\3.0	1593	142
6×[25.4]\6.0	1567	144			

4.33   [Copland *et al.*, 2005]

4.33.1   *General data on shields and projectiles*

Shields. Monolithic and layered (plates-in-contact and with air gaps) RHA shields; $c_{l,w} = 152.4\,mm$.

Projectiles. Tungsten alloy X-15C hemispherical-nose projectiles (Fig. 3.4.1); $D = 5.7\,mm$, $L = 142.2\,mm$, $m = 65\,g$.

4.33.2   *Test results*

Table 1 Test results for shields having different structure.

Shield structure	v_s (m/s)	P (mm)	v_s (m/s)	P (mm)	v_s (m/s)	P (mm)
1×[150.0]	1574	122.5	1576	122.0	1599	123.5
	1571	123.0	1574	123.0	1581	125.5
	1602	116.0	1606	127.5	1649	134.0
	1621	133.5	1592	123.5	1595	119.0
7×[25.0]	1607	142.0	1628	146.5	1595	141.0
	1620	143.5	1609	139.0		
7×[25.0]\1.5	1598	143.5	1609	144.5	1610	145.0
	1597	142.0	1620	146.5	1595	141.0
7×[25.0]\3.0	1607	145.0	1612	145.5	1608	141.0
	1614	147.0	1600	142.0		

4.34   [Corbett and Reid, 1993]

4.34.1   *General data on shields and projectiles*

Shields. Monolithic steel shields; $\sigma_y = 182\,MPa$, $\tilde{\sigma}_y = 230\,MPa$, $\sigma_u = 320\,MPa$, $\varepsilon_f = 40\%$, $c_l = 315\,mm$, $c_w = 300\,mm$.

Projectiles. Hemispherical-nose projectiles (Fig. 3.4.1) with $m = 70\,g$ and cylindrical projectiles (Fig. 3.1.1) with $m = 67\,g$. All projectiles with $D = 12.7\,mm$ were made from hardened and tempered tool steel and showed no signs of deformation during testing.

4.34.2 *Test results*

Table 1. Test results for hemispherical-nose projectiles.

T (mm)	W_{perf} (J)	T (mm)	W_{perf} (J)	T (mm)	W_{perf} (J)
1.0	97	3.0	458	5.0	1026
6.0	1380	8.0	2280	10.0	2850

Table 2. Test results for cylindrical projectiles.

T (mm)	W_{perf} (J)	T (mm)	W_{perf} (J)	T (mm)	W_{perf} (J)
1.0	46	3.0	366	5.0	673
6.0	863	8.0	1102	10.0	1412

4.35 [Corran *et al.*, 1983a]

Additional information can be also found in [Corran et al., 1983b].

4.35.1 *General data on shields and projectiles*

Shields. Monolithic stainless steel (SS, $\sigma_y = 290\,MPa$, $\sigma_u = 660\,MPa$, $\varepsilon_f = 65\%$), aluminum alloy (AA, $\sigma_y = 460\,MPa$, $\sigma_u = 480\,MPa$, $\varepsilon_f = 10\%$), and mild steel (see Table 1) shields.

Table 1. Material properties of mild steel shields vs. T.

T (mm)	σ_y (MPa)	σ_u (MPa)	ε_f (%)
1.3	220	315	37
2.0	300	360	32
3.0	270	350	50
5.9	220	335	47
6.4	362	500	20

Projectiles. Cylindrical projectiles (Fig. 3.1.1) with $m = 35\,g$ and projectiles having spherical segmented nose (Fig. 3.8.1) with $m = 34\,g$. For all steel projectiles, $D = 12.5\,mm$.

4.35.2 *Test results*

Table 2#. Test results for spherical projectiles having segmented nose against mild steel shields with T=1.3 *mm*.

ρ (mm)	w_{perf} (J)	ρ (mm)	w_{perf} (J)	ρ (mm)	w_{perf} (J)
∞	77	21.75	142	12.74	198
9.37	215	6.25	126		

Note. $\rho = \infty$ and $\rho = 6.25$ correspond to flat nose and hemispherical-nose, respectively.

Table 3#. Test results for cylindrical projectiles.

Shield material	T (mm)	v_{bl} (m/s)	T (mm)	v_{bl} (m/s)	T (mm)	v_{bl} (m/s)
SS	1.3	115	2.0	147	2.7	170
	3.4	180	4.2	168	6.5	200
AA	1.3	27	2.7	87	4.0	129
	6.6	142				

4.36 [Corran *et al.*, 1983b]

Additional information can be also found in [Corran *et al.*, 1983a].

4.36.1 *General data on shields and projectiles*

Shields. Monolithic and layered (plates-in-contact and with air gaps) steel shields ($D^{sh} = 300\,mm$). Material properties (values in parenthesis are valid for some plates with $T = 6.4\,mm$): $\sigma_y = 260 \pm 15\% (360)\,MPa$, $\sigma_u = 337 \pm 7\% (500)\,MPa$.

Projectiles. Steel (HV 320-850) cylindrical projectiles (Fig. 3.1.1); $D = 12.5\,mm$, $L_{cyl} = 50\,mm$, $m = 35\,g$. Steel hardness was sufficient to prevent from gross plastic deformations of the missile.

4.36.2 *Test results for monolithic shields*

Table 1#. BLV vs. shield thickness.

T (mm)	v_{bl} (m/s)	T (mm)	v_{bl} (m/s)	T (mm)	v_{bl} (m/s)
1.3	67	2.0	90	2.7	135
3.0	155	5.0	186	6.4	187

Table 2#. v_r vs. v_s for shields having different T.

T (mm)	v_s (m/s)	v_r (m/s)	v_s (m/s)	v_r (m/s)	v_s (m/s)	v_r (m/s)	v_s (m/s)	v_r (m/s)
1.2	72	10	100	76	115	92	153	147
	175	171	195	192	215	216		
2.0	99	18	111	61	123	86	128	96
	147	122	166	148	182	167	209	201
3.0	146	25	163	73	173	92	181	103
	195	120	204	133	219	165		

4.36.3 Test results for monolithic and layered shields

Table 3#. Test results on w_{perf}.

T_{sum} (mm)	Shield structure	w_{perf} (J)	T_{sum} (mm)	Shield structure	w_{perf} (J)
1.18	1×[1.18]	70	1.96	1×[1.96]	150
2.36	2×[1.18]	168	2.36	2×[1.18]\12.0	180
2.7	1×[2.7]	302	3.0	1×[3.0]	405
3.54	3×[1.18]	372	3.54	3×[1.18]\12.0	328
3.92	2×[1.96]	395	3.92	2×[1.96]\12.0	436
4.2	[3.0]+[1.2]	481	4.2	[1.2]+[3.0]	701
5.0	1×[5.0]	583	5.88	3×[1.96]	878
5.88	3×[1.96]\12.0	801	6.0	2×[3.0]	940
6.0	2×[3.0]\12.0	817	6.4	1×[6.4]	586

4.37 [Crouch *et al.*, 1990]

4.37.1 General data on shields and projectiles

Shields. Monolithic 2024-T351 AA shields, $\widetilde{D}^{sh} = 37.5\,mm$.

Table 1. Shield material properties.

T (mm)	Rolling direction			Long transverse direction		
	σ_y (MPa)	σ_u (MPa)	ε_f (%)	σ_y (MPa)	σ_u (MPa)	ε_f (%)
1.59	320	520	13	285	508	16

T	Rolling direction			Long transverse direction		
(mm)	σ_y (MPa)	σ_u (MPa)	ε_f (%)	σ_y (MPa)	σ_u (MPa)	ε_f (%)
3.10	320	535	17	290	512	15
4.81	335	545	17	295	540	17
6.48	370	595	17	315	580	17

Projectiles. Non-deformable steel cylindrical projectiles (Fig. 3.1.1); $D=12.7\,mm$, $m=25\,g$.

4.37.2 Test results

Table 2. Test results on BLV.

T (mm)	v_{bl} (m/s)	T (mm)	v_{bl} (m/s)
1.59	62±1	3.10	103±2
4.81	151±3	6.48	154±6

4.38   [Ćwik *et al.*, 2012]

4.38.1 *General data on shields and projectiles*

Shields. Monolithic Armox 370T and Armox 440T shields; $c_{l,w}=300\,mm$; $\gamma=7845\,kg/m^3$.

Table 1. Parameters of shields (average values).

Material	T (mm)	m^{sh} (kg)	σ_u (MPa)	ε_f (%)	E (GPa)	v
Armox 370T	3.8	3.2	1232	2.3	202	0.42
Armox 440T	4.7	2.6	1375	2.1	198	0.39

Projectiles. Steel 20-mm FSPs (Fig. 3.10.1, Table 3.10.1; $m=53.1\pm0.15\,g$) and copper ($\gamma=8930\,kg/m^3$, $E=110\,GPa$, $\sigma_v=70\,MPa$) FSPs having the same geometry with $m=60.25\pm0.25\,g$.

4.38.2 *Test results*

Table 2. Test results for different combinations of shield/projectile.

| Material | | v_s | v_r | v_s | v_r |
Shield	Projectile	(m/s)	(m/s)	(m/s)	(m/s)
Armox 370T	Steel	319#	195#	371	217
		523#	399#	656	512
		827	675	1008#	841#
		1021	886		
Armox 370T	Copper	238	0	463#	209#
		568	439	795	653
		994	878		
Armox 440T	Steel	330	71	373	190
		714	597	1016	892
Armox 440T	Copper	253	0	328	0
		559	449	989	932

4.39 [Dean *et al.*, 2009]

4.39.1 *General data on shields and projectiles*

Shields. Monolithic 304 stainless steel shields; $\widetilde{D}^{sh} = 60\,mm$, $T = 0.4\,mm$.

Projectiles. Hardened steel spherical projectiles (Fig. 3.5.1); $D = 8\,mm$, $m = 2\,g$.

4.39.2 *Test results*

Table 1. Test results.

| v_s | v_r | m^{fr} | Failure mode |
(m/s)	(m/s)	(g)	
176	110	0	Hinged cap
179	122	0	4 Petals
181	113	0	Hinged cap
185	121	0	Hinged cap
190	128	0	Hinged cap
215	179	0	6 Petals

v_s (m/s)	v_r (m/s)	m^{fr} (g)	Failure mode
218	172	0	7 Petals
221	173	0	4 Petals
222	188	0.098	Plugging
228	189	0	Hinged cap
310	286	0.156	Plugging/petaling
316	290	0.250	Plugging/petaling
318	292	0.199	Plugging/petaling
577	542	0.222	Petaling/fragments
580	544	0.202	Petaling/fragments
583	546	0.237	Petaling/fragments
592	560	0.114	Petaling/fragments

4.40   [Delsasso *et al.*, 1943]

4.40.1   *General data on shields and projectiles*

Shields. Monolithic STS armor shields, $c_{l.w} = 762\,mm$.

Table 1. Properties of shields' materials.

Designation	σ_y (MPa)	σ_u (MPa)	δ (%)	A_{red} (%)	HB
Sh-1	917	975	19.0	51	294
Sh-2	843	992	18.5	54	283
Sh-3	764	865	22.0	67	264
Sh-4	737	876	21.0	61	272
Sh-5	739	891	22.0	62	257
Sh-6	732	867	22.0	63	271
Sh-7	741	899	22.0	65	289

Table 2. Projectiles.

Label	Projectile	D (mm)
Projectile-1	0.244 Caliber smoothbore projectile	6.20
Projectile-2	0.50 Caliber E6 projectile	12.57
Projectile-3	Standard 37 mm M51 projectiles with removed cap, windshield and tracer	36.7

4.40.2 *Test results*

Table 3. Test results for Projectile-1.

Shield	T (mm)	m (g)	v_{bl} (m/s)	Shield	T (mm)	m (g)	v_{bl} (m/s)
Sh-1	6.5	5.25	454*	Sh-2	9.5	5.26	579

* Bullets slightly blunted.

Table 4. Test results for Projectile-2.

Shield	T (mm)	m (g)	v_{bl} (m/s)	Shield	T (mm)	m (g)	v_{bl} (m/s)
Sh-1	6.5	29.45	393	Sh-2	9.4	29.85	477
Sh-3	15.1	29.02	588	Sh-4	24.8	29.02	807

Table 5. Test results for Projectile-3.

Shield	T (mm)	m (g)	v_{bl} (m/s)	Shield	T (mm)	m (g)	v_{bl} (m/s)
Sh-3	15.1	754.2	303	Sh-4	24.8	755.6	396
Sh-5	31.1	756.4	447	Sh-6	36.4	759.8	489
Sh-7	49.1	728.9*	596				

* Without band.

4.41 [Deng *et al.*, 2012]

4.41.1 *General data on shields and projectiles*

Shields. Monolithic and layered (plates-in-contact and with air gaps) Q235 low carbon steel shields; $\widetilde{D}^{sh} = 170\,mm$, $\sigma_y = 229\,MPa$.

Projectiles. Hardened steel 38CrSi (HRC 58) hemispherical-nose projectiles (Fig. 3.4.1); $D = 12.62mm$, $L_{cyl} = 33.6\,mm$, $m = 34.5\,g$.

4.41.2 *Test results for* $T_{sum} = 2\,mm$

Table 1. Test results for 1×[2.0] shield.

v_s (m/s)	v_r (m/s)	v_s (m/s)	v_r (m/s)	v_s (m/s)	v_r (m/s)	v_s (m/s)	v_r (m/s)
128	0	134	20	135	27	138	45
139	55	148	74	151	81	153	81
159	97	159	97				

$v_{bl} = 133.8\,m/s$; $a = 0.9$, $p = 2.6$.

Table 2. Test results for 2×[1.0] shield.

v_s (m/s)	v_r (m/s)	v_s (m/s)	v_r (m/s)	v_s (m/s)	v_r (m/s)	v_s (m/s)	v_r (m/s)
125	0	129	27	132	40	132	48
132	48	134	53	139	72	141	74
144	79	151	93	151	93	152	95
156	104						

$v_{bl} = 127.8\, m/s$; $a = 0.96,\ p = 2.56$.

Table 3. Test results for [0.5]+[1.5] shield.

v_s (m/s)	v_r (m/s)	v_s (m/s)	v_r (m/s)	v_s (m/s)	v_r (m/s)	v_s (m/s)	v_r (m/s)
121	0	125	26	125	30	127	49
131	53	137	71	138	73	139	82
150	95	152	101	154	104		

$v_{bl} = 123.5\, m/s$; $a = 0.95,\ p = 2.53$.

Table 4. Test results for [1.5]+[0.5] shield.

v_s (m/s)	v_r (m/s)	v_s (m/s)	v_r (m/s)	v_s (m/s)	v_r (m/s)	v_s (m/s)	v_r (m/s)
124	0	127	37	127	29	131	43
133	55	136	62	139	71	149	87
156	107	158	100				

$v_{bl} = 124.8\, m/s$; $a = 0.96,\ p = 2.29$.

Table 5. Test results for 4×[0.5] shield.

v_s (m/s)	v_r (m/s)	v_s (m/s)	v_r (m/s)	v_s (m/s)	v_r (m/s)	v_s (m/s)	v_r (m/s)
129	0	132	25	135	43	136	44
138	50	145	64	146	74	147	76
152	88	156	97				

$v_{bl} = 130.68\, m/s$; $a = 0.97,\ p = 2.33$.

Table 6. Test results for 2×[1.0]\6.0 shield.

v_s (m/s)	v_r (m/s)	v_s (m/s)	v_r (m/s)	v_s (m/s)	v_r (m/s)	v_s (m/s)	v_r (m/s)
111	0	115	28	119	48	129	66
139	95	154	118	174	142		

$v_{bl} = 112.8\, m/s$; $a = 0.98,\ p = 2.45$.

4.41.3 *Test results for* $T_{sum} = 3\,mm$

Table 7. Test results for $1\times[3.0]$ shield.

v_s (m/s)	v_r (m/s)	v_s (m/s)	v_r (m/s)	v_s (m/s)	v_r (m/s)	v_s (m/s)	v_r (m/s)
184	0	191	43	191	40	191	39
192	52	196	61	200	77		

$v_{bl} = 187.5\,m/s$; $a = 0.96,\ p = 2.2$.

Table 8. Test results for $6\times[0.5]$ shield.

v_s (m/s)	v_r (m/s)	v_s (m/s)	v_r (m/s)	v_s (m/s)	v_r (m/s)	v_s (m/s)	v_r (m/s)
158	0	159	17	162	41	164	41
165	46	168	64	176	88	182	100

$v_{bl} = 158.5\,m/s$; $a = 0.98,\ p = 2.3$.

4.41.4 *Test results for* $T_{sum} = 6\,mm$

Table 9. Test results for $1\times[6.0]$ shield.

v_s (m/s)	v_r (m/s)	v_s (m/s)	v_r (m/s)	v_s (m/s)	v_r (m/s)	v_s (m/s)	v_r (m/s)
255	0	273	60	290	109	310	163
336	225	372	269	373	270	423	327

$v_{bl} = 275\,m/s$; $a = 0.94,\ p = 2.48$.

Table 10. Test results for $3\times[2.0]$ shield.

v_s (m/s)	v_r (m/s)	v_s (m/s)	v_r (m/s)	v_s (m/s)	v_r (m/s)	v_s (m/s)	v_r (m/s)
271	0	318	188	323	188	340	214
341	218	383	284	416	308	435	338

$v_{bl} = 285\,m/s$; $a = 0.92,\ p = 2.63$.

4.42 [Deng *et al.*, 2013]

4.42.1 *General data on shields and projectiles*

Shields. Monolithic and layered (plates in-contact and with air gaps) Q235 low carbon steel shields; $\widetilde{D}^{sh} = 170\,mm$, $\sigma_y = 229\,MPa$.

Projectiles. Hardened steel 38CrSi (HRC 58) ogive-nose projectiles (Fig. 3.3.1); $D = 12.62\,mm$, $L_{cyl} = 26.32\,mm$, $L_{nose} = 21.57\,mm$, $\psi = 3$, $m = 34.5\,g$.

4.42.2 Test results for $T_{sum} = 2\,mm$

Table 1. Test results for 1×[2.0] shield.

v_s (m/s)	v_r (m/s)	v_s (m/s)	v_r (m/s)	v_s (m/s)	v_r (m/s)
116	0	118	26	118	26
119	31	121	38	123	44
126	54	129	60	135	69

$v_{bl} = 115.8\,m/s;\ a = 1,\ p = 2.08$.

Table 2. Test results for 2×[1.0] shield.

v_s (m/s)	v_r (m/s)	v_s (m/s)	v_r (m/s)	v_s (m/s)	v_r (m/s)	v_s (m/s)	v_r (m/s)
104	0	110	33	110	31	112	30
115	57	120	69	124	67	135	82
140	98	144	101	151	119	179	145

$v_{bl} = 106.8\,m/s;\ a = 1,\ p = 2.18$.

Table 3. Test results for [0.5]+[1.5] shield.

v_s (m/s)	v_r (m/s)	v_s (m/s)	v_r (m/s)	v_s (m/s)	v_r (m/s)	v_s (m/s)	v_r (m/s)
107	0	108	0	113	38	114	39
115	39	118	50	122	55	129	68
134	74	143	92	151	105		

$v_{bl} = 108.1\,m/s;\ a = 1,\ p = 2$.

Table 4. Test results for [1.5]+[0.5] shield.

v_s (m/s)	v_r (m/s)	v_s (m/s)	v_r (m/s)	v_s (m/s)	v_r (m/s)	v_s (m/s)	v_r (m/s)
99	0	102	8	106	26	112	49
112	40	114	49	118	56	121	59
126	70	131	81	136	81	141	91
145	97						

$v_{bl} = 101.3\,m/s;\ a = 1,\ p = 1.84$.

Table 5. Test results for 4×[0.5] shield.

v_s (m/s)	v_r (m/s)	v_s (m/s)	v_r (m/s)	v_s (m/s)	v_r (m/s)	v_s (m/s)	v_r (m/s)
89	0	95	7	97	26	101	39
102	37	105	47	105	42	112	57
116	69	117	69	124	79	127	83
132	92						

$v_{bl} = 94.3 m/s$; $a = 1$, $p = 2.01$.

Table 6. Test results for 2×[1.0]\6.0 shield.

v_s (m/s)	v_r (m/s)	v_s (m/s)	v_r (m/s)	v_s (m/s)	v_r (m/s)
93	0	99	10	113	57
126	78	142	110	163	137

$v_{bl} = 98.5 m/s$; $a = 1$, $p = 2.2$.

4.42.3 *Test results for* $T_{sum} = 6 mm$

Table 7. Test results for 1×[6.0] shield.

v_s (m/s)	v_r (m/s)	v_s (m/s)	v_r (m/s)	v_s (m/s)	v_r (m/s)	v_s (m/s)	v_r (m/s)
241	0	248	26	286	140	308	188
327	222	327	205	338	240	370	273
397	316	496	417				

$v_{bl} = 245 m/s$; $a = 1$, $p = 2$.

Table 8. Test results for 3×[2.0] shield.

v_s (m/s)	v_r (m/s)	v_s (m/s)	v_r (m/s)	v_s (m/s)	v_r (m/s)	v_s (m/s)	v_r (m/s)
234	0	269	83	273	95	295	169
300	182	341	257	371	308		

$v_{bl} = 265 m/s$; $a = 1$, $p = 2.65$.

4.43 [Deng *et al.*, 2014a]

4.43.1 *General data on shields and projectiles*

Shields. Layered (plates-in-contact) shields, $\widetilde{D}^{sh} = 170 mm$, $T_{sum} = 12 mm$. Layers materials: 45 mild carbon alloy steel after HT ($\sigma_y = 714 MPa$,

HRC 29, hereafter *HS*-material) and low carbon alloy steel Q235 ($\sigma_y = 229\,MPa$, hereafter *LS*-material).

Projectiles. Hardened steel 38CrSi (HRC 58) projectiles with $D = 12.62\,mm$ and $m = 34.5\,g$: cylindrical projectiles (Fig. 3.1.1, $L = 37.8\,mm$) and ogive-nose projectiles (Fig. 3.3.1, $L_{cyl} = 26.4\,mm$, $L_{nose} = 20.93\,mm$, $\rho = 37.86\,mm$).

4.43.2 *Test results for cylindrical projectiles*

Table 1. Test results for [(HS) 6.0]+ [(LS) 6.0] shields.

v_s (m/s)	v_r (m/s)	m_r (g)	L_r (mm)	D_r (mm)
571	0	30.01	29.2	13.6
602	0	28.89	29.2	13.3
616	0	26.12	30.0	13.6
650	0	25.26	30.1	13.6
681	48	28.48	30.8	13.5
695	127	25.62	30.5	13.3
698	177	24.57	29.5	13.6
705	182	25.56	30.6	13.2
733	271	24.74	30.1	13.3
774	344	24.81	29.4	13.2

$v_{bl} = 679\,m/s$; $a = 0.88$, $p = 2.1$.

Table 2. Test results for [(LS) 6.0]+ [(HS) 6.0] shields.

v_s (m/s)	v_r (m/s)	m_r (g)	L_r (mm)	D_r (mm)
625	0	26.12	29.5	13.1
628	0	25.98	29.2	13.3
643	110	25.89	31.6	13.1
643	121	25.38	31.2	13.1
652	169	26.33	28.5	13.1
702	272	24.64	30.2	13.5
708	270	24.28	29.4	13.3
752	356	23.72	29.4	13.3

$v_{bl} = 632\,m/s$; $a = 0.77$, $p = 2.29$.

4.43.3 *Test results for ogive-nose projectiles*

Table 3. Test results for [(HS) 6.0]+ [(LS) 6.0] shields.

v_s (m/s)	v_r (m/s)	m_r (g)	L_r (mm)	D_r (mm)
373	0	34.40	46.2	12.6
354	0	34.34	46.2	12.6
427	0	34.42	46.7	12.7
450	142	34.45	46.0	12.7
465	176	34.56	46.3	12.6
470	189	34.53	45.7	12.7
518	299	34.50	46.2	12.6
582	392	34.46	45.7	12.7
604	439	34.52	45.6	12.7

$v_{bl} = 428\,m/s$; $a = 1$, $p = 2$.

Table 4. Test results for [(LS) 6.0]+ [(HS) 6.0] shields.

v_s (m/s)	v_r (m/s)	m_r (g)	L_r (mm)	D_r (mm)
404	0	34.29	46.5	12.7
433	84	34.36	46.4	12.7
449	163	34.26	46.0	12.7
494	244	34.46	46.1	12.7
549	335	34.12	46.3	12.7
675	494	34.20	45.3	12.7

$v_{bl} = 419\,m/s$; $a = 0.95$, $p = 2$.

4.44 [Deng *et al.*, 2014b]

4.44.1 *General data on shields and projectiles*

Shields. Monolithic and layered (plates-in-contact) shields; $\widetilde{D}^{sh} = 170\,mm$, $T_{sum} = 6\,mm$. Material of layers: 45 mild carbon alloy steel after HT ($\sigma_y = 714\,MPa$, HRC 29).

Projectiles. *Low strength* (HRC 19.1) and *high strength* (HRC 53.1) steel 38CrSi ogive-nose projectiles (Fig. 3.3.1); $D = 12.62mm$, $L_{cyl} = 26.4mm$, $L_{nose} = 20.93mm$, $\rho = 37.86mm$, $m = 34.5\,g$.

4.44.2 *Test results for high strength projectiles*

Table 1. Test results for 1×[6.0] shield.

v_s (m/s)	v_r (m/s)	v_s (m/s)	v_r (m/s)	v_s (m/s)	v_r (m/s)	v_s (m/s)	v_r (m/s)
317	0	341	93	347	117	365	153
380	166	394	192	441	237		

$v_{bl} = 325\,m/s;\ a = 0.7,\ p = 2.45$.

Table 2. Test results for 2×[3.0] shield.

v_s (m/s)	v_r (m/s)	v_s (m/s)	v_r (m/s)	v_s (m/s)	v_r (m/s)
291	0	328	128	356	171
420	254	433	266	481	321

$v_{bl} = 294.6\,m/s;\ a = 0.84,\ p = 2.05$.

Table 3. Test results for 3×[2.0] shield.

v_s (m/s)	v_r (m/s)	v_s (m/s)	v_r (m/s)	v_s (m/s)	v_r (m/s)	v_s (m/s)	v_r (m/s)
261	0	294	141	300	110	373	239
416	294	446	330	474	330	561	468

$v_{bl} = 276.5\,m/s;\ a = 0.96,\ p = 1.98$.

4.44.3 *Test results for low strength projectiles*

Table 4. Test results for 1×[6.0] shield.

v_s (m/s)	v_r (m/s)	m_r (g)	L_r (mm)	D_r (mm)
435	0	33.57	33.5	15.0
487	53	32.09	33.5	15.2
507	163	32.45	33.6	15.6
522	145	32.00	32.5	15.5
531	163	32.30	33.2	16.0
543	205	32.30	33.8	16.2
553	217	31.88	32.6	16.3
559	217	31.75	32.8	16.2
616	274	31.78	32.5	17.1

$v_{bl} = 482.5\,m/s;\ a = 0.6,\ p = 2.61$.

Table 5. Test results for 2×[3.0] shield.

v_s (m/s)	v_r (m/s)	m_r (g)	L_r (mm)	D_r (mm)
348	0	33.61	39.6	12.9
362	41	33.55	39.3	12.9
377	81	33.48	39.1	13.0
384	61	33.30	38.8	13.0
395	123	33.60	38.3	13.3
440	155	33.58	38.6	13.2
466	208	33.43	38.9	13.3
507	281	34.28	38.6	13.5

$v_{bl} = 358.5\,m/s$; $a = 0.76$, $p = 2.61$.

4.45 [Deng *et al.*, 2014c]

4.45.1 *General data on shields and projectiles*

Shields. Monolithic and layered (plates-in-contact) shields; $\widetilde{D}^{sh} = 170\,mm$, $T_{sum} = 6\,mm$. Material of layers: 45 mild carbon alloy steel after HT ($\sigma_y = 714\,MPa$, HRC 29).

Projectiles. *Low strength* (HRC 19.1) and *high strength* (HRC 53.1) steel 38CrSi cylindrical projectiles (Fig. 3.1.1); $D = 12.7\,mm$, $L = 37.3\,mm$, $m = 34.5\,g$.

4.45.2 *Test results for high strength projectiles*

Table 1. Test results for 1×[6.0] shield.

v_s (m/s)	v_r (m/s)	v_s (m/s)	v_r (m/s)	v_s (m/s)	v_r (m/s)	v_s (m/s)	v_r (m/s)
189	0	232	76	256	114	302	182
320	196	354	239	364	245	420	298

$v_{bl} = 215\,m/s$; $a = 0.82$, $p = 2.1$.

Table 2. Test results for 2×[3.0] shield.

v_s (m/s)	v_r (m/s)	v_s (m/s)	v_r (m/s)	v_s (m/s)	v_r (m/s)	v_s (m/s)	v_r (m/s)
212	0	244	69	255	112	279	141
305	174	372	244	401	275	537	391

$v_{bl} = 237.5\,m/s$; $a = 0.76$, $p = 2.95$.

Table 3. Test results for 3×[2.0] shield.

v_s (m/s)	v_r (m/s)	v_s (m/s)	v_r (m/s)	v_s (m/s)	v_r (m/s)
230	0	255	57	273	114
278	136	280	138	318	189
390	260	417	284	477	337

$v_{bl} = 252\,m/s;\ a = 0.72,\ p = 2.95$.

4.45.3 *Test results for low strength projectiles*

Table 4. Test results for 1×[6.0] shield.

v_s (m/s)	v_r (m/s)	m_r (g)	L_r (mm)	D_r (mm)
476	0	32.31	33.5	17.2
507	169	31.10	33.0	15.6
521	183	31.45	33.1	15.7
585	305	31.84	32.1	15.6
591	320	30.96	31.6	15.0

$v_{bl} = 485.4\,m/s;\ a = 0.8,\ p = 2.45$.

Table 5. Test results for 2×[3.0] shield.

v_s (m/s)	v_r (m/s)	m_r (g)	L_r (mm)	D_r (mm)
410	0	34.08	35.1	15.5
426	36	34.29	35.0	15.6
440	116	33.96	34.6	15.3
447	174	33.69	35.1	15.1
484	198	34.01	35.0	14.9
550	277	33.86	34.8	14.4
580	316	34.01	34.9	14.5
609	348	33.68	33.7	14.3

$v_{bl} = 424\,m/s;\ a = 0.71,\ p = 2.4$.

4.46 [Deniz, 2010]

See also [Deniz and Yildirim, 2013].

4.46.1 *General data on shields and projectiles*

Shields. Monolithic   AISI 4340 shields, HT to different hardness, $D^{sh} = 70\,mm$.

Projectiles. 7.62 mm AP projectiles.
Impact velocity. $v_s = 782 \pm 5.4\, m/s$.

4.46.2 *Test results*

Table 1. Test results for residual velocity, v_s (m/s).

HRC	T (mm)				
	7.2	9.0	10.8	12.7	14.4
39.5	570	500	400	0	0
49.5	470	0	0	0	0
52.5	395	260	0	0	0
58.5	265	0	0	0	0

4.47 [Deshpande and Gupta, 2000]

4.47.1 *General data on shields and projectiles*

Shields. Monolithic mild steel, *armor and duralumin* shields.
Projectiles. Steel spherical projectiles (Fig. 3.5.1) machined of tungsten carbide and mild steel.

4.47.2 *Test results for tungsten carbide projectiles ($D = 6\,mm$) against mild steel shields*

Table 1#. Test results for $T = 4\ mm$, $m = 1.832 - 1.930\ g$.

v_s (m/s)	v_r (m/s)	v_s (m/s)	v_r (m/s)	v_s (m/s)	v_r (m/s)	v_s (m/s)	v_r (m/s)
710	399	808	510	955	651	1053	751

Table 2#. Test results for $T = 6\ mm$, $m = 1.690 - 1.875\ g$.

v_s (m/s)	v_r (m/s)	v_s (m/s)	v_r (m/s)	v_s (m/s)	v_r (m/s)	v_s (m/s)	v_r (m/s)
706	290	808	359	958	500	1006	571

Table 3#. Test results for $T = 8\ mm$, $m = 1.730 - 1.780\ g$.

v_s (m/s)	v_r (m/s)	v_s (m/s)	v_r (m/s)	v_s (m/s)	v_r (m/s)	v_s (m/s)	v_r (m/s)
906	257	956	320	1103	466	1203	534

4.47.3 *Test results for mild steel projectiles against mild steel shields*

Table 4#. Test results for $T = 4$ *mm*, $D = 6$ *mm*, $m = 1.050$ *g*.

v_s (m/s)	v_r (m/s)	v_s (m/s)	v_r (m/s)	v_s (m/s)	v_r (m/s)
803	180	883	251	922	260
1001	351	1100	382		

Table 5#. Test results for $T = 6$ *mm*, $D = 16$ *mm*, $m = 16.400 - 16.420$ *g*.

v_s (m/s)	v_r (m/s)	v_s (m/s)	v_r (m/s)	v_s (m/s)	v_r (m/s)
497	298	598	310	697	338
800	443	999	575		

Table 6#. Test results for $T = 8$ *mm*, $D = 16$ *mm*, $m = 16.430 - 16.600$ *g*.

v_s (m/s)	v_r (m/s)	v_s (m/s)	v_r (m/s)	v_s (m/s)	v_r (m/s)	v_s (m/s)	v_r (m/s)
702	290	803	364	1002	496	1102	547

Table 7#. Test results $T = 10$ *mm*, $D = 16$ *mm*, $m = 16.590 - 16.620$ *g*.

v_s (m/s)	v_r (m/s)	v_s (m/s)	v_r (m/s)	v_s (m/s)	v_r (m/s)
650	111	701	182	802	235
904	291	1101	411		

4.47.4 *Test results for mild steel projectiles against armor shields*

Table 8#. Test results for $T = 10$ *mm*, $D = 16$ *mm*, $m = 16.360$ *g*.

v_s (m/s)	v_r (m/s)	v_s (m/s)	v_r (m/s)	v_s (m/s)	v_r (m/s)
809	195	857	170	982	253
983	267	987	237	996	244
1011	261	1017	273		

Table 9#. Test results for $T = 12$ *mm*, $D = 16$ *mm*, $m = 16.500$ *g*.

v_s (m/s)	v_r (m/s)	v_s (m/s)	v_r (m/s)	v_s (m/s)	v_r (m/s)
943	96	959	101	961	173
964	123	964	160	976	131

v_s (m/s)	v_r (m/s)	v_s (m/s)	v_r (m/s)	v_s (m/s)	v_r (m/s)
982	112	1002	187	1019	214
1034	200	1061	164		

Table 10#. Test results for $T = 4$ *mm*, $D = 6.3$ *mm*, $m = 1.028$ *g*.

v_s (m/s)	v_r (m/s)	v_s (m/s)	v_r (m/s)	v_s (m/s)	v_r (m/s)
663	111	668	76	669	101
685	118	699	156	704	168
722	183	857	171		

Table 11#. Test results for $T = 6$ *mm*, $D = 6.3$ *mm*, $m = 1.028$ *g*.

v_s (m/s)	v_r (m/s)	v_s (m/s)	v_r (m/s)	v_s (m/s)	v_r (m/s)
854	154	1127	107	1146	111
1146	119	1159	115	1165	124

4.47.5 *Test results for mild steel projectiles against duralumin shields*

Table 12#. Test results for $T = 10$ *mm*, $D = 6.3$ *mm*, $m = 1.028$ *g*.

v_s (m/s)	v_r (m/s)	v_s (m/s)	v_r (m/s)	v_s (m/s)	v_r (m/s)
787	98	798	119	804	131
809	184	816	133	816	164
817	185	819	169	825	136
828	203	835	193	836	191
840	231				

4.48 [Dey *et al.*, 2004]

4.48.1 *General data on shields and projectiles*

Shields. Monolithic shields ($\widetilde{D}^{sh} = 500\,mm$) manufactured of Weldox 460 E ($\sigma_y = 499\,MPa$), Weldox 700 E ($\sigma_y = 859\,MPa$) and Weldox 900 E ($\sigma_y = 992\,MPa$).

Projectiles. Arne oil hardened steel (HRC 51-53) projectiles with $D = 20\,mm$: cylindrical projectiles (Fig. 3.1.1; $L = 80\,mm$), truncated cone-nose projectiles (Fig. 3.6.1; $L_{cyl} = 68\,mm$, $L_{nose} = 30\,mm$, $d = 2\,mm$), ogive-nose projectiles (Fig. 3.3.1; $L_{cyl} = 62\,mm$, $L_{nose} = 33\,mm$, $\psi = 3$).

Table 1. Notations in column Note in tables below.

Notation	Description
^	Measurement is not available.
(1)	Test was performed using the material from other supply.
(2)	Projectile is lost in rag box.
(3)	Projectile is embedded in shield.
(4)	Projectile is shattered during impact.

4.48.2 *Test results for cylindrical projectiles*

Table 2. Test results for Weldox 460 E shields.

v_s (m/s)	m (g)	T (mm)	v_r (m/s)	v^{pl} (m/s)	m^{pl} (g)	T^{pl} (mm)	D^{pl}_{fr} (mm)	D^{pl}_{re} (mm)	ΔD (mm)	ΔL (mm)
173.7	196.1	12.1	0.0	0.0	–	–	–	–	0.14	0.22
177.3	197.0	12.1	0.0	0.0	–	–	–	–	0.15	0.13
179.4	196.8	12.0	0.0	0.0	–	–	–	–	0.16	0.13
181.5	196.7	12.1	0.0	0.0	–	–	–	–	0.11	0.24
184.3	196.8	12.1	30.8	45.3	27.8	11.7	20.5	21.3	0.17	0.20
184.8	^	^	0.0	0.0	^	^	^	–	0.17	0.00
188.8	196.9	12.1	43.2	66.9	27.8	11.7	20.6	21,3	0.19	0.27
189.2	196.7	12.1	40.1	66.8	28.0	11.7	20.4	21.4	0.21	0.34
189.6	196.9	12.1	43.7	71.8	27.9	11.8	20.4	21.3	0.19	0.28
189.6	196.7	12.1	42.0	64.0	27.7	11.7	20.4	21.2	0.17	0.22
199.1	196.9	12.1	67.3	104.0	27.8	11.6	20.4	21.3	0.17	0.22
200.4	197.0	12.0	71.4	103.7	27.8	11.6	20.4	21.4	0.29	0.28
224.7	196.9	12.0	113.7	169.0	27.3	11.5	20.4	21.2	0.49	0.47
244.2	196.9	12.1	132.6	187.7	28.1	11.6	20.7	21.7	0.50	0.53
285.4	197.3	12.0	181.1	224.7	27.6	11.4	20.7	21.7	0.77	0.70
303.5	196.6	12.1	199.7	242.3	27.6	11.4	21.0	21.9	1.01	0.90
399.6	196.9	11.9	291.3	–	36.0	11.5	–	–	–	–

$v_{bl} = 184.5\,m/s$; $a = 0.79$, $p = 2.24$.

Table 3. Test results for Weldox 700 E shields.

v_s (m/s)	m (g)	T (mm)	v_r (m/s)	v^{pl} (m/s)	m^{pl} (g)	T^{pl} (mm)	D_{fr}^{pl} (mm)	D_{re}^{pl} (mm)	ΔD (mm)	ΔL (mm)	Note
152.4	197.0	11.9	0.0	0.0	–	–	–	–	0.10	0.40	(1)
159.2	197.0	12.1	0.0	0.0	–	–	–	–	0.12	0.11	(1)
159.5	196.6	12.1	0.0	0.0	–	–	–	–	0.13	0.11	(1)
160.6	196.8	12.0	0.0	^	28.6	11.8	20.7	19.9	0.16	0.24	(1)
161.0	196.4	12.1	0.0	0.0	–	–	–	–	0.10	2.03	–
165.1	196.2	12.2	0.0	0.0	–	–	–	–	0.08	0.44	–
170.8	196.9	12.2	79.0	^	28.8	11.3	20.7	19.7	0.14	0.22	–
171.2	196.4	12.1	20.0	40.0	28.7	11.9	20.3	19.1	0.11	0.38	–
174.1	196.4	12.1	0.0	70.7	28.7	12.0	20.2	19.4	–	–	(3)
176.3	196.5	12.1	22.0	15.0	28.8	12.0	21.6	21.4	0.05	0.43	–
176.8	196.4	12.1	47.0	63.1	28.5	11.9	20.4	20.1	0.07	0.35	–
178.4	196.8	11.9	75.8	79.0	27.6	11.7	20.4	20.0	0.19	0.23	(1)
182.2	196.8	11.9	85.4	96.7	28.1	11.7	20.3	20.2	0.18	0.36	(1)
186.3	196.8	11.9	87.5	102.5	28.2	11.7	20.3	20.1	0.21	0.39	(1)
194.1	196.9	11.9	96.0	107.9	27.2	11.8	20.4	19.8	0.22	0.37	(1)
200.4	196.6	12.1	92.0	115.1	28.4	11.9	20.2	19.4	0.26	20.8	–
235.3	196.7	11.9	143.3	148.1	26.8	11.6	20.4	19.3	0.41	0.50	(1)
249.4	196.5	12.1	141.0	157.0	28.2	11.9	20.3	19.6	0.67	2.59	–
294.1	196.8	11.9	192.1	231.2	25.3	11.7	20.2	19.3	0.69	0.89	(1)
305.9	196.5	12.1	195.0	216.0	28.7	11.8	20.8	20.1	1.24	3.27	–
356.7	195.0	12.1	228.0	274.0	28.5	11.7	20.9	19.7	–	–	(2)

$v_{bl} = 168.0\,m/s$; $a = 0.70, p = 2.55$.

Table 4. Test results for Weldox 900 E shields.

v_s (m/s)	m (g)	T (mm)	v_r (m/s)	v^{pl} (m/s)	m^{pl} (g)	T^{pl} (mm)	D_{fr}^{pl} (mm)	D_{re}^{pl} (mm)	ΔD (mm)	ΔL (mm)	Note
156.8	194.8	12.1	0.0	0.0	–	–	–	–	0.00	0.29	–
165.1	194.9	12.0	^	25.3	29.4	12.0	20.1	19.7	0.15	0.50	–
172.2	194.8	12.0	54.0	73.0	29.1	12.0	20.1	19.3	0.09	0.38	–
202.1	197.0	12.0	95.0	123.0	28.7	11.8	20.4	18.6	0.42	2.44	–
246.0	196.2	12.1	140.6	175.0	28.4	11.8	20.4	18.6	0.56	1.00	–
307.4	196.3	12.0	193.0	248.0	28.4	11.8	20.8	18.9	0.98	1.65	–
359.6	195.0	12.0	^	139.3	43.3	11.8	25.9	23.2	–	–	(4)

$v_{bl} = 161.0\,m/s$; $a = 0.70, p = 2.34$.

4.48.3 *Test results for truncated cone-nose projectiles*

Table 5. Test results for Weldox 460 E shields (plug is not formed).

v_s (m/s)	m (g)	T (mm)	v_r (m/s)	ΔD (mm)	ΔL (mm)
206.9	196.6	12.1	0.0	0.01	0.01
248.7	196.7	12.3	0.0	0.06	0.02
280.9	196.5	12.2	0.0	0.02	0.04
300.3	196.7	12.3	110.3	0.03	0.02
317.9	196.7	12.4	155.8	0.02	0.01
355.6	196.7	12.1	232.3	0.02	0.05
405.7	196.3	12.3	312.0	0.01	0.19

$v_{bl} = 290.6\,m/s$; $a \equiv 1.00$, $p = 2.37$.

Table 6. Test results for Weldox 700 E shields.

v_s (m/s)	m (g)	T (mm)	v_r (m/s)	v^{pl} (m/s)	m^{pl} (g)	T^{pl} (mm)	D_{fr}^{pl} (mm)	D_{re}^{pl} (mm)	ΔD (mm)	ΔL (mm)	Note
277.9	196.2	12.1	0.0	^	4.8	^	^	^	0.00	4.63	–
304.9	196.3	12.2	0.0	^	7.5	9.7	12.4	15.4	–	–	(3)
323.4	196.4	12.1	0.0	^	3.7	^	^	^	–	–	(3)
346.6	196.4	12.1	122.0	^	7.0	8.6	13.5	11.5	0.24	14.12	–
370.2	196.2	12.2	177.0	^	8.0	8.0	13.8	13.9	0.17	14.32	–

$v_{bl} = 335.0\,m/s$; $a = 0.72$, $p = 3.19$.

Table 7. Test results for Weldox 900 E shields.

v_s (m/s)	m (g)	T (mm)	v_r (m/s)	v^{pl} (m/s)	m^{pl} (g)	T^{pl} (mm)	D_{fr}^{pl} (mm)	D_{re}^{pl} (mm)	ΔD (mm)	ΔL (mm)	Note
302.5	196.0	12.1	0.0	228.0	11.5	10.4	14.0	13.3	0.23	14.72	–
327.8	196.4	12.0	0.0	93.0	8.5	9.7	12.5	12.0	–	–	(3)
330.9	196.2	12.0	0.0	286.0	6.7	8.9	11.4	10.4	–	–	(3)
352.4	196.1	12.0	^	77.3	4.8	9.4	10.0	8.9	0.24	13.31	–
365.3	196.1	12.0	85.0	^	10.3	^	14.3	14.2	0.16	15.73	–

$v_{bl} = 340.1\,m/s$; $a = 0.47$, $p = 4.31$.

4.48.4 *Test results for ogive-nose projectiles (plug is not formed)*

Table 8. Test results for Weldox 460 E shields.

v_s (m/s)	m (g)	T (mm)	v_r (m/s)	ΔD (mm)	ΔL (mm)	Note
274.0	196.3	12.1	0.0	0.01	–	(3)
293.5	196.4	11.9	0.0	0.01	–	(3)

v_s (m/s)	m (g)	T (mm)	v_r (m/s)	ΔD (mm)	ΔL (mm)	Note
298.2	196.7	12.1	91.2	0.01	0.06	–
324.6	196.1	12.1	160.3	0.01	0.06	–
346.6	196.3	12.1	193.3	0.02	0.00	–
365.0	196.3	12.1	239.9	0.00	0.00	–

$v_{bl} = 295.9\,m/s$; $a \equiv 1.00, p = 2.35$.

Table 9. Test results for Weldox 700 E shields.

v_s (m/s)	m (g)	T (mm)	v_r (m/s)	ΔD (mm)	ΔL (mm)	Note
299.7	196.2	12.2	0.0	–	–	(3)
307.3	196.3	12.2	0.0	–	–	(3)
309.9	196.5	12.2	0.0	–	–	(3)
315.0	196.5	12.3	0.0	–	–	(3)
321.1	196.6	12.2	63.5	0.00	0.11	(2)
333.5	196.4	12.2	99.4	0.00	–	–
354.7	196.4	12.2	159.7	^	^	–
366.2	196.4	12.2	174.9	0.01	0.32	–

$v_{bl} = 318.1\,m/s$; $a \equiv 1.00, p = 2.02$.

Table 10. Test results for Weldox 900 E shields.

v_s (m/s)	m (g)	T (mm)	v_r (m/s)	ΔD (mm)	ΔL (mm)	Note
293.1	196.5	12.1	0.0	–	–	(3)
310.6	196.4	12.1	0.0	–	–	(3)
313.4	196.4	12.2	0.0	–	–	(3)
319.3	196.3	12.1	0.0	–	–	(3)
320.8	196.2	12.1	0.0	–	–	(3)
323.5	196.4	12.2	82.9	0.02	0.26	–
330.4	196.5	12.2	105.2	0.00	0.52	–
330.5	196.4	12.2	132.4	0.00	4.69	–
335.0	196.3	12.1	115.9	0.00	1.01	–
351.0	196.6	12.1	165.0	0.02	0.32	–

$v_{bl} = 322.2\,m/s$; $a \equiv 1.00, p = 2.44$.

4.49 [Dey *et al.*, 2007]

4.49.1 *General data on shields and projectiles*

Shields. Monolithic and layered (plates-in-contact and with air gaps) Weldox 700 E steel alloy shields, $\widetilde{D}^{sh} = 500\,mm$.

Projectiles. Hardened tool steel (HRC 52) projectiles with $D = 20\,mm$ and $m = 197\,g$: cylindrical projectiles (Fig. 3.1.1; $L = 80\,mm$) and ogive-nose projectiles (Fig. 3.3.1; $L_{cyl} = 62\,mm$, $L_{nose} = 33\,mm$, $\psi = 3$).

Table 1. Notations in column Note in tables below.

Notation	Description
(1)	Projectile lost in rag box.
(2)	Projectile embedded in target.
(3)	Some yaw.
(4)	No penetration.
(5)	A (part of a) ring is ejected.
(6)	Projectile shatters.
(7)	Projectile slides along the plate.

Note. A number of indicators in tables for two-layer shields are presented in the form of a fraction whereby numerator and denominator are related to the first and second plate, correspondingly.

4.49.2   *Test results for cylindrical projectiles*

Table 2. Test results for 1× [6.0] shield.

v_s (m/s)	v_r (m/s)	m^{pl} (g)	D_{fr}^{pl} (mm)	D_{re}^{pl} (mm)	T^{pl} (mm)	ΔD (mm)	ΔL (mm)
141	5	13.9	20.2	19.7	6.2	0.0	0.0
152	55	14.0	20.8	20.0	6.3	0.1	0.2
158	116	14.3	20.5	19.5	6.2	0.1	0.0
182	138	14.2	20.2	19.6	6.2	0.1	0.1
197	155	14.6	20.5	20.1	6.2	0.1	0.1
246	204	14.7	20.8	20.4	6.3	0.2	0.2

$v_{bl} = 140.8\,m/s$; $a = 0.92, p = 2.88$.

Table 3. Test results for 1× [12.0] shield.

v_s (m/s)	v_r (m/s)	m^{pl} (g)	D_{fr}^{pl} (mm)	D_{re}^{pl} (mm)	T^{pl} (mm)	ΔD (mm)	ΔL (mm)	Note
152	0	0.0	–	–	–	0.1	0.4	–
159	0	0.0	–	–	–	0.1	0.1	–
160	0	0.0	–	–	–	0.1	0.1	–
161	0	0.0	–	–	–	0.1	2.0	–

v_s (m/s)	v_r (m/s)	m^{pl} (g)	D_{fr}^{pl} (mm)	D_{re}^{pl} (mm)	T^{pl} (mm)	ΔD (mm)	ΔL (mm)	Note
161	–	28.6	20.7	19.9	11.8	0.2	0.2	–
165	0	0.0	–	–	–	0.1	0.4	–
171	20	28.7	20.3	19.1	11.9	0.1	0.4	–
171	79	28.8	20.7	19.7	11.9	0.1	0.2	–
174	0	28.7	20.2	19.4	12.0	–	–	(2)
176	22	28.8	21.6	21.4	12.0	0.1	0.4	–
177	47	28.5	20.4	20.1	11.9	0.1	0.3	–
178	76	27.6	20.4	20.0	11.7	0.2	0.2	–
182	85	28.1	20.5	20.2	11.7	0.2	0.4	–
186	88	28.2	20.5	20.1	11.7	0.2	0.4	–
194	96	27.2	20.4	19.8	11.8	0.2	0.4	–
200	92	28.4	20.2	19.4	11.9	0.3	2.1	–
235	143	26.8	20.4	19.5	11.6	0.4	0.5	–
249	141	28.2	20.3	19.6	11.9	0.7	2.6	–
294	192	25.3	20.2	19.5	11.7	0.7	0.9	–
306	195	28.7	20.8	20.1	11.8	1.2	3.3	–
357	228	28.5	20.9	19.7	11.7	–	–	(1)

$v_{bl} = 168.0\,m/s$; $a = 0.70, p = 2.55$.

Table 4. Test results for 2× [6.0] shield.

v_s (m/s)	v_r (m/s)	m^{pl} (g)	D_{fr}^{pl} (mm)	D_{re}^{pl} (mm)	T^{pl} (mm)	ΔD (mm)	ΔL (mm)	Note
243	0	–/–	–	–	–	–	–	(2)
252	34	13.8/11.0	21.1/20.2	20.2/7.8	6.8/7.5	0.6	1.6	–
257	98	12.4/10.5	20.2/19.4	19.5/17.5	6.6/7.1	0.3	0.8	(5)
279	138	12.0/10.5	20.2/19.7	19.5/18.2	6.2/7.6	0.2	0.8	(5)
307	172	12.2/10.6	20.5/19.7	19.7/18.6	6.3/7.8	0.4	0.9	–
355	215	13.4/12.0	21.6/21.1	20.9/19.7	6.6/8.3	0.7	1.4	–
375	0	15.9/–	22.6/_	22.4/–	7.6/–	0.7	–	(3), (4)

$v_{bl} = 247.3\,m/s$; $a = 0.74, p = 2.70$.

Table 5. Test results for 2× [6.0]\24.0 shield.

v_s (m/s)	v_r (m/s)	m^{pl} (g)	D_{fr}^{pl} (mm)	D_{re}^{pl} (mm)	T^{pl} (mm)	ΔD (mm)	ΔL (mm)	Note
225	0	14.6/–	22.2/–	22.6/–	7.0/–	–	–	(2)
244	70	14.7/15.7	22.2/22.4	22.5/22.2	6.1/7.1	0.3	0.4	–
249	102	14.5/15.9	22.3/22.4	22.5/21.5	6.0/7.2	0.2	0.5	–
252	0	14.6/–	22.0/–	22.7/–	6.7/–	0.1	0.2	(6)
260	73	14.6/15.2	22.5 /23.0	22.6/21.4	6.2/7.2	0.1	9.7	(6)
269	128	14.6/17.4	22.3/23.1	22.5/22.3	6.1/7.5	0.0	10.1	(6)

v_s (m/s)	v_r (m/s)	m^{pl} (g)	D_{fr}^{pl} (mm)	D_{re}^{pl} (mm)	T^{pl} (mm)	ΔD (mm)	ΔL (mm)	Note
271	97	14.4/16.8	23.0/23.5	23.3/23.3	6.8/7.5	0.3	0.6	–
283	86	14.2/16.6	22.9/23.4	23.8/24.1	7.1/7.9	0.3	0.5	–
297	156	14.6/17.5	22.8/23.4	22.8/22.1	5.4/8.2	–	12.5	(6)
297	98	14.4/17.7	22.5/23.2	22.8/23.2	6.3/7.9	–	34.5	(6)
309	90	14.4/16.4	22.9/24.3	24.1/25.0	7.6/8.4	0.4	0.7	–
351	190	10.9/ 9.4	21.3/20.8	20.3/20.0	5.5/7.1	0.6	0.9	(5)

$v_{bl} = 234.6\,m/s$; $a = 0.62, p = 2.87$.

4.49.3 *Test results for ogive-nose projectiles (plug is not formed)*

Table 6. Test results for $1\times$ [6.0] shield.

v_s (m/s)	v_r (m/s)	ΔD (mm)	ΔL (mm)	Note
179	0	0.0	0.1	–
195	0	–	–	(2)
202	19	0.0	0.0	–
206	51	0.0	0.0	–
216	0	0.0	7.4	(3), (7), (2)
219	65	0.0	0.2	–
252	159	0.0	0.1	–
309	244	0.0	0.1	–
370	318	0.0	0.1	–

$v_{bl} = 198.0\,m/s$; $a \equiv 1.00, p = 1.93$.

Table 7. Test results for $1\times$ [12.0] shield.

v_s (m/s)	v_r (m/s)	ΔD (mm)	ΔL (mm)	Note
300	0	–	–	(2)
307	0	–	–	(2)
310	0	–	–	(2)
315	0	–	–	(2)
321	64	0.0	0.1	–
334	99	0.0	–	–
355	160	–	–	(1)
366	175	0.0	0.3	–

$v_{bl} = 318.1\,m/s$; $a \equiv 1.00, p = 2.02$.

Table 8. Test results for 2× [6.0] shield.

v_s (m/s)	v_r (m/s)	ΔD (mm)	ΔL (mm)	Note
286	0	–	–	(1)
290	72	0.0	0.6	–
299	91	0.0	0.4	–
305	142	–	–	(1)
310	144	–	–	(1)
314	133	–	–	(1)
330	157	0.0	0.1	–
347	211	0.0	0.3	–

$v_{bl} = 288.3\,m/s$; $a \equiv 1.00, p = 2.25$.

Table 9. Test results for 2× [6.0]\24.0 shield.

v_s (m/s)	v_r (m/s)	ΔD (mm)	ΔL (mm)	Note
283	59	0.0	0.1	–
284	0	–	–	(2)
290	67	0.0	0.1	–
292	82	0.0	0.5	–
294	82	0.0	0.0	–
305	119	0.0	0.1	–
318	155	0.0	0.4	–
323	161	0.0	1.6	–
360	243	–	–	–

$v_{bl} = 280.0\,m/s$; $a \equiv 1.00, p = 2.03$.

Test results for 2× [6.0]\12.0 shield:
$v_s = 348\,m/s$, $v_r = 207\,m/s$.

4.50 [Dey *et al.*, 2011]

4.50.1 *General data on shields and projectiles*

Shields. Two-layer shields consisting of Weldox 700E and Armox 560T plates-in-contact; thickness of each plate is 6 mm; $\tilde{c}_{l.w} = 500\,mm$.
Projectiles. Hardened tool steel (HRC 52) projectiles with $D = 20\,mm$ and $m = 197\,g$: cylindrical projectiles (Fig. 3.1.1; $L = 80\,mm$) and ogive-nose projectiles (Fig. 3.3.1; $L_{cyl} = 62\,mm$, $L_{nose} = 33\,mm$, $\psi = 3$).

4.50.2 *Test results*

Table 1. Test results on BLV.

Projectile	Two-layered Shield	v_{bl} (m/s)
Cylindrical	[(Weldox 700E) 6.0]+[(Armox 560T) 6.0]	291
Cylindrical	[(Armox 560T) 6.0]+[(Weldox 700E) 6.0]	>354
Ogive nose	[(Weldox 700E) 6.0]+[(Armox 560T) 6.0]	294
Ogive nose	[(Armox 560T) 6.0]+[(Weldox 700E) 6.0]	>341

4.51 [Dikshit and Sundararajan, 1992a]

4.51.1 *General data on shields and projectiles*

Shields. Monolithic steel (HV 350) shields; $c_{l,w} = 450\,mm$, $T = 20\,mm$, $m^{sh} = 31.8\,\mathrm{kg}$.

Projectiles. Steel (HV 650) ogive-nose projectiles (Fig. 3.3.1); $D = 20mm$, $m = 108\,\mathrm{g}$

Note. All Vickers Hardness values (HVs) at 30 kg=294 N load.

4.51.2 *Shield holder*

Shield holder was fabricated from a 5 mm thick mild steel angled iron. The base of the shield holder was massive and was further anchored firmly in the hard ground. The arm plate was clamped firmly to this holder and one set of experiments was carried out which simulated rigid-clamping. Another set of experiments was carried out with the shield plate hanging from the holder using a high tension 500 mm long steel wire which simulated the 'non-rigid' clamping situation.

4.51.3 *Test results*

Table 1#. Rigid clamping.

v_s (m/s)	P (mm)	v_s (m/s)	P (mm)	v_s (m/s)	P (mm)	v_s (m/s)	P (mm)
312	12.3	346	14.0	378	15.3	446	18.9
466	20.3	532	25.8	581	31.3		

Table 2#. Non-rigid clamping.

v_s (m/s)	P (mm)	v_s (m/s)	P (mm)	v_s (m/s)	P (mm)
318	12.7	361	14.2	398	16.3
492	21.5	539	26.0		

4.52 [Dikshit and Sundararajan, 1992b]

See also [Sundararajan and Dikshit, 2009].

4.52.1 *General data on shields and projectiles*

Shields. Monolithic RHA steel shields, $c_{l.w} = 1\,m$.

Table 1. Mechanical properties of shield materials.

T (mm)	HB	HV	σ_y (MPa)	σ_u (MPa)	ε_u (%)	ε_f (%)
20	320±10	350±10	1080	1160	5.0	15
40	300±10	315±10	920	1000	6.5	17
80	290±10	295±10	860	960	7.0	19

Projectiles. Steel (HV 600) ogive-nose projectiles (Fig. 3.3.1); $D = 20\,mm$, $L_{cyl} = 34\,mm$, $L_{nose} = 20\,mm$, $m = 110\,g$.

Notes.
- Parameters L_{nose} и L_{cyl} were determined approximately from the figure showing the dependence of the volume of a projectile between the nose and the plane normal to the projectile axis vs. the distance from the nose to this plane.
- Vickers Hardness (HV) at 30 kg = 294 N load.

4.52.2 *Test results*

Table 2#. Test results for shields with $T = 20\,mm$.

v_s (m/s)	P (mm)	v_s (m/s)	P (mm)	v_s (m/s)	P (mm)	v_s (m/s)	P (mm)
304	11.7	311	12.5	345	14.1	378	15.2
404	15.2	428	17.3	441	19.0	468	20.1
499	19.8	531	22.5	574	29.0	582	30.0
591	31.9						

Table 3#. Test results for shields with $T = 40\,mm$.

v_s (m/s)	P (mm)	v_s (m/s)	P (mm)	v_s (m/s)	P (mm)	v_s (m/s)	P (mm)
289	12.0	328	13.5	408	15.0	445	17.5
456	18.1	525	21.0	542	22.4	571	23.0
596	24.1	605	24.9	628	26.5	676	28.1

Table 4#. Test results for shields with $T = 80\,mm$.

v_s (m/s)	P (mm)	v_s (m/s)	P (mm)	v_s (m/s)	P (mm)	v_s (m/s)	P (mm)
280	12.3	344	14.2	393	16.1	417	17.0
451	19.1	499	21.1	546	22.6	635	26.6
672	28.0	715	29.5	793	32.0		

4.53   [Dikshit *et al.*, 1995]

4.53.1   *General data on shields and projectiles*

Shields. Monolithic RHA steel shields having various hardness, $c_{l.w} = 450\,mm$.

Projectiles. Steel (HV 600) ogive-nose projectiles (Fig. 3.3.1); $D = 20mm$, $L_{cyl} = 34\,mm$, $L_{nose} = 20\,mm$, $m = 110\,g$.

Notes.

- Parameters L_{cyl} and L_{nose} were determined from the figure in [Dikshit and Sundararajan, 1992b] showing the dependence of the volume of a projectile between the nose and the plane normal to the projectile axis vs. the distance from the nose to this plane.
- HV is here Vickers hardness at 30 kg = 294 N load.

4.53.2   *Test results*

Table 1#. Test results for shields with $T = 20\,mm$.

HV	v_s (m/s)	P (mm)	v_s (m/s)	P (mm)	v_s (m/s)	P (mm)	v_s (m/s)	P (mm)
300	319	13.4	356	15.0	405	16.9	443	18.2
	495	23.7	526	27.2	562	30.0		
350	311	12.5	347	14.0	378	15.1	428	17.1

HV	v_s (m/s)	P (mm)	v_s (m/s)	P (mm)	v_s (m/s)	P (mm)	v_s (m/s)	P (mm)
	470	20.0	502	19.5	534	22.1	578	28.8
	585	29.7						
440	319	11.1	354	11.2	399	12.6	442	14.4
	485	16.5	521	25.7				
520	307	10.9	348	11.9	394	13.9	437	16.3
	454	17.3	503	21.9				

Table 2#. Test results for shields with $T = 80\,mm$.

HV	v_s (m/s)	P (mm)	v_s (m/s)	P (mm)	v_s (m/s)	P (mm)	v_s (m/s)	P (mm)
295	414	16.9	448	19.1	499	21.0	545	22.5
	632	26.4	672	28.0	716	29.4	800	31.9
440	456	15.4	506	18.7	567	20.0	664	22.4
	791	22.8						
520	496	11.3	559	11.8	677	12.7	760	11.8
	800	11.2						

4.54 [Dikshit, 1998a]

4.54.1 *General data on shields and projectiles*

Shields. Monolithic AISI-4330 steel shields in different hot-rolled condition, $T = 20\,mm$.

Projectiles. Steel (HV 750) cone-nose projectiles (Fig. 3.2.1); $D = 6.1\,mm$, $L_{cyl} = 18.9\,mm$, $L_{nose} = 8.5\,mm$, $m = 5.2\,g$.

Note. All Vickers Hardness values (HVs) at 30 kg = 294 N load.

4.54.2 *Test results*

Table 1#. Test results for shields having different hardness.

HV	v_s (m/s)	P (mm)	v_s (m/s)	P (mm)	v_s (m/s)	P (mm)	v_s (m/s)	P (mm)
350	198	4.7	309	6.6	361	7.0	510	9.0
	596	10.1	643	11.0				
450	233	3.5	307	5.0	426	7.5	559	9.0
	646	7.0						
550	242	2.1	356	5.0	450	7.0	643	4.0

4.55 [Dikshit, 1998b]

4.55.1 *General data on shields and projectiles*

Shields. Monolithic RHA (HV 370±10) shields; $c_l = 450\,mm$, $c_w = 300\,mm$, $T = 10\,mm$.

Projectiles. Steel (HV 600) ogive-nose projectiles (Fig. 3.3.1); $D = 20mm$, $L = 114.6mm$, $m = 110\,g$.

4.55.2 *Test results*

Table 1#. Test results for different θ_s

θ_s (deg)	v_s (m/s)	P (mm)	v_s (m/s)	P (mm)	v_s (m/s)	P (mm)
15	298	14.1	351	17.8		
30	302	9.4	345	10.7	390	14.7
45	298	2.4	351	3.5	386	4.6
	443	7.1	486	8.8	533	11.2

4.56 [Dikshit, 1998c]

4.56.1 *General data on shields and projectiles*

Shields. Monolithic RHA steel (HV 350) shields; $c_{l,w} = 450\,mm$, $T = 20\,mm$.

Projectiles. Steel (HV 600) ogive-nose projectiles (Fig. 3.3.1); $D = 20mm$, $L = 114.6mm$, $m = 110\,g$.

Note. All Vickers Hardness values (HVs) are at 30 kg = 294 N load.

4.56.2 *Test results*

Table 1#. Test results for different θ_s.

θ_s (deg)	v_s (m/s)	P (mm)	v_s (m/s)	P (mm)	v_s (m/s)	P (mm)	v_s (m/s)	P (mm)
0	306	11.8	349	13.8	388	15.2	449	18.8
	466	19.8	539	25.3	583	30.8		

θ_s (deg)	v_s (m/s)	P (mm)	v_s (m/s)	P (mm)	v_s (m/s)	P (mm)	v_s (m/s)	P (mm)
15	318	10.6	372	12.2	415	14.8	444	15.9
	495	20.5	568	26.4	579	27.8		
30	295	4.1	325	5.8	355	6.1	414	7.8
	439	8.5	486	10.2	530	10.0	546	10.4
	638	12.3	717	13.7	773	14.4		
45	318	2.3	368	3.5	421	3.2	445	3.6
	644	10.9	705	9.9	772	13.6		

4.57 [Dikshit, 1999]

4.57.1 *General data on shields and projectiles*

Shields. Monolithic RHA steel shields; $T = 10\,mm$.
Projectiles. Steel cone-nose projectiles (Fig. 3.2.1); $D = 6.1mm$, $L_{cyl} = 18.9\,mm$, $L_{nose} = 8.5\,mm$, $m = 5.2\,g$.
Note. Plate-to-projectile Vickers Hardness (at 30 kg = 294 N load) ratio is 0.51.

4.57.2 *Test results*

Table 1#. Test results for different θ_s.

θ_s (deg)	v_s (m/s)	P (mm)	v_s (m/s)	P (mm)	v_s (m/s)	P (mm)
0	328	5.5	376	7.6	476	9.1
	514	9.7	555	10.2	609	12.3
15	310	3.0	384	4.0	446	6.6
	537	5.6	644	5.8		
30	327	1.7	394	2.3	438	2.1
	580	3.7	691	4.3	724	4.5
45	341	0.6	412	1.6	560	2.3
	763	2.7	823	3.8		

4.58 [Dikshit, 2000]

4.58.1 *General data on shields and projectiles*

Shields. Monolithic tempered steel shields; $T = 20\,mm$.

Projectiles. Steel (HV 650) ogive-nose projectiles (Fig. 3.3.1); $D = 20\,mm$, $L_{cyl} = 34\,mm$, $L_{nose} = 20\,mm$, $m = 110\,g$.

Notes.

- Parameters L_{cyl} and L_{nose} were determined from the figure in [Dikshit and Sundararajan, 1992b] showing the dependence of the volu1`me of a projectile between the nose and the plane normal to the projectile axis vs. distance from the nose to this plane.
- HV is the value of Vickers hardness at 30 kg = 294 N load.
- Plugging velocity, v^{pl}, is a striking velocity of a projectile at which a plate begins to deform and a plug is formed.

4.58.2 *Test results*

Table 1#. Test results.

HV	v_{bl} (m/s)	v^{pl} (m/s)	HV	v_{bl} (m/s)	v^{pl} (m/s)	HV	v_{bl} (m/s)	v^{pl} (m/s)
300	639	–	350	558	–	400	536	575
440	561	561	470	526	495	520	526	419

4.59   [Dolinski and Rittel, 2015]

4.59.1 *General data on shields and projectiles*

Shields. Monolithic thick RHA (HRC 39) shield; $c_w = 300\,mm$, $c_l = 250\,mm$.

Projectiles. FSP (Fig. 3.10.1, $D = 20\,mm$) having a trimmed nose (or a cylinder with small rear "skirt") made of 4340 steel (HRC 30).

4.59.2 *Test results*

Table 1. Test results.

L (mm)	m (g)	v_s (m/s)	P (mm)	L (mm)	m (g)	v_s (m/s)	P (mm)
20.7	46.7	1920	*	21.1	47.7	1715	*
20.9	46.6	1460	*	20.5	46.4	1466	*
20.7	46.9	1256	14.1	20.7	46.7	1248	13.7
20.0	45.9	1322	16.8	20.6	46.5	1394	*
20.3	46.8	1151	11.1				

* Perforation.

4.60   [Durmuş *et al.*, 2011]

4.60.1   *General data on shields and projectiles*

Shields. Low carbon cold rolled H320LA (DIN EN 10268–99) steel monolithic and layered (plates-in-contact) shields; $\tilde{c}_{l.w} = 164\,mm$.

Projectiles. 9 mm × 19 Parabellum cartridge. The mass of the full metal jacketed projectile is 8 g and the core and the jacket are made from 2 wt.% antimony-lead alloy and brass, respectively.

Table 1. Projectile failure types (PFTs).

Notation	Description	Notation	Description	Notation	Description
F	Flattening	M	Mushrooming	S	Separation from the jacket

4.60.2   *Test results*

Table 2. Test results for 1×[1.0] shields.

v_s (m/s)	v_r (m/s)	m^{pl} (g)	PFT	L_r (mm)	D_r (mm)
71	–	–	F	15.01	4.62
78	–	–	F	14.91	4.94
97	–	–	F	14.70	5.23
137	35	0.2	F	14.94	5.74
153	79	0.2	F	14.70	5.94
211	158	0.2	M	13.64	9.12
238	192	0.3	M	13.58	9.22
260	216	0.3	M	13.38	9.33
272	229	0.3	M	13.63	9.19
300	262	0.4	M	13.48	9.35
311	270	0.3	M	13.60	9.30
329	291	0.3	M	13.60	9.41
341	311	0.3	M	13.62	9.40
370	336	0.3	M	13.44	9.50
377	339	0.4	M	13.44	9.53

$v_{bl} = 97\,m/s$.

Table 3. Test results for 1×[2.0] shields.

v_s (m/s)	v_r (m/s)	m^{pl} (g)	PFT	L_r (mm)	D_r (mm)
69	–	–	F	13.81	6.81
93	–	–	M	13.13	9.05

v_s (m/s)	v_r (m/s)	m^{pl} (g)	PFT	L_r (mm)	D_r (mm)
103	–	–	M	12.96	9.24
116	–	–	M	12.70	9.65
218	–	–	S	–	–
232	–	–	S	–	–
259	–	–	S	–	–
266	–	–	S	–	–
287	–	–	S	–	–
307	–	–	S	–	–
325	–	–	S	–	–
338	125	0.5	S	–	–
363	209	0.6	S	–	–
371	224	0.6	S	–	–

$v_{bl} = 332\, m/s$.

Table 4. Test results for 2×[1.0] shields.

v_s (m/s)	v_r (m/s)	Front plate m^{pl} (g)	Back plate m^{pl} (g)	PFT	L_r (mm)	D_r (mm)
90	–	–	–	F	13.60	7.54
94	–	–	–	F	13.49	7.65
117	–	–	–	M	12.84	9.50
160	–	–	–	M	11.64	11.53
211	–	–	–	M	9.62	14.26
235	–	–	–	M	9.93	14.30
251	–	–	–	M	11.82	15.75
263	–	–	–	M	10.19	14.66
300	–	–	–	M	11.53	15.31
301	–	–	–	M	12.72	14.98
306	–	–	–	M	12.26	15.06
324	104	0.4	0.5	S	–	–
335	120	0.4	0.4	S	–	–
363	183	0.4	0.4	S	–	–
373	222	0.4	0.4	S	–	–

$v_{bl} = 306\, m/s$.

4.61   [Edwards and Man, 2000]

4.61.1   *General data on shields and projectiles*

Shields. Monolithic shields, $c_{l,w} = 300\, mm$. *Shield-1* was hardened and tempered to HV 510 low alloy steel, $T = 5\, mm$; *Shield-2* was hardened and tempered to HV 345 low alloy steel, $T = 6\, mm$.

Projectiles. 7.62 mm FFV AP projectiles consist of a tungsten carbide core (HV 1200) enclosed in a low carbon steel jacket, coated in gilding metal, and an aluminum cup (*Projectile-1, m* = 8.3 g). 7.62 mm P80 armor piercing projectiles consist of a steel core (HV 700) enclosed in a copper jacket (*Projectile-2, m* = 9.7 g) .

4.61.2 *Test results*

Table 1. Test results for Projectile-1 against Shield-1.

v_s (m/s)	v_r (m/s)	v_s (m/s)	v_r (m/s)	v_s (m/s)	v_r (m/s)
736	576	783	640	796	654
820	681	846	707	869	734
945	805				

Table 2. Test results for Projectile-2 against Shield-2.

v_s (m/s)	v_r (m/s)	v_s (m/s)	v_r (m/s)	v_s (m/s)	v_r (m/s)
670	338	679	342	706	451
729	475	760	523	779	550
818	609				

4.62 [Edwards and Mathewson, 1997]

4.62.1 *General data on shields and projectiles*

Shields. Monolithic steel shields having various hardness.

Table 1. Projectiles.

Notation	Projectile	Shape	Core Material	Core HV	m (g)
Kalashnikov round	7.62 mm Kalashnikov steel cored ball round	Fig. 3.11.5	Fine grained ferrite-pearlite steel	236	3.56
FFV round	7.62 mm FFV tungsten carbide AP round	Fig. 3.11.6	Tungsten carbide	1200	5.91

4.62.2 *Test results*

Table 2. Test results for Kalashnikov round and shields with $T = 5\ mm$.

Shield material	HV	v_{bl} (m/s)	Shield material	HV	v_{bl} (m/s)
Tool steel	386	707	Tool steel	200	578
Mild steel	133	466			

Table 3. Test results for FFV round and for shields with $T = 10\ mm$.

Shield material	HV	v_{bl} (m/s)	Shield material	HV	v_{bl} (m/s)
Tool steel	364	588	Tool steel	200	502
RHA *	340	520			

* [Lowery, 1990].

4.63 [Elek *et al.*, 2005]

4.63.1 *General data on shields and projectiles*

Shields. Monolithic and layered (plates-in-contact) steel (HB 80, $\sigma_y = 320\ MPa$) shields; $c_{l,w} = 400\ mm$.

Projectiles. Steel (HB 243) cylindrical projectiles (Fig. 3.1.1); $D = 5.71mm$, $m = 1.41\,g$.

4.63.2 *Test results*

Table 1. Test results for shields of different structure.

Shield structure	v_s (m/s)	v_r (m/s)	v_s (m/s)	v_r (m/s)
1×[1.25]#	450	332	607	437
	731	564	949	745
1×[2.2]#	534	174	608	265
	736	413	943	631
[1.25]+[2.2]	915	405		
[2.2]+[1.25]	923	508		

4.64 [Erice *et al.*, 2010]

4.64.1 *General data on shields and projectiles*

Shields. Monolithic FV535 steel shields; $c_{l,w} = 101\ mm$, $T = 2\ mm$.

Projectiles. AISI 52100 bearing steel spherical projectiles (Fig. 3.5.1); $D = 5.55mm$.

Shield damage types (SDTs). P – Perforation, C – Containment.

4.64.2 *Test results*

Table 1. Test results for $T_{room} = 24\,°C$.

v_s (m/s)	v_r (m/s)	SDT	v_s (m/s)	v_r (m/s)	SDT	v_s (m/s)	v_r (m/s)	SDT
430	0	C	451	10	P	461	40	P
474	128	P	508	210	P	510	242	P
626	330	P	658	–	P	710	390	P
735	451	P	780	455	P			

$v_{bl} = 440\,m/s$.

Table 2. Test results for $T_{room} = 400\,°C$.

v_s (m/s)	v_r (m/s)	SDT	v_s (m/s)	v_r (m/s)	SDT	v_s (m/s)	v_r (m/s)	SDT
422	0	C	457	0	C	466	0	C
473	0	C	512	159	P	530	211	P
555	–	P	600	298	P	661	320	P
672	331	P	709	–	P	754	–	P
795	456	P						

$v_{bl} = 430\,m/s$.

Table 3. Test results for $T_{room} = 700\,°C$.

v_s (m/s)	v_r (m/s)	SDT	v_s (m/s)	v_r (m/s)	SDT	v_s (m/s)	v_r (m/s)	SDT
350	0	P	380	0	P	386	0	C
387	0	C	406	0	C	425	107	P
461	168	P	490	215	P	596	319	P
720	449	P						

$v_{bl} = 385\,m/s$.

4.65 [Favorsky *et al.*, 2011]

4.65.1 *General data on shields and projectiles*

Shields. Monolithic aluminum 7075-T7351 plates, $c_l = 200\,mm$, $c_w = 300\,mm$.

Projectiles. Hardened steel 0.3" (7.62 mm) ogive-tip AP projectiles.
Notations in tables. * − penetration, ** − penetration and ricochet.

4.65.2 *Test results*

Table 1. Test results for $T = 25\ mm$, $\theta_s = 0°$.

v_s (m/s)	v_r (m/s)	v_s (m/s)	v_r (m/s)
863	579	862	576

Table 2. Test results for shields with $T = 32\ mm$.

θ_s (deg)	v_s (m/s)	P (mm)	θ_s (deg)	v_s (m/s)	P (mm)
60	850	2.9	55	841	3.9
50	863	5.1	45	835	5.6
40	833	6.9			

Table 3. Test results for shields with $T = 19\ mm$.

θ_s (deg)	Shot 1		Shot 2		Shot 3	
	v_s (m/s)	P (mm)	v_s (m/s)	P (mm)	v_s (m/s)	P (mm)
60	837	3.4	830	3.1	831	3.6
55	845	4.6	836	3.9	834	3.9
50	850	5.3	838	5.5	830	4.7
45	818	5.7	825	5.7	847	6.1
40	829	*	834	*	814	6.4

Table 4. Test results for shields with $T = 15.9\ mm$.

θ_s (deg)	Shot 1		Shot 2		Shot 3	
	v_s (m/s)	P (mm)	v_s (m/s)	P (mm)	v_s (m/s)	P (mm)
60	845	3.2	832	3.1	835	3.4
55	840	4.3	834	4.2	822	3.7
50	821	4.8	846	5.3	849	5
45	845	6.8	845	6.3	819	6.1
40	826	*	835	*	853	*

Table 5. Test results for shields with $T = 9.5\ mm$.

θ_s	Shot 1		Shot 2		Shot 3	
(deg)	v_s (m/s)	P (mm)	v_s (m/s)	P (mm)	v_s (m/s)	P (mm)
60	835	**	837	**	821	**
55	829	**	836	**	830	**
50	815	**	823	**	828	*
45	835	*	832	*	844	*

4.66 [Forrestal and Hanchak, 1999]

4.66.1 *General data on shields and projectiles*

Shields. Monolithic HY-100 steel ($\sigma_y = 739\ MPa$, $\sigma_u = 831\ MPa$) shields; $\widetilde{D}^{sh} = 305\ mm$.

Projectiles. Cylindrical projectiles (Fig. 3.1.1) made of Maraging T-250 steel ($\sigma_y = 1720\ MPa$, $\sigma_u = 1790\ MPa$) and 4340 steel ($\sigma_y = 1170\ MPa$, $\sigma_u = 1240\ MPa$, HRC 38).

Table 1. Additional initial data for experimental sets.

Set	Projectile material	D (mm)	L (mm)	m (kg)	T (mm)
Set 1	Maraging T-250 steel	30	282	1.58	10.5
Set 2	4340 steel	30	282	1.56	10.5
Set 3	4340 steel	30.8	268	1.56	5.3
Set 4	4340 steel	30.8	89.1	0.52	10.5

4.66.2 *Test results*

Table 2. Test results for Set 1.

v_s (m/s)	v_r (m/s)	v^{pl} (m/s)	m^{pl} (g)	D_r (mm)
78	0	0	0	30.0
127	98	112	56	30.1
162	142	164	56	30.4
258	232	268	66	32.8
370	333	389	–	Rod spalled

Table 3. Test results for Set 2.

v_s (m/s)	v_r (m/s)	v^{pl} (m/s)	m^{pl} (g)	D_r (mm)	v_s (m/s)	v_r (m/s)	v^{pl} (m/s)	m^{pl} (g)	D_r (mm)
82	0	0	0	30.3	105	0	0	0	30.3
114	0	0	0	30.3	125	80	80	54	30.5
126	84	104	55	30.5	163	138	163	57	30.7
165	141	163	57	30.7	207	168	210	59	31.3
261	233	269	64	32.4	263	234	273	65	33.3
318	284	335	63	33.3	322	289	335	67	33.4

Table 4. Test results for Set 3.

v_s (m/s)	v_r (m/s)	v^{pl} (m/s)	m^{pl} (g)	D_r (mm)	v_s (m/s)	v_r (m/s)	v^{pl} (m/s)	m^{pl} (g)	D_r (mm)
84	0	0	0	31.0	99	63	71	27.3	31.2
106	75	94	27.0	31.2	130	104	117	27.6	31.1
202	185	234	30.6	31.8	246	229	260	33.6	32.3
294	270	323	35.0	33.0	331	313	341	33.6	32.8

Table 5. Test results for Set 4.

v_s (m/s)	v_r (m/s)	v^{pl} (m/s)	m^{pl} (g)	D_r (mm)	v_s (m/s)	v_r (m/s)	v^{pl} (m/s)	m^{pl} (g)	D_r (mm)
149	0	0	0	31.8	169	0	0	0	32.0
178	69	80	62.1	32.1	183	102	106	63.9	32.4
196	122	130	65.3	32.6	241	166	166	69.4	33.5
288	203	229	69.6	34.0	349	256	292	70.0	34.7

4.67   [Forrestal and Piekutowski, 2000]

4.67.1 *General data on shields and projectiles*

Shields. Monolithic 6061-T6511 Al shields, $D^{sh} = 250\,mm$.

Projectiles. VAR 4340 steel hemispherical-nose projectiles (Fig. 3.4.1); $D = 7.11\,mm$, $L_{cyl} = 71.14\,mm$, $m = 22.8\,g$.

4.67.2 *Test results*

Table 1. Test results for HRC 36.6 projectiles.

T (mm)	v_s (m/s)	P (mm)	ΔL (mm)	T (mm)	v_s (m/s)	P (mm)	ΔL (mm)
216	720	67.8	3.3	152	806	74.7	6.6

T (mm)	v_s (m/s)	P (mm)	ΔL (mm)	T (mm)	v_s (m/s)	P (mm)	ΔL (mm)
178	892	84.1	9.4	216	1042	41.6	–
241	1216	50.7	–	241	2479	137.9	–

Table 2. Test results for HRC 39.5 projectiles.

T (mm)	v_s (m/s)	P (mm)	ΔL (mm)	T (mm)	v_s (m/s)	P (mm)	ΔL (mm)
114	496	37.6	0.8	121	572	48.1	1.2
124	781	72.7	4.6	152	821	84.3	4.0
178	841	91.4	2.5	178	932	96.5	6.6
178	967	94.4	10.2	178	1037	64.6	–
175	1193	50.7	–	297	1337	61.8	–
178	1515	76.0	–	190	1802	94.3	–
216	2052	113.9	–	241	2204	124.6	–
241	2777	147.0	–	241	3075	151.0	–

Table 3. Test results for HRC 46.2 projectiles.

T (mm)	v_s (m/s)	P (mm)	ΔL (mm)	T (mm)	v_s (m/s)	P (mm)	ΔL (mm)
184	909	109.6	0.5	197	1086	126.3	6.4
203	1174	67.5	–	208	1174	66.5	–
156	1284	78.8	–	151	1411	106.1	–
254	1813	120.0	–	214	2255	137.4	–
240	2787	159.0	–				

4.68 [Forrestal *et al.*, 1987]

4.68.1 *General data on shields and projectiles*

Shields. Monolithic 6061-T6 Al ($\sigma_y \approx 300\,MPa$) shields; $D^{sh} = 304\,mm$, $T = 25.4\,mm$.

Projectiles. Maraging T-200 steel spherical blunted cone-nose projectiles (Fig. 3.2.2a and Fig. 3.2.2b), $m = 78\,g$.

Table 1. Geometries of projectiles.

Label	Shape	D (mm)	D_{cyl} (mm)	L (mm)	L_{nose} (mm)	ρ_1 (mm)	ρ_0 (mm)
Pr.1	Fig. 3.2.2a	9.53	–	146	14.3	0.89	–
Pr.2	Fig. 3.2.2b	10.2	9.53	146	14.3	0.89	3.18

4.68.2 *Test results*

Table 2#. Test results (projectiles were not deformed).

Projectile	v_s (m/s)	v_r (m/s)	Projectile	v_s (m/s)	v_r (m/s)
Pr.2	217	0	Pr.2	252	79
Pr.2	273	141	Pr.2	314	209
Pr.2	346	259	Pr.1	535	497
Pr.1	685	640	Pr.1	685	640

4.69 [Forrestal *et al.*, 1988]

See also [Anderson *et al.*, 1992].

4.69.1 *General data on shields and projectiles*

Shields. Monolithic 6061-T651 Al ($\sigma_y \approx 400\,MPa$) shields, $D^{sh} = 152\,mm$.

Table 1. Projectiles properties, $D = 7.1\ mm$, $m \approx 24\ g$.

Projectile	Shape	D_{cyl} (mm)	L_{nose} (mm)	L (mm)	ρ (mm)	ψ	Material	σ_y (MPa)
Ogive-nose projectiles	Fig. 3.3.1	–	–	82.9	21.3	3.0	T-200 maraging steel	1380
Hemispherical-nose projectiles	Fig. 3.4.1	–	–	74.7	–	–	T-300 maraging steel	2070
Mushroom-shaped projectiles	Fig. 3.4.2	6.4	–	95.0	–	–	T-300 maraging steel	2070
Cone-nose projectiles	Fig. 3.2.1	–	10.7	81.8	–	–	T-300 maraging steel	2070

4.69.2 *Test results*

Table 2. Test results for ogive-nose projectiles.

v_s (m/s)	P (mm)	v_s (m/s)	P (mm)	v_s (m/s)	P (mm)
440	44	730	94	790	105
1000	163	1260	253	1460	333

Table 3. Test results for hemispherical-nose projectiles.

v_s (m/s)	P (mm)	v_s (m/s)	P (mm)	v_s (m/s)	P (mm)
450	30	760	72	960	115
1020	112	1120	152	1160	159

Table 4. Test results for mushroom-shaped projectiles.

v_s (m/s)	P (mm)	v_s (m/s)	P (mm)	v_s (m/s)	P (mm)
480	36	770	84	1000	119

Table 5. Test results for cone-nose projectiles.

v_s (m/s)	P (mm)	v_s (m/s)	P (mm)	v_s (m/s)	P (mm)	v_s (m/s)	P (mm)
510	40	660	66	940	121	1000	145
1060	153	1150	196	1260	201	1370	273

4.70 [Forrestal *et al.*, 1990]

4.70.1 *General data on shields and projectiles*

Shields. Monolithic 5083-H131 Al shields; $c_{l,w} = 304\,mm$.

Projectiles. 94% tungsten alloy cone-nose projectiles (Fig. 3.2.1); $D = 8.31\,mm$, $L_{cyl} = 20.7\,mm$, $L_{nose} = 14.83\,mm$, $m = 26\,g$.

4.70.2 *Test results*

Table 1. Test results for shields with $T = 12.7\ mm$.

v_s (m/s)	v_r (m/s)	v_s (m/s)	v_r (m/s)	v_s (m/s)	v_r (m/s)	v_s (m/s)	v_r (m/s)
220	0	221	0	278	142	290	177
479	412	532	464	726	684	892	859
1023	989	1195	1156				

Table 2. Test results for shields with $T = 50.8\ mm$.

v_s (m/s)	v_r (m/s)	v_s (m/s)	v_r (m/s)	v_s (m/s)	v_r (m/s)
506	0	513	0	573	242

578	250	590	255	655	378
810	595	1037	863	1176	1022

Table 3. Test results for shields with $T = 76.2$ *mm*.

v_s (m/s)	v_r (m/s)	v_s (m/s)	v_r (m/s)	v_s (m/s)	v_r (m/s)	v_s (m/s)	v_r (m/s)
623	0	635	0	690	169	709	248
792	420	959	667	1021	767	1168	889

Table 4#. Test results for BLV.

T (mm)	v_{bl} (m/s)	T (mm)	v_{bl} (m/s)	T (mm)	v_{bl} (m/s)
12.7	224	50.8	515	76.2	639

4.71   [Forrestal *et al.*, 1991]

See also [Anderson *et al.*, 1992].

4.71.1   *General data on shields and projectiles*

Shields. Monolithic 6061-T651 Al shields; $D^{sh} = 152$ *mm* .
Projectiles. T-200 maraging steel ($\gamma = 8000 \, kg/m^3$) hemispherical-nose projectiles (Fig. 3.4.1).

Table 1. Projectiles details.

Label	D (mm)	L_{cyl} (mm)	m (g)
Projectile 1	7.11	71.12	23.38±0.06
Projectile 2	5.08	71.12	11.79±0.07
Projectile 3	7.11	35.56	12.05±0.05

4.71.2   *Test results*

Table 2. Test results for Projectile 1.

v_s (m/s)	P (mm)	T (mm)	v_s (m/s)	P (mm)	T (mm)	v_s (m/s)	P (mm)	T (mm)
359	21	90	430	32	89	490	40	82
519	41	114	673	73	178	792	85	178
959	109	198	1009	129	216			

Table 3. Test results for Projectile 2.

v_s (m/s)	P (mm)	T (mm)	v_s (m/s)	P (mm)	T (mm)	v_s (m/s)	P (mm)	T (mm)
253	12	90	357	20	107	425	27	102
457	33	92	599	56	140	740	80	178
866	102	203	980	117	203			

Table 4. Test results for Projectile 3.

v_s (m/s)	P (mm)	T (mm)	v_s (m/s)	P (mm)	T (mm)	v_s (m/s)	P (mm)	T (mm)
318	12	90	473	23	102	634	38	140
743	47	89	806	56	127	945	68	152

4.72 [Forrestal *et al.*, 1992]

4.72.1 *General data on shields and projectiles*

Shields. Monolithic 7075-T651 Al ($E = 73.1\,GPa$, $v = 0.33$, $\gamma = 2710\,kg/m^3$, $\sigma_y = 448\,MPa$) shields, $D^{sh} = 152\,mm$.

Projectiles. T-200 maraging steel ogive-nose projectiles (Fig. 3.3.1); $D = 7.11\,mm$, $L_{nose} = 11.79\,mm$, $L = 71.12\,mm$, $\psi = 3$, $\gamma = 8020\,kg/m^3$, $m = 24.75\,g$.

4.72.2 *Test results*

Table 1. Test results on DOP.

v_s (m/s)	P (mm)	v_s (m/s)	P (mm)	v_s (m/s)	P (mm)
372	26	695	70	978	127
1067	147	1258	209		

4.73 [Forrestal *et al.*, 2010]

Shields. 7075-T651 Al ($\sigma_u = 582\,MPa$, $\tilde{\sigma}_y = 520\,MPa$, $\varepsilon_f = 11\%$) monolithic and layered shields (plates-in-contact).

Projectiles. 0.30 Caliber AP M2 (the whole) and 0.30 Caliber AP M2 (core only) projectiles; Fig. 3.11.1, Table 3.11.1.

4.73.1   *Test results for 0.30 caliber AP M2 projectiles*

Table 1. Test results for 1× [20.0] shields.

v_s (m/s)	v_r (m/s)	v_s (m/s)	v_r (m/s)	v_s (m/s)	v_r (m/s)	v_s (m/s)	v_r (m/s)
611	0	616	0	629	51	641	151
650	204	662	233	694	315	782	509
825	562	875	651				

$v_{bl} = 628\,m/s; a = 1.00, p = 2.17.$

Table 2. Test results for 2× [20.0] shields.

v_s (m/s)	v_r (m/s)	v_s (m/s)	v_r (m/s)	v_s (m/s)	v_r (m/s)	v_s (m/s)	v_r (m/s)
886	0	899	0	901	0	909	0
921	181	930	174	932	0	934	193

$v_{bl} = 909\,m/s; a = 1.00, p = 1.98.$

4.73.2   *Test results for 0.30 caliber AP M2 projectiles (core only)*

Table 3. Test results for 1×[20.0] shields.

v_s (m/s)	v_r (m/s)	v_s (m/s)	v_r (m/s)	v_s (m/s)	v_r (m/s)	v_s (m/s)	v_r (m/s)
598	0	614	0	626	0	641	129
642	77	654	141	667	199	670	191
716	357	868	580				

$v_{bl} = 633\,m/s; a = 1.00, p = 1.95.$

Table 4. Test results for 2×[20.0] shields.

v_s (m/s)	v_r (m/s)	v_s (m/s)	v_r (m/s)	v_s (m/s)	v_r (m/s)
984	0	989	184	1009	293
1026	220	1094	466		

$v_{bl} = 984\,m/s; a = 1.00, p = 2.16.$

4.74   [Fras *et al.*, 2015a]

See also [Fras *et al.*, 2015b].

4.74.1 *General data on shields and projectiles*

Shields. Monolithic AA7020-T651 ($E = 71\,GPa$, $G = 25\,GPa$ $\gamma = 2770\,kg/m^3$, $v = 0.3$, HV 133) shields; $T = 40\,mm$.
Projectiles. Steel 20-mm FSPs (Fig. 3.10.1, Table 3.10.1).

4.74.2 *Test results*

Table 1#. Test results.

v_s (m/s)	v_r (m/s)	v_s (m/s)	v_r (m/s)	v_s (m/s)	v_r (m/s)	v_s (m/s)	v_r (m/s)
884	0	889	0	906	63	921	79
931	89	933	119	934	70	940	86
1051	215	1083	224	1087	271	1246	419
1282	478	1422	652	1425	656		

$v_{bl} = 890\,m/s$; $a = 0.85$, $p = 1.23$.

4.75 [Fugelso and Bloedow, 1966]

4.75.1 *General data on shields and projectiles*

Shields. Monolithic 7075-T6 AA and Ti 5 Al 2.5 Sn TA shields.

Table 1. 7075-T6 AA shields. Material properties.

T (mm)	E (GPa)	σ_u (MPa)	σ_y (MPa)	HV	HRA
0.25	67.6	538	496	177	53.5*
1.27	69.0	572	510	188	55.5
2.29	69.6	583	514	195	57.0
6.35	70.3	576	521	190	56.0
Mean values	69.0	565	510	188	55.5

* Not experimental data.

Table 2. Ti 5 Al 2.5 Sn TA shields. Material properties.

T (mm)	E (GPa)	σ_u (MPa)	σ_y (MPa)	HV*	HRA
0.84	111.0	914	814	334	67.5
1.24	115.8	896	841	309	66.0
1.93	110.3	841	796	325	67.0
2.54	113.8	827	803	317	66.5
Mean values	113.1	872	810	320	66.75

* Not experimental data.

Projectiles. 4340 Steel normalized (HRC 42, HB 388, $\sigma_y = 1241 MPa$) cylindrical projectiles (Fig. 3.1.1).

Table 3. Characteristics of projectiles.

Notation	D (mm)	L (mm)	m (g)
Projectile-1	5.54	6.35	1.18
Projectile-2	5.54	9.53	1.79
Projectile-3	7.59	8.66	3.07

4.75.2 *Test results for Projectile-1*

Table 4#. Test results for 7075-T6 AA shields with $T = 0.25$ *mm*.

v_s (m/s)	v_r (m/s)	v_s (m/s)	v_r (m/s)	v_s (m/s)	v_r (m/s)	v_s (m/s)	v_r (m/s)
189	103	198	129	246	174	270	201
356	296	397	305	514	426	676	596
877	786	1179	1049	1207	1054		

$v_{bl} = 152 \pm 30 m/s$.

Table 5#. Test results for 7075-T6 AA shields with $T = 1.27$ *mm*.

v_s (m/s)	v_r (m/s)	v_s (m/s)	v_r (m/s)	v_s (m/s)	v_r (m/s)	v_s (m/s)	v_r (m/s)
172	10	220	156	265	113	271	149
312	160	351	185	376	216	432	278
547	363	825	600	903	637	903	669

$v_{bl} = 198 \pm 15 m/s$.

Table 6#. Test results for 7075-T6 AA shields with $T = 2.29$ *mm*.

v_s (m/s)	v_r (m/s)	v_s (m/s)	v_r (m/s)	v_s (m/s)	v_r (m/s)	v_s (m/s)	v_r (m/s)
201	8	322	154	393	251	552	378
619	442	731	530	818	605	861	626
976	723	1123	804				

$v_{bl} = 229 \pm 15 m/s$.

Table 7#. Test results for 7075-T6 AA shields with $T = 6.35$ *mm*.

v_s (m/s)	v_r (m/s)	v_s (m/s)	v_r (m/s)	v_s (m/s)	v_r (m/s)	v_s (m/s)	v_r (m/s)
344	9	371	44	415	49	506	123
519	148	553	196	583	284	595	211

v_s (m/s)	v_r (m/s)	v_s (m/s)	v_r (m/s)	v_s (m/s)	v_r (m/s)	v_s (m/s)	v_r (m/s)
636	260	868	431	868	376	881	415
1030	542	1076	565	1218	639	1237	675
1351	698						

$v_{bl} = 442 \pm 15\,m/s$.

Table 8#. Test results for Ti 5 Al 2.5 Sn TA shields with $T = 0.81$ *mm*.

v_s (m/s)	v_r (m/s)	v_s (m/s)	v_r (m/s)	v_s (m/s)	v_r (m/s)	v_s (m/s)	v_r (m/s)
284	105	329	206	429	303	433	291
667	537	681	516	958	798	989	766

$v_{bl} = 244 \pm 30\,m/s$.

Table 9#. Test results for Ti 5 Al 2.5 Sn TA shields with $T = 1.22$ *mm*.

v_s (m/s)	v_r (m/s)	v_s (m/s)	v_r (m/s)	v_s (m/s)	v_r (m/s)	v_s (m/s)	v_r (m/s)
241	10	267	92	350	201	372	209
595	414	597	375	708	504	862	635
999	758	1121	855	1161	928		

$v_{bl} = 274 \pm 15\,m/s$.

Table 10#. Test results for Ti 5 Al 2.5 Sn TA shields with $T = 1.93$ *mm*.

v_s (m/s)	v_r (m/s)	v_s (m/s)	v_r (m/s)	v_s (m/s)	v_r (m/s)	v_s (m/s)	v_r (m/s)
246	6	280	16	467	246	478	251
580	357	698	442	769	534	879	595
924	674	1084	781	1165	864		

$v_{bl} = 320 \pm 15\,m/s$.

Table 11#. Test results for Ti 5 Al 2.5 Sn TA shields with $T = 2.54$ *mm*.

v_s (m/s)	v_r (m/s)	v_s (m/s)	v_r (m/s)	v_s (m/s)	v_r (m/s)	v_s (m/s)	v_r (m/s)
313	9	347	47	382	71	644	373
729	430	843	495	847	559	915	555
993	610	1042	678				

$v_{bl} = 366 \pm 15\,m/s$.

4.75.3 *Test results for Projectile-2*

Table 12#. Test results for 7075-T6 AA shields with $T = 1.27$ *mm*.

v_s (m/s)	v_r (m/s)	v_s (m/s)	v_r (m/s)	v_s (m/s)	v_r (m/s)	v_s (m/s)	v_r (m/s)
155	10	203	99	213	110	216	125
237	138	255	140	295	191	481	364
626	498	639	542	653	565		

$v_{bl} = 168 \pm 15 m/s$.

Table 13#. Test results for 7075-T6 AA shields with $T = 2.29$ *mm*.

v_s (m/s)	v_r (m/s)	v_s (m/s)	v_r (m/s)	v_s (m/s)	v_r (m/s)
182	20	222	117	230	90
242	120	257	161	308	155
457	299	597	420	710	530

$v_{bl} = 198 \pm 15 m/s$.

4.75.4 *Test results for Projectile-3*

Table 14#. Test results for 7075-T6 AA shields with $T = 1.27$ *mm*.

v_s (m/s)	v_r (m/s)	v_s (m/s)	v_r (m/s)	v_s (m/s)	v_r (m/s)	v_s (m/s)	v_r (m/s)
140	8	169	58	183	94	185	81
186	105	214	79	238	134	393	307
393	266	403	287	410	362	696	526
862	758						

$v_{bl} = 152 \pm 15 m/s$.

Table 15#. Test results for 7075-T6 AA shields with $T = 2.29$ *mm*.

v_s (m/s)	v_r (m/s)	v_s (m/s)	v_r (m/s)	v_s (m/s)	v_r (m/s)	v_s (m/s)	v_r (m/s)
172	9	175	53	224	141	241	104
264	138	377	253	403	330	413	252
438	320	473	292	636	453	684	515
762	595						

$v_{bl} = 198 \pm 15 m/s$.

4.76   [Gallardy, 2012]

See also [Gallardy, 2011].

4.76.1 *General data on shields and projectiles*

Shields. Monolithic 7085-T7E01 and 7085-T7E02 shields.
Projectiles. References to projectile description are given in titles of subsections.
Note. Brinell hardness for 29420 N = 3000 kg load.

Table 1. Notes in tables below.

Note label	Description
(1)	Unpublished data from W. Gooch (U.S. Army Research Laboratory: Aberdeen Proving Ground, MD)
(2)	Partial penetration at maximum projectile velocity
(3)	Unpublished data collected by the U.S. Army Aberdeen Test Center (Gessleman, D. U.S. Army Aberdeen Test Center: Aberdeen Proving Ground, MD)

4.76.2 *Test results for 0.30 caliber APM2 projectile (Fig. 3.11.1 and Table 3.11.1)*

Table 2. Test results for 7085-T7E01 shields and $\theta_s = 0°$.

T (mm)	HB	v_{bl} (m/s)	Note	T (mm)	HB	v_{bl} (m/s)	Note
18.52	–	598	(3)	18.57	179	598	–
25.4	170	772	–	37.95	179	908	–
40.51	174	938	(1)	50.83	–	1020	(1), (2), (3)

Table 3. Test results for 7085-T7E01 shields and $\theta_s = 30°$.

T (mm)	HB	v_{bl} (m/s)	Note	T (mm)	HB	v_{bl} (m/s)	Note
12.52	–	511	(3)	12.60	183	511	–
18.52	–	652	(3)	18.57	179	668	–

Table 4. Test results for 7085-T7E02 shields and $\theta_s = 0°$.

T (mm)	HB	v_{bl} (m/s)	Note	T (mm)	HB	v_{bl} (m/s)	Note
18.34	–	563	(3)	18.52	153	567	–
25.45	149	664	–	25.45	–	671	(3)
37.97	143	842	–	40.49	149	872	(1)
52.76	–	993	(1), (2), (3)				

Table 5. Test results for 7085-T7E02 shields and $\theta_s = 30°$.

T (mm)	HB	v_{bl} (m/s)	Note	T (mm)	HB	v_{bl} (m/s)	Note
12.62	149	481	–	18.42	–	608	(3)
18.52	153	614	–				

4.76.3 *Test results for normal impact of 0.50 caliber APM2 projectile (Fig. 3.13.1 and Table 3.13.1)*

Table 6. Test results for 7085-T7E01 shields.

T (mm)	HB	v_{bl} (m/s)	Notes	T (mm)	HB	v_{bl} (m/s)	Notes
37.95	179	665	–	50.11	–	782	(3)
52.83	–	809	(1), (3)	56.31	–	843	(3)
56.52	166	853	–	63.35	174	907	–
74.00	170	990	(2)				

Table 7. Test results for 7085-T7E02 shields.

T (mm)	HB	v_{bl} (m/s)	Notes	T (mm)	HB	v_{bl} (m/s)	Notes
37.97	143	635	–	50.39	–	738	(3)
50.83	–	737	(3)	50.93	–	732	(3)
51.05	–	734	(3)	51.18	–	734	(3)
51.41	–	735	(3)	52.76	146	753	(1)
56.26	–	783	(3)	56.31	146	792	–
63.60	146	847	–	75.31	–	938	(3)
77.47	–	962	(3)				

4.76.4 *Test results for normal impact of 14.5 mm BS41 projectile (Fig. 3.14.1 and Table 3.14.1)*

Table 8. Test results for 7085-T7E01 shields.

T (mm)	HB	v_{bl} (m/s)	Notes	T (mm)	HB	v_{bl} (m/s)	Notes
63.35	174	838	–	74.00	170	937	(2)

4.76.5 *Test results for normal impact of 0.50 caliber FSP (Fig.3.10.1 and Table 3.10.1)*

Table 9. Test results for 7085-T7E01 shields.

T (mm)	HB	v_{bl} (m/s)	Notes	T (mm)	HB	v_{bl} (m/s)	Notes
18.52	–	568	(3)	18.57	179	631	–
25.35	170	1026	–	25.40	170	1061	–

Table 10. Test results for 7085-T7E02 shields.

T (mm)	HB	v_{bl} (m/s)	Notes	T (mm)	HB	v_{bl} (m/s)	Notes
18.36	–	589	(3)	18.52	153	573	–
25.32	–	960	(3)	25.40	149	982	–
25.45	149	957	–				

4.76.6 *Test results for normal impact of 20 mm FSP (Fig.3.10.1 and Table 3.10.1)*

Table 11. Test results for 7085-T7E01 shields.

T (mm)	HB	v_{bl} (m/s)	T (mm)	HB	v_{bl} (m/s)	T (mm)	HB	v_{bl} (m/s)
25.35	170	461	25.40	170	448	37.95	179	835

Table 12. Test results for 7085-T7E02 shields.

T (mm)	HB	v_{bl} (m/s)	Notes	T (mm)	HB	v_{bl} (m/s)	Notes
25.40	149	457	–	25.43	–	418	(3)
25.45	149	430	–	37.97	143	806	–

4.77 [Goel *et al.*, 1988]

4.77.1 *General data on shields and projectiles*

Shields. Monolithic hardened steel (HB 360) shields, $T = 55\,mm$.
Projectiles. Two types of ogive-nose (Fig. 3.3.1) AP projectiles with $\psi = 1$: tungsten carbide projectiles ($D = 17.0\,mm$, $L = 52.8\,mm$, $m = 145.6\,g$) and hardened steel projectiles ($D = 19.9\,mm$, $L = 54.5\,mm$, $m = 108.0\,g$).

4.77.2 *Test results*

Table 1. Test results for tungsten carbide projectiles, $v_s = 685\,m/s$.

θ_s (deg)	P (mm)	θ_s (deg)	P (mm)	θ_s (deg)	P (mm)
0	50.7	10	51.0	20	50.0
30	45.8	40	23.0	50	11.0
60	8.2	70	3.8	80	0.8

Table 2 Test results for hardened steel projectiles, $v_s = 677\,m/s$.

θ_s (deg)	P (mm)	θ_s (deg)	P (mm)	θ_s (deg)	P (mm)
0	35.5	5	33.5	25	33.0
35	18.2	55	8.6	65	6.2
75	3.8	80	2.5	85	0.8

4.78 [Gogolewski and Morgan, 1999]

4.78.1 *General data on shields and projectiles*

Shields. Monolithic 2024-T3 AA ($\sigma_y = 317\,MPa$, $\sigma_u = 464\,MPa$, $\delta = 18\%$) shields, $D^{sh} = 152\,mm$.
Projectiles. 6AL-4V titanium alloy ($\tilde{\sigma}_y = 979\,MPa$, $\sigma_u = 1110\,MPa$, $\delta = 15\%$, HRC 37) FSPs (Fig. 3.10.3; $D = 12.7\,mm$, $\alpha = 39°$, $m = 7.93\,g$) and two types of cylindrical projectiles (Fig. 3.1.1) with $D = 12.7\,mm$: *CP-1* ($L = 12.7\,mm$, $m = 8.17\,g$) and *CP-2* ($L = 2.54\,mm$, $m = 1.43\,g$).
Shield damage types (SDTs). Pl – plugging, Pet – petaling.

4.78.2 *Test results*

Table 1. Test results for FSPs.

T (mm)	v_{bl} (m/s)	SDT	T (mm)	v_{bl} (m/s)	SDT	T (mm)	v_{bl} (m/s)	SDT
3.81	192±14	Pl	2.54	162±3	Pet	1.27	126±3	Pet

Table 2. Test results for cylindrical projectiles.

Projectile	T (mm)	v_{bl} (m/s)	SDT	Projectile	T (mm)	v_{bl} (m/s)	SDT
CP-1	3.81	189±9	Pl	CP-2	1.27	269±1	Pl

4.79 [Gogolewski *et al.*, 1996]

4.79.1 *General data on shields and projectiles*

Shields. Monolithic and layered (plates in-contact and with air gaps) 6061-T6 AA ($\sigma_v = 276\,MPa$, $\sigma_u = 310\,MPa$, $\varepsilon_f = 12\%$) shields; $c_{l,w} = 83\,mm$.
Projectiles. Hard steel (HRC 38) cylindrical projectiles (Fig. 3.1.1; $D = 7.62mm$, $m = 6.8\,g$).

4.79.2 *Test results*

Table 1. Test results on BLV.

Shield structure	v_{bl} (m/s)	Shield structure	v_{bl} (m/s)
1×[6.70]	233 ± 4	3×[2.23]	287 ± 12
3×[2.23]*	268 ± 9	3×[2.23]\2.5	228 ± 8
3×[2.23]\7.5	217 ± 6		

* Frictionless interface between layers

4.80 [Goldsmith and Finnegan, 1971]

See also [Calder and Goldsmith, 1971].

4.80.1 *General data on shields and projectiles*

Shields. Monolithic shields, $c_{l,w} = 305\,mm$.

Table 1. Properties of shields materials.

Label	Material	E (GPa)	ν	σ_p (MPa)	σ_u (MPa)	σ_{ss} (MPa)	HRB/ HRC/ HRH Value	Scale
Al-1	2024-0 Aluminum	73	0.33	88	234	131	88	H
Al-2	2024-T3 Aluminum	73	0.33	365	482	282	76	B
Al-3	2024-T4 Aluminum	73	0.33	365	477	282	78	B
St-1	SAE 1020 Steel, large-grained	204	0.29	441	489	311	90	B
St-2	SAE 1020 Steel, small-grained	204	0.29	331	410	261	77	B
St-3	SAE 4130 Steel, quenched and tempered	204	0.29	1171	1302	782	40	C

Projectiles. AISI 52-100 hard steel spherical projectiles (Fig. 3.5.1).

4.80.2 *Test results Al-1 shields*

4.80.2.1 *Shields with* $T = 1.27\,mm$ *and projectiles with* $D = 6.35\,mm$

Table 2. Results for v_r.

v_s (m/s)	v_r (m/s)	v_s (m/s)	v_r (m/s)	v_s (m/s)	v_r (m/s)
201	73	869	797	2606	2438

Table 3#. Results for T^{pl}.

v_s (m/s)	T^{pl} (m/s)	v_s (m/s)	T^{pl} (m/s)	v_s (m/s)	T^{pl} (m/s)
231	1.0	553	0.6	918	0.4
1227	0.4	1256	0.5		

4.80.3 *Test results Al-2 shield*

4.80.3.1 *Shields with* $T = 3.17\,mm$ *and projectiles with* $D = 6.35\,mm$

Table 4. Results for v_r.

v_s (m/s)	v_r (m/s)	v_s (m/s)	v_r (m/s)
869	686	2606	2225

Table 5#. Results for T^{pl}.

v_s (m/s)	T^{pl} (m/s)	v_s (m/s)	T^{pl} (m/s)	v_s (m/s)	T^{pl} (m/s)
346	2.4	844	1.6	1291	0.8

4.80.3.2 *Shields with $T = 1.27\,mm$ and projectiles with $D = 6.35\,mm$*

Table 6. Results for v_r.

v_s (m/s)	v_r (m/s)	v_s (m/s)	v_r (m/s)	v_s (m/s)	v_r (m/s)
853	777	2569	2409	2393	2301

Table 7#. Results for T^{pl}.

v_s (m/s)	T^{pl} (m/s)	v_s (m/s)	T^{pl} (m/s)	v_s (m/s)	T^{pl} (m/s)
613	1.0	806	0.9	1004	0.8

4.80.3.3 *Shields with $T = 1.27\,mm$ and projectiles with $D = 9.53\,mm$*

Results for v_r: $v_s = 2393\,m/s$, $v_r = 2301\,m/s$.

Table 8#. Results for T^{pl}.

v_s (m/s)	T^{pl} (m/s)	v_s (m/s)	T^{pl} (m/s)	v_s (m/s)	T^{pl} (m/s)
844	1.2	1123	0.9	1498	0.6

4.80.4 *Test results Al-3 shield*

4.80.4.1 *Shields with $T = 6.35\,mm$ and projectiles with $D = 6.35\,mm$*

Table 9. Results for v_r.

v_s (m/s)	v_r (m/s)	v_s (m/s)	v_r (m/s)
869	457	2545	1899

Table 10#. Results for T^{pl}.

v_s (m/s)	T^{pl} (m/s)	v_s (m/s)	T^{pl} (m/s)	v_s (m/s)	T^{pl} (m/s)
477	4.1	557	3.8	822	2.3
956	2.2	1060	1.7	1230	1.7

4.80.4.2 *Shields with $T = 6.35\ mm$ and projectiles with $D = 9.53\ mm$*

Table 11. Results for v_r.

v_s (m/s)	v_r (m/s)	v_s (m/s)	v_r (m/s)
904	689	2414	1774

Table 12#. Results for T^{pl}.

v_s (m/s)	T^{pl} (m/s)	v_s (m/s)	T^{pl} (m/s)	v_s (m/s)	T^{pl} (m/s)
598	3.7	861	2.9	1115	2.0

4.80.5 *Test results for St-1 shield with $T = 6.35\ mm$ and spherical projectiles with $D = 6.35\ mm$*

Table 13. Results for v_r.

v_s (m/s)	v_r (m/s)	v_s (m/s)	v_r (m/s)
1474	219	2115	721

Table 14#. Results for T^{pl}.

v_s (m/s)	T^{pl} (m/s)	v_s (m/s)	T^{pl} (m/s)	v_s (m/s)	T^{pl} (m/s)
894	3.8	2.93	3.2	4.09	2.9
1551	1.7	2048	1.6	2196	1.7

4.80.6 *Test results for St-2 shield*

4.80.6.1 *Shields with $T = 6.35\ mm$ and projectiles with $D = 6.35\ mm$*

Table 15. Results for v_r .

v_s (m/s)	v_r (m/s)	v_s (m/s)	v_r (m/s)
899	*	2621	914

* Penetration without perforation.

Table 16#. Results for T^{pl} .

v_s (m/s)	T^{pl} (m/s)	v_s (m/s)	T^{pl} (m/s)	v_s (m/s)	T^{pl} (m/s)
848	3.6	1460	1.9	2080	1.8

4.80.6.2 *Shields with* $T = 1.57\ mm$ *and projectiles with* $D = 6.35\ mm$

Table 17. Results for v_r .

v_s (m/s)	v_r (m/s)	v_s (m/s)	v_r (m/s)	v_s (m/s)	v_r (m/s)	v_s (m/s)	v_r (m/s)
853	640	1314	1003	1696	1320	2669	2068

Table 18#. Results for T^{pl} .

v_s (m/s)	T^{pl} (m/s)	v_s (m/s)	T^{pl} (m/s)	v_s (m/s)	T^{pl} (m/s)
305	1.4	389	1.3	464	1.2
870	0.9	1295	0.6		

4.80.6.3 *Shields with* $T = 1.57\ mm$ *and projectiles with* $D = 9.53\ mm$

Table 19. Results for v_r .

v_s (m/s)	v_r (m/s)	v_s (m/s)	v_r (m/s)	v_s (m/s)	v_r (m/s)
853	704	1207	1023	2384	2048

Table 20#. Results for T^{pl} .

v_s (m/s)	T^{pl} (m/s)	v_s (m/s)	T^{pl} (m/s)	v_s (m/s)	T^{pl} (m/s)	v_s (m/s)	T^{pl} (m/s)
599	1.1	845	1.0	1136	0.8	1500	0.5

4.80.7   *Test results for St-3 shield with* $T = 6.35\ mm$ *and projectiles with* $D = 6.35\ mm$

Table 21. Results for v_r.

v_s (m/s)	v_r (m/s)	v_s (m/s)	v_r (m/s)	v_s (m/s)	v_r (m/s)	v_s (m/s)	v_r (m/s)
884	*	1531	210	1622	355	1890	536
2146	655	2490	863	2637	892		

* Penetration without perforation.

Table 22#. Results for T^{pl}.

v_s (m/s)	T^{pl} (m/s)	v_s (m/s)	T^{pl} (m/s)	v_s (m/s)	T^{pl} (m/s)	v_s (m/s)	T^{pl} (m/s)
918	5.3	1387	4.0	1485	3.9	1510	3.6
1610	3.1	1636	3.3	1688	2.7	1738	3.1
1898	2.8	1936	2.7	2026	2.7	2218	2.5
2469	2.4						

4.81   [Goldsmith and Finnegan, 1985]

4.81.1   *General data on shields and projectiles*

Shields. Monolithic shields ($\widetilde{D}^{sh} = 114\ mm$) made from different materials: 2024-0 Al (HB 48), mild steel (HB 130), medium carbon steel (HB 220), and hard steel (HRC 53–60).

Projectiles. Hard steel cone-nose projectiles (Fig. 3.2.1, $\beta = 30°$) and cylindrical projectiles (Fig. 3.1.1, $D = 12.3 mm$). For more details see Tables 1 and 2.

Table 1. Test sets description for cone-nose projectiles.

Set	m (g)	D (mm)	Shield material	T (mm)
Set-1	29.8	12.5	2024-0 Al	3.175
Set-2	Varied	12.3	2024-0 Al	Varied
Set-3	Varied	12.3	Mild steel	Varied
Set-4	Varied	12.3	Medium carbon steel	6.35

Table 2. Test sets description for cylindrical projectiles.

Set	Projectile material	θ_s (deg)	Shield material	T (mm)
Set-5	Hard steel	Varied	2024-0 Al	Varied
Set-6	Hard steel	Varied	Mild steel	Varied
Set-7	Hard steel	Varied	Medium carbon steel	6.35
Set-8	2024-0 Al	0	2024-0 Al	Varied
Set-9	2024-0 Al	0	Mild steel	3.175
Set-10	2024-0 Al	0	Medium carbon steel	6.35

Table 3. Notes in tables.

Note number	Description
(1)	No velocity decreases within experimental error
(2)	At or below BLV
(3)	Slightly above BLV
(4)	BLV
(5)	Projectile missed the multi-beam array
(6)	Blind hole in the shield
(7)	Projectile was embedded in shield
(8)	Ricochet
(9)	Very close to BLV
(10)	Projectile broke during penetration
(11)	Projectile collapsed and mushroomed. Shield and projectile fragments widely scattered
(12)	Plug separated and petals were formed
(13)	Final velocity of cloud
(14)	Curved plug
(15)	Projectile shattered
(16)	3.175 mm thick mild steel shields
(17)	1.58 mm thick 2024-0 aluminum plate
(18)	Final velocity was determined from photographs before complete exit of projectile from shield

4.81.2 *Test conditions and results for cone-nose projectiles*

Table 4. Test results for Set-1.

θ_s (deg)	v_s (m/s)	θ_r (deg)	v_r (m/s)	Note
0	96.5	–	0	(4)
0	102.9	0	36.2	–
0	126.1	0	83.6	–

θ_s (deg)	v_s (m/s)	θ_r (deg)	v_r (m/s)	Note
0	154.1	0	123.8	–
0	155.7	0	152.4	(17)
0	156.6	0	128.3	–
0	157.2	0	127.2	–
0	157.7	0	127.4	–
0	159.0	0	128.0	–
0	160.7	–	–	(6),(16)
0	170.7	–	–	(6), (16)
0	182.9	–	–	(7), (16)
10	91.6	–	0	(2)
10	98.9	3.7	48.3	–
10	107.5	–7.4	49.5	–
10	122.0	–4.7	77.5	–
10	150.0	–	141.3	(17)
10	150.6	3	120.7	–
10	151.2	2.2	387.7	–
10	156.4	2.9	79.0	–
10	160.1	11.3	121.3	–
20	99.5	–	0	(2)
20	121.8	–8.0	53.9	–
20	122.3	-6.9	53.9	–
20	123.2	13.0	67.4	–
20	151.0	6.7	120.2	–
20	151.0	7.3	115.6	–
20	159.8	13.6	–	–
20	160.1	9.1	118.0	–
20	162.8	9.9	120.7	–
30	102.8	–	–	(8)
30	103.7	–	~0	(3)
30	121.4	–21.5	–	–
30	~122.0	2.1	52.7	–
30	122.6	-5.0	–	(5)
30	123.2	2.3	50.3	–
30	133.1	9.3	99.2	–
30	151.2	8.0	105.2	–
30	160.1	7.0	100.6	–
30	161.3	15.0	117.7	–
40	~122.0	2.9	–	(5)
40	155.3	~0	69.1	–
40	159.8	17.8	103.0	–
40	160.0	16.1	105.8	–
50	160.4	37.0	82.6	–

Table 5. Test results for Set-2, $T = 3.175$ *mm*.

m (g)	θ_s (deg)	v_s (m/s)	θ_r (deg)	v_r (m/s)	Note
29.1	0	229.5	0	161.9	(18)
29.0	0	293.9	0	245.1	–
29.0	0	382.6	0	247.9	–
30.7	0	587.8	0	567.0	–
28.9	0	604.9	0	576.8	–
29.1	0	960.4	0	718.1	(18)
29.0	10	333.8	8.5	282.3	–
30.7	10	600.6	10.0	600.6	–
29.3	20	337.8	19.0	286.7	(18)
31.0	20	608.2	20.0	608.2	–
30.3	30	334.4	26.0	304.9	–
29.1	30	356.4	26.0	255.7	(18)
29.2	30	596.0	32.0	568.9	(18)
30.8	40	300.6	39.0	281.4	–
30.4	40	323.8	32.9	–	–
29.2	40	332.3	36.0	313.1	–
30.4	40	605.2	40.0	530.4	–
29.1	50	301.8	48.0	176.8	(18)
30.8	50	315.5	43.0	266.4	–
30.6	50	326.8	45.4	–	–
29.0	50	605.8	46.5	591.0	(18)

Table 6. Test results for Set-2, $T = 6.35$ *mm*.

m (g)	θ_s (deg)	v_s (m/s)	θ_r (deg)	v_r (m/s)	Note
29.1	0	605.2	0	523.5	–
30.9	0	960.4	0	960.4	(1)
29.2	10	326.2	8.0	235.5	(18)
30.6	10	614.3	8.0	588.4	(18)
30.6	10	931.4	8.0	931.4	–
29.3	20	332.0	18.0	256.1	(18)
30.5	20	612.8	17.0	<612.8	(1), (18)
30.6	20	923.8	21.0	897.0	–
30.5	30	312.3	25.0	306.1	–
29.3	30	326.2	26.0	267.0	–
29.0	30	605.2	31.0	558.6	(18)
30.5	30	930.0	35.0	930.0	(1)
30.4	40	304.9	36.0	240.8	–
29.2	40	328.4	47.0	217.1	–
30.9	40	613.4	39.0	<609.8	(18)
31.1	40	920.7	40.0	749.7	–

| m | θ_s | v_s | θ_r | v_r | Note |
(g)	(deg)	(m/s)	(deg)	(m/s)	
29.2	50	332.3	42.5	244.2	–
29.0	50	601.5	47.0	537.6	(18)
30.4	50	936.0	50.0	~915.0	–

Table 7. Test results for Set-3, $T = 3.175$ *mm*.

| m | θ_s | v_s | θ_r | v_r | Note |
(g)	(deg)	(m/s)	(deg)	(m/s)	
29.0	0	337.2	0	191.8	–
29.3	0	607.0	0	503.4	–
30.9	0	608.5	0	582.3	–
30.5	0	930.0	0	875.6	–
29.2	10	362.8	8.0	248.1	–
30.6	10	609.8	8.0	557.9	(18)
30.6	10	926.8	10.0	926.8	(1)
29.2	20	332.3	16.0	252.5	(18)
30.5	20	612.8	23.0	579.3	–
31.0	20	965.9	21.0	879.0	–
29.3	30	336.9	24.0	254.9	(18)
29.0	30	593.0	31.0	553.9	(18)
30.9	30	923.8	30.0	922.6	–
30.9	40	317.7	27.0	223.5	–
29.2	40	334.5	29.0	232.0	–
30.9	40	586.9	42.0	497.0	(18)
30.3	40	904.0	41.5	875.3	–
29.1	50	332.3	48.5	209.5	(18)
29.0	50	594.3	44.5	521.3	(18)
30.9	50	~918.3	53.0	918.3	(1)

Table 8. Test results for Set-3, $T = 6.35$ *mm*.

| m | θ_s | v_s | θ_r | v_r | Note |
(g)	(deg)	(m/s)	(deg)	(m/s)	
31.2	0	320.1	–	0	(7)
31.3	0	607.0	0	490.9	–
30.6	0	933.9	0	812.5	–
30.7	10	316.5	–	0	(7)
30.7	10	325.6	15.0	230.5	(18)
30.5	10	622.0	8.0	503.0	–
30.6	20	640.3	20.0	512.2	–
30.9	40	644.0	36.0	515.2	–

Table 9. Test results for Set-4.

m (g)	θ_s (deg)	v_s (m/s)	θ_r (deg)	v_r (m/s)	Note
29.1	0	597.6	0	564.9	(18)
29.0	10	334.5	–	60.4	(18)
30.6	10	1025.6	9.5	873.8	–
29.1	20	332.3	–1.5	81.3	(18)
30.4	20	940.0	18.0	902.4	–
29.2	26	339.0	24.0	65.8	(9)
29.2	30	~325.0	–	–	(8)
29.0	30	603.7	–	399.2	(18)
30.8	30	932.5	28.0	857.9	–
30.9	40	932.9	–	784.4	(10)
29.4	50	482.1	–	–	(8)
29.3	50	604.5	–	511.6	–
30.6	50	999.2	–	749.4	(10)

4.81.3 *Test conditions and results for cylindrical projectiles*

Table 10. Test results for Set-5.

T (mm)	m (g)	θ_s (deg)	v_s (m/s)	θ_r (deg)	v_r (m/s)	Note
3.175	~34.8	30	331.7	26.5	327.7	–
3.175	34.9	50	360.4	46.0	300.3	–
6.35	~34.8	0	328.4	0	284.6	–
6.35	37.1	0	610.5	0	~580.0	–
6.35	36.6	0	690.5	0	566.2	–
6.35	34.6	0	882.6	0	700.9	–
6.35	36.3	0	985.2	0	~807.9	–
6.35	36.3	40	707.1	41.0	544.5	–
6.35	~34.8	50	330.8	44.0	240.9	–
12.7	36.3	0	324.2	0	291.8	–
12.7	36.6	0	593.3	0	559.0	–
12.7	36.5	0	594.0	0	547.0	–
12.7	36.4	0	880.1	0	814.6	–
12.7	36.2	40	600.0	38.0	445.4	–
15.88	36.6	0	325.9	0	221.6	–
19.05	36.3	0	338.1	0	127.1	–
19.05	36.5	0	670.6	0	449.3	(14)
19.05	36.2	0	887.2	0	738.4	(10)
25.4	36.5	0	332.9	–	0	(7)

Table 11. Test results for Set-6.

T (mm)	m (g)	θ_s (deg)	v_s (m/s)	θ_r (deg)	v_r (m/s)	Note
3.175	~34.8	0	341.2	0	278.8	–
3.175	34.7	50	342.4	42.0	218.3	–
12.7	36.5	0	884.1	0	554.9	(15)
12.7	36.5	40	750.0	24.0	75.0	–
19.05	36.3	0	~884.1	0	314.3	(13)

Table 12. Test results for Set-7.

m (g)	θ_s (deg)	v_s (m/s)	θ_r (deg)	v_r (m/s)	Note
34.8	0	326.2	0	196.1	–
36.2	0	606.7	0	481.1	–
36.3	0	861.3	0	689.0	(10)
34.7	0	882.6	0	690.5	(18)
34.7	26	328.7	25.0	183.2	–
36.2	40	514.5	31.5	334.4	–
34.8	50	556.4	–	275.3	(18)

Table 13. Test results for Set-8.

T (mm)	m (g)	v_s (m/s)	v_r (m/s)	Note
6.35	14.2	275.0	0	(7)
6.35	14.2	368.0	226.8	–
6.35	14.7	661.6	442.7	–
6.35	14.2	917.7	658.2	–
12.7	14.3	678.0	372.9	–
12.7	15.2	813.2	300.9	–
12.7	13.8	1013.7	691.1	–
19.05	14.1	929.0	0	(7), (13)
25.4	14.2	920.7	0	(7)

Table 14. Test results for Set-9.

m (g)	v_s (m/s)	v_r (m/s)
14.0	652.4	411.0

Table 15. Test results for Set-10.

m (g)	v_s (m/s)	v_r (m/s)	Note
14.0	801.8	146.6	(11)
14.6	923.8	288.4	–

4.82   [Goldsmith and Finnegan, 1986]

See also [Goldsmith and Finnegan, 1985].

4.82.1   *General data on shields and projectiles*

Shields. Monolithic 2024–0 Al and mild steel shields, $\widetilde{D}^{sh} = 114\,mm$.
Projectiles. Hard steel cone-nose projectiles (Fig. 3.2.1; $\beta = 30°$, HRC 53-60, $L = 38.1\,mm$) and hard steel and 2024–0 Al cylindrical projectiles (Fig. 3.1.1) with $D = 12.3\,mm$.

4.82.2   *Test results for cone-nose projectiles*

Table 1. Test results for 2024-0 Al shields.

T (mm)	m (g)	θ_s (deg)	v_{bl} (m/s)
3.175	28.9	0	95
3.175	28.9	0	88-90
3.175	30.9	30	104
6.35	30.6	0	150

Table 2. Test results for mild steel shields and $\theta_s = 0°$.

T (mm)	m (g)	v_{bl} (m/s)	T (mm)	m (g)	v_{bl} (m/s)
3.175	31.0	195	6.35	31.2	320

4.82.3   *Test results for cylindrical projectiles*

Table 3. Test results for hard steel projectiles against 2024–0 Al shields.

T (mm)	m (g)	θ_s (deg)	v_{bl} (m/s)	T (mm)	m (g)	θ_s (deg)	v_{bl} (m/s)
3.175	37.1	0	105	3.175	37.7	0	88
12.7	37.5	40	610	19.05	37.5	0	305
25.4	36.5	0	333				

Table 4. Test results for 2024–0 Al projectiles against 2024-0 Al shields and $\theta_s = 0°$.

T (mm)	m (g)	v_{bl} (m/s)	T (mm)	m (g)	v_{bl} (m/s)
6.35	14.1	275	25.4	14.2	921

4.83   [Golubev and Medvedkin, 2001]

4.83.1 *General data on shields and projectiles*

Shields. Monolithic low-carbon steel ($\tilde{\sigma}_y = 203\,MPa$, $\sigma_u = 458\,MPa$, $\delta = 24.5\%$, HB 129) shields; $D^{sh} = 160\,mm$, $T = 70\,mm$.

Projectiles. Steel (HRC 60-64)   projectiles with $D = 20\,mm$ and $L_{cyl} = 70\,mm$; cone-nose projectiles (Fig. 3.2.1; $L_{nose} = 10\,mm$, $\beta = 45°$, $m = 179\,g$) and hemispherical-nose projectiles (Fig. 3.4.1; $m = 187\,g$).

4.83.2 *Test results*

Table 1. Test results for cone-nose projectiles.

v_s (m/s)	P (mm)	v_s (m/s)	P (mm)	v_s (m/s)	P (mm)	v_s (m/s)	P (mm)
390	22.0	410	24.0	530	34.0	540	36.0

Table 2. Test results for hemispherical-nose projectiles.

v_s (m/s)	P (mm)	v_s (m/s)	P (mm)	v_s (m/s)	P (mm)	v_s (m/s)	P (mm)
364	19.5	425	23.0	435	24.0	524	33.5

4.84   [Gooch *et al.*, 2007]

4.84.1 *General data on shields and projectiles*

Shields. Monolithic shields made from different aluminum alloys.
Projectiles. 0.30 Caliber AP M2 projectiles (Fig. 3.11.1, Table 3.11.1) and 20 mm FSPs (Fig. 3.10.1, Table 3.10.1).

4.84.2 *Test results for 0.30 caliber AP M2 projectiles*

Table 1#. Test results for AA6061-T651 shields.

T (mm)	v_{bl} (m/s)	T (mm)	v_{bl} (m/s)	T (mm)	v_{bl} (m/s)	T (mm)	v_{bl} (m/s)
25.6	596	25.9	583	38.8	753	51.2	883

Test results for AA6061-Extrusion shields#. $T = 57.4\,mm$, $v_{bl} = 929\,m/s$.

Table 2#. Test results for AA5059 shields.

T (mm)	v_{bl} (m/s)	T (mm)	v_{bl} (m/s)	T (mm)	v_{bl} (m/s)
25.1	587	30.4	661	34.6	704
37.9	762	47.8	859		

4.84.3 *Test results for 20 mm FSP*

Table 3#. Test results for AA6061-T651 shields.

T (mm)	v_{bl} (m/s)	T (mm)	v_{bl} (m/s)	T (mm)	v_{bl} (m/s)
25.7	444	38.8	762	51.2	1217

Test results for AA6061-Extrusion shields#. $T = 57.2\,mm$, $v_{bl} = 1365\,m/s$.

Table 4#. Test results for AA5059 shields.

T (mm)	v_{bl} (m/s)	T (mm)	v_{bl} (m/s)	T (mm)	v_{bl} (m/s)
25.2	419	30.4	580	34.7	681
37.9	800	47.8	1212		

4.85 [Grabarek, 1971]

Data from [Anderson *et al.*, 1992].

4.85.1 *General data on shields and projectiles*

Shields. Monolithic RHA ($\gamma = 7860\,kg/m^3$) shields (except for several experiments).

Projectiles. Cylindrical projectiles (Fig. 3.1.1).

Note. More detailed data on shields and projectiles is given in tables below.

4.85.2 *Test results*

Table 1. Test results for $\theta_s = 0°$.

| Material | Projectile | | | | | | Shield | | v_{bl} |
	γ (kg/m^3)	HB	δ (%)	L (mm)	D (mm)	m (g)	T (mm)	HB	(m/s)
U	19150	226	6.0	25.4	6.4	10.60	12.7	353	701
U	19150	226	6.0	25.4	6.4	10.60	19.1	381	919
U	19150	226	6.0	49.4	4.1	10.05	22.2	409	975
U	19150	226	6.0	49.4	4.1	10.95	22.2	409	978
U	19150	226	6.0	44.7	3.7	8.85	22.2	409	1039
U	19150	226	6.0	63.0	5.1	23.00	29.2	353	943
U	19150	226	6.0	81.3	2.8	9.10	31.8	344	1067
U	19150	226	6.0	76.2	3.8	15.70	31.8	344	989
U	19150	226	6.0	76.2	3.8	15.70	47.0	409	1298
U	19150	226	6.0	76.2	3.8	15.70	31.8	400	1036
U	19150	226	6.0	25.4	6.4	13.70	22.2	409	1082
U	19150	226	6.0	25.4	6.4	13.65	31.8	344	1213
U	19150	226	6.0	25.4	6.4	13.65	15.9	371	799
U	19150	226	6.0	25.4	6.4	13.65	12.7	353	721
U	19150	226	6.0	25.4	6.4	13.65	19.1	381	931
U	19150	226	6.0	25.4	6.4	13.65	25.4	344	1152
U	19150	226	6.0	13.6	1.8	0.54	12.7	390	1303
U	19150	226	6.0	13.6	1.8	0.54	19.2	381	1827
U	19150	226	6.0	21.1	1.8	0.83	19.2	381	1294
U	19150	226	6.0	21.1	1.8	0.83	12.7	344	1023
U	19150	226	6.0	77.0	3.6	12.40	25.4	381	893
U	19150	226	6.0	77.0	3.6	12.40	66.0	301	1573
U (Depl)	18700	243	9.0	80.3	4.0	19.44	38.1	327	1064
U (Depl)	18700	243	9.0	80.3	4.0	19.44	50.8	327	1179
M-3000	17000	253	15.0	15.6	3.1	1.94	6.3	381	689
M-3000	17000	253	15.0	15.6	3.1	1.94	12.7	381	1181
M-3000	17000	253	15.0	19.6	3.9	3.89	12.7	381	1024
M-3000	17000	253	15.0	24.8	5.0	7.78	12.7	381	831
U-8.5 Mo	17300	253	10.0	82.8	4.1	19.44	38.1	327	1055
U-8.5 Mo	17300	253	10.0	82.8	4.1	19.44	50.8	327	1190
M3000	17000	271	15.0	25.4	6.4	12.15	15.9	362	855
SD170	17000	279	–	28.2	7.0	16.00	25.5	371	1045
SD170	17000	279	–	28.2	7.0	16.00	19.1	381	913
SD170	17000	279	–	28.2	7.0	16.00	22.2	409	1052
SD170	17000	279	–	28.2	7.0	16.00	15.9	390	859
W-2	18600	286	6.0	80.8	4.0	19.44	38.1	327	1158
W-2	18600	286	6.0	80.8	4.0	19.44	50.8	327	1260
W-10	17000	294	–	83.3	4.2	19.44	38.1	327	1176
W-10	17000	294	–	83.3	4.2	19.44	50.8	327	1281
DU-2 Mo	18140	390	–	108.0	5.2	36.87	38.1	443	925

	Projectile						Shield		v_{bl}
Material	γ (kg/m^3)	HB	δ (%)	L (mm)	D (mm)	m (g)	T (mm)	HB	(m/s)
DU-2 Mo	18140	390	–	108.0	5.2	36.87	50.8	390	1052
DU-2 Mo	18140	390	–	108.0	5.2	36.87	38.1	319	932
DU-2 Mo	18140	390	–	108.0	5.2	36.87	50.8	319	1097
U-2 Mo	18400	455	–	75.7	6.4	42.40	31.8	344	853
U-2 Mo	18400	455	–	75.7	6.4	42.40	47.0	432	1143
U-2 Mo	18400	455	–	25.4	6.4	13.35	15.9	390	748
U-2 Mo	18400	455	–	25.4	6.4	13.35	22.2	409	1032
U-2 Mo	18400	455	–	25.4	6.4	13.35	31.8	390	1346
U-2 Mo	18400	455	–	25.4	6.4	13.35	22.2	409	1070
U-2 Mo	18400	455	–	34.9	6.4	18.55	22.2	409	939
U-2 Mo	18400	455	–	50.8	6.4	27.55	22.2	409	841
U-2 Mo	18400	455	–	50.8	6.4	27.55	47.0	432	1358
U-2 Mo	18400	455	–	73.0	6.4	40.90	47.0	432	1157
U-Quad	17700	512	4.0	82.3	4.1	19.44	38.1	327	1036
U-Quad	17700	512	4.0	82.3	4.1	19.44	50.8	327	1149
Bearcat	7800	560	10.0	20.2	4.0	1.94	12.7	381	1432
Bearcat	7800	560	10.0	203.5	20.4	453.6	25.4	327	381
Bearcat	7800	560	10.0	203.5	20.4	453.6	50.8	258	671
Bearcat	7800	560	10.0	311.4	15.8	453.6	152.4	271	1547
Ketos	7800	722	–	57.2	6.4	13.40	25.4	381	1265
Ketos	7800	722	–	31.8	6.4	6.45	17.5	344	1170
Ketos	7800	722	–	31.8	6.4	6.45	12.5	390	917
52100	7800	739	–	108.2	5.4	19.44	38.1	327	1352
52100	7800	739	–	108.2	5.4	19.44	50.8	327	1495
WC	14200	800	–	423.7	14.1	907.2	50.8	258	431
WC	14200	800	–	88.9	4.4	19.44	38.1	327	1065
WC	14200	800	–	88.9	4.4	19.44	50.8	327	1275
WC(K95)	14600	800	–	25.4	6.4	10.40	22.2	409	1073*

* $\theta_s = 25°$.

Table 2. Test results for $\theta_s = 30°$.

	Projectile						Shield		v_{bl}
Material	γ (kg/m^3)	HB	δ (%)	L (mm)	D (mm)	m (g)	T (mm)	HB	(m/s)
U	19150	226	6.0	63.0	5.1	23.00	29.2	353	1055
U	19150	226	6.0	25.4	6.4	13.70	22.2	409	1161
U	19150	226	6.0	21.1	1.8	0.83	12.7	344	1125
U (Depl)	18700	243	9.0	80.3	4.0	19.44	38.1	327	1113
U (Depl)	18700	243	9.0	80.3	4.0	19.44	50.8	327	1281
U-8.5 Mo	17300	253	10.0	82.8	4.1	19.44	38.1	327	1106
U-8.5 Mo	17300	253	10.0	82.8	4.1	19.44	50.8	327	1278
W-2	18600	286	6.0	80.8	4.0	19.44	38.1	327	1166

| Projectile | | | | | | Shield | | v_{bl} |
Material	γ (kg/m^3)	HB	δ (%)	L (mm)	D (mm)	m (g)	T (mm)	HB	(m/s)
W-2	18600	286	6.0	80.8	4.0	19.44	50.8	327	1419
W-10	17000	294	–	83.3	4.2	19.44	38.1	327	1204
W-10	17000	294	–	83.3	4.2	19.44	50.8	327	1372
U-2 Mo	18400	455	–	78.7	6.4	44.35	31.8	344	881
U-2 Mo	18400	455	–	25.4	6.4	13.35	22.2	409	1134
U-2 Mo	18400	455	–	73.0	6.4	40.90	47.0	432	1265
U-Quad	17700	512	4.0	82.3	4.1	19.44	38.1	327	1083
52100	7800	739	–	108.2	5.4	19.44	38.1	327	1423
52100	7800	739	–	108.2	5.4	19.44	50.8	327	1617
WC	14200	800	–	88.9	4.4	19.44	38.1	327	1166
WC	14200	800	–	88.9	4.4	19.44	50.8	327	1431
WC(K95)	14600	800	–	25.4	6.4	10.40	22.2	409	1174
WC(K95)	14600	800	–	42.2	5.3	11.95	13.2	362	806
WC(K95)	14600	800	–	50.8	6.4	21.40	22.2	409	875
WC(K95)	14600	800	–	81.3	6.4	33.40	15.9	390	475
WC(K95)	14600	800	–	81.3	6.4	33.40	22.2	409	768

Table 3. Test results for $\theta_s = 45°$.

| Projectile | | | | | | Shield | | v_{bl} |
Material	γ (kg/m^3)	HB	δ (%)	L (mm)	D (mm)	m (g)	T (mm)	HB	(m/s)
U	19150	226	6.0	49.4	4.1	10.05	22.2	409	1271
U	19150	226	6.0	49.4	4.1	10.95	22.2	409	1228
U	19150	226	6.0	44.7	3.7	8.85	22.2	409	1219
U	19150	226	6.0	44.7	3.7	9.05	22.2	409	1244
U	19150	226	6.0	81.3	2.8	9.10	12.7	381	1165
U	19150	226	6.0	76.2	3.8	15.70	22.2	409	1012
U	19150	226	6.0	25.4	6.4	13.70	22.2	409	1321
U	19150	226	6.0	21.1	1.8	0.83	12.7	344	1253
U(Depl)	18700	243	9.0	80.3	4.0	19.44	25.4	327	1041
U(Depl)	18700	243	9.0	80.3	4.0	19.44	38.1	327	1207
U-8.5 Mo	17300	253	10.0	82.8	4.1	19.44	25.4	327	1031
U-8.5 Mo	17300	253	10.0	82.8	4.1	19.44	38.1	327	1199
M3000	17000	271	15.0	40.6	10.2	50.00	19.1	381	869
SD170	17000	279	–	28.2	7.0	16.00	22.2	409	1335
W-2	18600	286	6.0	80.8	4.0	19.44	25.4	327	1059
W-2	18600	286	6.0	80.8	4.0	19.44	38.1	327	1308
W-10	17000	294	–	83.3	4.2	19.44	25.4	327	1096
W-10	17000	294	–	83.3	4.2	19.44	38.1	327	1381
U-2 Mo	18400	455	–	78.7	6.4	44.35	31.8	344	1006
U-2 Mo	18400	455	–	25.4	6.4	13.35	22.2	409	1343

| | Projectile | | | | | | Shield | | v_{bl} |
Material	γ (kg/m^3)	HB	δ (%)	L (mm)	D (mm)	m (g)	T (mm)	HB	(m/s)
U-2 Mo	18400	455	–	25.4	6.4	13.35	22.2	409	1315
U-2 Mo	18400	455	–	25.4	6.4	13.35	12.7	344	730
U-Quad	17700	512	4.0	82.3	4.1	19.44	25.4	327	1002
52100	7800	739	–	108.2	5.4	19.44	25.4	327	1312
52100	7800	739	–	108.2	5.4	19.44	38.1	327	1547
WC	14200	800	–	88.9	4.4	19.44	25.4	327	1106
WC	14200	800	–	88.9	4.4	19.44	38.1	327	1379
WC(K95)	14600	800	–	25.4	6.4	10.40	22.2	409	1417
WC(K95)	14600	800	–	42.2	5.3	12.15	12.7	362	905
WC(K95)	14600	800	–	42.2	5.3	12.15	13.2	362	975
WC(K95)	14600	800	–	42.2	5.3	11.95	13.2	362	992
WC(K95)	14600	800	–	50.8	6.4	21.40	12.7	390	658
WC(K95)	14600	800	–	50.8	6.4	21.40	15.9	390	817
WC(K95)	14600	800	–	25.4	6.4	9.60	15.9	390	1091
WC(K95)	14600	800	–	81.3	6.4	33.40	15.9	390	735

Table 4. Test results for $\theta_s = 60°$.

| | Projectile | | | | | | Shield | | v_{bl} |
Material	γ (kg/m^3)	HB	δ (%)	L (mm)	D (mm)	m (g)	T (mm)	HB	(m/s)
U	19150	226	6.0	25.4	6.4	10.60	12.7	353	1097
U	19150	226	6.0	63.0	5.1	23.00	12.7	344	901
U	19150	226	6.0	63.0	5.1	23.00	13.2	344	922
U	19150	226	6.0	81.3	2.8	9.10	19.1	390	1202
U	19150	226	6.0	76.2	3.8	15.70	19.1	371	1100
U	19150	226	6.0	76.2	3.8	15.70	25.4	371	1324
U	19150	226	6.0	76.2	3.8	15.70	22.2	409	1250
U	19150	226	6.0	25.4	6.4	13,65	12.7	344	1032
U	19150	226	6.0	25.4	6.4	13.65	19.1	353	1372
U	19150	226	6.0	25.4	6.4	13.65	12.7	353	1023
U	19150	226	6.0	77.0	3.6	12.40	15.9	381	988
U	19150	226	6.0	77.0	3.6	12.40	25.4	381	1311
U	19150	226	6.0	77.0	3.6	12.40	31.8	344	1570
U (Depl)	18700	243	9.0	80.3	4.0	19.44	25.4	327	1191
U (Depl)	18700	243	9.0	80.3	4.0	19.44	38.1	327	1440
U-8.5 Mo	17300	253	10.0	82.8	4.1	19.44	25.4	327	1177
M-3950	18000	264	9.0	76.2	7.6	61.48	25.4	371	1167
M-3950	18000	264	9.0	78.7	7.9	61.50	25.4	381	1167
M3000	17000	271	15.0	25.4	6.4	12.15	15.9	362	1306
M3000	17000	271	15.0	55.9	6.4	25.65	12.7	353	945

	Projectile						Shield		v_{bl}
Material	γ (kg/m^3)	HB	δ (%)	L (mm)	D (mm)	m (g)	T (mm)	HB	(m/s)
M3000	17000	271	15.0	55.9	6.4	25.65	19.1	371	1132
SD170	17000	279	–	28.2	7.0	16.00	15.9	390	1248
SD170	17000	279	–	28.2	7.0	16.00	12.7	353	1035
M-3000	17000	286	15.0	303.8	15.4	907.2	101.6	258	1585
W-2	18600	286	6.0	77.0	7.7	64.80	25.4	381	1112
W-2	18600	286	6.0	80.8	4.0	19.44	25.4	327	1206
W-10	17000	294	–	83.3	4.2	19.44	25.4	327	1259
W-10	17000	294	–	83.3	4.2	19.44	38.1	327	1514
E592-U	18750	353	7.5	78.8	7.9	71.00	25.4	381	1020
E591-U	18890	371	16.0	78.8	7.9	68.40	25.4	381	1003
F584-U	17900	421	7.0	79.6	8.0	66.50	25.4	381	1112
F583-U	17400	432	1.0	79.3	7.9	59.50	25.4	371	1143
F583-U	17400	432	1.0	76.2	7.6	59.50	25.4	381	1143
U-2 Mo	18400	455	–	78.7	6.4	44.35	25.4	381	1091
U-2 Mo	18400	455	–	25.4	6.4	13.35	12.7	344	972
U-2 Mo	18400	455	–	25.4	6.4	13.35	19.1	381	1432
U-2 Mo	18400	455	–	34.9	6.4	18.55	19.1	381	1240
U-2 Mo	18400	455	–	34.9	6.4	18.55	19.1	390	1390
U-2 Mo	18400	455	–	50.8	6.4	27.55	22.2	409	1303
U-2 Mo	18400	455	–	73.3	6.4	40.90	22.2	409	1107
E593-U	18570	481	3.0	78.8	7.9	67.00	25.4	381	966
F583-U	17400	512	1.0	79.3	7.9	67.00	25.4	381	1044
U-Quad	17700	512	4.0	82.3	4.1	19.44	25.4	327	1147
581-U	17680	525	1.5	79.4	7.9	67.00	25.4	381	1105
Bearcat	7800	560	10.0	203.5	20.4	453.6	76.2	301	1742
Bearcat	7800	560	10.0	101.6	10.2	64.80	25.4	381	1356
1095 St	7800	654	–	406.2	13.7	453.6	101.6	258	1884
1095 St	7800	654	–	391.7	19.9	907.2	101.6	258	1697
Ketos	7800	722	–	57.2	6.4	13.40	12.7	353	1282
52100	7800	739	–	108.2	5.4	19.44	25.4	327	1463
52100	7800	739	–	108.2	5.4	19.44	38.1	327	1817
WC	14600	800	–	518.2	25.9	3629	101.6	258	1094
WC	14200	800	–	259.1	13.0	453.6	101.6	258	1609
WC	14200	800	–	259.1	13.0	453.6	127.0	258	2062
WC	14200	800	–	259.1	13.0	453.6	152.4	247	2085
WC	14200	800	–	338.3	11.3	453.6	101.6	258	1701
WC	14200	800	–	325.1	16.3	907.2	25.4	371	524
WC	14200	800	–	325.1	16.3	907.2	101.6	258	1489
WC	14200	800	–	325.1	16.3	907.2	127.0	258	1803
WC	14200	800	–	325.1	16.3	907.2	152.4	247	1919
WC	14200	800	–	423.7	14.1	907.2	101.6	258	1399
WC	14200	800	–	83.9	8.4	64.80	25.4	381	1245

| Projectile | | | | | | Shield | | v_{bl} |
Material	γ (kg/m^3)	HB	δ (%)	L (mm)	D (mm)	m (g)	T (mm)	HB	(m/s)
WC	14200	800	–	88.9	4.4	19.44	25.4	327	1175
WC	14200	800	–	88.9	4.4	19.44	38.1	327	1558
WC(K95)	14600	800	–	79.5	4.0	13.50	15.9	390	975
WC(K95)	14600	800	–	79.5	4.0	13.50	15.9	400	1097
WC(K95)	14600	800	–	79.5	4.0	13.50	25.4	253	1234
WC(K95)	14600	800	–	79.5	4.0	13.50	15.9	253	1006
WC(K95)	14600	800	–	50.8	6.4	21.40	12.7	390	870
WC(K95)	14600	800	–	50.8	6.4	21.40	15.9	400	1085
WC(K95)	14600	800	–	50.8	6.4	21.40	22.2	409	1294
WC(K95)	14600	800	–	25.4	6.4	9.60	12.7	400	1180
WC(K95)	14600	800	–	25.4	6.4	9.60	15.9	390	1474
WC(K95)	14600	800	–	81.3	6.4	33.40	15.9	400	977
WC(K95)	14600	800	–	36.8	7.1	16.40	19.1	381	1367*

* $\theta_s = 64°$.

4.86 [Graves and Beatty, 1994]

4.86.1 *General data on shields and projectiles*

Shields. Monolithic shields made from AerMet 100 Steel subjected to various ageing treatments.

Table 1. Shield materials properties.

Label	Treatment	HRC	Label	Treatment	HRC
Steel-1	468°C, 1 hour	55	Steel-2	468°C, 5 hours	54
Steel-3	482°C, 1 hour	56	Steel-4	482°C, 5 hours	53

Projectiles. U.S. 0.30 (7.62mm) caliber AP M2 and U.S. 0.50 (12.7 mm) caliber AP M2 projectiles.

Additional notation. Parameter N_{test} in tables below indicates the number of test firings used to calculate the BLV: velocities from $N_{test}/2$ complete penetrations and $N_{test}/2$ partial penetrations were used to calculate the BLV.

4.86.2 *Test results*

Table 2#. Test results for 0.30 caliber AP M2 projectiles.

Material	T (mm)	N_{test}	v_{bl} (m/s)	Material	T (mm)	N_{test}	v_{bl} (m/s)
Steel-1	4.9	10	386	Steel-2	3.4	6	100
Steel-2	4.8	10	266	Steel-3	3.7	12	314
Steel-4	3.9	14	201	Steel-4	3.9	8	179

Table 3#. Test results for 0.50 caliber AP M2 projectiles.

Material	T (mm)	N_{test}	v_{bl} (m/s)	Material	T (mm)	N_{test}	v_{bl} (m/s)
Steel-1	8.8	8	333	Steel-1	11.7	8	381
Steel-2	5.7	4	69	Steel-2	9.0	4	332
Steel-2	9.1	6	338	Steel-2	12.2	6	402
Steel-2	12.4	4	391	Steel-3	7.7	2	334
Steel-3	11.9	8	375	Steel-4	5.4	6	78
Steel-4	8.2	8	286	Steel-4	9.7	6	311
Steel-4	11.5	8	357	Steel-4	12.4	4	348

4.87 [Griffin, 1959]

4.87.1 *General data on shields and projectiles*

Shields. Monolithic 5083 AA shields.

Table 1. Projectiles.

Projectile	m (g)	HRC	Projectile	m (g)	HRC
0.3 Caliber FDP	2.85	29–31	0.3 Caliber M2 Ball	9.72	–
0.5 Caliber M2 Ball	45.55	–	0.3 Caliber AP M2	10.63	–
0.5 Caliber AP M2	46.01	–			

4.87.2 *Test results*

Table 2. Test results for 0.3 caliber M2 ball.

T (mm)	θ_s (deg)	v_{bl} (m/s)	T (mm)	θ_s (deg)	v_{bl} (m/s)	T (mm)	θ_s (deg)	v_{bl} (m/s)
13.5	0	477	13.5	30	580	13.5	45	662
13.1	60	828	19.5	0	522	19.5	30	699
19.5	45	820	20.2	60	>1096	26.0	0	677

T (mm)	θ_s (deg)	v_{bl} (m/s)	T (mm)	θ_s (deg)	v_{bl} (m/s)	T (mm)	θ_s (deg)	v_{bl} (m/s)
26.0	30	906	26.0	45	1044	38.0	0	1036
38.0	30	>1045	44.9	0	>1062			

Table 3. Test results for 0.3 caliber AP M2 projectiles.

T (mm)	θ_s (deg)	v_{bl} (m/s)	T (mm)	θ_s (deg)	v_{bl} (m/s)	T (mm)	θ_s (deg)	v_{bl} (m/s)
13.3	0	418	13.3	30	457	13.3	45	528
13.1	60	776	13.3	60	850	19.5	60	974
26.0	0	608	26.0	30	677	26.0	45	839
26.0	60	>1057	38.0	30	813	44.9	30	864

Table 4. Test results for 0.3 caliber FDPs.

T (mm)	θ_s (deg)	v_{bl} (m/s)	T (mm)	θ_s (deg)	v_{bl} (m/s)
13.3	0	726	20.2	0	>1065

Table 5. Test results for 0.5 caliber M2 ball.

T (mm)	θ_s (deg)	v_{bl} (m/s)	T (mm)	θ_s (deg)	v_{bl} (m/s)
19.5	30	397	38.0	30	592
38.0	60	>874	44.9	30	641

Table 6. Test results for 0.5 caliber AP M2 projectiles.

T (mm)	θ_s (deg)	v_{bl} (m/s)	T (mm)	θ_s (deg)	v_{bl} (m/s)	T (mm)	θ_s (deg)	v_{bl} (m/s)
19.5	30	417	26.0	60	>946	38.0	30	584
38.0	60	>950	44.9	30	600			

4.88   [Gupta and Madhu, 1997]

See also [Gupta and Madhu, 1992], [Madhu *et al.*, 2003].

4.88.1 *General data on shields and projectiles*

Shields. Monolithic and layered (plates in-contact and with air gaps) shields made from various materials; $D^{sh} \approx 200\,mm$.

Table 1. Shield materials properties.

Label	Material	γ (kg/m^3)	HV
MS-1	Mild Steel	7810	110-115
MS-2	Mild Steel	7830	150-155
MS-3	Mild Steel	7830	140-145
RHA	RHA steel	8130	280-300
AL-1	Aluminum	2590	90-92
AL-2	Aluminum	2710	95-100
AL-3	Aluminum	2320	30-35

Projectiles. Hard-steel alloy (HV 900) ogive-nose projectiles (the projectile core had a copper sheath such that the total diameter of the shot was 7.8 mm); $D = 6.2\,mm$, $L = 27.6\,mm$, $m = 5.2$ g for projectile core.

4.88.2 *Test results*

Table 2. Test results for MS-1 shield and $\theta_s = 0°$.

T_{sum} (mm)	Shield structure	v_s (m/s)	v_r (m/s)
4.7	1×[4.7]	821	759
9.4	2×[4.7]	818	711
14.1	3×[4.7]	827	637
14.1	3×[4.7]\30.0	825	652
14.1	3×[4.7]\40.0	826	675
18.8	4×[4.7]	826	607
23.5	5×[4.7]	820	510
28.2	6×[4.7]	812	303

Table 3. Test results for MS-2 shield and $\theta_s = 0°$.

T_{sum} (mm)	Shield structure	v_s (m/s)	v_r (m/s)
6.0	1×[6.0]	866	792
12.0	2×[6.0]	870	725
18.0	3×[6.0]	864	613
24.0	4×[6.0]	875	489
30.0	5×[6.0]	867	378
36.0	6×[6.0]	863	0

Table 4. Test results for monolithic RHA shield and $\theta_s = 0°$.

T (mm)	v_s (m/s)	v_r (m/s)	T (mm)	v_s (m/s)	v_r (m/s)
8.0	819	644	12.0	821	487
16.0	813	0	20.0	819	0

Table 5. Test results for monolithic MS-3 shields.

T (mm)	v_s (m/s)	θ_s (deg)	v_r (m/s)	θ_r (deg)
10.0	828	0	702	–
10.0	815	15	690	22.5
10.0	826	30	654	42.6
10.0	790	45	500	57.0
10.0	820	60	494	56.3
10.0	828	61.5	294	–
10.0	821	62*	0	–
12.0	818	0	662	–
12.0	843	15	672	12.5
12.0	802	30	598	35.3
12.0	808	45	555	49.4
12.0	809	57	369	62.8
12.0	815	59*	0	–
16.0	820	0	562	–
16.0	817	15	544	12.8
16.0	818	30	496	39.9
16.0	806	45	0	49.4
16.0	819	51*	0	–
16.0	–	55	–	65.0
20.0	821	0	405	–
20.0	826	15	296	8.9
20.0	818	30	152	43.6
20.0	797	45	0	–
20.0	813	51*	0	–
25.0	842	0	108	–
25.0	800	15	0	–
25.0	819	50*	0	–

* Critical ricochet angle.

Table 6. Test results for two-layer MS-3 shields and $\theta_s = 0°$.

T_{sum} (mm)	Shield structure	v_s (m/s)	v_r (m/s)
20.0	2×[10.0]	827	410
26.0	[10.0]+[16.0]	820	113

T_{sum} (mm)	Shield structure	v_s (m/s)	v_r (m/s)
26.0	[16.0]+[10.0]	833	117
30.0	[10.0]+[20.0]	821	E
30.0	[20.0]+[10.0]	812	R
36.0	[16.0]+[20.0]	812	E
36.0	[20.0]+[16.0]	806	R
37.0	[25.0]+[12.0]	810	R

E – embedded projectile, R – rebounded projectile.

Table 7. Test results for AL-1 shield and $\theta_s = 0°$.

T_{sum} (mm)	Shield structure	v_s (m/s)	v_r (m/s)
6.1	1×[6.1]	855	786
12.2	2×[6.1]	837	744
18.3	3×[6.1]	835	728
18.3	3×[6.1]\20.0	830	779
18.3	3×[6.1]\30.0	838	801
24.4	4×[6.1]	820	687
30.5	5×[6.1]	833	635
36.6	6×[6.1]	829	576

Table 8. Test results for monolithic AL-2 shield with $T = 20$ *mm*.

v_s (m/s)	θ_s (deg)	v_r (m/s)	θ_r (deg)	v_s (m/s)	θ_s (deg)	v_r (m/s)	θ_r (deg)
815	0	614	0	815	30	596	20.54
826	45	455	6.4	835	52	0	–

Table 9. Test results for monolithic AL-3 shields.

T (mm)	v_s (m/s)	θ_s (deg)	v_r (m/s)	θ_r (deg)
10.0	837	0	774	–
10.0	831	30	722	23.0
10.0	840	45	654	33.0
10.0	–	53	–	39.0
10.0	831	60	558	44.0
20.0	835	0	762	0
20.0	827	30	662	25.1
20.0	843	45	612	35.1
20.0	853	60	512	52.8
30.0	836	0	662	0
30.0	838	30	619	26.0

T (mm)	v_s (m/s)	θ_s (deg)	v_r (m/s)	θ_r (deg)
30.0	833	45	568	40.5
30.0	834	60	482	59.8
40.0	833	0	593	0
40.0	831	30	548	34.0
40.0	827	45	520	47.4
40.0	820	60	359	62.0

4.89 [Gupta *et al.*, 2001]

4.89.1 *General data on shields and projectiles*

Shields. Monolithic commercial purity Al (HRC 47-52) shields; $\widetilde{D}^{sh} = 205\,mm$.

Table 1. Shield material properties depending on T .

T (mm)	$\widetilde{\sigma}_y$ (MPa)	σ_u (MPa)	T (mm)	$\widetilde{\sigma}_y$ (MPa)	σ_u (MPa)	T (mm)	$\widetilde{\sigma}_y$ (MPa)	σ_u (MPa)
0.5	120	125	1.74	130	134	1.0	124	130
1.5	115	120	2.0	120	126			

Projectiles. Hardened steel ogive-nose projectiles (Fig. 3.3.1; $\psi = 2$).

Table 2. Geometrical parameters and mass of projectiles.

Label	D (mm)	L_{cyl} (mm)	L (mm)	m (g)
Projectile 1	10	10	23.24	10.35
Projectile 2	10	20	33.30	16.7
Projectile 3	10	30	43.32	23.0
Projectile 4	10	40	53.24	28.8
Projectile 5	15	30	49.60	55.0

4.89.2 *Test results for Projectiles 1,3,4*

Table 3#. Test results for shields with $T = 1.00\ mm$.

Projectile	v_{bl} (m/s)	Projectile	v_{bl} (m/s)	Projectile	v_{bl} (m/s)
Projectile 1	62	Projectile 3	43	Projectile 4	38

4.89.3 *Test results for Projectile 2*

Table 4#. Test results for shields with $T = 0.74$ *mm*.

v_s (m/s)	v_r (m/s)	v_s (m/s)	v_r (m/s)	v_s (m/s)	v_r (m/s)	v_s (m/s)	v_r (m/s)
38	0	43	4	45	21	49	24
53	27	54	28	55	36	56	30
57	29	60	39	65	51	67	54
73	56	73	61				

Table 5#. Test results for shields with $T = 1.00$ *mm*.

v_s (m/s)	v_r (m/s)	v_s (m/s)	v_r (m/s)	v_s (m/s)	v_r (m/s)	v_s (m/s)	v_r (m/s)
47	0	57	16	61	29	65	37
66	39	68	36	69	47	80	63

$v_{bl} = 47 m/s$.

4.89.4 *Test results for Projectile 5*

Table 6#. Test results for shields with $T = 0.50$ *mm*.

v_s (m/s)	v_r (m/s)	v_s (m/s)	v_r (m/s)	v_s (m/s)	v_r (m/s)	v_s (m/s)	v_r (m/s)
18	0	21	10	26	19	31	26
32	26	39	33	54	49	60	57
65	59	68	67	76	75		

$v_{bl} = 18 m/s$.

Table 7#. Test results for shields with $T = 0.74$ *mm*.

v_s (m/s)	v_r (m/s)	v_s (m/s)	v_r (m/s)	v_s (m/s)	v_r (m/s)	v_s (m/s)	v_r (m/s)
25	0	27	10	29	13	30	12
30	14	31	16	32	18	34	21
37	25	37	24	39	27	44	33
46	34	47	35	47	36	51	42
54	46	56	50	58	50	58	50
76	67	77	70				

$v_{bl} = 25 m/s$.

Table 8#. Test results for shields with $T = 1.00$ *mm*.

v_s (m/s)	v_r (m/s)	v_s (m/s)	v_r (m/s)	v_s (m/s)	v_r (m/s)	v_s (m/s)	v_r (m/s)
32	0	33	9	36	14	40	23

v_s (m/s)	v_r (m/s)	v_s (m/s)	v_r (m/s)	v_s (m/s)	v_r (m/s)	v_s (m/s)	v_r (m/s)
43	27	48	34	49	35	54	41
56	45	63	53	65	56	67	58
78	68						

$v_{bl} = 32 m/s$.

Table 9#. Test results for shields with $T = 1.50$ *mm.*

v_s (m/s)	v_r (m/s)	v_s (m/s)	v_r (m/s)	v_s (m/s)	v_r (m/s)	v_s (m/s)	v_r (m/s)
45	0	46	7	47	13	52	27
55	33	61	38	65	45	70	52
72	54	83	63				

$v_{bl} = 45 m/s$.

Table 10#. Test results for shields with $T = 2.00$ *mm.*

v_s (m/s)	v_r (m/s)	v_s (m/s)	v_r (m/s)	v_s (m/s)	v_r (m/s)	v_s (m/s)	v_r (m/s)
64	0	65	13	68	30	69	35
72	35	72	40	76	34	77	47
77	42	78	46	81	54	83	44

$v_{bl} = 64 m/s$.

4.90 [Gupta *et al.*, 2006]

4.90.1 *General data on shields and projectiles*

Shields. Monolithic 1100-H12 AA shields; $\tilde{D}^{sh} = 230\, mm$, $T = 1\, mm$.
Projectiles. EN-24 steel (HRC 47-52) cylindrical projectiles (Fig. 3.1.1; $m = 52.5\, g$) and hemispherical-nose projectiles (Fig. 3.4.1; $m = 47\, g$) with L=50.8 *mm*, D=19 *mm*.

4.90.2 *Test results*

Table 1. Test results for cylindrical projectiles.

v_s (m/s)	v_r (m/s)	m^{pl} (g)	v_s (m/s)	v_r (m/s)	m^{pl} (g)
115.6	92.9	1.0	104.0	80.2	1.5
102.5	79.2	1.0	92.5	67.5	0.9
87.5	58.3	1.0	73.9	43.8	–

$v_{bl} = 61.3 m/s$.

Table 2. Test results for hemispherical-nose projectile.

v_s (m/s)	v_r (m/s)	v_s (m/s)	v_r (m/s)	v_s (m/s)	v_r (m/s)
114.7	92.3	109.7	86.8	106.1	80.1
105.5	79.0	98.3	71.0	92.3	63.5

$v_{bl} = 85.9\,m/s$.

4.91   [Gupta *et al.*, 2007]

See also [Iqbal and Gupta, 2011].

4.91.1   *General data on shields and projectiles*

Shields. Monolithic 1100-H12 AA shields.

Projectiles. Hard steel cylindrical (Fig. 3.1.1), ogive-nose (Fig. 3.3.1; $\rho = 30\,mm$) and hemispherical-nose (Fig. 3.4.1) projectiles with $L = 50.8\,mm$, $D = 19\,mm$, $m = 52.35\,g$.

Note. In tables below, the values of v_s corresponding to $v_r = 0$ were interpreted in [Iqbal and Gupta, 2011] as v_{bl}.

4.91.2   *Test results for cylindrical projectiles.*

Table 1. Test results for shields with $T = 0.5$ *mm*.

v_s (m/s)	v_r (m/s)	v_s (m/s)	v_r (m/s)	v_s (m/s)	v_r (m/s)	v_s (m/s)	v_r (m/s)
35.6	0.0	47.9	10.9	59.9	27.1	65.8	38.6
73.7	49.7	80.4	59.3	88.8	71.6	91.0	74.9
101.2	85.8	104.3	89.7	109.8	96.7	113.4	102.8

Table 2. Test results for shields with $T = 0.71$ *mm*.

v_s (m/s)	v_r (m/s)	v_s (m/s)	v_r (m/s)	v_s (m/s)	v_r (m/s)
42.8	0.0	51.4	12.3	64.5	28.3
76.2	43.8	82.2	56.8	97.5	71.3
98.3	73.4	101.0	78.3	114.3	93.2

Table 3. Test results for shields with $T = 1.0$ *mm*.

v_s (m/s)	v_r (m/s)	v_s (m/s)	v_r (m/s)	v_s (m/s)	v_r (m/s)	v_s (m/s)	v_r (m/s)
61.3	0.0	74.0	43.8	87.5	58.3	92.5	67.5
102.5	79.2	104.0	80.2	115.6	93.0		

Table 4. Test results for shields with $T = 1.5$ *mm*.

v_s (m/s)	v_r (m/s)	v_s (m/s)	v_r (m/s)	v_s (m/s)	v_r (m/s)	v_s (m/s)	v_r (m/s)
64.1	0.0	71.4	31.8	85.4	51.0	96.3	63.4
104.8	74.6	107.5	78.9	112.4	86.9		

Table 5. Test results for shields with $T = 2.0$ *mm*.

v_s (m/s)	v_r (m/s)	v_s (m/s)	v_r (m/s)	v_s (m/s)	v_r (m/s)	v_s (m/s)	v_r (m/s)
66.7	0.0	83.0	38.3	87.9	49.1	97.4	62.4
108.4	75.2	116.3	84.9	121.8	91.6		

Table 6. Test results for shields with $T = 2.5$ *mm*.

v_s (m/s)	v_r (m/s)	v_s (m/s)	v_r (m/s)	v_s (m/s)	v_r (m/s)	v_s (m/s)	v_r (m/s)
74.4	0.0	87.3	32.6	99.4	50.2	105.7	61.6
112.0	72.0	116.4	81.1	121.1	88.0		

Table 7. Test results for shields with $T = 3.0$ *mm*.

v_s (m/s)	v_r (m/s)	v_s (m/s)	v_r (m/s)	v_s (m/s)	v_r (m/s)	v_s (m/s)	v_r (m/s)
87.3	0.0	93.6	30.9	101.3	44.2	108.2	57.8
114.6	64.9	115.8	67.4	118.2	73.6		

4.91.3 *Test results for ogive-nose projectiles.*

Table 8. Test results for shields with $T = 0.5$ *mm*.

v_s (m/s)	v_r (m/s)	v_s (m/s)	v_r (m/s)	v_s (m/s)	v_r (m/s)	v_s (m/s)	v_r (m/s)
33.7	0.0	40.7	8.0	47.5	17.2	54.1	27.6
62.5	38.8	69.6	48.1	77.4	58.2	83.9	68.4
92.5	81.8	97.9	88.7	107.8	99.4		

Table 9. Test results for shields with $T = 0.71$ *mm*.

v_s (m/s)	v_r (m/s)	v_s (m/s)	v_r (m/s)	v_s (m/s)	v_r (m/s)	v_s (m/s)	v_r (m/s)
38.4	0.0	44.3	8.5	51.2	19.6	65.0	38.9
72.3	48.6	88.3	70.0	97.9	83.4	103.0	90.4
111.1	99.4	114.9	103.7				

Table 10. Test results for shields with $T = 1.0$ *mm*.

v_s (m/s)	v_r (m/s)	v_s (m/s)	v_r (m/s)	v_s (m/s)	v_r (m/s)
45.3	0.0	51	8.7	57.3	17.9
65.8	29.7	73.3	44.4	81.9	58.2
83.0	61.6	97.2	78.3	112.7	99.1

Table 11. Test results for shields with $T = 1.5$ *mm*.

v_s (m/s)	v_r (m/s)	v_s (m/s)	v_r (m/s)	v_s (m/s)	v_r (m/s)
54.3	0.0	62.9	23.5	74.6	39.5
88.7	58.2	96.3	70.0	100.2	75.7
106.2	84.0	108.9	87.9	112.0	97.0

Table 12. Test results for shields with $T = 2.0$ *mm*.

v_s (m/s)	v_r (m/s)	v_s (m/s)	v_r (m/s)	v_s (m/s)	v_r (m/s)	v_s (m/s)	v_r (m/s)
67.2	0.0	83.7	30.0	95.8	46.2	105.3	58.0
111.3	69.4	116.7	77.4	126.5	91.4		

Table 13. Test results for shields with $T = 2.5$ *mm*.

v_s (m/s)	v_r (m/s)	v_s (m/s)	v_r (m/s)	v_s (m/s)	v_r (m/s)
79.4	0.0	96.7	40.6	103.1	57.0
111.4	69.6	117.8	79.0	123.0	88.1

Table 14. Test results for shields with $T = 3.0$ *mm*.

v_s (m/s)	v_r (m/s)	v_s (m/s)	v_r (m/s)	v_s (m/s)	v_r (m/s)
90.4	0.0	96.9	20.8	100.2	27.6
105.6	36.7	112.4	60.4	118.1	72.7

4.91.4 *Test results for hemispherical-nose projectiles*

Table 15. Test results for shields with $T = 0.5$ *mm*.

v_s (m/s)	v_r (m/s)	v_s (m/s)	v_r (m/s)	v_s (m/s)	v_r (m/s)	v_s (m/s)	v_r (m/s)
42.4	0.0	48.5	12.8	57.4	25.3	65.2	36.5
70.3	44.9	78.5	56.3	82.4	61.8	87.5	67.9
94.2	75.3	100.6	85.5	112.4	99.3	118.6	107.5

Table 16. Test results for shields with $T = 0.71$ *mm*.

v_s (m/s)	v_r (m/s)	v_s (m/s)	v_r (m/s)	v_s (m/s)	v_r (m/s)
48.9	0.0	54.6	14.6	65.4	30.8
71.6	42.2	79.4	51.6	90.4	70.9
98.5	81.8	112.0	97.4	115.3	102.7

Table 17. Test results for shields with $T = 1.0$ *mm*.

v_s (m/s)	v_r (m/s)	v_s (m/s)	v_r (m/s)	v_s (m/s)	v_r (m/s)
62.2	0.0	72.4	33.8	80.7	45.2
91.9	60.3	98.3	69.8	105.3	78.4
108.1	83.8	112.4	88.8	116.5	95.4

Table 18. Test results for shields with $T = 1.5$ *mm*.

v_s (m/s)	v_r (m/s)	v_s (m/s)	v_r (m/s)	v_s (m/s)	v_r (m/s)
69.5	0.0	75.5	18.8	83.6	31.4
98.3	50.9	106.3	65.1	110.4	78.9

Table 19. Test results for shields with $T = 2.0$ *mm*.

v_s (m/s)	v_r (m/s)	v_s (m/s)	v_r (m/s)	v_s (m/s)	v_r (m/s)
81.4	0.0	96.4	42.7	102.9	54.6
109.7	66.0	113.6	71.7	119.8	82.0

Table 20. Test results for shields with $T = 2.5$ *mm*.

v_s (m/s)	v_r (m/s)	v_s (m/s)	v_r (m/s)	v_s (m/s)	v_r (m/s)
95.9	0.0	106.8	54.3	118.0	76.2

Table 21. Test results for shields with $T = 3.0$ *mm*.

v_s (m/s)	v_r (m/s)	v_s (m/s)	v_r (m/s)
108.2	0.0	116.6	38.1

4.92 [Gupta *et al.*, 2008]

See also [Iqbal and Gupta, 2011].

4.92.1 *General data on shields and projectiles*

Shields. Layered (plates-in-contact) 1100-H12 AA shields, $\tilde{D}^{sh} = 230\,mm$

.

Projectiles. Oil quenched EN-24 steel (HRC 47-52) cylindrical (Fig. 3.1.1), ogive-nose (Fig. 3.3.1; $\rho = 30\,mm$) and hemispherical-nose (Fig. 3.4.1) projectiles with $L = 50.8\ mm$, $D = 19\ mm$, $m = 52.5\,g$.

Note. Experimental results for monolithic shields with thicknesses 0.5, 0.71, 1.0 and 1.5 mm are given in [Gupta *et al.*, 2007].

4.92.2 *Test results for cylindrical projectiles*

Table 1. Test results for 2×[0.5] shield.

v_s (m/s)	v_r (m/s)	v_s (m/s)	v_r (m/s)	v_s (m/s)	v_r (m/s)	v_s (m/s)	v_r (m/s)
51.2	0.0	66.9	29.6	81.3	49.6	88.5	60.4
92.3	66.9	98.2	75.2	107.7	86.4	113.9	95.5
116.1	99.5	120.6	105.3				

Table 2. Test results for 3×[0.5] shield.

v_s (m/s)	v_r (m/s)	v_s (m/s)	v_r (m/s)	v_s (m/s)	v_r (m/s)
56.4	0.0	67.6	21.3	79.2	40.1
84.7	46.9	90.2	55.9	91.6	64.4
104.4	80.8	112.7	90.9	118.2	98.8

Table 3. Test results for 4×[0.5] shield.

v_s (m/s)	v_r (m/s)	v_s (m/s)	v_r (m/s)	v_s (m/s)	v_r (m/s)
63.3	0.0	73.8	23.6	85.9	41.4
92.6	53.7	97.4	61.9	102.7	69.8
109.5	79.4	111.3	82.2	119.0	91.6

Table 4. Test results for 5×[0.5] shield.

v_s (m/s)	v_r (m/s)	v_s (m/s)	v_r (m/s)	v_s (m/s)	v_r (m/s)	v_s (m/s)	v_r (m/s)
70.2	0.0	83.4	30.5	95.7	53.0	102.5	64.0
105.2	69.4	107.0	72.5	113.4	81.3	118.3	90.0

Table 5. Test results for 2×[0.71] shield.

v_s (m/s)	v_r (m/s)	v_s (m/s)	v_r (m/s)	v_s (m/s)	v_r (m/s)	v_s (m/s)	v_r (m/s)
56.5	0.0	63.4	12.8	69.4	21.0	78.6	32.2
86.1	45.6	91.6	56.2	94.2	62.2	103.3	74.6
110.1	83.3	112.4	86.9	117.3	93.8		

Table 6. Test results for 3×[0.71] shield.

v_s (m/s)	v_r (m/s)	v_s (m/s)	v_r (m/s)	v_s (m/s)	v_r (m/s)	v_s (m/s)	v_r (m/s)
64.5	0.0	69.2	18.0	76.4	28.8	82.7	39.4
89.7	48.0	93.3	55.7	99.7	63.7	107.9	75.6
113.8	83.1	120.1	91.8				

Table 7. Test results for 4×[0.71] shield.

v_s (m/s)	v_r (m/s)	v_s (m/s)	v_r (m/s)	v_s (m/s)	v_r (m/s)	v_s (m/s)	v_r (m/s)
73.7	0.0	84.4	25.0	89.3	34.0	96.2	45.1
104.8	55.7	109.4	63.7	113.0	70.8	120.8	82.5

Table 8. Test results for 2×[1.0] shield.

v_s (m/s)	v_r (m/s)	v_s (m/s)	v_r (m/s)	v_s (m/s)	v_r (m/s)
65.8	0.0	71.9	21.3	77.5	31.4
88.6	45.4	95.1	54.8	98.7	59.3
103.1	66.6	111.4	78.6	116.1	86.4

Table 9. Test results for 3×[1.0] shield.

v_s (m/s)	v_r (m/s)	v_s (m/s)	v_r (m/s)	v_s (m/s)	v_r (m/s)	v_s (m/s)	v_r (m/s)
83.5	0.0	91.8	32.2	96.5	40.3	101.7	55.6
107.6	63.8	113.0	72.9	121.1	87.5		

Table 10. Test results for 2×[1.5] shield.

v_s (m/s)	v_r (m/s)	v_s (m/s)	v_r (m/s)	v_s (m/s)	v_r (m/s)
86.5	0.0	93.2	31.9	98.1	40.2
109.9	57.7	114.5	67.9	121.5	80.7

4.92.3 *Test results for ogive-nose projectiles*

Table 11. Test results for 2×[0.5] shield.

v_s (m/s)	v_r (m/s)	v_s (m/s)	v_r (m/s)	v_s (m/s)	v_r (m/s)	v_s (m/s)	v_r (m/s)
39.2	0.0	54.8	12.0	64.5	25.4	77.4	43.3
85.6	56.9	89.6	64.5	95.3	74.0	103.5	85.9
110.4	96.9	116.2	105.0				

Table 12. Test results for 3×[0.5] shield.

v_s (m/s)	v_r (m/s)	v_s (m/s)	v_r (m/s)	v_s (m/s)	v_r (m/s)	v_s (m/s)	v_r (m/s)
51.0	0.0	66.8	18.6	72.2	30.6	86.1	50.0
93.2	62.6	97.6	70.6	105.1	82.0	110.6	91.2
113.7	95.5	119.4	102.6				

Table 13. Test results for 4×[0.5] shield.

v_s (m/s)	v_r (m/s)	v_s (m/s)	v_r (m/s)	v_s (m/s)	v_r (m/s)
61.7	0.0	67.5	18.7	79.5	36.1
87.7	47.6	98.3	60.4	104.9	71.3
106.8	74.3	115.0	85.3	122.5	97.9

Table 14. Test results for 5×[0.5] shield.

v_s (m/s)	v_r (m/s)	v_s (m/s)	v_r (m/s)	v_s (m/s)	v_r (m/s)	v_s (m/s)	v_r (m/s)
66.0	0.0	76.3	24.7	83.7	36.5	94.9	54.1
98.4	60.4	105.5	69.6	113.9	82.2	117.7	88.2

Table 15. Test results for 2×[0.71] shield.

v_s (m/s)	v_r (m/s)	v_s (m/s)	v_r (m/s)	v_s (m/s)	v_r (m/s)	v_s (m/s)	v_r (m/s)
51.2	0.0	57.1	9.7	64.8	21.1	73.5	34.8
79.3	43.5	87.6	56.1	92.7	65.5	94.6	70.8
100.4	79.6	105.1	85.2	112.0	95.0		

Table 16. Test results for 3×[0.71] shield.

v_s (m/s)	v_r (m/s)	v_s (m/s)	v_r (m/s)	v_s (m/s)	v_r (m/s)	v_s (m/s)	v_r (m/s)
62.6	0.0	69.2	14.8	75.4	25.2	81.5	35.2
87.2	45.1	94.7	54.9	96.1	61.5	103.7	72.0
108.7	80.8	114.6	91.1				

Table 17. Test results for 4×[0.71] shield.

v_s (m/s)	v_r (m/s)	v_s (m/s)	v_r (m/s)	v_s (m/s)	v_r (m/s)
69.3	0.0	76.4	16.8	83.3	28.8
92.2	41.7	97.7	49.3	104.4	62.7
107.7	68.0	111.6	77.8	118.7	91.6

Table 18. Test results for 2×[1.0] shield.

v_s (m/s)	v_r (m/s)	v_s (m/s)	v_r (m/s)	v_s (m/s)	v_r (m/s)
62.9	0.0	67.4	13.6	76.5	27.3
89.0	43.4	96.2	53.8	102.4	62.6
109.7	75.1	112.1	82.9	119.9	95.3

Table 19. Test results for 3×[1.0] shield.

v_s (m/s)	v_r (m/s)	v_s (m/s)	v_r (m/s)	v_s (m/s)	v_r (m/s)	v_s (m/s)	v_r (m/s)
74.2	0.0	81.4	28.2	86.6	36.4	98.8	51.4
105.3	61.5	107.6	66.0	115.4	78.0	122.6	90.4

Table 20. Test results for 2×[1.5] shield.

v_s (m/s)	v_r (m/s)	v_s (m/s)	v_r (m/s)	v_s (m/s)	v_r (m/s)	v_s (m/s)	v_r (m/s)
78.6	0.0	84.2	27.5	89.2	34.8	97.8	47.6
106.7	60.7	110.1	67.4	117.3	80.8		

4.92.4 *Test results for hemispherical-nose projectiles*

Table 21. Test results for 2×[0.5] shield.

v_s (m/s)	v_r (m/s)	v_s (m/s)	v_r (m/s)	v_s (m/s)	v_r (m/s)	v_s (m/s)	v_r (m/s)
53.7	0.0	65.3	13.8	76.6	29.9	83.3	44.1
90.4	57.9	97.8	68.0	106.2	80.9	107.1	82.3
112.4	89.7	119.8	101.4				

Table 22. Test results for 3×[0.5] shield.

v_s (m/s)	v_r (m/s)	v_s (m/s)	v_r (m/s)	v_s (m/s)	v_r (m/s)
65.2	0.0	78.5	24.6	85.2	37.9
93.6	53.5	97.6	58.7	100.6	62.8
109.7	75.8	114.8	84.2	121.4	95.0

Table 23. Test results for 4×[0.5] shield.

v_s (m/s)	v_r (m/s)	v_s (m/s)	v_r (m/s)	v_s (m/s)	v_r (m/s)	v_s (m/s)	v_r (m/s)
75.2	0.0	84.8	30.6	91.5	44.2	95.8	50.9
98.2	55.0	106.8	68.8	111.7	76.9	114.6	83.3

Table 24. Test results for 5×[0.5] shield.

v_s (m/s)	v_r (m/s)	v_s (m/s)	v_r (m/s)	v_s (m/s)	v_r (m/s)	v_s (m/s)	v_r (m/s)
79.4	0.0	86.2	24.5	92.3	35.7	97.6	43.4
105.6	57.4	108.3	65.5	116.3	79.2	122.6	87.8

Table 25. Test results for 2×[0.71] shield.

v_s (m/s)	v_r (m/s)	v_s (m/s)	v_r (m/s)	v_s (m/s)	v_r (m/s)
63.3	0.0	72.5	20.7	80.7	33.4
87.8	44.7	93.0	51.5	97.2	57.4
102.6	67.0	107.6	73.8	115.3	87.9

Table 26. Test results for 3×[0.71] shield.

v_s (m/s)	v_r (m/s)	v_s (m/s)	v_r (m/s)	v_s (m/s)	v_r (m/s)
69.3	0.0	76.5	19.2	78.6	24.3
85.5	35.3	93.9	47.7	99.7	56.2
109.2	70.2	114.8	79.0	120.3	92.9

Table 27. Test results for 4×[0.71] shield.

v_s (m/s)	v_r (m/s)	v_s (m/s)	v_r (m/s)	v_s (m/s)	v_r (m/s)
80.2	0.0	85.1	17.7	93.5	32.1
98.6	40.9	105.4	54.2	107.5	60.7
111.1	69.6	116.9	78.2	122.8	93.1

Table 28. Test results for 2×[1.0] shield.

v_s (m/s)	v_r (m/s)	v_s (m/s)	v_r (m/s)	v_s (m/s)	v_r (m/s)	v_s (m/s)	v_r (m/s)
83.1	0.0	87.7	27.5	94.5	41.9	102.8	54.9
109.2	66.9	111.9	70.2	118.8	80.1		

Table 29. Test results for 3×[1.0] shield.

v_s (m/s)	v_r (m/s)	v_s (m/s)	v_r (m/s)	v_s (m/s)	v_r (m/s)	v_s (m/s)	v_r (m/s)
91.7	0.0	98.2	28.4	107.8	47.4	113.3	60.5
120.5	75.9						

Table 30. Test results for 2×[1.5] shield.

v_s (m/s)	v_r (m/s)	v_s (m/s)	v_r (m/s)	v_s (m/s)	v_r (m/s)	v_s (m/s)	v_r (m/s)
96.8	0.0	106.1	42.7	115.6	58.7	119.0	69.7

4.93 [Haight *et al.*, 2012]

4.93.1 *General data on shields and projectiles*

Shields. Monolithic A36 steel shields with $T = 6.35\,mm$ and various $c_{l.w}$.

Table 1. Projectiles.

Designation	Description	D (mm)	m (g)
M80	Standard bullet with full metal jacket and a lead core	7.62	9.39
M193	Standard bullet with full metal jacket and a lead core	5.56	3.56
M855	Bullet with a steel penetrator tip	5.56	4.01

4.93.2 *Test results*

Table 2. Test results on BLV, $v_{bl}\ (m/s)$.

$c_{l,w}$ (mm)	Projectile		
	M80	M193	M855
75	590	700	633
150	585	683	607

4.94 [Hanchak *et al.*, 1993]

4.94.1 *General data on shields and projectiles*

Shields. Monolithic HY-100 steel shields; $\widetilde{D}^{sh} = 305\,mm$, $T = 10.4\,mm$.
Projectiles. 4340 steel cylindrical projectiles (Fig. 3.1.1); $D = 30\,mm$, $L = 281\,mm$, $m = 1.58\,kg$.

4.94.2   *Test results*

Table 1. Test results.

v_s (m/s)	v_r (m/s)	v^{pl} (m/s)	m^{pl} (g)	v_s (m/s)	v_r (m/s)	v^{pl} (m/s)	m^{pl} (g)
78	0	0	0	82	0	0	0
105	0	0	0	114	0	0	0
125	80	80	54	126	84	104	55
127	98	112	56	162	142	164	56
163	138	163	57	165	141	163	57
207	168	210	59	258	232	268	66
261	233	269	64	263	234	273	65
318	284	335	63	322	289	335	67
370	333	389	62				

4.95   [Holmen *et al.*, 2013]

4.95.1   *General data on shields and projectiles*

Shields. Monolithic shields made from 6070 AA with different tempers (described in original paper), $c_{l,w} = 300\,mm$, $T = 20\,mm$.

Table 1. Properties of shield material.

Material	Orientation with respect to the rolling direction (deg)	$\widetilde{\sigma}_y$ (MPa)	σ_u (MPa)	True peak stress (MPa)	True failure strain
AA6070-O	0	51	139	243	0.79
AA6070-O	45	51	136	231	0.76
AA6070-O	90	50	138	225	0.66
AA6070-T4	0	187	320	487	0.52
AA6070-T4	45	186	328	474	0.46
AA6070-T4	90	187	328	457	0.39
AA6070-T6	0	373	393	475	0.26
AA6070-T6	45	379	396	423	0.07
AA6070-T6	90	387	399	416	0.04
AA6070-T7	0	341	354	419	0.32
AA6070-T7	45	346	360	396	0.14
AA6070-T7	90	340	356	391	0.13

Projectiles. 0.30 Caliber AP M2 projectiles (Fig. 3.11.1, Table 3.11.1).

4.95.2 *Test results*

Table 2#. Test results for AA6070-O shields.

v_s (m/s)	v_r (m/s)	v_s (m/s)	v_r (m/s)	v_s (m/s)	v_r (m/s)
345	0	377	166	432	264
550	416	659	506	901	824

$v_{bl} = 348.0\,m/s; a = 1, p = 1.94.$

Table 3#. Test results for AA6070-T4 shields.

v_s (m/s)	v_r (m/s)	v_s (m/s)	v_r (m/s)	v_s (m/s)	v_r (m/s)	v_s (m/s)	v_r (m/s)
468	0	503	0	509	62	513	32
520	117	604	341	894	780		

$v_{bl} = 506.2\,m/s; a = 1, p = 2.05.$

Table 4#. Test results for AA6070-T6 shields.

v_s (m/s)	v_r (m/s)	v_s (m/s)	v_r (m/s)	v_s (m/s)	v_r (m/s)
537	0	562	0	582	170
589	199	614	294	623	312
646	336	663	378	903	742

$v_{bl} = 562.5\,m/s; a = 1, p = 2.21.$

Table 5#. Test results for AA6070-T7 shields.

v_s (m/s)	v_r (m/s)	v_s (m/s)	v_r (m/s)	v_s (m/s)	v_r (m/s)
510	0	527	0	539	123
587	277	658	418	901	771

$v_{bl} = 529.1\,m/s; a = 1, p = 2.20.$

4.96 [Holt *et al.*, 1993]

4.96.1 *General data on shields and projectiles*

Shields. Monolithic Ti6A14V alloy ($\tilde{\sigma}_y = 840\,MPa$, $\sigma_u = 941\,MPa$, $\delta = 10\%$) shields; $D^{sh} = 34\,mm$, $T = 9.52\,mm$.

Projectiles. Hardened steel (HRC 50-58) cylindrical projectiles (Fig. 3.1.1); $D = 7.94mm$, $L = 71.9\,mm$.

4.96.2 *Test results*

Table 1. Test results for plug displacement.

v_s (m/s)	P^{pl} (mm)	v_s (m/s)	P^{pl} (mm)	v_s (m/s)	P^{pl} (mm)	v_s (m/s)	P^{pl} (mm)
219	0.89	234	2.28	253	2.53	262	4.36
283	5.70	290	7.46	361	*	456	*

* Separated from shield.

4.97 [Huang *et al.*, 2011]

4.97.1 *General data on shields and projectiles*

Shields. Monolithic   steel   ($\sigma_y = 685\,MPa$, $E = 204\,GPa$, $v = 0.33$, $\gamma = 7800\,kg/m^3$) shields, $T = 50\,mm$.

Projectiles. Slightly modified steel (HRC 52) cone-nose projectiles (Fig. 3.2.2c); $m = 4.25\,kg$.

4.97.2 *Test results*

Table 1. Test results.

v_s (m/s)	P (mm)	v_s (m/s)	P (mm)	v_s (m/s)	P (mm)	v_s (m/s)	P (mm)
277	64	300	70	316	75	317.6	74
331	79						

4.98 [Hutchings *et al.*, 1981]

4.98.1 *General data on shields and projectiles*

Shields. Monolithic mild steel (BS 970, En1A, 0.15% C, cold rolled, HV 235) plates with a plane polished surface.

Projectiles. Hard steel (HV 850) spherical projectiles (Fig. 3.5.1); $D = 9.5\,mm$.

4.98.2 *Test results*

Note. v_r is a rebound velocity after ricochet.

Table 1#. Test results for $\theta_s = 0°$.

v_s (m/s)	v_r (m/s)	v_s (m/s)	v_r (m/s)	v_s (m/s)	v_r (m/s)	v_s (m/s)	v_r (m/s)
47	11	99	17	144	16	210	17
209	25	245	22	275	7	276	22
277	25	305	19	339	22		

Table 2#. Test results for $\theta_s = 15°$.

v_s (m/s)	v_r (m/s)	θ_r (deg)	v_s (m/s)	v_r (m/s)	θ_r (deg)
209	21	149	276	33	149

Table 3#. Test results for $\theta_s = 30°$.

v_s (m/s)	v_r (m/s)	θ_r (deg)	v_s (m/s)	v_r (m/s)	θ_r (deg)	v_s (m/s)	v_r (m/s)	θ_r (deg)
82	43	120	141	60	123	205	71	129
211	89	126	275	70	134	275	79	139

Table 4#. Test results for $\theta_s = 45°$.

v_s (m/s)	v_r (m/s)	θ_r (deg)	v_s (m/s)	v_r (m/s)	θ_r (deg)	v_s (m/s)	v_r (m/s)	θ_r (deg)
208	109	115	245	133	117	275	144	117

Table 5#. Test results for $\theta_s = 55°$.

v_s (m/s)	v_r (m/s)	θ_r (deg)	v_s (m/s)	v_r (m/s)	θ_r (deg)	v_s (m/s)	v_r (m/s)	θ_r (deg)
209	144	111	243	172	108	274	189	108

Table 6#. Test results for $\theta_s = 60°$.

v_s (m/s)	v_r (m/s)	θ_r (deg)	v_s (m/s)	v_r (m/s)	θ_r (deg)	v_s (m/s)	v_r (m/s)	θ_r (deg)
95	76	109	95	79	–	140	118	–
140	106	110	140	–	110	178	145	111
178	152	–	207	–	108	210	163	108
210	160	108	210	169	–	226	175	–

v_s (m/s)	v_r (m/s)	θ_r (deg)	v_s (m/s)	v_r (m/s)	θ_r (deg)	v_s (m/s)	v_r (m/s)	θ_r (deg)
268	197	–	273	200	110	273	205	110
273	211	109	277	205	–	325	226	110
325	220	109						

Table 7#. Test results for $\theta_s = 70°$.

v_s (m/s)	v_r (m/s)	θ_r (deg)	v_s (m/s)	v_r (m/s)	θ_r (deg)	v_s (m/s)	v_r (m/s)	θ_r (deg)
94	91	102	137	129	104	209	176	104
275	211	104	275	230	103	277	234	104

Table 8#. Test results for $\theta_s = 75°$.

v_s (m/s)	v_r (m/s)	θ_r (deg)	v_s (m/s)	v_r (m/s)	θ_r (deg)	v_s (m/s)	v_r (m/s)	θ_r (deg)
208	193	102	275	232	102	280	254	102

Table 9#. Test results for $\theta_s = 80°$.

v_s (m/s)	v_r (m/s)	θ_r (deg)	v_s (m/s)	v_r (m/s)	θ_r (deg)	v_s (m/s)	v_r (m/s)	θ_r (deg)
211	200	100	213	–	99	275	244	100

4.99   [Ipson *et al.*, 1973]

4.99.1   *General data on shields and projectiles*

Shields. Monolithic 4130 steel (HB 230) shields.
Projectiles.   HT 4340   Steel (HRC 50-55)   cone-nose   projectiles (Fig. 3.2.1) and cylindrical projectiles (Fig. 3.1.1) with $D = 12.7mm$.

4.99.2   *Test results for cone-nose projectiles*

4.99.2.1   *Test results for shields with T=3.2 mm and projectiles with 26.6 g*

Table 1. Test results on BLV.

θ_s (deg)	β (deg)	v_{bl} (m/s)	θ_s (deg)	β (deg)	v_{bl} (m/s)	θ_s (deg)	β (deg)	v_{bl} (m/s)
0	30	211	0	60	219	30	30	238
30	60	240	60	30	430			

Table 2#. Test results on v_r for $\theta_s = 0°, \beta = 30°$.

v_s (m/s)	v_r (m/s)	v_s (m/s)	v_r (m/s)	v_s (m/s)	v_r (m/s)
367	280	471	401	590	524
655	590	794	745		

Table 3#. Test results on v_r for $\theta_s = 0°, \beta = 60°$.

v_s (m/s)	v_r (m/s)	v_s (m/s)	v_r (m/s)	v_s (m/s)	v_r (m/s)	v_s (m/s)	v_r (m/s)
336	275	489	426	573	500	793	740

4.99.2.2 *Test results for shields with T=6.4 mm*

Table 4. Test results on BLV.

θ_s (deg)	m (g)	β (deg)	v_{bl} (m/s)	θ_s (deg)	m (g)	β (deg)	v_{bl} (m/s)
0	28.8	30	393	0	29.2	60	275
30	26.6	30	433	30	29.2	60	376
60	28.8	60	676				

Table 5#. Test results on v_r for $m = 26.6\,g$, $\theta_s = 0°, \beta = 30°$.

v_s (m/s)	v_r (m/s)	v_s (m/s)	v_r (m/s)	v_s (m/s)	v_r (m/s)
409	136	510	327	528	167
624	443	897	670		

Table 6#. Test results on v_r for $m = 26.6\,g$, $\theta_s = 0°, \beta = 60°$.

v_s (m/s)	v_r (m/s)	v_s (m/s)	v_r (m/s)	v_s (m/s)	v_r (m/s)	v_s (m/s)	v_r (m/s)
481	340	589	425	668	467	767	516

Table 7#. Test results on v_r and θ_r for $m = 26.6$ g, $\theta_s = 30°$, $\beta = 30°$.

v_s (m/s)	θ_r (deg)	v_s (m/s)	θ_r (deg)	v_s (m/s)	θ_r (deg)	v_s (m/s)	θ_r (deg)
85	105	100	105	152	105	157	106
183	100	234	97	290	105	329	125

Table 8#. Test results on v_r and θ_r for $m = 26.6$ g, $\theta_s = 30°$, $\beta = 60°$.

v_s (m/s)	θ_r (deg)	v_s (m/s)	θ_r (deg)	v_s (m/s)	θ_r (deg)	v_s (m/s)	θ_r (deg)
149	99	158	100	163	100	195	110
237	132	275	135	279	130	345	153

4.99.2.3 *Test results for shields with T=9.6 mm*

Table 9. Test results on BLV for $m = 26.6$ g, $\theta_s = 0°$.

β (deg)	v_{bl} (m/s)	β (deg)	v_{bl} (m/s)
30	468	60	364

4.99.3 *Test results for cylindrical projectiles with m = 26.6 g*

Table 10. Test results on BLV for $\theta_s = 0°$.

T (mm)	v_{bl} (m/s)	T (mm)	v_{bl} (m/s)	T (mm)	v_{bl} (m/s)
3.2	198	6.4	272	9.5	395

Table 11#. Test results on θ_r for $T = 6.4$ mm, $\theta_s = 30°$.

v_s (m/s)	θ_r (deg)	v_s (m/s)	θ_r (deg)	v_s (m/s)	θ_r (deg)
168	111	179	115	266	121
267	112	268	112	280	145

4.100 [Iqbal *et al.*, 2006]

4.100.1 *General data on shields and projectiles*

Shields. Monolithic 1100-H12 AA ($E = 66\,GPa, \gamma = 2700\,kg/m^3$, $\nu = 0.3$) shields; $D^{sh} = 230\,mm$, $T = 1\,mm$.

Projectiles. Steel (HRC 52) ogive-nose projectiles (Fig. 3.3.1); $D = 15\,mm$, $L_{cyl} = 30\,mm$, $\rho = 30\,mm$, $m = 55\,g$.

4.100.2 *Test results*

Table 1. Test results.

v_s (m/s)	v_r (m/s)	h^{def} (mm)	v_s (m/s)	v_r (m/s)	h^{def} (mm)
65.8	52.7	8.0	73.3	59.4	7.9
81.9	70.4	7.2	83.0	71.2	6.9
97.2	85.7	6.1	103.7	92.1	5.7

4.101 [Iqbal *et al.*, 2013]

4.101.1 *General data on shields and projectiles*

Shields. Monolithic 1100-H12 AA ($\sigma_y = 103\,MPa$, $E = 69\,GPa$, $\gamma = 2713\,kg/m^3$ $v = 0.33$) shields; $\widetilde{D}^{sh} = 255\,mm$.

Projectiles. EN-31 steel (HRC 47-50) projectiles ($m = 29.7\,g$), see Fig. 3.7.1b, Fig. 3.6.4a, and Fig. 3.6.3 with $L_{cyl} = 26.25\,mm$, $L_{nose} = 12.80\,mm$, $D = 12.80\,mm$, $d = 6.40\,mm$.

4.101.2 *Test results for projectiles shown in Fig. 3.7.1b*

Table 1. Test results for shields with $T = 0.82\ mm$.

v_s (m/s)	v_r (m/s)	h^{def} (mm)	N	v_s (m/s)	v_r (m/s)	h^{def} (mm)	N
39.4	13.4	9.4	4	50.8	32.2	9.1	5
63.2	51.1	8.8	4	65.6	54.1	8.8	4
72.5	63.1	8.6	6	87.0	79.5	8.5	4
92.0	85.0	8.4	4	100.0	93.8	8.2	4
112.3	106.7	7.9	4				

$v_{bl} = 31.4\ m/s$.

Table 2. Test results for shields with $T = 1.82\ mm$.

v_s (m/s)	v_r (m/s)	h^{def} (mm)	N	v_s (m/s)	v_r (m/s)	h^{def} (mm)	N
72.4	34.0	11.1	5	80.6	47.3	10.9	4

v_s (m/s)	v_r (m/s)	h^{def} (mm)	N	v_s (m/s)	v_r (m/s)	h^{def} (mm)	N
91.4	63.1	10.7	5	100.0	75.2	10.3	4
104.2	80.0	9.9	4	113.6	91.9	9.6	4

$v_{bl} = 60.9\ m/s$.

4.101.3 *Test results for projectiles shown in Fig. 3.6.3*

Table 3. Test results for shields with $T = 0.82$ *mm*.

v_s (m/s)	v_r (m/s)	h^{def} (mm))	v_s (m/s)	v_r (m/s)	h^{def} (mm)	v_s (m/s)	v_r (m/s)	h^{def} (mm)
55.8	38.2	10.6	71.4	59.6	10.3	86.2	77.1	10.1
92.6	85.0	10.0	100.0	94.1	9.7	103.6	97.9	9.7
109.3	103.7	9.5	113.6	108.1	9.3			

$v_{bl} = 34.8\ m/s$.

Table 4. Test results for shields with $T = 1.82$ *mm*.

v_s (m/s)	v_r (m/s)	h^{def} (mm))	v_s (m/s)	v_r (m/s)	h^{def} (mm)	v_s (m/s)	v_r (m/s)	h^{def} (mm)
73.5	36.6	12.2	80.6	57.4	11.9	92.6	73.9	11.8
100.0	82.1	11.6	108.7	90.2	11.4	113.6	94.5	11.3

$v_{bl} = 63.2\ m/s$.

4.101.4 *Test results for projectiles shown in Fig. 3.6.4a*

Table 5. Test results for shields with $T = 0.82$ *mm*.

v_s (m/s)	v_r (m/s)	h^{def} (mm))	v_s (m/s)	v_r (m/s)	h^{def} (mm)	v_s (m/s)	v_r (m/s)	h^{def} (mm)
54.5	44.2	12.6	69.4	61.9	12.6	83.3	77.0	12.6
92.4	85.8	12.5	100.0	94.0	12.4	106.6	101.5	12.4
109.6	104.6	12.3						

$v_{bl} = 25.6\ m/s$.

Table 6. Test results for shields with $T = 1.82$ *mm*.

v_s (m/s)	v_r (m/s)	h^{def} (mm))	v_s (m/s)	v_r (m/s)	h^{def} (mm)	v_s (m/s)	v_r (m/s)	h^{def} (mm)
73.5	52.1	14.9	78.1	58.3	14.8	90.1	73.2	14.8

v_s (m/s)	v_r (m/s)	h^{def} (mm))	v_s (m/s)	v_r (m/s)	h^{def} (mm)	v_s (m/s)	v_r (m/s)	h^{def} (mm)
98.2	82.7	14.7	103.4	88.0	14.7	108.7	94.0	14.7

$v_{bl} = 60.1\, m/s$.

4.102   [Iqbal *et al.*, 2015]

4.102.1   *General data on shields and projectiles*

Shields. Monolithic 1100-H12 Al ($E = 65.7\,GPa$, $\gamma = 2700\,kg/m^3$, $\sigma_y = 148\,MPa$) sheets with $T = 1\,mm$ and different $\widetilde{D}^{sh}$.

Projectiles. Ogive-nose (Fig. 3.3.1) and cylindrical (Fig. 3.1.1) projectiles ($D = 19\,mm$, $m = 52.5\,g$) machined through EN-24 steel, oil quenched to HRC 47-52 and annealed for releasing internal stresses.

4.102.2   *Test results for cylindrical projectiles*

Table 1. Test results for $\widetilde{D}^{sh} = 68\,mm$.

v_s (m/s)	v_r (m/s)	v_s (m/s)	v_r (m/s)	v_s (m/s)	v_r (m/s)	v_s (m/s)	v_r (m/s)
32.0	–	32.5	0.0	42.0	21.0	52.9	34.6
65.6	44.1	70.0	54.5	80.0	58.3	108.1	94.6

$v_{bl} = 37.2\, m/s$.

Table 2. Test results for $\widetilde{D}^{sh} = 95\,mm$.

v_s (m/s)	v_r (m/s)	v_s (m/s)	v_r (m/s)	v_s (m/s)	v_r (m/s)	v_s (m/s)	v_r (m/s)
34.2	–	34.5	–	37.3	0.0	47.7	22.5
52.9	27.0	75.0	55.3	106.7	93.3	120.0	105.0

$v_{bl} = 42.4\, m/s$.

Table 3. Test results for $\widetilde{D}^{sh} = 150\,mm$.

v_s (m/s)	v_r (m/s)	v_s (m/s)	v_r (m/s)	v_s (m/s)	v_r (m/s)	v_s (m/s)	v_r (m/s)
41.6	0	44.8	–	45.0	–	48.5	15.5
50.6	22.7	65.6	40.9	83.2	61.3	97.1	83.2
109.2	92.6						

$v_{bl} = 46.8\, m/s$.

Table 4. Test results for $\widetilde{D}^{sh} = 190mm$.

v_s (m/s)	v_r (m/s)	v_s (m/s)	v_r (m/s)	v_s (m/s)	v_r (m/s)	v_s (m/s)	v_r (m/s)
46.1	0	52.5	12.7	53.9	–	54.0	–
60.0	30.0	77.6	59.1	110.4	96.6	115.01	101.2

$v_{bl} = 49.2\,m/s$.

Table 5. Test results for $\widetilde{D}^{sh} = 285mm$.

v_s (m/s)	v_r (m/s)	v_s (m/s)	v_r (m/s)	v_s (m/s)	v_r (m/s)	v_s (m/s)	v_r (m/s)
48.5	0	56.5	–	56.8	–	60.0	20.0
90	74.3	107.1	94.7	116.8	100.1		

$v_{bl} = 51.6\,m/s$.

4.102.3 *Test results for ogive-nose projectiles*

Table 6. Test results for $\widetilde{D}^{sh} = 68mm$.

v_s (m/s)	v_r (m/s)	v_s (m/s)	v_r (m/s)	v_s (m/s)	v_r (m/s)	v_s (m/s)	v_r (m/s)
38.5	–	40.5	0	45.0	25.7	64.3	51.7
70.6	55.6	92.3	78.9	100	85.7	109.1	100

$v_{bl} = 42.7\,m/s$.

Table 7. Test results for $\widetilde{D}^{sh} = 95mm$.

v_s (m/s)	v_r (m/s)	v_s (m/s)	v_r (m/s)	v_s (m/s)	v_r (m/s)	v_s (m/s)	v_r (m/s)
40.0	–	41.2	0	46.0	22.2	57.5	38.3
66.7	50	83.1	72.7	108.1	101.3	115.8	108.1

$v_{bl} = 43.6\,m/s$.

Table 8. Test results for $\widetilde{D}^{sh} = 150mm$.

v_s (m/s)	v_r (m/s)	v_s (m/s)	v_r (m/s)	v_s (m/s)	v_r (m/s)	v_s (m/s)	v_r (m/s)
40.1	0	42.5	–	50.0	26.5	55.3	32.9
64.3	42.9	70.0	50.0	79.4	65.5	97.1	85.6
104.0	97.1						

$v_{bl} = 45.0\,m/s$.

Table 9. Test results for $\widetilde{D}^{sh} = 190\,mm$.

v_s (m/s)	v_r (m/s)	v_s (m/s)	v_r (m/s)	v_s (m/s)	v_r (m/s)	v_s (m/s)	v_r (m/s)
42.9	0	43.0	–	48.0	20.8	83.1	72.5
85.7	77.1	106.2	101.2	115.0	108.4		

$v_{bl} = 45.4\,m/s$.

Table 10. Test results for $\widetilde{D}^{sh} = 285\,mm$.

v_s (m/s)	v_r (m/s)	v_s (m/s)	v_r (m/s)	v_s (m/s)	v_r (m/s)	v_s (m/s)	v_r (m/s)
43.5	0	44.0	–	46	–	49.4	20.6
57.1	35.2	90.0	80.0	101.5	93.3	113.3	105.0

$v_{bl} = 46.4\,m/s$.

4.103 [Jagannathan, 2010]

4.103.1 *General data on shields and projectiles*

Shields. Monolithic hot rolled A36 mild steel shields.
Projectiles. Winchester 147 grains full metal jacket bullet (7.62 mm caliber) which consists of a soft lead inner core enclosed into a copper jacket, $m = 9.53\,g$.

4.103.2 *Test results*

Table 1#. Test results for shields with $T = 9.5$ *mm*.

v_s (m/s)	P (mm)	v_s (m/s)	P (mm)	v_s (m/s)	P (mm)	v_s (m/s)	P (mm)
704.6	11.1	717.7	11.8	734.5	11.9	737.2	12.4
744.8	8.9	752.1	11.5	752.7	11.5	754.3	12.4
776.6	13.0	809.5	12.0	822.7	15.8		

Table 2#. Test results for shields with $T = 15.9$ *mm*.

v_s (m/s)	P (mm)	v_s (m/s)	P (mm)	v_s (m/s)	P (mm)
737.0	6.0	740.0	4.6	748.3	4.7
762.0	6.4	780.6	8.6	792.5	5.4
805.3	8.9	817.8	10.5	839.4	11.2

Table 3#. Test results for shields with $T = 19.1\ mm$.

v_s (m/s)	P (mm)	v_s (m/s)	P (mm)	v_s (m/s)	P (mm)	v_s (m/s)	P (mm)
724.5	4.4	737.0	3.4	742.1	4.8	748.6	5.5
759.6	6.9	763.5	3.7	766.0	6.9	775.7	6.7
777.0	6.8	778.5	7.0	783.4	6.7	791.9	7.2
796.2	7.5	798.9	7.3	811.4	4.6		

4.104 [Jena *et al.*, 2010a]

4.104.1 *General data on shields and projectiles*

Shields. Monolithic shields made from ultra-high strength armor steel that was austenitized at $910^\circ C$ followed by tempering at different temperatures (T^{temp}); $c_l = 150mm$, $c_w = 100mm$, $T = 25\,mm$.

Table 1. Shields material properties.

T^{temp} (°C)	σ_y (MPa)	σ_u (MPa)	HV	δ (%)	Charpy (J)
0	1367	1900	586	9	19
200	1417	1808	555	12	31
300	1463	1700	518	7.5	22
400	1433	1587	490	10	26
500	1286	1409	442	12	38
600	1146	1247	381	16	85

Projectiles. Hard steel alloy (HV 930) 7.62 mm projectiles having a mass of $m = 5.34\,\mathrm{g}$.

Impact velocity. $v_s = 840 \pm 15\,m/s$.

4.104.2 *Test results*

Table 2. DOP vs T^{temp}.

T^{temp} (°C)	P (mm)	T^{temp} (°C)	P (mm)	T^{temp} (°C)	P (mm)
0	16.5	200	12.6	300	15.5
400	16.0	500	17.6	600	20.5

4.105 [Jena *et al.*, 2010b]

4.105.1 *General data on shields and projectiles*

Shields. Monolithic 7017 AA ($\sigma_y = 458\,MPa$, $\sigma_u = 508\,MPa$, $\delta = 13\%$, $\gamma = 2780\,kg/m^3$, $E = 75\,GPa$) shields; $c_l = 300\,mm$, $c_w = 100\,mm$, $T = 15\,mm$.

Projectiles. Hard steel alloy (HV 930) 7.62 mm AP projectiles; $D = 6.1\,mm$, $m = 5.34\,g$.

Impact velocity. $v_s = 840 \pm 15\,m/s$.

4.105.2 *Test results*

Table 1#. Test results.

θ_s (deg)	δ_v	θ_s (deg)	δ_v	θ_s (deg)	δ_v
0	0.079	15	0.105	30	0.134
45	0.168	50	0.249		

4.106 [Jenq *et al.*, 1988]

4.106.1 *General data on shields and projectiles*

Shields. Monolithic 2024-0 Al ($E = 71\,GPa$, $\gamma = 2700\,kg/m^3$, $v = 0.33$, $\sigma_y = 75.8\,MPa$, $\sigma_u = 186\,MPa$, HB 43-56) shields; $D^{sh} = 140\,mm$.

Projectiles. Steel (HRC 60) cylindrical projectiles (Fig. 3.1.1); $D = 12.7\,mm$, $L = 38.1\,mm$, $m = 35.5\,g$.

4.106.2 *Test results*

Table 1. Test results for shields with $T = 3.2$ *mm*.

v_s (m/s)	v_r (m/s)	h^{def} (mm)	v_s (m/s)	v_r (m/s)	h^{def} (mm)	v_s (m/s)	v_r (m/s)	h^{def} (mm)
77	0	13.2	86	0	18.2	93	0	20.2
107	0	15.2	150	127	14.9	172	157	9.3

Table 2. Test results for shields with $T = 6.4$ *mm*.

v_s (m/s)	v_r (m/s)	h^{def} (mm)	v_s (m/s)	v_r (m/s)	h^{def} (mm)	v_s (m/s)	v_r (m/s)	h^{def} (mm)
104	0	12.6	123	0	14.4	126	–	13.4
110	13.9	15.2	343	300	8.0	416	382	7.6

4.107 [Johnson *et al.*, 1959]

4.107.1 *General data on shields and projectiles*

Shields. Monolithic shields made from different metals.

Table 1. Properties of shield materials.

Material	σ_{ss} (MPa)	σ_y (MPa)	Material	σ_{ss} (MPa)	σ_y (MPa)
Lead	12	10	Aluminum	65	102
Magnesium	130	210	Zinc	131	158
Copper	158	335	4140 Steel	550	760
Silver	–	–			

Projectiles. Spherical projectiles (Fig. 3.5.1; $D = 4.76mm$, $m = 0.44\,g$) made from a high-carbon, chrome alloy steel through-hardened to HRC 64-66. A number of balls were annealed at $650°C$.

4.107.2 *Test results for impact by non-annealed projectile*

Table 2. Test results for copper shields.

v_s (km/s)	P (mm)	v_s (km/s)	P (mm)	v_s (km/s)	P (mm)
0.42	2.17	0.45	2.53	0.57	3.70
0.76	1.48	0.82	4.60	0.99	4.49
1.10	4.92	1.19	4.95	1.27	4.90
1.36	5.00	1.64	5.80	1.65	6.03
1.65	5.60	1.70	6.25	1.73	6.20
1.73	6.10	1.73	5.80	1.79	6.20
1.87	6.68	1.95	6.64	1.96	7.11
1.98	6.90	2.01	7.62	2.06	6.96
2.12	7.08	2.13	6.91	2.16	7.16

v_s (km/s)	P (mm)	v_s (km/s)	P (mm)	v_s (km/s)	P (mm)
2.23	7.31	2.27	7.40	2.27	8.46
2.30	7.50	2.44	7.70	2.50	7.75

Table 3. Test results for 4140 steel shields.

v_s (km/s)	P (mm)	v_s (km/s)	P (mm)	v_s (km/s)	P (mm)
0.44	1.16	0.52	1.22	0.56	1.30
0.63	1.47	0.63	1.61	0.74	1.68
0.97	1.91	0.98	1.93	1.28	2.58
1.32	2.77	1.34	3.03	1.49	3.07
1.51	3.42	1.60	3.56	1.62	3.39
1.62	3.38	1.68	4.05	1.75	4.39
1.76	3.89	1.80	4.25	1.89	4.00
1.92	4.35	1.97	4.24	1.97	4.23
2.02	4.23	2.02	4.31	2.02	4.06
2.04	3.45	2.08	4.45	2.09	4.61
2.10	4.81	2.10	4.30	2.11	4.50
2.12	5.05	2.14	4.35	2.17	4.47
2.19	4.93	2.19	4.43	2.20	5.15
2.30	4.51				

Table 4. Test results for lead shields.

v_s (km/s)	P (mm)	v_s (km/s)	P (mm)	v_s (km/s)	P (mm)
0.44	10.30	0.47	11.10	0.56	12.90
0.63	14.40	0.97	12.30	1.28	10.76
1.30	10.87	1.40	11.24	1.48	11.45
1.49	10.88	1.51	11.60	1.53	11.90
1.66	12.59	1.69	12.51	1.83	12.93
1.88	13.20	1.99	13.18	2.02	12.70
2.17	13.10	2.24	13.45	2.27	11.35
2.38	14.05				

Table 5. Test results for aluminum shields.

v_s (km/s)	P (mm)	v_s (km/s)	P (mm)	v_s (km/s)	P (mm)
0.12	0.93	0.12	0.83	0.23	1.90
0.43	2.90	0.57	6.70	0.66	8.30
1.01	14.70	1.06	17.00	1.33	22.90
1.38	23.40	1.49	24.70	1.59	27.70
1.61	27.90	1.69	27.10	1.83	25.20

v_s (km/s)	P (mm)	v_s (km/s)	P (mm)	v_s (km/s)	P (mm)
1.93	24.30	2.01	21.60	2.03	22.70
2.07	19.30	2.23	17.00	2.28	16.40
2.56	17.80				

Table 6. Test results for zinc shields.

v_s (km/s)	P (mm)	v_s (km/s)	P (mm)	v_s (km/s)	P (mm)
0.37	2.01	0.45	2.65	0.57	3.50
0.72	4.43	0.74	4.40	0.91	5.70
1.26	5.32	1.26	5.75	1.46	6.00
1.58	6.54	1.61	6.80	1.63	6.60
1.65	6.43	1.67	6.67	1.70	6.53
1.74	6.91	1.75	6.75	1.89	7.13
1.90	7.61	2.03	7.28	2.51	7.83

Table 7. Test results for silver shields.

v_s (km/s)	P (mm)	v_s (km/s)	P (mm)	v_s (km/s)	P (mm)
0.50	4.40	0.64	5.40	0.65	6.00
0.70	6.10	0.79	6.40	1.00	6.58
1.02	5.94	1.22	6.39	1.25	6.44
1.67	7.24	1.73	7.63	1.78	7.61
1.85	7.64	1.87	7.39	2.54	9.24

Table 8. Test results for magnesium shields.

v_s (km/s)	P (mm)	v_s (km/s)	P (mm)	v_s (km/s)	P (mm)
0.39	1.50	0.63	5.50	0.85	8.80
0.98	11.30	1.11	14.00	1.39	19.60
1.52	22.90	1.62	24.30	1.89	30.10
1.96	32.30	2.27	33.00	2.38	30.20

4.107.3 *Test results for impact by annealed projectile*

Table 9. Test results for aluminum shields.

v_s (km/s)	P (mm)	v_s (km/s)	P (mm)	v_s (km/s)	P (mm)
0.55	6.10	0.83	10.00	0.86	12.10
1.23	17.80	1.35	16.20	1.44	18.20
1.55	18.00	1.94	16.60	1.95	16.80
2.15	16.60	2.18	16.20	2.33	18.40

Table 10. Test results for copper shields.

v_s (km/s)	P (mm)	v_s (km/s)	P (mm)	v_s (km/s)	P (mm)
0.74	2.44	0.95	3.87	1.05	4.30
1.38	5.27	1.49	5.28	1.58	6.19
1.62	6.13	1.85	5.88	1.97	6.52
2.09	6.72				

4.108 [Jones and Placzankis, 2011]

4.108.1 *General data on shields and projectiles*

Shields. Monolithic aluminum and magnesium alloys shields.

Table 1. Properties of shield materials.

Notation	Material	σ_u (MPa)	σ_y (MPa)	δ (%)	Form	γ (kg/m^3)
Mg-1	AZ31B-H24 Mg	235	125	7	Rolled	1770
AA	5083-H131 AA	391	282	13	Rolled	2660
Mg-2	E675-T5 Mg	410	310	9	Extruded	1950

Projectiles. 0.30 Caliber AP M2 projectiles (Fig. 3.11.1, Table 3.11.1); 0.50 caliber FSPs and 20 mm FSPs (Fig. 3.10.1, Table 3.10.1).

4.108.2 *Test results*

Table 2. Test results.

T (mm)	Projectile	Shield material	v_{bl} (m/s)
38.1	0.30 Caliber AP M2	Mg-1	582
63.5	0.30 Caliber AP M2	Mg-1	824
76.2	0.30 Caliber AP M2	Mg-1	906
38.1	0.30 Caliber AP M2	AA	586
63.5	0.30 Caliber AP M2	AA	792
76.2	0.30 Caliber AP M2	AA	931
38.1	0.30 Caliber AP M2	Mg-2	749*
63.5	0.30 Caliber AP M2	Mg-2	931*
76.2	0.30 Caliber AP M2	Mg-2	>982**
40.4	0.50 Caliber FSP	Mg-1	849
40.4	0.50 Caliber FSP	AA	1027

T (mm)	Projectile	Shield material	v_{bl} (m/s)
40.4	0.50 Caliber FSP	Mg-2	1156
59.7	20 mm FSP	Mg-1	781
59.7	20 mm FSP	AA	911
59.7	20 mm FSP	Mg-2	976

* Half aged; ** Normally aged.

4.109 [Jones *et al.*, 2007]

4.109.1 *General data on shields and projectiles*

Shields. Monolithic shields.

Table 1. Properties of shield materials.

Material	T (mm)	σ_u (MPa)	σ_y (MPa)	δ (%)
AZ31B-O	7.6	254	153	21.5
AZ31B-O	31.5	258	151	11.5
AZ31B-H24	7.7	265	179	19.0
AZ31B-H24	76.5	262	169	9.5

Projectiles. 0.30 Caliber AP M2 projectiles (Fig. 3.11.1, Table 3.11.1), 0.50 Caliber AP M2 projectiles (Fig. 3.13.1, Table 3.13.1); 0.22 Caliber FSP, 0.50 Caliber FSP, 20 mm FSP (Fig. 3.10.1, Table 3.10.1).

4.109.2 *Test results*

Table 2. Test results for AZ31B-O shields.

T (mm)	Projectile	v_{bl} (m/s)	T (mm)	Projectile	v_{bl} (m/s)
31.5	0.30 Caliber AP M2	511	76.5	0.50 Caliber AP M2	649
7.6	0.22 Caliber FSP	417	31.5	0.50 Caliber FSP	639

Table 3. Test results for AZ31B-H24shields.

T (mm)	Projectile	v_{bl} (m/s)	T (mm)	Projectile	v_{bl} (m/s)
76.5	0.30 Caliber AP M2	863	7.7	0.22 Caliber FSP	421
76.5	20 mm FSP	897			

4.110 [Jones *et al.*, 2008]

4.110.1 *General data on shields and projectiles*

Shields. Monolithic Mg AZ31B-H24 (HB 55-61 at 500 kg = 4.9 kN load) shields.

Table 1. Properties of shield materials.

T (mm)	σ_u (MPa)	σ_y (MPa)	δ (%)
19.1–25.4	265	165	13.1
25.4–38.1	263	167	12.5
38.1–63.5	264	170	11.9
63.5–88.9	261	166	11.1

Projectiles. 0.30 caliber AP M2 projectiles (Fig. 3.11.1, Table 3.11.1), 0.50 Caliber AP M2 projectiles (Fig. 3.13.1, Table 3.13.1); 0.50 caliber FSP and 20 mm FSPs (Fig. 3.10.1, Table 3.10.1).

4.110.2 *Test results*

Table 2. Test results for FSPs.

Caliber	T (mm)	v_{bl} (m/s)	Caliber	T (mm)	v_{bl} (m/s)	Caliber	T (mm)	v_{bl} (m/s)
0.50"	25.0	507	0.50"	38.7	742	20 mm	38.7	477
20 mm	49.7	576	20 mm	63.1	735			

Table 3. Test results for AP M2 projectiles.

Caliber	T (mm)	v_{bl} (m/s)	Caliber	T (mm)	v_{bl} (m/s)	Caliber	T (mm)	v_{bl} (m/s)
0.30"	38.8	579	0.30"	49.7	687	0.30"	63.5	787
0.50"	76.5	650	0.50"	88.9	688	0.50"	102.0	746

4.111 [Khan and Ansari, 2003]

4.111.1 *General data on shields and projectiles*

Shields. Monolithic commercially available Al ($\sigma_y = 110 MPa$) shields; $D^{sh} = 255 mm$.

Projectiles. Hardened (HRC 58) cylindrical projectiles (Fig. 3.1.1); $D = 12.8\,mm$, $L = 25.6\,mm$, $m = 25.08\,g$.

4.111.2 *Test results*

Table 1#. Test results on residual velocities for $\theta_s = 15°$.

T (mm)	v_s (m/s)	v_r (m/s)	v_s (m/s)	v_r (m/s)	v_s (m/s)	v_r (m/s)
0.81	24.4	0	47.0	31.4	66.9	58.1
	79.2	70.4	90.5	74.3		
1.52	40.3	0	49.0	5.4	66.9	21.5
	79.3	33.9	90.5	52.5	105.4	103.5
1.91	50.8	0	66.9	54.0	79.3	74.5
	90.5	83.3	105.4	95.5		

Table 2#. Test results on residual velocities for $\theta_s = 30°$.

T (mm)	v_s (m/s)	v_r (m/s)	v_s (m/s)	v_r (m/s)	v_s (m/s)	v_r (m/s)
0.81	23.4	0	45.4	23.0	57.7	53.0
	79.3	78.6	90.6	86.2	105.3	103.0
1.52	39.3	0	52.9	25.2	63.4	62.6
	84.5	81.7	90.5	89.4	105.4	84.0
1.91	55.7	0	63.5	25.0	79.3	70.4
	90.5	72.7				

Table 3#. Test results on BLV.

T (mm)	θ_s (deg)	v_{bl} (m/s	θ_s (deg)	v_{bl} (m/s)	θ_s (deg)	v_{bl} (m/s)
0.81	0	28.8	15	24.4	30	23.4
	45	37.3	60	48.8		
1.52	0	45.3	15	40.3	30	39.3
	45	45.3	60	52.9		
1.91	0	51,0	15	50.8	30	55.7
	45	57.0	60	79.3		

4.112 [Khan *et al.*, 2014]

4.112.1 *General data on shields and projectiles*

Shields. Monolithic 1100-H14 aluminum plates, $\widetilde{D}^{sh} = 230\,mm$, $T = 0.82\,mm$.

Projectiles. Cone-nose projectiles (Fig. 3.2.1); $L_{cyl} = 25\,mm$, $L_{nose} = 15\,mm$, $\beta = 22.5°$ and cylindrical projectiles (Fig. 3.1.1); $L = 30\,mm$. Projectiles were made from EN-31 alloy steel; $D = 12.8\,mm$, $m = 29.7\,g$.

4.112.2 *Test results*

Table 1. Test results for cone-nose projectiles.

v_s (m/s)	v_r (m/s)	v_s (m/s)	v_r (m/s)	v_s (m/s)	v_r (m/s)
28.2	2.1	35.0	15.9	39.4	23.4
50.1	38.2	83.3	75.5	100.0	93.1

Table 2. Test results for cylindrical projectiles.

v_s (m/s)	v_r (m/s)	v_s (m/s)	v_r (m/s)	v_s (m/s)	v_r (m/s)
34.5	8.5	42.2	26.2	65.8	57.2
83.3	77.4	104.2	101.0		

4.113 [Khan, 2010]

Note. Data from [Ansari, 1998].

4.113.1 *General data on shields and projectiles*

Shields. Monolithic commercial aluminum sheets.
Projectiles. Ogive-nose projectiles (Fig. 3.3.1) made of hardened steel; $D = 15\,mm$, $L_{cyl} = 30\,mm$, $\rho = 30\,mm$.

4.113.2 *Test results*

Table 1#. Test results for shields with $T = 0.5\ mm$.

v_s (m/s)	v_r (m/s)	v_s (m/s)	v_r (m/s)	v_s (m/s)	v_r (m/s)
20.9	9.8	25.7	18.9	31.0	25.9
32.0	26.5	38.3	33.5	53.7	48.7
59.8	57.0	64.7	59.1	68.2	67.0
76.2	74.6				

Table 2#. Test results for shields with $T = 0.74$ *mm*.

v_s (m/s)	v_r (m/s)	v_s (m/s)	v_r (m/s)	v_s (m/s)	v_r (m/s)	v_s (m/s)	v_r (m/s)
27.1	10.0	29.6	12.4	29.1	12.9	29.5	14.0
30.7	15.9	31.9	18.1	32.2	20.9	34.1	21.5
36.7	25.4	37.3	24.1	38.3	27.4	44.1	33.1
45.6	33.8	46.4	35.3	46.6	36.6	47.3	37.3
50.4	42.4	53.9	46.4	56.0	49.9	57.5	50.5
57.8	49.6	62.5	54.0	67.9	60.4	75.5	67.2
76.9	70.3	78.9	75.6				

Table 3#. Test results for shields with $T = 1$ *mm*.

v_s (m/s)	v_r (m/s)	v_s (m/s)	v_r (m/s)	v_s (m/s)	v_r (m/s)	v_s (m/s)	v_r (m/s)
33.1	9.8	35.2	14.4	39.5	23.5	42.9	27.1
47.7	34.5	48.6	35.6	53.7	40.8	55.8	45.0
62.2	52.9	65.4	56.2	67.2	58.6	77.9	68.7

Table 4#. Test results for shields with $T = 1.5$ *mm*.

v_s (m/s)	v_r (m/s)	v_s (m/s)	v_r (m/s)	v_s (m/s)	v_r (m/s)
45.2	7.3	47.0	13.1	51.5	26.9
54.7	32.8	60.3	37.6	64.7	44.4
70.0	51.4	71.7	53.5	83.0	62.7

4.114 [Khoda-rahmi *et al.*, 2006]

4.114.1 *General data on shields and projectiles*

Shields. Monolithic steel shields; $c_{l,w} = 300\,mm$, $T = 25\,mm$.
Projectiles. Steel ogive-nose projectiles (Fig. 3.3.1); $D = 6.7\,mm$, $L_{cyl} = 34.4\,mm$, $m = 7.7\,g$.

Table 1. Data on sets of experiments.

Set	Projectile		Shield	
	$\tilde{\sigma}_y$ (MPa)	σ_u (MPa)	$\tilde{\sigma}_y$ (MPa)	σ_u (MPa)
Set 1	308	496	382	543
Set 2	533	603	308	462

4.114.2 *Test results*

Table 2. Test results on set 1.

v_s (m/s)	P (mm)	v_s (m/s)	P (mm)	v_s (m/s)	P (mm)
858	8.50	849	7.46	851	7.51
758	6.89	750	6.15	746	5.43
650	4.32	656	4.08	630	4.61

Table 3. Test results on set 2.

v_s (m/s)	P (mm)	v_s (m/s)	P (mm)	v_s (m/s)	P (mm)
858	10.65	850	9.62	846	9.15
768	8.84	760	6.46	780	9.38
648	2.99	650	3.45	660	4.97

4.115 [Kpenyigba *et al.*, 2013]

See also [Rodríguez-Martínez *et al.*, 2013], [Kpenyigba *et al.*, 2015], [Jankowiak *et al.*, 2014], [Jankowiak and Rusinek, 2015].

4.115.1 *General data on shields and projectiles*

Shields. Monolithic and layered (plates-in-contact) shields, $c_{l,w} = 100\,mm$, made from rolled low-carbon ferritic steel, $\sigma_y = 154\,MPa$, $\sigma_u = 347\,MPa$. Thickness of a single sheet (layer) is 1 *mm*.

Projectiles. Cylindrical projectiles (Fig. 3.1.1, $L = 29\,mm$, *Projectile-1*), hemispherical-nose projectiles (Fig. 3.4.1, $L_{cyl} = 25\,mm$, *Projectile-2*) and cone-nose projectiles (Fig. 3.2.1) with different sizes (see Table 1). Projectile material is maraging steel subjected to heat treatment, $\sigma_y = 2\,GPa$. $D = 13mm$ and $m \approx 30\,g$ for all projectiles.

Table 1. Cone-nose projectiles.

Notation	β (deg)	L_{cyl} (mm)	L_{nose} (mm)	Notation	β (deg)	L_{cyl} (mm)	L_{nose} (mm)
Projectile-3	20	22.98	17.98	Projectile-4	36	25	8.95
Projectile-5	50	27.14	5.5	Projectile-6	60	27.71	3.79

4.115.2 *Test results for monolithic shields*

Table 2#. Test results for Projectile-1.

v_s (m/s)	v_r (m/s)	v_s (m/s)	v_r (m/s)	v_s (m/s)	v_r (m/s)	v_s (m/s)	v_r (m/s)
81	37	88	46	101	67	121	94
141	117	171	148	180	158		

$v_{bl} = 72\,m/s; a = 1, p = 1.880.$

Table 3#. Test results for Projectile-2.

v_s (m/s)	v_r (m/s)	v_s (m/s)	v_r (m/s)	v_s (m/s)	v_r (m/s)	v_s (m/s)	v_r (m/s)
80	0	88	23	100	54	122	89
141	112	171	146	181	160		

$v_{bl} = 83.5\,m/s; a = 1, p = 1.940.$

Table 4#. Test results for Projectile-4.

v_s (m/s)	v_r (m/s)	v_s (m/s)	v_r (m/s)	v_s (m/s)	v_r (m/s)	v_s (m/s)	v_r (m/s)
72	0	80	37	87	52	101	67
121	92	141	118	172	148	180	156

$v_{bl} = 72\,m/s; a = 1, p = 1.823.$

Table 5#. Test results for cone-nose projectiles.

Projectile	v_s (m/s)	$w_r - w_s$ (J)	Projectile	v_s (m/s)	$w_r - w_s$ (J)
Projectile-4	72	76	Projectile-4	80	75
Projectile-4	88	82	Projectile-4	101	85
Projectile-4	121	87	Projectile-4	141	92
Projectile-4	171	108	Projectile-4	180	111
Projectile-3	98	73	Projectile-3	99	76
Projectile-3	128	79	Projectile-3	128	86
Projectile-3	128	91	Projectile-3	171	83
Projectile-3	172	93	Projectile-5	174	133
Projectile-5	174	130	Projectile-6	128	108
Projectile-6	128	103	Projectile-6	128	101
Projectile-6	128	110			

4.115.3 *Test results for layered shields*

Table 6#. Test results for Projectile-4 ($m = 29.4\,g$, $v_s = 172\,m/s$).

Shield	v_r (m/s)	Shield	v_r (m/s)	Shield	v_r (m/s)
2×[1.0]	128.5	3×[1.0]	82.0	4×[1.0]	No perforation

4.116 [Kpenyigba *et al.*, 2015]

See also [Kpenyigba *et al.*, 2013], [Rodríguez-Martínez *et al.*, 2013].

4.116.1 *General data on shields and projectiles*

Shields. Monolithic interstitial-free shields, $c_{l,w} = 100\,mm$, $T = 1\,mm$.

Projectiles. Hemispherical-nose projectiles (Fig. 3.4.1, $L_{cyl} = 23.5\,mm$, *Projectile-1*), cone-nose projectile (Fig. 3.2.1, $L_{cyl} = 26\,mm$, $L_{nose} = 8.95\,mm$, $\beta = 36°$, *Projectile-2*) and two projectiles having more complicated shapes, (Fig. 3.7.1c, *Projectile-3*) and (Fig. 3.7.1d, *Projectile-4)*. Projectile material is maraging steel subjected to heat treatment, $\sigma_y = 2\,GPa$; $D = 13\,mm$ and $m = 30\,g$ for all projectiles.

4.116.2 *Test results*

Table 1#. Test results for Projectile-1.

v_s (m/s)	v_r (m/s)	v_s (m/s)	v_r (m/s)	v_s (m/s)	v_r (m/s)
88	0	93	14	101	45
128	104	159	146	176	169

$v_{bl} = 90\,m/s; a = 1, p = 2.39.$

Table 2#. Test results for Projectile-2.

v_s (m/s)	v_r (m/s)	v_s (m/s)	v_r (m/s)	v_s (m/s)	v_r (m/s)
86.5	0	93	48	99	60
110	77	128	98	147	121
159	137	173	160	178	164

$v_{bl} = 86.5\,m/s; a = 1, p = 1.22.$

Table 3#. Test results for Projectile-3.

v_s (m/s)	v_r (m/s)	v_s (m/s)	v_r (m/s)	v_s (m/s)	v_r (m/s)	v_s (m/s)	v_r (m/s)
79	0	86	39	99	64	128	108
158	147	172	161				

$v_{bl} = 79\,m/s; a = 1, p = 2.28.$

Table 4#. Test results for Projectile-4.

v_s (m/s)	v_r (m/s)	v_s (m/s)	v_r (m/s)	v_s (m/s)	v_r (m/s)	v_s (m/s)	v_r (m/s)
78	0	88	39	92	51	99	65
126	108	158	146	172	161		

$v_{bl} = 78\,m/s; a = 1, p = 2.25.$

4.117 [Kymer and Fatzinger, 1957]

4.117.1 *General data on shields and projectiles*

Shields. Monolithic shields made from 2024-T4 AA (hereafter labeled as *Sh-1*) and 7075-T6 AA (hereafter labeled as *Sh-2*).

Projectiles. 20 mm AP T33, 20 mm AP M95, and 0.50 caliber (12.7 mm) AP M2 projectiles.

4.117.2 *Test results for 20 mm AP T33 projectiles*

Table 1. Test results for shields with $T = 12.7\ mm$.

θ_s (deg)	$v_{bl}\ (m/s)$ Sh-1	Sh-2	θ_s (deg)	$v_{bl}\ (m/s)$ Sh-1	Sh-2	θ_s (deg)	$v_{bl}\ (m/s)$ Sh-1	Sh-2
60	442	343	65	506	341	70	581	419

Table 2. Test results for shields with $T = 19.1\ mm$.

θ_s (deg)	$v_{bl}\ (m/s)$ Sh-1	Sh-2	θ_s (deg)	$v_{bl}\ (m/s)$ Sh-1	Sh-2
45	457	424	60	597	536
65	689	588	70	837	677

Table 3. Test results for shields with $T = 25.4$ *mm*.

θ_s	v_{bl} (m/s)		θ_s	v_{bl} (m/s)		θ_s	v_{bl} (m/s)	
(deg)	Sh-1	Sh-2	(deg)	Sh-1	Sh-2	(deg)	Sh-1	Sh-2
0	342	314	30	424	392	45	567	547
60	832	732	65	942	829	70	1091	919

Table 4. Test results for shields with $T = 38.1$ *mm*.

θ_s	v_{bl} (m/s)		θ_s	v_{bl} (m/s)		θ_s	v_{bl} (m/s)	
(deg)	Sh-1	Sh-2	(deg)	Sh-1	Sh-2	(deg)	Sh-1	Sh-2
0	509	433	30	581	572	45	719	750
55	911	927	60	1084	1070			

4.117.3 *Test results for 20 mm AP M95 projectiles*

Table 5 Test results for shields with $T = 12.7$ *mm*.

θ_s	v_{bl} (m/s)		θ_s	v_{bl} (m/s)		θ_s	v_{bl} (m/s)	
(deg)	Sh-1	Sh-2	(deg)	Sh-1	Sh-2	(deg)	Sh-1	Sh-2
60	472	375	65	520	421	70	596	443

Table 6. Test results for shields with $T = 19.1$ *mm*.

θ_s	v_{bl} (m/s)		θ_s	v_{bl} (m/s)	
(deg)	Sh-1	Sh-2	(deg)	Sh-1	Sh-2
45	460	430	60	562	524
65	593	575	70	671	733

Table 7. Test results for shields with $T = 25.4$ *mm*.

θ_s	v_{bl} (m/s)		θ_s	v_{bl} (m/s)		θ_s	v_{bl} (m/s)	
(deg)	Sh-1	Sh-2	(deg)	Sh-1	Sh-2	(deg)	Sh-1	Sh-2
0	399	296	30	443	431	45	594	590
60	652	709	65	792	786	70	866	968

4.117.4 *Test results for Caliber 0.50 AP M2 projectiles*

Table 8. Test results for shields with $T = 3.17$ *mm*.

θ_s	v_{bl} (m/s)		θ_s	v_{bl} (m/s)		θ_s	v_{bl} (m/s)	
(deg)	Sh-1	Sh-2	(deg)	Sh-1	Sh-2	(deg)	Sh-1	Sh-2
60	305	273	70	344	332	80	480	436

Table 9. Test results for shields with $T = 6.35$ *mm*.

θ_s	v_{bl} (m/s)		θ_s	v_{bl} (m/s)		θ_s	v_{bl} (m/s)	
(deg)	Sh-1	Sh-2	(deg)	Sh-1	Sh-2	(deg)	Sh-1	Sh-2
60	450	405	70	556	523	80	826	808

Table 10. Test results for shields with $T = 12.7$ *mm*.

θ_s	v_{bl} (m/s)		θ_s	v_{bl} (m/s)		θ_s	v_{bl} (m/s)	
(deg)	Sh-1	Sh-2	(deg)	Sh-1	Sh-2	(deg)	Sh-1	Sh-2
45	459	468	55	553	619	60	680	677
70	960	875	80	>1280	>1356			

Table 11. Test results for shields with $T = 19.1$ *mm*.

θ_s	v_{bl} (m/s)		θ_s	v_{bl} (m/s)		θ_s	v_{bl} (m/s)	
(deg)	Sh-1	Sh-2	(deg)	Sh-1	Sh-2	(deg)	Sh-1	Sh-2
0	416	384	30	411	433	45	529	658
750	817	–	60	898	884			

Table 12. Test results for shields with $T = 25.4$ *mm*.

θ_s	v_{bl} (m/s)		θ_s	v_{bl} (m/s)		θ_s	v_{bl} (m/s)	
(deg)	Sh-1	Sh-2	(deg)	Sh-1	Sh-2	(deg)	Sh-1	Sh-2
0	509	517	30	506	555	45	696	771
55	916	1036	60	1103	1143			

Table 13. Test results for shields with $T = 38.1$ *mm*.

θ_s	v_{bl} (m/s)		θ_s	v_{bl} (m/s)	
(deg)	Sh-1	Sh-2	(deg)	Sh-1	Sh-2
0	619	632	30	678	727
45	939	975	55	1234	>1277

4.118　[Lambert, 1978]

See also [Anderson *et al.*, 1992].

4.118.1　*General data on shields and projectiles*

Shields. Monolithic RHAs. More details are given in Table 2.
Projectiles. Steel hemispherical-nose projectiles (Fig. 3.4.1). More details are presented in Tables 1 and 2.

Table 1. Properties of projectile materials.

Series No.	Material	HRC	Remark
1-16,19,21,22	AISI-S7 tool steel	55	–
17	Steel	55	–
18,20	Steel	55	Stem
18,20	Steel	42	Cap

Table 2. Test series characteristics.

Series No.	Projectile				Shield	
	m	D	L	θ_s	T	HB
	(g)	(mm)	(mm)	(deg)	(mm)	
1	64.8	13.0	64.8	0	19.1	364
2	64.9	13.0	64.8	0	25.4	340
3	64.8	13.0	64.8	0	31.8	321
4	64.8	13.0	64.8	0	38.1	321
5	64.8	13.0	64.8	0	50.8	286
6	64.8	13.0	64.8	60	25.4	340
7	63.7	10.2	101.6	0	19.1	364
8	63.8	10.2	101.6	0	25.4	340
9	63.7	10.2	101.6	0	38.1	321
10	63.7	10.2	101.6	0	51.0	288
11	64.0	10.2	101.6	60	25.4	340
12	64.7	8.1	161.9	0	25.4	340
13	64.7	8.1	161.9	0	38.1	321
14	64.6	8.1	161.9	0	50.8	291
15	64.5	8.1	161.9	60	25.4	340
16	64.2	10.2	101.6	45	25.5	364
17	64.1	10.2	101.6	0	38.3	331
18	64.4	9.7	96.8	0	38.2	331
19	64.3	10.2	101.6	60	25.4	358
20	64.4	9.7	96.8	60	25.3	372
21	64.7	8.1	161.9	0	19.1	364
22	64.7	8.1	161.9	0	31.8	321
23	63.7	10.2	101.6	0	31.8	321

Table 3. Symbols in tables below and their description.

Notation	Description
–	Not applicable
^	Applicable but unknown
(1)	Total yaw exceeds $2.5°$, round is not used in determining v_{bl}
(2)	Perforation but v_r is not available, round is not used in determining v_{bl}
(3)	Non-perforated, small bulge in rear surface of shield

Notation	Description
(4)	Non-perforated, large bulge in rear surface of shield.
(5)	Non-perforated, rear surface of shield is fractured.
(6)	Rod slightly bent at launch.
(7)	Perforated, penetrator is severely shattered.

4.118.2 *Test results*

Table 4. Test results for Series 1.

v_s (m/s)	v_r (m/s)	m_r (g)	Δm^{sh} (g)	Notes	v_s (m/s)	v_r (m/s)	m_r (g)	Δm^{sh} (g)	Notes
692	0	–	20	(3)	690	0	–	15	–
721	81	^	50	–	743	364	^	40	–
763	429	^	^	–	909	609	^	55	–
917	548	^	^	(1)	1053	730	^	^	(1)

Table 5. Test results for Series 2.

v_s (m/s)	v_r (m/s)	m_r (g)	Δm^{sh} (g)	Notes	v_s (m/s)	v_r (m/s)	m_r (g)	Δm^{sh} (g)	Notes
858	0	–	29	(3)	879	249	22.3	45	(1)
913	213	19.7	60	–	942	363	17.8	45	–
966	367	28.1	103	–	987	474	25.4	62	–
1022	525	^	69	–	1124	624	^	55	–
1272	797	^	124	–					

Table 6. Test results for Series 3.

v_s (m/s)	v_r (m/s)	m_r (g)	Δm^{sh} (g)	Notes	v_s (m/s)	v_r (m/s)	m_r (g)	Δm^{sh} (g)	Notes
1111	0	–	^	(4)	1130	36	22.0	^	–
1137	65	^	^	–	1164	223	22.9	^	–
1185	314	18.8	65	(1)	1241	465	^	65	–
1304	661	^	46	–	1352	849	^	52	–

Table 7. Test results for Series 4.

v_s (m/s)	v_r (m/s)	m_r (g)	Δm^{sh} (g)	Notes	v_s (m/s)	v_r (m/s)	m_r (g)	Δm^{sh} (g)	Notes
1259	0	–	^	(4)	1283	0	–	^	(4), (5)
1306	199	^	^	–	1330	514	21.3	22	–

Table 8. Test results for Series 5.

v_s (m/s)	v_r (m/s)	m_r (g)	Δm^{sh} (g)	Notes	v_s (m/s)	v_r (m/s)	m_r (g)	Δm^{sh} (g)	Notes
1455	0	–	40	(1),(3)	1470	0	–	^	(1), (3)
1503	0	–	^	(4),(5	1512	0	–	^	(1), (4)
1526	0	–	^	(4)	1553	0	–	37	–
1555	0	–	18	(1)	1566	0	–	^	–
1567	0	–	^	(1),(4),(5)	1567	162	^	20	–
1585	0	–	^	(1),(5)	1591	^	^	^	(1), (2)
1595	0	–	0	(1)	1595	143	^	27	–
1598	0	–	19	(1),(4)	1611	298	^	^	(1)
1618	0	–	^	(1)	1620	0	–	60	–
1620	^	^	55	(2)	1629	^	^	45	(1), (2)
1630	44	^	24	(1)	1630	474	^	84	(1)
1632	478	^	45	(1)	1634	0	–	40	(1), (5)
1638	^	^	45	(1),(2)	1639	262	^	59	(1)
1642	0	–	4	(1),(4),(5)	1644	308	^	58	–
1646	128	^	40	(1)	1650	373	^	38	(1)
1652	517	^	^	(1)	1667	391	^	106	–
1679	532	^	40	(7)	1832	973	^	65	–
1941	1283	^	40	–					

Table 9. Test results for Series 6.

v_s (m/s)	v_r (m/s)	m_r (g)	Δm^{sh} (g)	Notes	v_s (m/s)	v_r (m/s)	m_r (g)	Δm^{sh} (g)	Notes
1349	0	–	40	(1)	1448	0	–	44	(1)
1451	0	–	70	–	1461	277	11.6	115	(1)
1462	197	^	119	–	1520	398	12.5	150	–
1546	568	^	165	(1)	1558	351	^	195	–
1565	0	–	88	(1)	1575	453	^	180	–
1576	470	^	140	–	1581	^	^	112	(1),(2)
1712	704	^	175	(1),(7)	1753	901	^	215	(1)
1841	948	^	235	–					

Table 10. Test results for Series 7.

v_s (m/s)	v_r (m/s)	m_r (g)	Δm^{sh} (g)	Notes	v_s (m/s)	v_r (m/s)	m_r (g)	Δm^{sh} (g)	Notes
625	0	–	7	–	671	0	–	11	–
738	0	–	17	–	751	0	–	21	(3)
783	0	–	5		786	0	–	^	(3)
789	0	–	10	–	792	0	–	10	–
800	0	–	^	(3)	801	379	30.8	^	–

v_s (m/s)	v_r (m/s)	m_r (g)	Δm^{sh} (g)	Notes	v_s (m/s)	v_r (m/s)	m_r (g)	Δm^{sh} (g)	Notes
811	0	–	17	–	828	493	^	46	–
852	467	^	42	–	1002	712	^	40	–
1361	1229	^	65	–					

Table 11. Test results for Series 8.

v_s (m/s)	v_r (m/s)	m_r (g)	Δm^{sh} (g)	Notes	v_s (m/s)	v_r (m/s)	m_r (g)	Δm^{sh} (g)	Notes
915	0	–	41	(3)	930	0	–	33	(1),(3)
937	0	–	14	(1),(3)	955	432	23.2	50	–
956	246	19.3	60	(1)	984	317	24.7	48	(1)
1053	594	25.8	50	–	1103	690	32.7	65	–
1137	753	31.6	67	(1)	1219	910	^	82	–

Table 12. Test results for Series 9.

v_s (m/s)	v_r (m/s)	m_r (g)	Δm^{sh} (g)	Notes	v_s (m/s)	v_r (m/s)	m_r (g)	Δm^{sh} (g)	Notes
1195	0	–	15	(4),(5)	1211	128	19.0	31	(1)
1237	313	12.1	30	–	1243	501	18.5	20	–
1366	907	20.5	26	–	1384	868	24.4	47	(1)

Table 13. Test results for Series 10.

v_s (m/s)	v_r (m/s)	m_r (g)	Δm^{sh} (g)	Notes	v_s (m/s)	v_r (m/s)	m_r (g)	Δm^{sh} (g)	Notes
1294	0	–	4	(1)	1312	0	–	17	(1)
1318	0	–	^	(1)	1374	0	–	^	(1)
1383	0	–	^	–	1402	0	–	^	(3)
1405	0	–	^	(1)	1410	^	^	^	(2)
1414	290	^	78	–	1416	0	–	^	(4),(5)
1418	0	–	5	(4),(5)	1422	255	^	33	–
1422	375	^	40	–	1439	^	^	30	(2)
1450	251	^	^	–	1454	628	^	49	–
1464	0	–	^	(2)	1528	0	–	^	(1)
1531	772	^	61	–	1765	1286	^	70	–

Table 14. Test results for Series 11.

v_s (m/s)	v_r (m/s)	m_r (g)	Δm^{sh} (g)	Notes	v_s (m/s)	v_r (m/s)	m_r (g)	Δm^{sh} (g)	Notes
1268	0	–	34	(4)	1294	0	–	23	(1)

v_s (m/s)	v_r (m/s)	m_r (g)	Δm^{sh} (g)	Notes	v_s (m/s)	v_r (m/s)	m_r (g)	Δm^{sh} (g)	Notes
1304	367	16.3	78	(4)	1307	0	–	41	(4)
1309	0	–	50	(3)	1310	0	–	55	(1)
1323	581	18.7	98	-	1329	∧	∧	73	(2)
1333	0	–	40	–	1339	0	–	42	(4)
1339	427	∧	80	–	1352	586	∧	98	(7)
1358	∧	∧	58	–	1361	151	∧	76	(1)
1361	212	∧	65	–	1365	0	–	∧	(1)
1366	∧	∧	85	–	1370	171	∧	155	–
1396	178	∧	70	–	1504	907	∧	111	-
1538	∧	∧	118	(1)					

Table 15. Test results for Series 12.

v_s (m/s)	v_r (m/s)	m_r (g)	Δm^{sh} (g)	Notes	v_s (m/s)	v_r (m/s)	m_r (g)	Δm^{sh} (g)	Notes
990	0	-	28	–	998	125	19.8	32	–
1011	428	26.7	28	–	1090	776	∧	40	–
1154	886	∧	58	(6)					

Table 16. Test results for Series 13.

v_s (m/s)	v_r (m/s)	m_r (g)	Δm^{sh} (g)	Notes	v_s (m/s)	v_r (m/s)	m_r (g)	Δm^{sh} (g)	Notes
1130	0	–	38	(3)	1148	0	–	∧	(4)
1160	338	16.3	20	–	1234	834	29.5	17	–
1327	1025	34.2	42	–					

Table 17. Test results for Series 14.

v_s (m/s)	v_r (m/s)	m_r (g)	Δm^{sh} (g)	Notes	v_s (m/s)	v_r (m/s)	m_r (g)	Δm^{sh} (g)	Notes
1270	0	–	∧	–	1296	0	–	∧	(3)
1296	0	–	20	(4),(5)	1297	498	∧	40	–
1315	215	∧	28	(2)	1316	∧	∧	60	–
1318	383	∧	10	–	1322	∧	∧	23	–
1326	234	∧	40	–	1327	502	∧	5	–
1328	378	∧	10	–	1329	∧	∧	0	–
1331	∧	∧	21	–	1347	796	∧	10	–
1361	∧	∧	0	–	1364	842	21.0	50	–
1400	948	∧	43	–	1500	∧	∧	30	–
1566	1183	∧	∧	–					

Table 18. Test results for Series 15.

v_s (m/s)	v_r (m/s)	m_r (g)	Δm^{sh} (g)	v_s (m/s)	v_r (m/s)	m_r (g)	Δm^{sh} (g)
1227	0	–	38	1234	0	–	26
1242	707	22.3	46	1260	735	21.3	48
1261	0	–	32	1263	564	^	69
1271	725	18.8	61				

Table 19. Test results for Series 16.

v_s (m/s)	v_r (m/s)	m_r (g)	Δm^{sh} (g)	v_s (m/s)	v_r (m/s)	m_r (g)	Δm^{sh} (g)
1061	0	–	29	1096	199	8.1	51
1138	536	11.8	0	1187	744	16.7	94
1205	815	19.1	99	1213	857	17.7	69
1431	1166	30.3	120	1433	1161	29.5	113
1542	1280	30.0	129	1828	1575	32.2	206

Table 20. Test results for Series 17.

v_s (m/s)	v_r (m/s)	m_r (g)	Δm^{sh} (g)	v_s (m/s)	v_r (m/s)	m_r (g)	Δm^{sh} (g)
1196	0	–	53	1217	139	9.9	26
1244	507	12.5	55	1246	530	12.1	^
1468	1200	22.0	86	1809	1548	31.0	125

Table 21. Test results for Series 18.

v_s (m/s)	v_r (m/s)	m_r (g)	Δm^{sh} (g)	v_s (m/s)	v_r (m/s)	m_r (g)	Δm^{sh} (g)
1095	0	–	^	1097	54	8.6	1
1097	^	8.1	9	1125	478	11.6	17
1185	868	19.8	37	1257	1033	25.3	45
1258	1045	23.1	34	1269	1047	29.6	54
1377	1194	29.2	70	1627	1485	31.7	107
1789	1586	32.4	111				

Table 22. Test results for Series 19.

v_s (m/s)	v_r (m/s)	m_r (g)	Δm^{sh} (g)	v_s (m/s)	v_r (m/s)	m_r (g)	Δm^{sh} (g)
1201	0	–	14	1227	204	8.6	68
1282	509	13.6	97	1360	768	20.7	96
1471	997	18.5	98	1489	903	19.6	118
1647	1145	16.8	156	1800	1285	13.6	171

Table 23. Test results for Series 20.

v_s (m/s)	v_r (m/s)	m_r (g)	Δm^{sh} (g)	v_s (m/s)	v_r (m/s)	m_r (g)	Δm^{sh} (g)
1201	0	–	59	1331	898	21.4	91
1758	1399	22.9	176				

Table 24. Test results for Series 21.

v_s (m/s)	v_r (m/s)	m_r (g)	Δm^{sh} (g)	Notes	v_s (m/s)	v_r (m/s)	m_r (g)	Δm^{sh} (g)	Notes
856	0	–	^	(4)	865	0	–	126	–
882	445	^	34	(6)	886	555	^	^	–

Table 25. Test results for Series 22.

v_s (m/s)	v_r (m/s)	m_r (g)	Δm^{sh} (g)	Notes	v_s (m/s)	v_r (m/s)	m_r (g)	Δm^{sh} (g)	Notes
1022	0	–	20	(4)	1025	0	–	^	(6)
1038	0	–	^	(4)	1045	0	–	^	–
1061	503	25.0	^	–	1081	^	24.0	35	(1),(2)
1087	597	^	^	–					

Table 26. Test results for Series 23.

v_s (m/s)	v_r (m/s)	m_r (g)	Δm^{sh} (g)	Notes	v_s (m/s)	v_r (m/s)	m_r (g)	Δm^{sh} (g)	Notes
1146	566	26.9	^	–	1185	684	^	25	(1)
1198	623	20.8	^	(1)	1201	638	^	35	(1)
1214	662	^	^	(1)					

4.118.3 *Approximations*

Table 27. Parameters of approximation.

Series No.	v_{bl} (m/s)	a	p	Series No.	v_{bl} (m/s)	a	p
1	720	0.94	2.9	2	906	0.70	3.3
3	1130	1.00	2.0	4	1304	1.00	2.9
5	1568	1.00	1.8	6	1451	0.80	2.0
7	798	1.00	2.7	8	916	0.83	3.2
9	1234	0.93	3.8	10	1411	1.00	2.2
11	1296	1.00	1.9	12	997	1.00	3.3
13	1157	1.00	3.7	14	1296	1.00	2.8
15	1234	1.00	3.1	16	1095	0.94	3.4

Series No.	v_{bl} (m/s)	a	p	Series No.	v_{bl} (m/s)	a	p
17	1216	1.00	3.0	18	1097	1.00	3.4
19	1225	0.80	3.3	20	1229	0.82	5.4
21	879	1.00	5.8	22	1045	0.90	3.7

4.119 [Landgrov and Sarkisyan, 1984]

4.119.1 *General data on shields and projectiles*

Shields. Monolithic shields made from lead ($\gamma = 1135\,kg/m^3$, $v^{sound} = 1200\,m/s$, HM 5.8) and aluminum ($\gamma = 2700\,kg/m^3$, $v^{sound} = 5080\,m/s$, HM 36).

Projectiles. Steel spherical projectiles (Fig. 3.5.1).

4.119.2 *Test results for lead shields*

Table 1#. Test results for $D = 10.3$ *mm*.

T (mm)	v_{bl} (m/s)	T (mm)	v_{bl} (m/s)	T (mm)	v_{bl} (m/s)	T (mm)	v_{bl} (m/s)
0.9	71	2.4	117	3.4	167	4.5	188
6.0	211	7.6	241	10.9	311	11.5	326
13.0	369	13.9	389	19.0	501		

Table 2#. Test results for $D = 6.37$ *mm*.

T (mm)	v_{bl} (m/s)	T (mm)	v_{bl} (m/s)	T (mm)	v_{bl} (m/s)	T (mm)	v_{bl} (m/s)
1.9	75	2.3	87	4.5	132	5.1	137
6.0	145	7.6	178	10.0	205	11.2	220
13.0	252	16.4	299	19.3	331		

4.119.3 *Test results for aluminum shields*

Note. Projectile penetrated through a predrilled cylindrical hole in a shield. Tests were made for balls with $D = 6.37\,mm$ and shield with $T = 4.85\,mm$; $\hat{v}_{bl}$ and v_{bl} denote BLVs of the shield with the hole and without the hole, correspondingly.

Table 3#. Test results.

$\hat{D}/D$	$\hat{v}_{bl}/v_{bl}$	$\hat{D}/D$	$\hat{v}_{bl}/v_{bl}$	$\hat{D}/D$	$\hat{v}_{bl}/v_{bl}$
0	1	0.29	0.9	0.58	0.75
0.79	0.57	0.92	0.34	1	0

4.120   [Landkof and Goldsmith, 1985]

4.120.1   *General data on shields and projectiles*

Shields. Monolithic 2024-0 Al (HB 50) shields; $D^{sh} = 139.7\,mm$ $T = 3.175\,mm$.

Projectiles. Slightly spherically blunted cone-nose hard steel (HRC 60) projectiles (Fig. 3.2.2a, $\beta = 30°$). Other details regarding projectiles are given in Table 1 and in the caption of Table 2.

4.120.2   *Test results*

Table 1. Test results on penetration into non-predrilled shields.

m (g)	D (mm)	L (mm)	ρ_1 (mm)	v_s (m/s)	v_r (m/s)	Failure mode
29.5	12.45	38.1	0	154.1	124.0	Only petals
29.5	12.45	38.1	0	157.8	127.4	Only petals
27.8	12.32	35.55	1.27	153.7	118.3	Only petals
27.8	12.32	35.55	2.54	151.4	109.6	Petals and plug
29.6	12.32	35.55	3.81	151.3	112.7	Petals and plug
30.7	12.32	35.55	5.08	150.6	–	Plug and small cracks
28.7	12.32	33.02	6.35	151.6	114.5	Plug and small cracks

Table 2. Test results on penetration into predrilled shields by impactor with $m = 29.5\,g$, $D = 12.5\,mm$, $\rho = 0$ and $v_s = 150\,m/s$.

$\dfrac{\hat{D}}{D}$	h^{lip} (mm)	$\dfrac{\hat{D}}{D}$	h^{lip} (mm)	$\dfrac{\hat{D}}{D}$	h^{lip} (mm)	$\dfrac{\hat{D}}{D}$	h^{lip} (mm)
0	11.3	0.26	6.1	0.32	7.5	0.38	3.1
0.45	2.2	0.51	1.8	0.75	0.0		

4.121   [Liss and Goldsmith, 1984]

4.121.1   *General data on shields and projectiles*

Shields. Monolithic 2024-0 Al.
Projectiles. Cylindrical projectiles (Fig. 3.1.1).

Table 1. Parameters of projectiles.

Type	D (mm)	L (mm)	Material	HB	HRC	m (g)
Non-deformable	12.5	73	Steel	–	56-59	–
Deformable	12.3	38.1	2024-0 Al	47	–	12.6

4.121.2 *Test results*

Table 2. Test results for non-deformable projectiles.

Shield		Projectile			Plug			
T (mm)	HB	m (g)	v_s (m/s)	v_r (m/s)	m^{pl} (g)	T^{pl} (mm)	v^{pl} (m/s)	Notes*
3.2	49	39.8	55.3	0	–	3.0	–	N
3.2	49	39.8	66.7	0	–	3.0	–	N
3.2	49	39.8	86.1	0	–	3.0	–	N
3.2	49.5	39.7	86.8	0	–	3.0	–	N
3.2	50.5	39.7	92.9	0	–	3.0	–	N
3.2	49	39.7	93.0	0	–	3.0	–	N
3.2	42.5	40.2	105.6	–	0.85	2.9	–	P
3.2	44.5	39.7	110.4	68.1	0.86	2.9	70.1	P
3.2	43.2	39.7	121.8	–	0.90	2.8	92.9	P
3.2	43	39.8	132.2	102.8	0.86	2.8	109.8	P
3.2	46	34.4	284	263	0.9	2.7	–	P
6.4	48	39.8	96.8	0	–	5.9	–	N
6.4	46.5	39.8	107.2	0	–	5.8	–	N
6.4	48	39.8	109.5	0	–	5.9	–	N
6.4	48	39.8	119.8	0	–	5.8	–	N
6.4	49	39.8	132.6	0	–	5.7	–	N
6.4	48	39.8	142.4	–	1.5	5.8	21.2	N
6.4	48	39.8	155.6	105.0	1.6	5.9	111.5	P
6.4	53	40.0	175.1	135.9	1.7	5.9	151.6	P
6.4	50	33.4	342.9	300	1.6	5.1	–	P
6.4	54	34.4	356.6	304.8	1.7	4.9	–	P
6.4	50	33.4	582.1	532.0	0.7	–	–	P
6.4	54	34.4	594.3	490.7	1.2	4.1	–	P
9.8	45.5	39.8	150.5	0	–	8.5	–	N
9.8	44.9	40.0	185.4	121.3	2.5	8.5	137.0	P
9.8	123	33.4	180	75.3	2.8	9.3	–	Y
12.75	51	39.8	130.1	0	–	–	–	N
12.75	56	33.4	212.4	0	–	–	–	N
12.75	56	33.4	350.5	224	2.0	9.6	–	P
12.75	56	33.4	582.1	452	1.6	8.0	–	P

* P – Perforated, N – Not perforated.

Table 3. Test results for deformable projectiles; HB 47.5 of shield material.

T (mm)	v_s (m/s)	v_r (m/s)	m^{pl} (g)	T^{pl} (mm)	v^{pl} (m/s)	Notes*
6.4	228.4	0	–	5.9	–	N
6.4	240.3	0	–	5.8	–	N
6.4	241.1	0	–	5.8	–	N
3.2	240.3	153.6	9.1	2.8	186.6	P
12.7	240.1	0	–	12.0	–	N

* P – Perforated, N – Not perforated.

4.122 [Liss *et al.*, 1983]

4.122.1 *General data on shields and projectiles*

Shields. Monolithic 2024-T3 Al (HB 120) shields; $T = 6.35\,mm$.
Projectiles. Hard steel cylindrical projectiles (Fig. 3.1.1); $D = 7.82\,mm$, $m = 7.3\,g$.

4.122.2 *Test results*

Table 1#. Test results.

v_s (m/s)	v_r (m/s)	v^{pl} (m/s)	v_s (m/s)	v^{pl} (m/s)	v_s (m/s)	v^{pl} (m/s)	v_s (m/s)
237	66	321	181	335	156	408	225
460	145	514	381	531	407	560	434
561	419	572	446	590	466	679	506
881	713						

4.123 [Liu and Stronge, 2000]

4.123.1 *General data on shields and projectiles*

Shields. Monolithic shields.

Table 1. Shield characteristics.

Shield label	Material	T (mm)	γ (kg/m^3)	σ_y (MPa)	σ_u (MPa)	ε_u (%)	ε_f (%)
Sh-1	Mild steel	1.15	7800	175	288	0.28	1.36
Sh-2	Mild steel	2.98	7800	371	485	0.17	1.00
Sh-3	Mild steel	6.10	7800	365	481	0.24	0.98
Sh-4	2014A TB AA	3.25	2700	302	464	0.18	0.40

Shield label	Material	T (mm)	γ (kg/m^3)	σ_y (MPa)	σ_u (MPa)	ε_u (%)	ε_f (%)
Sh-5	2014A TB AA	6.27	2700	288	309	0.06	0.38
Sh-6	1200 Al	1.22	2700	106	112	0.03	1.05
Sh-7	1200 Al	3.03	2700	115	127	0.06	1.07
Sh-8	1200 Al	6.42	2700	114	125	0.04	1.20

Projectiles. Cylindrical projectiles (Fig. 3.1.1) made from hardened steel and ductile materials.

Table 2. Ductile projectiles materials.

Material	γ (kg/m^3)	σ_y (MPa)	Material	γ (kg/m^3)	σ_y (MPa)
1200 Al	2700	60	6063 TF	2700	150
6061 T6	2700	320	Teflon (PTFE)	2170	8

4.123.2 *Test results on hardened steel projectiles*

Table 3. Test results for mild steel shields and projectiles having $m = 23.6\,g$.

Shield	D^{sh} (mm)	T (mm)	D (mm)	v_{bl} (m/s)
Sh-1	95	1.15	12.5	81 ± 4
Sh-2	95	2.98	12.5	186 ± 4
Sh-3	95	6.10	12.5	287 ± 19
Sh-1	190	1.15	12.5	84 ± 2
Sh-2	190	2.98	12.5	181 ± 9
Sh-1	95	1.15	25.0	134 ± 2
Sh-2	95	2.98	25.0	314 ± 9

Table 4. Test results for 2014A TB AA shields with $D^{sh} = 95\,mm$ and projectiles having D=12.5 *mm* and $m = 23.6$ *g*.

Shield	v_{bl} (m/s)	Shield	v_{bl} (m/s)
Sh-4	122 ± 5	Sh-5	147 ± 3

Table 5. Test results for 1200 Al shields with $D^{sh} = 95\,mm$ and projectiles having $D = 12.5\,mm$.

m (g)	Shield	v_{bl} (m/s)	m (g)	Shield	v_{bl} (m/s)
9.7	Sh-6	56 ± 3	9.7	Sh-7	144.5 ± 3.5
9.7	Sh-8	218 ± 5	3.8	Sh-6	97 ± 3

m (g)	Shield	v_{bl} (m/s)	m (g)	Shield	v_{bl} (m/s)
3.8	Sh-7	229.5 ± 5.5	3.8	Sh-8	357.5 ± 5.5

4.123.3 *Test results on ductile projectiles*

Table 6. Test results.

Projectile material	Shield	v_{bl} (m/s)	Projectile material	Shield	v_{bl} (m/s)
Teflon (PTFE)	Sh-6	129 ± 3	Teflon (PTFE)	Sh-7	401.5 ± 1.5
1200 Al	Sh-7	354 ± 7	6063 TF	Sh-7	319 ± 7
6061 T6	Sh-7	247 ± 3			

4.124 [Loikkanen *et al.*, 2005]

4.124.1 *General data on shields and projectiles*

Shields. Monolithic 2024-T3 Al shields; $\widetilde{c}_{l,w} = 254\,mm$.
Projectiles. Chrome steel ($E = 207 GPa$, $\gamma = 7830\,kg/m^3$, $v = 0.3$, $\sigma_y = 470 MPa$) spherical projectiles (Fig. 3.5.1); $D = 12.7\,mm$.

4.124.2 *Test results*

Table 1#. Test results for shields with $T = 1.6\,mm$.

v_s (m/s)	v_r (m/s)	v_s (m/s)	v_r (m/s)	v_s (m/s)	v_r (m/s)	v_s (m/s)	v_r (m/s)
88	0	92	0	101	0	108	0
116	0	119	0	120	19	120	7
121	49	122	29	126	39	127	49
129	44	130	54	132	37	142	83
151	96	160	112	164	119	175	139
181	144	185	148	186	150	188	151
199	161	204	169	225	192	266	232
271	237	275	245	316	285		

Table 2#. Test results for shields with $T = 3.2\,mm$.

v_s (m/s)	v_r (m/s)	v_s (m/s)	v_r (m/s)	v_s (m/s)	v_r (m/s)	v_s (m/s)	v_r (m/s)
160	0	190	0	196	0	201	0
208	0	213	0	216	0	219	0

v_s (m/s)	v_r (m/s)	v_s (m/s)	v_r (m/s)	v_s (m/s)	v_r (m/s)	v_s (m/s)	v_r (m/s)
218	29	220	46	220	43	220	40
220	25	224	64	225	68	226	72
231	82	236	101	242	109	245	115
247	122	252	129	254	131	255	135
258	140	259	143	262	148	269	158
271	164	348	255				

Table 3#. Test results for shields with $T = 6.4\ mm$.

v_s (m/s)	v_r (m/s)	v_s (m/s)	v_r (m/s)	v_s (m/s)	v_r (m/s)	v_s (m/s)	v_r (m/s)
312	0	318	0	404	0	405	72
409	60	410	0	408	115	411	132
413	0	419	132	423	138	427	147
434	176	437	170	442	178	451	187
460	219	470	231	482	262	501	250
509	286	538	331	559	345	572	375

4.125   [Lou *et al.*, 2009]

4.125.1   *General data on shields and projectiles*

Shields. Monolithic NVE36 steel (after different surface treatment) shields, $c_{l,w} = 300\ mm$, $T = 12\ mm$. Basic properties of untreated material: $\sigma_y = 355\ MPa$, $\sigma_u = 490 - 630\ MPa$, $\delta = 21\%$.

Table 1. Shields materials.

Notation	Treatment conditions
Material-1	As received
Material-2	Boronized (untempered)
Material-3	Carburized + boronized (tempered at 150°C × 0.5 hour)
Material-4	Carburized (untempered)
Material-5	Carburized (tempered at 165°C × 2 hours)
Material-6	Carburized (tempered at 245°C × 2 hours)

Projectiles. 0.30 caliber AP M2 projectiles (Fig. 3.11.1, Table 3.11.1).

4.125.2 *Test results*

Table 2#. Test results for Material-1.

v_s (m/s)	v_r (m/s)	v_s (m/s)	v_r (m/s)	v_s (m/s)	v_r (m/s)	v_s (m/s)	v_r (m/s)
515	0	552	0	575	0	576	0
592	0	600	183	602	186	603	162
607	202	613	217	665	343	704	399
712	414	775	551	846	681		

$v_{bl} = 592\,m/s.$

Table 3#. Test results for Material-2.

v_s (m/s)	v_r (m/s)	v_s (m/s)	v_r (m/s)	v_s (m/s)	v_r (m/s)	v_s (m/s)	v_r (m/s)
558	52	573	185	587	237	611	310
712	477	724	512	743	539		

$v_{bl} = 557\,m/s.$

Table 4#. Test results for Material-3.

v_s (m/s)	v_r (m/s)	v_s (m/s)	v_r (m/s)	v_s (m/s)	v_r (m/s)	v_s (m/s)	v_r (m/s)
729	0	743	0	758	128	776	200
775	255	786	319	831	434		

$v_{bl} = 754\,m/s.$

Table 5#. Test results for Material-4.

v_s (m/s)	v_r (m/s)	v_s (m/s)	v_r (m/s)	v_s (m/s)	v_r (m/s)	v_s (m/s)	v_r (m/s)
728	0	755	0	759	166	771	0
780	0	784	165	852	400	889	468

$v_{bl} = 754\,m/s.$

Table 6#. Test results for Material-5.

v_s (m/s)	v_r (m/s)	v_s (m/s)	v_r (m/s)	v_s (m/s)	v_r (m/s)
699	0	711	0	716	99
732	136	742	0	749	206
791	351	824	321	887	516

$v_{bl} = 711\,m/s.$

Table 7#. Test results for Material-6.

v_s (m/s)	v_r (m/s)	v_s (m/s)	v_r (m/s)	v_s (m/s)	v_r (m/s)	v_s (m/s)	v_r (m/s)
761	0	766	0	779	106	783	124
785	93	818	248	874	326	927	492

$v_{bl} = 766\,m/s$.

4.126 [Ma and Liew, 2013]

4.126.1 *General data on shields and projectiles*

Shields. Monolithic mild steel A36 ($\sigma_y = 250\,MPa$ $\sigma_u = 525\,MPa$) and high strength steel XAR-450 ($E = 216\,GPa, \sigma_u = 1285\,MPa$, $\widetilde{\sigma}_y = 1068\,MPa$); $c_{l,w} = 200\,mm$.

Projectiles. 9 mm caliber round nose full metal jacket bullet (Fig. 3.12.1, $m = 8\,g$; hereafter, *Bullets*) and AISI4340 steel FSPs (Fig. 3.10.3, $E = 200\,GPa$, $\gamma = 7858\,kg/m^3$, $v = 0.3$, $\sigma_y = 968\,MPa$, $\sigma_u = 1091\,MPa$; hereafter, *FSPs*).

4.126.2 *Test results for bullets*

Table 1. Test results for XAR-450 steel shield with $T = 6\,mm$.

v_s (m/s)	P (mm)	v_s (m/s)	P (mm)	v_s (m/s)	P (mm)
425.5	0.20	425.5	0.19	429.2	0.14
434.8	0.18	434.8	0.21		

Test results for A36 steel shield with $T=8\,mm$: $v_s = 426\,m/s$, $P = 1.05\,mm$.

4.126.3 *Test results for the FSPs against shields with T=6 mm*

Table 2#. Test results for XAR-450 steel shields.

v_s (m/s)	P (mm)	v_s (m/s)	P (mm)	v_s (m/s)	P (mm)	v_s (m/s)	P (mm)
304.8	1.0	318.5	1.1	330.8	1.2	330.8	1.3
354.3	1.4	389.4	1.7	402.1	2.0	413.8	2.1
427.3	1.9	438.4	1.9	442.8	2.4	484.5	2.8
495.6	3.0	550.2	4.0	560.8	4.0	562.6	4.3
563.7	4.5						

Table 3#. Test results for A36 steel shields.

v_s (m/s)	P (mm)	v_s (m/s)	P (mm)	v_s (m/s)	P (mm)
290.5	2.0	315.7	2.3	351.7	3.2
397.0	3.4	412.6	4.3	439.0	3.9

4.127 [Manes *et al.*, 2014]

4.127.1 *General data on shields and projectiles*

Shields. Monolithic 6061-T6 Al shields, $c_l = 300\,mm$, $c_w = 600\,mm$.

Projectiles. 7.62 mm AP projectiles: *Projectile-1* (Fig. 3.11.2a-c, $m = 9.28\,g$) and *Projectile-2* (Fig. 3.11.3a-c, $m = 11.6\,g$, $m^{core} = 5.1\,g$).

4.127.2 *Test results*

Table 1. Test results for Projectile-1 and $v_s \approx 771\,m/s$.

T (mm)	P (mm)	v_r (m/s)	T (mm)	P (mm)	v_r (m/s)	T (mm)	P (mm)	v_r (m/s)
76.2	46.35	–	101.6	46.45	–	25.0	–	~500

Table 2. Test results for Projectile-2 and $v_s \approx 742\,m/s$.

T (mm)	P (mm)	v_r (m/s)	T (mm)	P (mm)	v_r (m/s)	T (mm)	P (mm)	v_r (m/s)
76.2	38.11	–	101.6	41.29	–	25.0	–	~200

4.128 [Marom and Bodner, 1979]

4.128.1 *General data on shields and projectiles*

Shields. Monolithic and layered (plates-in-contact and with air gaps) shields; $c_l = 40\,mm$, $\tilde{c}_w = 230\,mm$.

Projectiles. 0.22 caliber lead bullets having approximately spherical shape, $D = 5.56\,mm$, $m = 2.6\,g$.

Table 1. Material properties of layers.

Shield material	T (mm)	σ_y (MPa)	σ_u (MPa)	σ_{ss} (MPa)
Al 1100-H14	1.0	122	132	76
Al 1100-H16	2.0	137	162	82
Al 1100-H16	2.5	137	162	82
Al 1100-H16	3.0	137	162	82
Al 6061-T6	1.0	235	274	207
Al 6061-T6	1.7	235	274	207
Al 6061-T6	2.1	274	325	207
Al 6061-T6	3.6	255	309	207
Al 6061-T6	4.8	274	339	207

4.128.2 *Test results for shields composed of Al 1100-H14 layers*

Table 2. Test results on v_r.

Exp. No.	T_{sum} (mm)	Shield structure	v_s (m/s)	v_r (m/s)
1	1.0	1×[1.0]	387	369
2	2.0	2×[1.0]\13.0	374	344
3	3.0	3×[1.0]\13.0	374	326
4	4.0	4×[1.0]\13.0	377	318
5	5.0	5×[1.0]\13.0	373	284
6	6.0	6×[1.0]\13.0	380	249
7	7.0	7×[1.0]\13.0	376	236
8	8.0	8×[1.0]\13.0	377	207
9	9.0	9×[1.0]\13.0	375	139
10	10.0	10×[1.0]\13.0	376	0
11	2.0	2×[1.0]	385	356
12	3.0	3×[1.0]	362	297
13	4.0	4×[1.0]	369	267
14	5.0	5×[1.0]	367	190
15	6.0	6×[1.0]	359	115
16	7.0	7×[1.0]	366	0

Table 3. Test results on maximum deflection of layers (mm).

Exp. No.	Layer No.									
	1	2	3	4	5	6	7	8	9	10
1	0.7	–	–	–	–	–	–	–	–	–
2	0.7	0.8	–	–	–	–	–	–	–	–

Exp.	Layer No.									
No.	1	2	3	4	5	6	7	8	9	10
3	0.7	0.8	0.9	–	–	–	–	–	–	–
4	0.7	0.8	0.9	1.0	–	–	–	–	–	–
5	0.7	0.8	0.9	1.0	1.5	–	–	–	–	–
6	0.7	0.8	0.9	1.0	1.5	1.9	–	–	–	–
7	0.7	0.8	0.9	1.0	1.5	1.9	2.8	–	–	–
8	0.7	0.8	0.9	1.0	1.5	1.9	2.8	4.0	–	–
9	0.7	0.8	0.9	1.0	1.5	1.9	2.8	4.0	12.0	–
10	0.7	0.8	0.9	1.0	1.5	1.9	2.8	4.0	12.0	12.2
11	0.2	1.0	–	–	–	–	–	–	–	–
12	1.5	2.0	2.5	–	–	–	–	–	–	–
13	1.1	2.4	3.5	3.5	–	–	–	–	–	–
14	1.0	2.2	5.4	8.3	8.3	–	–	–	–	–
15	1.1	2.0	7.8	1.0	12.0	12.0	–	–	–	–
16	1.0	2.5	9.7	9.7	9.7	9.7	9.7	–	–	–

4.128.3 *Test results for shields composed of Al 1100-H16 layers*

Table 4. Test results on v_r and deflection of layers.

T_{sum} (mm)	Shield structure	v_s (m/s)	v_r (m/s)	Maximum deflection of layers (mm) Layer No.				
				1	2	3	4	5
2.0	1×[2.0]	385	346	1.4	–	–	–	–
4.0	2×[2.0]\13.0	383	280	1.4	3.4	–	–	–
6.0	3×[2.0]\13.0	370	153	1.4	3.4	8.9	–	–
8.0	4×[2.0]\13.0	375	0	1.4	3.4	8.9	7.9	–
4.0	2×[2.0]	369	256	1.1	3.6	–	–	–
6.0	3×[2.0]	377	45	1.5	9.2	11	–	–
2.5	1×[2.5]	381	341	2.1	–	–	–	–
5.0	2×[2.5]\13.0	369	225	2.1	4.6	–	–	–
7.5	3×[2.5]\13.0	376	0	2.1	4.6	13	–	–
5.0	2×[2.5]	379	241	1.4	5.5	–	–	–
7.5	3×[2.5]	375	0	1.2	8.7	9.0	–	–
3.0	1×[3.0]	397	332	1.2	–	–	–	–
6.0	2×[3.0]\13.0	371	120	1.2	6.0	–	–	–
9.0	3×[3.0]\13.0	375	0	1.2	6.0	11	–	–
6.0	2×[3.0]	365	110	1.5	7.0	–	–	–
9.0	3×[3.0]	370	0	1.5	5.0	5.0	–	–

4.128.4 *Test results for shields composed of Al 6061-T6 layers*

Table 5. Test results on v_r.

Exp. No.	T_{sum} (mm)	Shield structure	v_s (m/s)	v_r (m/s)
1	1.0	1×[1.0]	379	357
2	2.0	2×[1.0]\13.0	359	321
3	3.0	3×[1.0]\13.0	357	302
4	4.0	4×[1.0]\13.0	368	294
5	5.0	5×[1.0]\13.0	357	258
6	6.0	6×[1.0]\13.0	363	234
7	7.0	7×[1.0]\13.0	368	201
8	8.0	8×[1.0]\13.0	360	115
9	2.0	2×[1.0]	367	311
10	3.0	3×[1.0]	355	230
11	4.0	4×[1.0]	362	146
12	5.0	5×[1.0]	369	0
13	1.7	1×[1.7]	378	332
14	3.4	2×[1.7]\16.0	356	244
15	5.1	3×[1.7]\16.0	371	146
16	3.4	2×[1.7]	370	0
17	2.1	1×[2.1]	357	293
18	4.2	2×[2.1]\13.0	367	200
19	6.3	3×[2.1]\13.0	352	0
20	4.2	2×[2.1]	360	0
21	3.6	1×[3.6]	365	249
22	7.2	2×[3.6]\13.0	370	0
23	7.2	2×[3.6]	368	0

Table 6. Test results on maximum deflection of layers (mm).

Exp. No.	Layer No. 1	2	3	4	5	6	7	8	9
1	1.3	–	–	–	–	–	–	–	–
2	1.3	1.6	–	–	–	–	–	–	–
3	1.3	1.6	1.8	–	–	–	–	–	–
4	1.3	1.6	1.8	1.8	–	–	–	–	–
5	1.3	1.6	1.8	1.8	2.4	–	–	–	–
6	1.3	1.6	1.8	1.8	2.4	2.8	–	–	–
7	1.3	1.6	1.8	1.8	2.4	2.8	7.0	–	–
8	1.3	1.6	1.8	1.3	2.4	2.8	7.0	16	–
9	3.1	4.4	–	–	–	–	–	–	–
10	5.5	5.6	7.0	–	–	–	–	–	–
11	3.9	6.9	8.5	9.1	–	–	–	–	–

Exp.	Layer No.								
No.	1	2	3	4	5	6	7	8	9
12	5.7	7.3	7.6	8.5	8.4	–	–	–	–
13	2.9	–	–	–	–	–	–	–	–
14	2.9	3.9	–	–	–	–	–	–	–
15	2.9	3.9	9.2	–	–	–	–	–	–
16	9.2	11	–	–	–	–	–	–	–
17	3.6	–	–	–	–	–	–	–	–
18	3.6	6.1	–	–	–	–	–	–	–
19	3.6	6.1	3.0	–	–	–	–	–	–
20	7.2	7.2	–	–	–	–	–	–	–
21	3.6	–	–	–	–	–	–	–	–
22	3.6	3.3	–	–	–	–	–	–	–
23	3.2	3.2	–	–	–	–	–	–	–

4.129 [Martineau *et al.*, 2004]

4.129.1 *General data on shields and projectiles*

Shields. Monolithic high-strength low-alloy (HSLA-100) steel ($E = 197\,GPa$, $\gamma = 7842\,kg/m^3$, $v = 0.29$, $G = 76.3GPa$, $\sigma_y = 730\,MPa$, $\varepsilon_f = 0.8$) shields; $\tilde{c}_l = 148\,mm$, $\tilde{c}_w = 457\,mm$, $T = 51\,mm$.

Projectiles. Tungsten carbide ($E = 683GPa$, $\gamma = 14900\,kg/m^3$, $G = 251GPa$, $\sigma_y = 3630\,MPa$, $\varepsilon_f = 0.015$) spherical projectiles (Fig. 3.5.1); $D = 6.4\,mm$.

4.129.2 *Test results*

Table 1. Test results on DOP.

v_s (m/s)	P (mm)	v_s (m/s)	P (mm)	v_s (m/s)	P (mm)	v_s (m/s)	P (mm)
830	4.57	970	5.59	980	5.46	1270	7.09
1280	6.86	1500	8.51	1810	8.79	1910	8.53
2150	8.41	2220	8.64	2460	9.50	2550	9.60

4.130 [Mileiko and Sarkisyan, 1981]

4.130.1 *General data on shields and projectiles*

Shields. Monolithic shields made from different materials (labels of materials correspond to SU standard classification).

Projectiles. Steel spherical projectiles (Fig. 3.5.1).

4.130.2   *Test results for AMg-6 AA shields*

4.130.2.1   *Results for spherical projectiles with* $D = 6.37\,mm$

Table 1#. Results for shields with $T = 1.97\,mm$.

$\bar{v}_s$	$\bar{v}_r$	$\bar{v}_s$	$\bar{v}_r$	$\bar{v}_s$	$\bar{v}_r$	$\bar{v}_s$	$\bar{v}_r$
1.09	0.36	1.10	0.31	1.32	0.75	1.43	0.89
1.47	0.92	1.52	0.93	1.65	1.01	1.64	1.11
1.80	1.34	1.82	1.23	2.00	1.68	2.30	1.87
2.89	2.51	3.00	2.60	3.21	2.60		

$v_{bl} = 250\,m/s$.

Table 2#. Results for shields with $T = 3.00\,mm$.

$\bar{v}_s$	$\bar{v}_r$	$\bar{v}_s$	$\bar{v}_r$	$\bar{v}_s$	$\bar{v}_r$	$\bar{v}_s$	$\bar{v}_r$
1.09	0.47	1.21	0.64	1.31	0.79	1.35	0.82
1.38	0.86	1.59	1.12	1.71	1.17	1.95	1.53
2.10	1.71	2.16	1.77				

$v_{bl} = 350\,m/s$.

Table 3#. Results for shields with $T = 4.00\,mm$.

$\bar{v}_s$	$\bar{v}_r$	$\bar{v}_s$	$\bar{v}_r$	$\bar{v}_s$	$\bar{v}_r$
1.01	0.30	1.07	0.23	1.12	0.53
1.16	0.53	1.29	0.72	1.65	1.11
1.85	1.37	1.89	1.45	2.05	1.57

$v_{bl} = 405\,m/s$.

4.130.2.2   *Results for spherical projectiles with* $D = 10.3\,mm$

Table 4#. Results for shields with $T = 3.37\,mm$.

$\bar{v}_s$	$\bar{v}_r$	$\bar{v}_s$	$\bar{v}_r$	$\bar{v}_s$	$\bar{v}_r$	$\bar{v}_s$	$\bar{v}_r$
1.22	0.55	1.39	0.85	1.44	0.84	1.65	1.08
1.89	1.29	2.01	1.48	2.09	1.53		

$v_{bl} = 228\,m/s$.

4.130.3   *Test results for VT-14 titanium alloy shields and spherical projectiles with* $D = 6.37\,mm$.

Table 5#. Results for shields with $T = 1.00\,mm$.

$\bar{v}_s$	$\bar{v}_r$	$\bar{v}_s$	$\bar{v}_r$	$\bar{v}_s$	$\bar{v}_r$	$\bar{v}_s$	$\bar{v}_r$
1.16	0.53	1.23	0.77	1.31	0.82	1.40	0.97
1.50	1.05	1.67	1.24	1.98	1.58	2.58	2.14

$v_{bl} = 240\,m/s$.

Table 6#. Results for shields with $T = 1.43$ *mm*.

$\overline{v}_s$	$\overline{v}_r$	$\overline{v}_s$	$\overline{v}_r$	$\overline{v}_s$	$\overline{v}_r$	$\overline{v}_s$	$\overline{v}_r$
1.09	0.48	1.15	0.59	1.21	0.68	1.26	0.68
1.28	0.76	1.34	0.90	1.38	0.89	1.81	1.45
2.02	1.64	2.12	1.76	2.32	1.95		

$v_{bl} = 335\,m/s$.

Table 7#. Results for shields with $T = 2.05$ *mm*.

$\overline{v}_s$	$\overline{v}_r$	$\overline{v}_s$	$\overline{v}_r$	$\overline{v}_s$	$\overline{v}_r$	$\overline{v}_s$	$\overline{v}_r$
1.04	0.10	1.05	0.26	1.08	0.43	1.47	0.96
1.54	1.12	1.60	1.14	1.75	1.34	1.83	1.37
1.88	1.45						

$v_{bl} = 423\,m/s$.

4.130.4 *Test results for copper shields with T = 1.50 mm.*

Table 8#. Results for spherical projectiles with $D = 6.37$ *mm*.

$\overline{v}_s$	$\overline{v}_r$	$\overline{v}_s$	$\overline{v}_r$	$\overline{v}_s$	$\overline{v}_r$	$\overline{v}_s$	$\overline{v}_r$
1.01	0.20	1.08	0.38	1.29	0.86	1.60	1.12
1.63	1.18	1.83	1.40	1.99	1.60	2.20	1.71
2.32	1.84						

$v_{bl} = 300\,m/s$.

Table 9#. Results for spherical projectiles with $D = 10.3$ *mm*.

$\overline{v}_s$	$\overline{v}_r$	$\overline{v}_s$	$\overline{v}_r$	$\overline{v}_s$	$\overline{v}_r$	$\overline{v}_s$	$\overline{v}_r$
1.35	0.82	1.43	1.06	1.87	1.46	2.04	1.62

$v_{bl} = 220\,m/s$.

4.131 [Mileiko *et al.*, 1979]

4.131.1 *General data on shields and projectiles*

Shields. Monolithic and predrilled shields made from different materials (labels of materials correspond to SU standard classification).
Projectiles. Steel spherical projectiles (Fig. 3.5.1).

4.131.2　*Test results for penetration into non-predrilled shields*

Table 1#. Test results for D16T AA shields and spherical projectiles with $D = 6.35$ *mm*.

T (mm)	v_{bl} (m/s)	T (mm)	v_{bl} (m/s)	T (mm)	v_{bl} (m/s)
1.8	238	3.8	404	4.8	492
7.8	686	9.9	801		

Table 2#. Test results for shields made from different materials and spherical projectiles with $D = 6.35$ *mm*.

Shield material	A (kg/m^2)	v_{bl} (m/s)	Shield material	A (kg/m^2)	v_{bl} (m/s)
VT14 titanium alloy	4.6	238	VT14 titanium alloy	13.6	504
VT14 titanium alloy	27.5	831	VT1-0 titanium alloy	13.9	477
OT4-1 titanium alloy	14.4	509	MA-8 magnesium alloy	14.7	486
01420 aluminum alloy	8.0	340			

Table 3#. Test results for titanium alloy shields and spherical projectiles with $D = 10.3$ *mm*.

Shield material	A (kg/m^2)	v_{bl} (m/s)	Shield material	A (kg/m^2)	v_{bl} (m/s)
VT14 titanium alloy	13.6	353	OT4-1 titanium alloy	14.4	385

Table 4. Test results on BLV for two materials.

Shield material	σ_f (MPa)	HRC	δ (%)	T (mm)	D (mm)	v_{bl} (m/s)
D16T aluminum alloy	–	–	–	4.85	6.35	495
KhNM steel	186	49	9.5	6.35	10.3	660

4.131.3　*Test results for penetration into a center of a predrilled cylindrical hole in D16T AA shield*

Note. $\hat{v}_{bl}$ and $v_{bl} = 495 m/s$ are BLVs of the shield with the hole and without the hole, correspondingly.

Table 5#. Test results for balls with $D = 6.35$ *mm* and shields with $T = 4.85$ *mm*.

$\hat{D}/D$	$\hat{v}_{bl}/v_{bl}$	$\hat{D}/D$	$\hat{v}_{bl}/v_{bl}$	$\hat{D}/D$	$\hat{v}_{bl}/v_{bl}$	$\hat{D}/D$	$\hat{v}_{bl}/v_{bl}$
0	1.00	0.32	1.00	0.47	0.95	0.63	0.89

4.132 [Mohotti *et al.*, 2015a]

See also [Mohotti *et al.*, 2015b].

4.132.1 *General data on shields and projectiles*

Shields. Two-layer (plates-in-contact) shields, $2\times[(AA5083–H116)\ 8.0]$; $c_{l,w}=200\,mm$ (data on shields having polyurea layers are not presented below).
Projectiles. NATO SS109 (metal jacket 5.56 mm × 45 mm) projectiles with $m = 62\ g$.
Impact velocity. $v_s = 945\,m/s$.

4.132.2 *Test results*

Table 1. Test results for the same impact velocity.

v_r (m/s)	v_r (m/s)	v_r (m/s)	v_r (m/s)	v_r (m/s)	v_r (m/s)
560	549	560	550	570	520

4.133 [Mullin *et al.*, 1991]

Data from [Ravid *et al.*, 1995].

4.133.1 *General data on shields and projectiles*

Shields. Layered (plates-in-contact) shields.

Table 1. Material properties of layers.

Material label	Material	T (mm)	γ (kg/m^3)	E (GPa)	ν	σ_y (MPa)	ε_f (%)
RHA	MIL-A-12560	9.53	7850	211	0.31	1245	11.8
MS	A36 Mild Steel	6.35	7800	206	0.31	447	32.5
Al	5083-H115	19.05	2750	710	0.32	264	51.1

Table 2. Cylindrical projectiles (Fig. 3.1.1).

D (mm)	L (mm)	Material	E (GPa)	ν	γ (kg/m^3)	σ_y (MPa)	HRC	ε_f (%)	m (g)
13	38	4340 Steel	211	0.31	7850	1764	51	10.5	39.7

4.133.2 *Test results*

Table 3. Test results for shields composed of plates in contact under zero pressure.

Shield structure	T_{sum} (mm)	v_s (m/s)	v_r (m/s)	m^{pl} (g)	v^{pl} (m/s)
[(RHA) 9.53]+[(AL) 19.05]	28.58	898	215	28.4	–
[(AL) 19.05]+ [(RHA) 9.53]	28.58	~914	0	16.7	165
[(AL) 19.05]+ [(RHA) 9.53]	28.58	924	0	22.8	–
[(MS) 6.35]+[(AL) 19.05]	25.40	581	0	32.1	–
[(MS) 6.35]+[(AL) 19.05]	25.40	642	0	16.5	–
[(MS) 6.35]+[(AL) 19.05]	25.40	709	133	31.6	–
[(AL) 19.05]+ [(MS) 6.35]	25.40	698	380	39.7	–

Table 4. Test results for shields consisting of plates in contact and clamped under pressure of about 133 MPa.

Shield structure	T_{sum} (mm)	v_s (m/s)	v_r (m/s)	m^{pl} (g)	v^{pl} (m/s)
[(RHA) 9.53]+[(AL) 19.05]	28.58	914	244	26.2	–
[(RHA) 9.53]+[(AL) 19.05]	28.58	900	232	28.7	–
[(AL) 19.05]+ [(RHA) 9.53]	28.58	912	190	27.7	–
[(AL) 19.05]+ [(RHA) 9.53]	28.58	894	0	34.1	–
[(MS) 6.35]+[(AL) 19.05]	25.40	713	145	29.0	171
[(MS) 6.35]+[(AL) 19.05]	25.40	824	368	28.1	364
[(AL) 19.05]+ [(MS) 6.35]	25.40	710	368	39.5	387

4.134 [Muzychenko and Postnov, 1985]

4.134.1 *General data on shields and projectiles*

Shields. Monolithic shields made from different materials.
Projectiles. ShKh15 steel spherical projectiles (Fig. 3.5.1).
Note. Labels of materials correspond to SU standard classification.

4.134.2 *Test results for VT-20 titanium alloy shields*

Table 1#. Test results for shields with $D = 6.37$ *mm*.

T (mm)	v_s (m/s)	v_r (m/s)	v_s (m/s)	v_r (m/s)	v_s (m/s)	v_r (m/s)	v_s (m/s)	v_r (m/s)
2.0	377	0	408	75	441	142	571	303
2.0	614	358						

T (mm)	v_s (m/s)	v_r (m/s)	v_s (m/s)	v_r (m/s)	v_s (m/s)	v_r (m/s)	v_s (m/s)	v_r (m/s)
4.0	617	0	641	51	713	228	792	364
6.0	751	0	763	31	812	147		

Table 2#. Test results for projectiles with $D = 10.3$ *mm*.

T (mm)	v_s (m/s)	v_r (m/s)	v_s (m/s)	v_r (m/s)	v_s (m/s)	v_r (m/s)	v_s (m/s)	v_r (m/s)
2.0	269	0	332	138	418	225	474	270
4.0	427	0	478	124	641	342	699	406
6.0	548	0	569	57	638	203	692	295

Table 3# Test results for projectiles with $D = 16.0$ *mm*.

T (mm)	v_s (m/s)	v_r (m/s)	v_s (m/s)	v_r (m/s)	v_s (m/s)	v_r (m/s)	v_s (m/s)	v_r (m/s)
2.0	290	0	332	97	436	216	578	343
6.0	400	0	437	94	552	270	596	312
8.0	460	0	480	47	606	279	680	358

4.134.3 *Test results for OT-4-1 titanium alloy shields*

Table 4# Test results for projectiles with $D = 6.37$ *mm*.

T (mm)	v_s (m/s)	v_r (m/s)	v_s (m/s)	v_r (m/s)	v_s (m/s)	v_r (m/s)	v_s (m/s)	v_r (m/s)
2.0	401	0	433	115	468	191	639	422
4.0	597	0	624	103	659	204	726	337
4.0	778	400						
6.0	717	0	738	113	790	242	819	295
6.0	836	354						

Table 5#. Test results for projectiles with $D = 10.3$ *mm*.

T (mm)	v_s (m/s)	v_r (m/s)	v_s (m/s)	v_r (m/s)	v_s (m/s)	v_r (m/s)	v_s (m/s)	v_r (m/s)
2.0	250	0	279	81	335	185	437	296
2.0	500	349	568	403				
4.0	428	0	482	149	535	251	741	496
6.0	500	0	521	89	598	264	705	420

Table 6#. Test results for projectiles with $D = 16.0$ *mm*.

T (mm)	v_s (m/s)	v_r (m/s)	v_s (m/s)	v_r (m/s)	v_s (m/s)	v_r (m/s)	v_s (m/s)	v_r (m/s)
2.0	284	0	318	91	434	263	596	412

T (mm)	v_s (m/s)	v_r (m/s)	v_s (m/s)	v_r (m/s)	v_s (m/s)	v_r (m/s)	v_s (m/s)	v_r (m/s)
4.0	354	0	392	113	504	295	553	349
4.0	592	391						
10.0	479	0	501	89	596	294	623	341

4.134.4 *Test results for D16T AA shields*

Table 7#. Test results for projectiles with $D = 6.37$ *mm*.

T (mm)	v_s (m/s)	v_r (m/s)	v_s (m/s)	v_r (m/s)	v_s (m/s)	v_r (m/s)	v_s (m/s)	v_r (m/s)
2.0	263	0	280	78	338	232	453	358
2.0	531	424	744	585				
4.0	428	0	455	180	497	287	789	621
6.0	554	0	605	279	645	372	773	586
6.0	817	649						

Table 8#. Test results for projectiles with $D = 10.3$ *mm*.

T (mm)	v_s (m/s)	v_r (m/s)	v_s (m/s)	v_r (m/s)	v_s (m/s)	v_r (m/s)	v_s (m/s)	v_r (m/s)
2.8	226	125	311	247	395	343		
4.2	324	0	357	167	439	317	586	449
4.2	594	489	653	505				
7.4	466	0	505	194	583	376	799	640

Table 9#. Test results for projectiles with $D = 16.0$ *mm*.

T (mm)	v_s (m/s)	v_r (m/s)	v_s (m/s)	v_r (m/s)	v_s (m/s)	v_r (m/s)	v_s (m/s)	v_r (m/s)
2.0	208	0	227	81	293	201	396	310
6.0	282	0	322	169	358	242		
12.7	409	0	428	121	509	333	586	424

4.135 [Neilson, 1985]

4.135.1 *General data on shields and projectiles*

Shields. Monolithic mild steel shields; additional data are given in Table 1.

Projectiles. Rigid cylindrical projectiles (Fig. 3.1.1); additional data are given in Table 1.

4.135.2 *Test results*

Table 1. Parameters of shields, projectiles and test results.

Projectile			Shield		v_{bl}
D (mm)	L (mm)	m (kg)	T (mm)	σ_u (MPa)	(m/s)
32	370	4.3	1.0	320	12.5
32	370	4.3	1.0	320	9.5
63	820	19.8	15	470	80
63	820	19.8	25	470	102
85	220	9.72	6	473	79
85	220	9.72	12	473	99
43	110	1.22	6	473	98
40	100	1.0	1.0	420	43
40	100	1.0	3.0	420	83

4.136 [Nia and Hoseini, 2011]

4.136.1 *General data on shields and projectiles*

Shields. Monolithic and layered (plates in-contact and with air gaps) Al 1100 shields; $D^{sh} = 220\,mm$, $T_{sum} = 3\,mm$.

Projectiles. Silver steel (HRC 56) hemispherical-nose projectiles (Fig. 3.4.1); $D = 8.7\,mm$, $L = 27.55\,mm$, $m = 12.15\,g$.

4.136.2 *Test results*

Table 1. Test results for 1×[3.0] shield.

v_s (m/s)	v_r (m/s)	v_s (m/s)	v_r (m/s)	v_s (m/s)	v_r (m/s)
124	0	127	7	129	26
133	47	139	67		

$v_{bl} = 125\,m/s$.

Table 2. Test results for [0.5]+[1.0]+[1.5] shield.

v_s (m/s)	v_r (m/s)	v_s (m/s)	v_r (m/s)	v_s (m/s)	v_r (m/s)
114	0	119	22	127	51
133	62	139	85		

$v_{bl} = 144\,m/s$.

Table 3. Test results for [0.5]+[1.5]+[1.0] shield.

v_s (m/s)	v_r (m/s)	v_s (m/s)	v_r (m/s)	v_s (m/s)	v_r (m/s)
112	0	119	22	122	36
127	52	133	75	139	85

$v_{bl} = 112\,m/s$.

Table 4. Test results for [1.0]+[0.5]+[1.5] shield.

v_s (m/s)	v_r (m/s)	v_s (m/s)	v_r (m/s)	v_s (m/s)	v_r (m/s)
114	0	119	19	122	35
127	49	133	62	139	76

$v_{bl} = 115\,m/s$.

Table 5. Test results for [1.0]+[1.5]+[0.5] shield.

v_s (m/s)	v_r (m/s)	v_s (m/s)	v_r (m/s)	v_s (m/s)	v_r (m/s)
114	0	119	18	127	46
133	74				

$v_{bl} = 115.5\,m/s$.

Table 6. Test results for [1.5]+[0.5]+[1.0] shield.

v_s (m/s)	v_r (m/s)	v_s (m/s)	v_r (m/s)	v_s (m/s)	v_r (m/s)
119	0	122	26	127	39
133	57	139	71		

$v_{bl} = 119\,m/s$.

Table 7. Test results for [1.5]+[1.0]+[0.5] shield.

v_s (m/s)	v_r (m/s)	v_s (m/s)	v_r (m/s)	v_s (m/s)	v_r (m/s)
119	0	122	24	127	41
133	61	139	69		

$v_{bl} = 120.5\,m/s$.

Table 8. Test results for [1.0]\1.0\[0.5]\5.0\[1.5] shield.

v_s (m/s)	v_r (m/s)	v_s (m/s)	v_r (m/s)	v_s (m/s)	v_r (m/s)
108	0	114	9	119	30
122	40	127	62		

$v_{bl} = 107.5\,m/s$.

Table 9. Test results for [1.0]\2.0\[0.5]\5.0\[1.5] shield.

v_s (m/s)	v_r (m/s)	v_s (m/s)	v_r (m/s)	v_s (m/s)	v_r (m/s)
102	0	114	13	119	33
122	44	127	64		

$v_{bl} = 105 \, m/s$.

Table 10. Test results for [1.0]\3.0\[0.5]\3.0\[1.5] shield.

v_s (m/s)	v_r (m/s)	v_s (m/s)	v_r (m/s)	v_s (m/s)	v_r (m/s)
99	0	108	7	114	17
119	39	122	48	127	70

$v_{bl} = 103 \, m/s$.

Table 11. Test results for [1.0]\4.0\[0.5]\2.0\[1.5] shield.

v_s (m/s)	v_r (m/s)	v_s (m/s)	v_r (m/s)	v_s (m/s)	v_r (m/s)
102	0	108	9	114	28
119	42	122	53	127	76

$v_{bl} = 101.5 \, m/s$.

Table 12. Test results for [1.0]\5.0\[0.5]\1.0\[1.5] shield.

v_s (m/s)	v_r (m/s)	v_s (m/s)	v_r (m/s)	v_s (m/s)	v_r (m/s)
101	0	105	9	114	31
119	45	122	54	127	76

$v_{bl} = 100.5 \, m/s$.

4.137 [Nishiwaki, 1951]

4.137.1 *General data on shields and projectiles*

Shields. Monolithic aluminum (HB 40-49) shields.

Projectiles. Steel cone-nose projectiles (Fig. 3.2.1); $D = 6.5 \, mm$, $m = 5.5 \, g$.

4.137.2 *Test results*

Table 1#. Test results for shields with $T = 3.2$ *mm* and projectiles with $\beta = 30°$.

v_s (m/s)	v_r (m/s)	v_s (m/s)	v_r (m/s)	v_s (m/s)	v_r (m/s)
393	358	470	444	504	483
587	562	749	717	804	775
807	786	924	899	948	925

Table 2#. Test results for shields with $T = 3.2$ *mm* and projectiles with $\beta = 45°$.

v_s (m/s)	v_r (m/s)	v_s (m/s)	v_r (m/s)	v_s (m/s)	v_r (m/s)	v_s (m/s)	v_r (m/s)
331	282	422	373	508	469	567	515
632	607	648	608	743	729	813	782
873	849	934	916				

Table 3#. Test results for shields with $T = 4.9$ *mm* and projectiles with $\beta = 30°$.

v_s (m/s)	v_r (m/s)	v_s (m/s)	v_r (m/s)	v_s (m/s)	v_r (m/s)	v_s (m/s)	v_r (m/s)
309	231	421	334	470	407	583	535
631	584	674	644	724	667	743	682
752	614	783	755	849	804	894	798
949	878	959	886				

Table 4#. Test results for shields with $T = 4.9$ *mm* and projectiles with $\beta = 45°$.

v_s (m/s)	v_r (m/s)	v_s (m/s)	v_r (m/s)	v_s (m/s)	v_r (m/s)	v_s (m/s)	v_r (m/s)
389	308	451	394	497	440	548	511
635	605	735	701	796	742	829	775
921	872	923	884				

4.138 [NPG Report 7-43, 1943]

4.138.1 *General data on shields and projectiles*

Shields. Monolithic armor shields.

Table 1. Properties of shield material.

Label	T (mm)	σ_u (MPa)	Label	T (mm)	σ_u (MPa)	Label	T (mm)	σ_u (MPa)
Sh-1	34.5	823	Sh-2	18.5	840	Sh-3	18.5	898

Projectiles. Projectiles with $D = 76.2\,mm$.

Table 2. Projectiles details.

Label	Description	Shape	m (kg)	Comments
FN-1	Cylindrical projectile	Fig. 3.1.1	3.4	FN supplied by FA
FN-2	Cylindrical projectile	Fig. 3.1.1	5.0	FN made by sawing a nose of M79 projectile
FN-3	Cylindrical projectiles	Fig. 3.1.1	6.8	FN supplied by FA
M79	M79 AP projectile	–	6.8	FA M79 AP projectile

Table 3. Special nomenclature in tables below.

Notation	Description
CP	Complete penetration
FA	Frankford Arsenal
FN	Flat-nose projectile
*	Incomplete penetration. Projectile is rejected
**	Projectile stuck in shield

4.138.2 *Test results*

Table 4. Ballistic data for M79 projectiles.

Shield	θ_s (deg)	v_s (m/s)	P (mm)	v_r (m/s)	Shield	θ_s (deg)	v_s (m/s)	P (mm)	v_r (m/s)
Sh-2/3	0	181	CP	15	Sh-2/3	0	185	CP	52
Sh-1	0	300	130.2*	–	Sh-1	0	307	CP	–
Sh-1	30	322	CP	91	Sh-1	30	344	CP	139
Sh-2/3	60	431	CP	316	Sh-2/3	60	402	*	–
Sh-2/3	60	416	*	–	Sh-2/3	60	431	CP	–

Table 5. Summary of ballistic data for M79 projectiles.

Shield	θ_s (deg)	v_{bl} (m/s)	Shield	θ_s (deg)	v_{bl} (m/s)
Sh-2/3	0	180	Sh-2/3	0	180
Sh-1	0	304	Sh-1	0	304
Sh-1	30	312	Sh-1	30	312
Sh-2/3	60	422	Sh-2/3	60	422

Table 6. Ballistic data for cylindrical projectiles.

Projectile	Shield	θ_s (deg)	v_s (m/s)	P (mm)	v_r (m/s)	m^{pl} (kg)	v^{pl} (m/s)
FN-2	Sh-2/3	0	200	6.4*	–	–	–
FN-2	Sh-2/3	0	207	CP	118	–	–
FN-2	Sh-2/3	0	237	CP	–	–	–
FN-1	Sh-1	0	282	*	–	–	–
FN-1	Sh-1	0	323	CP	43	1.32	135
FN-1	Sh-1	0	323	CP	-	1.26	138
FN-1	Sh-1	0	353	CP	-	1.27	169
FN-1	Sh-1	0	359	CP	69	1.50	155
FN-2	Sh-1	0	234	*	–	–	–
FN-2	Sh-1	0	249	CP	68	–	–
FN-2	Sh-1	0	290	CP	110	–	–
FN-2	Sh-1	0	311	CP	135	–	–
FN-2	Sh-1	0	340	CP	165	–	–
FN-3	Sh-1	0	175	25.4*	–	–	–
FN-3	Sh-1	0	194	CP	63	1.20	114
FN-3	Sh-1	0	237	CP	120	1.16	157
FN-3	Sh-1	0	294	CP	175	1.13	211
FN-2	Sh-1	30	280	**	–	–	–
FN-2	Sh-1	30	279	CP	–	–	–
FN-3	Sh-2/3	60	354	CP	–	–	–
FN-3	Sh-2/3	60	295	CP	–	–	–
FN-3	Sh-2/3	60	257	CP	~30	–	–
FN-3	Sh-2/3	60	243	CP	28	–	–
FN-3	Sh-2/3	60	222	CP	–	–	–
FN-3	Sh-2/3	60	195	**	–	0.96	86

Table 7. Summary of ballistic data for cylindrical projectiles.

Projectile	Shield	θ_s (deg)	v_{bl} (m/s)	Projectile	Shield	θ_s (deg)	v_{bl} (m/s)
FN-2	Sh-2/3	0	202	FN-1	Sh-1	0	304
FN-2	Sh-1	0	235	FN-3	Sh-1	0	176
FN-2	Sh-1	30	276	FN-2	Sh-2/3	60	196

4.139 [NPG Report 12-44, 1944]

4.139.1 *General data on shields and projectiles*

Shields. Monolithic steel shields and one spaced steel shield.

Table 1. Properties of shield materials.

Label	T (mm)	σ_u (MPa)	Label	T (mm)	σ_u (MPa)
Sh-1	15.2	848	Sh-2	38.1	841
Sh-3	38.1	876	Sh-4	50.8	876
Sh-5	50.8	800	Sh-6	61.0	876
Sh-7	76.2	827	Sh-8	81.3	855
Sh-9^	–	–			

^ 9.5 mm, 38.1 mm and 6.4 mm STS plates with 61 cm air gaps between layers.

Projectiles. HT WD4150 steel projectiles (HRC 55-60 and HRC 40 for body and base, respectively); $D = 76.2\,mm$, $m = 6.7\,kg$.

Table 2. Characteristics of projectiles.

Label	Description	Shape	Comments
FN	Cylindrical projectile	Fig. 3.1.1	No cap
FN-5w	Cylindrical projectile	Fig. 3.1.1	5% welded flat cap
FN-15w	Cylindrical projectiles	Fig. 3.1.1	15% welded flat cap
FN-5g	Cylindrical projectiles	Fig. 3.1.1	5% grooved flat cap
FN-15g	Cylindrical projectiles	Fig. 3.1.1	15% grooved flat cap
M79	M79 AP projectile	–	–

Note. Cap mass is expressed in percentages of the total projectile mass.

Table 3. Special nomenclature used in tables below.

Notation	Description
B	Projectile is broken into two or three pieces
BB	Projectile is bent
BS	Projectile is broken on secondary impact against butt structure
CP	Complete penetration
D	Projectile is deformed but not broken
E	Projectile is not deformed
FN	Flat-nose projectile
NC	Projectile nose is chipped
Nose offset	For M79 projectiles this is the distance at which the nose is displaced from the longitudinal axis of the projectile. For flat-nosed projectiles this is the maximum displacement of one side of the flat nose with respect of the other along the longitudinal axis of the projectile

Notation	Description
X	Projectile shattered
ΔD	Increase in diameter of the forward bourrelet of projectile in the result of impact
*	Incomplete penetration. Projectile is rejected
**	Projectile stuck in plate

4.139.2 *Test results*

Table 4. Ballistic data for M79 projectiles.

Shield	θ_s (deg))	v_s (m/s)	P (mm)	v_r (m/s)	ΔD (mm)	Nose offset (mm)	Projectile condition
Sh-1	0	152	203.2**	–	0	0	E
Sh-1	0	184	CP	–	0	0	E
Sh-1	45	175	50.8*	–	0	0	E
Sh-1	45	188	CP	53	0	0	E
Sh-1	45	191	CP	56	0	0	E
Sh-1	60	292	–*	–	0	0	BS
Sh-1	60	297	–*	–	0	0	E
Sh-1	60	303	–*	–	0	0	E
Sh-1	60	310	CP	–	0	0	E
Sh-2	0	305	76.2*	–	0	0	E
Sh-2	0	315	101.6*	–	0	0	E
Sh-2	0	320	CP	23	0	0	E
Sh-2	0	321	203.2**	–	–	–	E
Sh-2	30	323	152.4**	–	0	0	E
Sh-2	30	332	CP	69	0	0	E
Sh-2	45	408	38.1*	–	–	–	B
Sh-2	45	429	25.4*	–	–	–	B
Sh-2	45	449	–	–	–	–	B
Sh-2	55	594	CP	–	–	–	X
Sh-2	55	614	–	–	–	–	X
Sh-2	55	645	CP	–	–	–	X
Sh-4	0	386	CP	109	0	0	E
Sh-4	20	380	CP	47	0	0	E
Sh-4	30	397	CP	27	0.13	1.5	D
Sh-4	40	469	44.5*	–	0.74	12.7	D
Sh-4	40	485	44.5*	–	–	–	BS
Sh-4	40	504	CP	239	0.41	4.6	D
Sh-6	0	415	127.0*	–	0	0	E
Sh-6	0	429	CP	67	0	0	E
Sh-6	0	445	CP	–	0	0	E
Sh-7	0	484	CP	92	0	0	E

Table 5. Summary of ballistic data for M79 projectiles.

Shield	θ_s (deg)	v_{bl} (m/s)	Projectile condition	Shield	θ_s (deg)	v_{bl} (m/s)	Projectile condition
Sh-1	0	151	E	Sh-2/3	0	320	–
Sh-4/5	0	369	E	Sh-6	0	420	E
Sh-8	0	487	E	Sh-4/5	20	375	E
Sh-2/3	30	330	E	Sh-4/5	30	389	D
Sh-4/5	40	495	D	Sh-1	45	217	E
Sh-2/3	45	>446	X	Sh-1	60	304	E
Sh-2/3	60	>693	X	Sh-9	30	344	E

Table 6. Ballistic data for cylindrical projectiles without cap.

Shield	θ_s (deg))	v_s (m/s)	P (mm)	v_r (m/s)	ΔD (mm)	Nose offset (mm)	Projectile condition
Sh-1	0	155	12.7*	–	0.20	0	D
Sh-1	0	160	CP	37	0.20	0	D
Sh-1	45	148	CP	–	0.05	0	D
Sh-1	45	153	CP	–	0.08	0	D
Sh-1	45	173	CP	87	0.05	0	D
Sh-2	0	179	6.4*	–	0.51	0	D
Sh-2	0	189	9.5*	–	0.56	0	D
Sh-2	0	197	CP	–	1.02	0	D
Sh-2	0	199	CP	26	0.81	0	D
Sh-2	30	200	44.5*	–	0.18	0	D
Sh-2	30	210	25.4*	–	0.15	0	D
Sh-2	30	247	CP	56	0.30	0	D
Sh-2	45	317	177.8**	–	–	–	D
Sh-2	45	330	CP	–	0.25	1.5	D
Sh-2	60	415	–*	–	0.08–0.61	1.5	D
Sh-2	60	461	CP	–	0.23–1.50	3.3	D
Sh-2	60	519	CP	–	0.56–2.79	6.4	D
Sh-4	0	223	38.1*	–	1.60	0	D
Sh-4	0	245	CP	–	1.93	0	D
Sh-4	0	303	CP	115	3.63	0	D
Sh-4	20	264	25.4*	–	1.04	6.4	D
Sh-4	20	290	38.1*	–	1.17	6.4	D
Sh-4	20	304	50.8*	–	1.73	6.4	D
Sh-4	20	315	50.8*	–	–	–	–
Sh-4	20	325	69.9*	–	1.78	6.4	D
Sh-5	30	360	88.9**	–	1.78	6.4	D
Sh-5	30	382	CP	–	–	–	B
Sh-5	30	387	CP	–	2.16	6.4	D
Sh-5	30	437	CP	161	4.67	9.7	D

Shield	θ_s (deg))	v_s (m/s)	P (mm)	v_r (m/s)	ΔD (mm)	Nose offset (mm)	Projectile condition
Sh-5	40	419	57.2*	–	1.22	6.4	D
Sh-5	40	427	25.4*	–	–	–	X
Sh-5	40	482	–*	–	–	–	B
Sh-5	40	496	95.3**	–	–	–	D
Sh-6	0	276	9.3*	–	–	–	X
Sh-6	0	330	–*	–	–	–	X
Sh-6	0	347	76.2**	–	–	–	X
Sh-6	0	423	CP	–	9.14	–	D
Sh-8	0	328	9.53*	–	–	–	X
Sh-8	0	423	19.1*	–	–	–	X
Sh-8	0	500	158.8*	–	–	–	X
Sh-9	30	247	38.1*	–	–	–	–
Sh-9	30	287	CP	–	–	–	–

Table 7. Ballistic data for cylindrical projectiles with different caps.

Shield	Projectile	θ_s (deg))	v_s (m/s)	P (mm)	Projectile condition
Sh-2	FN-5w	30	264	19.1	NC
Sh-2	FN-5g	30	266	50.8	NC
Sh-2	FN-15w	30	275	19.1	NC
Sh-2	FN-15g	30	264	19.1	NC
Sh-4	FN-15g	20	331	19.1	BB
Sh-4	FN-15g	20	328	34.9	NC
Sh 8	FN-15w	0	426	9.5	X
Sh-8	FN-15g	0	424	25.4	X

Table 8. Summary of ballistic data for cylindrical projectiles.

Projectile	Shield	θ_s (deg)	v_{bl} (m/s)	Projectile condition
FN	Sh-1	0	154	D
FN	Sh-2/3	0	196	D
FN	Sh-4/5	0	228	D
FN	Sh-6	0	363	D
FN	Sh-8	0	>500	X
FN-15w	Sh-8	0	>426	X
FN-15g	Sh-8	0	>424	X
FN	Sh-4/5	20	322	D
FN-15w	Sh-4/5	20	>331	D
FN-15g	Sh-4/5	20	>328	D
FN	Sh-2/3	30	241	D
FN-5w	Sh-2/3	30	>264	NC

Projectile	Shield	θ_s (deg)	v_{bl} (m/s)	Projectile condition
FN-5g	Sh-2/3	30	>266	NC
FN-15w	Sh-2/3	30	>275	NC
FN-15g	Sh-2/3	30	>264	NC
FN	Sh-4/5	30	362	D
FN	Sh-4/5	40	501	D
FN	Sh-1	45	>146	D
FN	Sh-2/3	45	320	D
FN	Sh-2/3	60	430	D
FN	Sh-9	30	268	D

4.140 [NPG Report 16-43, 1943]

4.140.1 *General data on shields and projectiles*

Shields. Monolithic steel shields.

Table 1. Properties of shield materials.

Label	T (mm)	σ_u (MPa)	Label	T (mm)	σ_u (MPa)	Label	T (mm)	σ_u (MPa)	Label	T (mm)	σ_u (MPa)
Sh-1	18.5	898	Sh-2	49.3	848	Sh-3	81.0	827	Sh-4	82.6	–

Projectiles. Projectiles with $D = 76.2 mm$.

Table 2. Characteristics of projectiles.

Notation	m (kg)	Cap mass (%)	Shape
M79 AP	6.80	0	
CPr-1	6.58	6	
CPr-2 (hard cap, HB 485)	6.17	11	
CPr-3 (soft cap, HB 210)	6.17	11	
CPr-4	5.85	10	
CPr-5	6.31	12	

Notation	$\dfrac{m}{(\text{kg})}$	Cap mass (%)	Shape
CPr-6	6.65	16	
CPr-7	6.44	15	

Table 3. Special nomenclature in tables below.

Notation	Description
B2	Projectile is broken into two pieces
BD	Projectile base is dented
CP	Complete penetration
E	Projectile is not deformed
NB	Nose is shattered (less than half of the projectile is shattered)
X	Projectile is shattered (more than half of the projectile is shattered)
*	Incomplete penetration. Projectile is rejected
**	Projectile stuck in shield
^	Ricochet

4.140.2   *Test results*

Table 4. Ballistic data for M79 AP projectiles.

Shield	θ_s (deg)	v_s (m/s)	P (mm)	v_r (m/s)	Projectile condition
Sh-1	30	192	CP	14	E
Sh-1	60	402	25.4*	–	B2
Sh-1	60	416	19.1*	–	B2
Sh-1	60	431	CP	–	E
Sh-1	60	431	CP	316	E
Sh-2	0	377	CP	96	E
Sh-2	30	408	CP	49	E
Sh-3	0	512	139.7*	–	E

Table 5. Ballistic data on penetration of capped projectiles into Sh-1 shields; $\theta_s = 30°$ (projectiles are not deformed).

Projectile	v_s (m/s)	P (mm)	v_r (m/s)	Projectile	v_s (m/s)	P (mm)	v_r (m/s)
CPr-1	196	CP	–	CPr-2	190	44.5*	–
CPr-2	192	CP	–	CPr-3	197	3.2^	–

Projectile	v_s (m/s)	P (mm)	v_r (m/s)	Projectile	v_s (m/s)	P (mm)	v_r (m/s)
CPr-3	215	6.4^	–	CPr-5	198	44.5*	–
CPr-6	197	50.8*	–	CPr-6	198	CP	–
CPr-6	200	47.6*	–	CPr-7	204	CP	52
CPr-7	220	CP	83				

Table 6. Ballistic data on penetration of capped projectiles into Sh-1 shields; $\theta_s = 60°$.

Projectile	v_s (m/s)	P (mm)	v_r (m/s)	Projectile condition
CPr-1	412	44.5*	–	BD
CPr-2	440	CP	–	E
CPr-4	414	38.1*	–	BD
CPr-4	422	CP	–	E
CPr-5	380	155.6*	–	–
CPr-5	418	CP	–	E
CPr-6	420	CP	263	E
CPr-7	362	25.4*	–	BD
CPr-7	378	CP	148	E
CPr-7	379	CP	–	E
CPr-7	382	CP	–	E
CPr-7	397	CP	333	E
CPr-7	435	CP	–	E

Table 7. Ballistic data on penetration of capped projectiles into Sh-2 shields; $\theta_s = 0°$
(projectiles were not deformed).

Projectile	v_s (m/s)	P (mm)	v_r (m/s)	Projectile	v_s (m/s)	P (mm)	v_r (m/s)
CPr-1	373	76.2*	–	CPr-1	392	82.6*	–
CPr-2	417	76.2*	–	CPr-2	467	CP	191
CPr-3	444	CP	156	CPr-4	452	CP	164
CPr-5	445	CP	138	CPr-6	426	152.4**	–
CPr-7	440	CP	76				

Table 8. Ballistic data on penetration of capped projectiles into Sh-2 shields; $\theta_s = 30°$
(projectiles were not deformed).

Projectile	v_s (m/s)	P (mm)	v_r (m/s)	Projectile	v_s (m/s)	P (mm)	v_r (m/s)
CPr-1	455	CP	221	CPr-2	469	CP	187
CPr-3	442	69.9*	–	CPr-3	461	CP	97

Projectile	v_s (m/s)	P (mm)	v_r (m/s)	Projectile	v_s (m/s)	P (mm)	v_r (m/s)
CPr-4	467	CP	129	CPr-5	478	CP	121
CPr-6	477	CP	130	CPr-7	438	57.2*	–
CPr-7	469	CP	25	CPr-7	473	CP	90
CPr-7	489	CP	119	CPr-7	491	CP	142
CPr-7	542	CP	263				

Table 9. Ballistic data on penetration of capped projectiles into Sh-3 shields; $\theta_s = 0°$ (projectiles were not deformed).

Projectile	v_s (m/s)	P (mm)	v_r (m/s)	Projectile	v_s (m/s)	P (mm)	v_r (m/s)
CPr-1	525	108.0*	–	CPr-1	539	CP	80
CPr-2	590	CP	107	CPr-3	591	CP	105
CPr-5	583	CP	137	CPr-6	609	CP	213
CPr-7	596	CP	58				

Table 10. Ballistic data on penetration of capped projectiles into Sh-4 shields; $\theta_s = 20°$.

Projectile	v_s (m/s)	P (mm)	Projectile condition
CPr-1	624	CP	NB
CPr-2	625	–**	X
CPr-5	606	–*	X
CPr-5	626	CP	NB
CPr-6	621	CP	X
CPr-7	540	25.4*	NB
CPr-7	571	–*	X
CPr-7	586	CP	BD
CPr-7	623	CP	BD

4.141 [Nurick and Crowther, 1994]

4.141.1 *General data on shields and projectiles*

Shields. Monolithic cold-rolled mild steel shields; $T = 0.5mm$.

Projectiles. Steel (HRC 60) projectiles having $D = 12.6mm$ and $m = 35g$: cylindrical projectiles (Fig. 3.1.1; $L = 36.0mm$), hemispherical-nose projectiles (Fig. 3.4.1), and cone-nose projectiles (Fig. 3.2.1; $L_{cyl} = 32.35mm$, $L_{nose} = 10.9mm$, $\beta = 30°$).

Note. Data on length and diameter of projectiles are taken from [Nurick, 2013]

4.141.2 *Test results*

Table 1. Test results for cylindrical projectiles.

v_s (m/s)	v_r (m/s)	ω (deg)	v_s (m/s)	v_r (m/s)	ω (deg)	v_s (m/s)	v_r (m/s)	ω (deg)
82.1	62.7	11.6	99.2	83.0	8.9	70.9	47.9	21.0
81.4	60.2	17.5	90.0	79.4	11.5	74.2	52.0	19.8
83.7	70.1	8.7	59.9	45.1	16.0	72.5	60.2	14.3
89.4	84.0	6.8	100.2	94.0	7.9	70.9	–	14.4
88.5	86.5	13.4	83.8	–	15.2	99.2	–	11.3
113.8	–	7.0	103.4	–	3.4	98.1	–	6.5
81.4	88.5	16.4	84.1	81.1	2.2	108.2	111.7	8.2
113.8	–	6.9	57.5	–	18.0			

Table 2. Test results for hemispherical-nose projectiles.

v_s (m/s)	v_r (m/s)	ω (deg)	v_s (m/s)	v_r (m/s)	ω (deg)	v_s (m/s)	v_r (m/s)	ω (deg)
88.6	55.1	4.0	97.5	77.8	2.2	112.2	97.1	13.0
111.1	94.2	5.3	67.6	43.2	7.6	65.4	38.7	24.0
80.0	62.8	21.8	85.6	70.2	15.1	92.8	78.9	15.1
75.9	–	16.3	84.0	–	17.2			

Table 3. Test results for cone-nose projectiles.

v_s (m/s)	v_r (m/s)	ω (deg)	v_s (m/s)	v_r (m/s)	ω (deg)	v_s (m/s)	v_r (m/s)	ω (deg)
82.1	64.9	3.5	83.5	66.4	11.0	90.0	77.7	9.4
101.2	77.7	11.8	116.3	103.6	12.9	105.3	93.4	6.0
69.9	41.5	22.1	75.6	34.3	22.3	82.6	75.3	9.5
89.9	59.1	3.2	99.2	88.0	10.0	82.0	58.6	14.4
59.8	46.4	11.8	70.9	60.7	2.2	87.0	81.7	4.1
97.3	92.5	3.9	87.5	–	8.2	118.2	–	7.1
120.0	–	7.3	121.5	–	6.6	106.1	–	4.4
110.5	–	2.8	83.0	–	6.3	82.1	57.2	14.8
87.7	–	3.2	95.4	–	3.9			

4.142 [Nurick and Walters, 1990]

4.142.1 *General data on shields and projectiles*

Shields. Monolithic and layered (plates in-contact and with air gaps) shields consisting of cold-rolled mild steel plates.

Projectiles. High strength steel (HRC 60) projectiles with $D = 12.6\,mm$ and $m = 35\,g$: cylindrical projectiles (Fig. 3.1.1; $L = 36.0\,mm$) and two types of cone-nose projectiles (Fig. 3.2.1) with $L_{cyl} = 32.35\,mm$, $L_{nose} = 10.9\,mm$, $\beta = 30°$ and with $L_{cyl} = 34.75\,mm$, $L_{nose} = 3.65\,mm$, $\beta = 60°$, respectively.

Note. Data on length and diameter of projectiles are taken from [Nurick, 2013].

4.142.2 *Test results*

Table 1#. Test results (cone with $\beta = 90°$ is a cylinder).

T_{sum} (mm)	Shield structure	$\beta = 30°$ v_{bl} (m/s)	$\beta = 60°$ v_{bl} (m/s)	$\beta = 90°$ v_{bl} (m/s)
0.5	1×[0.5]	38	52	35
0.8	1×[0.8]	51	74	42
1.0	1×[1.0]	61	85	54
1.0	2×[0.5]	57	78	55
1.0	2×[0.5]\2.0	56	79	55
1.0	2×[0.5]\5.0	55	78	53
1.0	2×[0.5]\10.0	55	75	52
1.0	2×[0.5]\15.0	54	73	51
1.0	2×[0.5]\25.0	53	72	48

4.143 [O'Daniel *et al.*, 2011]

4.143.1 *General data on shields and projectiles*

Shields. Monolithic A36 grade steel ($\gamma = 7197\,kg/m^3$, $\sigma_y = 248\,MPa$, $\sigma_u = 310\,MPa$, $\varepsilon_f = 25\%$) shields.

Projectiles. 4340 steel ($\gamma = 7833\,kg/m^3$, $\sigma_y = 1482\,MPa$, $\sigma_u = 1579\,MPa$, $\varepsilon_f = 16\%$) 0.5 Caliber FSPs (Fig. 3.10.1, Table 3.10.1).

4.143.2 *Test results*

Table 1. Test results on v_r.

T (mm)	v_s (m/s)	v_r (m/s)	T (mm)	v_s (m/s)	v_r (m/s)	T (mm)	v_s (m/s)	v_r (m/s)
3.81	1111.9	755.6	6.35	1122.9	609.9	9.53	1123.5	346.3
12.7	1125.3	112.6						

4.144 [Ohte *et al.*, 1982]

4.144.1 *General data on shields and projectiles*

Shields. Monolithic SGV 49 carbon SA ($\sigma_y = 343\,MPa$, $\sigma_u = 490\,MPa$, $\delta = 40\%$, $\varepsilon_f = 17\%$, $A_{red} = 75\%$) shields; $\widetilde{c}_{l.w} = 1500\,mm$.

Projectiles. Austenitic stainless steel SUS 304 ($\sigma_y = 284\,MPa$, $\sigma_u = 637\,MPa$, $\delta = 80\%$, $A_{red} = 80\%$) cone-nose projectiles (Fig. 3.2.1).

Table 1. General data on test.

T (mm)	β (deg)	D (mm)	m (kg)	v_s (m/s)
7	20;45	87.5	3-18	25-170
10	45	66-160	3-50	50-180
20	10;15;20;45	66-160	3-50	50-150
30	20;45	66-160	4-50	6-150
38	20	99-160	24-45	93-142

Table 2. Shield damage types (SDT).

Notation	Description
P	Perforation (projectile penetrated through a shield)
F	Fracture (cracks or a small hole observed)
N	No failure (only plastic deformation, without crack)

4.144.2 *Test results*

Table 3#. Test results for shields with $T = 7$ *mm*.

β (deg)	w_s (kJ)	SDT	w_s (kJ)	SDT	w_s (kJ)	SDT	w_s (kJ)	SDT
20	6.2	F	6.2	F	6.9	F	13	F
20	24	P						
45	5.8	N	7.1	N	10	F	11	F
45	12	F	23	P	36	P		

Table 4#. Test results for shields with $T = 10$ *mm*.

β (deg)	w_s (kJ)	SDT	w_s (kJ)	SDT	w_s (kJ)	SDT	w_s (kJ)	SDT
45	12	N	37	N	42	F	51	P
45	101	P						

Table 5#. Test results for shields with $T = 20$ *mm*.

β (deg).	w_s (kJ)	SDT	w_s (kJ)	SDT	w_s (kJ)	SDT	w_s (kJ)	SDT
10	22	N	33	F				
15	22	N	31	N				
20	11	N	18	N	19	N	30	N
20	34	N	49	F	115	P	244	P
45	13	N	49	N	104	N	110	N
45	141	N	150	N	153	N	201	F
45	209	F	259	F	503	P		

Table 6#. Test results for shields with $T = 30$ *mm*.

β (deg).	w_s (kJ)	SDT	w_s (kJ)	SDT	w_s (kJ)	SDT	w_s (kJ)	SDT
20	14	N	28	N	44	N	80	N
20	103	N	154	F	251	F	336	P
20	518	P						
45	51	N	155	N	247	N	593	N

Table 7#. Test results for shields with $T = 38$ *mm*.

β (deg).	w_s (kJ)	SDT	w_s (kJ)	SDT	w_s (kJ)	SDT	w_s (kJ)	SDT
20	151	N	216	N	352	F	509	F

4.145 [Özşahin and Tolun, 2010]

4.145.1 *General data on shields and projectiles*

Shields. Monolithic coated (coating thickness 0.762 mm) and uncoated 6061 T651 Al shields; $c_{l,w} = 250\,mm$.

Table 1. Mechanical properties of uncoated 6061 T651 Al shield.

E (GPa)	ν	γ (kg/m^3)	σ_y (MPa)	σ_u (MPa)	σ_{ss} (MPa)	δ (%)
68.9	0.33	2700	276	310	207	17

Projectiles. Bullet from 9×19 mm FMJ Parabellum cartridge, $m = 8\,g$.

Note. DOP is measured with respect to the initial position of a frontal surface of a shield before penetration; consequently occasionally $P > T$.

4.145.2 *Test results for uncoated shields*

Table 2. Test results for shields with $T = 6.35$ *mm*.

v_s (m/s)	P (mm)	h^{def} (mm)	v_s (m/s)	P (mm)	h^{def} (mm)	v_s (m/s)	P (mm)	h^{def} (mm)
359	6.74	5.37	366	6.58	5.24	366	6.67	5.31
367	7.38	5.84	378	10.25	9.28	379	7.26	5.80
397	*	–						

* Perforation.

Table 3. Test results for shields with $T = 8.00$ *mm*.

v_s (m/s)	P (mm)	h^{def} (mm)	v_s (m/s)	P (mm)	h^{def} (mm)	v_s (m/s)	P (mm)	h^{def} (mm)
368	5.70	4.62	376	6.07	4.91	379	6.39	5.26
384	6.11	4.93	391	6.88	5.43	400	*	–
401	*	–						

* Perforation.

4.145.3 *Test results for shields coated by Metco 68F-NS-1 spray powder*

Table 4. Test results for shields with $T = 6.35$ *mm*.

v_s (m/s)	P (mm)	h^{def} (mm)	v_s (m/s)	P (mm)	h^{def} (mm)	v_s (m/s)	P (mm)	h^{def} (mm)
366	5.51	4.23	375	7.05	5.70	376	6.21	4.76
380	5.99	4.76	391	8.22	6.56	398	8.23	6.95
406	8.58	7.22						

Table 5. Test results for shields with $T = 8.00$ *mm*.

v_s (m/s)	P (mm)	h^{def} (mm)	v_s (m/s)	P (mm)	h^{def} (mm)	v_s (m/s)	P (mm)	h^{def} (mm)
365	3.52	2.99	374	4.42	3.70	381	4.41	3.66
393	5.91	4.34	399	5.53	4.67	402	5.69	4.77

4.145.4 *Test results for shields coated by Metco 201NS spray powder*

Table 6. Test results for shields with $T = 6.35$ *mm*.

v_s (m/s)	P (mm)	h^{def} (mm)	v_s (m/s)	P (mm)	h^{def} (mm)	v_s (m/s)	P (mm)	h^{def} (mm)
372	4.81	3.55	375	7.59	6.40	378	7.80	6.47
382	5.46	3.76	401	10.02	7.97	406	10.25	8.24

Table 7. Test results for shields with $T = 8.00$ *mm*.

v_s (m/s)	P (mm)	h^{def} (mm)	v_s (m/s)	P (mm)	h^{def} (mm)	v_s (m/s)	P (mm)	h^{def} (mm)
372	4.27	3.41	376	4.49	3.16	393	6.35	4.74
396	6.52	4.81	400	6.52	5.14	406	7.13	4.55

4.146 [Özşahin *et al.*, 2011]

4.146.1 *General data on shields and projectiles*

Shields. Monolithic 6061 T651 AA shields; $c_{l,w} = 250\,mm$.
Projectiles. FMJ Parabellum bullets with a brass cup and lead-antimony alloyed core projectile; $D = 9\,mm$, $m = 8 \pm 0.075\,g$.

4.146.2 *Test results*

Table 1#. Test results on DOP.

T (mm)	v_s (m/s)	P (mm)	v_s (m/s)	P (mm)	v_s (m/s)	P (mm)
6.35	359	6.7	367.0	7.3	378	10.2
8.00	368	5.7	376	6.1	379	6.4

4.147 [Partom *et al.*, 2001]

4.147.1 *General data on shields and projectiles*

Shields. Monolithic 6061-T6 Al ($\sigma_y = 200\,MPa$) shields.

Projectiles. Two types of 7.62 mm AP M2 projectiles: common "pointed" projectile and backward moving "blunt" projectile.

4.147.2 *Test results*

Table 1#. Test results for "blunt" projectiles and $T = 6.6\ mm$.

v_s (m/s)	v_r (m/s)	v_s (m/s)	v_r (m/s)	v_s (m/s)	v_r (m/s)	v_s (m/s)	v_r (m/s)
282	0	289	0	315	94	321	115
380	261	408	278	438	327	507	334
515	358	843	778	844	788		

Table 2#. Test results for "pointed" projectiles.

v_s (m/s)	P (mm)	v_s (m/s)	P (mm)	v_s (m/s)	P (mm)
617	30	707	38	770	43
825	49	860	51		

4.148 [Pereira and Lerch, 2001]

4.148.1 *General data on shields and projectiles*

Shields. Monolithic HT Inconel 718 ($\gamma = 8190\ kg/m^3$) shields after different heat treatments; $\tilde{c}_{l.w} = 152\ mm$.

Table 1 Mechanical properties of HT Inconel 718.

HT label	E (GPa)	v	$\tilde{\sigma}_y$ (MPa)	σ_u (MPa)	ε_f (%)	A_{red} (%)	HRB/HRC
A	198	0.29	331	821	67	46	84/–
B	198	0.28	483	924	55	43	96/–
C	205	0.26	1117	1338	20	30	114/45

Projectiles. Ti-6-4 cylindrical projectiles (Fig. 3.1.1); $D = 12.7\ mm$, $L = 25.4\ mm$.

4.148.2 *Test results*

Table 2. Test results on BLV.

HT	T (mm)	v_{bl} (m/s)	HT	T (mm)	v_{bl} (m/s)	HT	T (mm)	v_{bl} (m/s)
A	1.0	198	A	1.8	265	B	1.0	192
B	1.8	277	B	2.0	282	C	1.0	168
C	1.8	244						

4.149   [Pereira *et al.*, 2013]

See also [Pereira *et al.*, 2012], [Pereira *et al.*, 2014].

4.149.1   *General data on shields and projectiles*

Shields. Monolithic shields made from different materials; $\widetilde{c}_{l,w} = 254\,mm$.
Projectiles. Projectile shapes are close to cylindrical (Fig. 3.1.1) although nose bluntness is slightly different from plane. Projectiles have a relatively large nose radius of 70 mm, which allows slight deviation of 5° in the normal orientation of the projectile without a frontal edge impact. The edges of the frontal face are rounded by a circle having a 0.80 mm radius.

Table 1. Notations for materials.

Notation	Material
Al	Al-2024 T3/T351 (AMS 4037)
Ti	Ti-6Al-4V (AMS 4911)
St	A2 Tool Steel

Table 2. Description of series of experiments.

Series	Shield		Projectile				
	Material	T (mm)	Material	HRC	L (mm)	D (mm)	m (g)
Series 1	Al	3.2	Ti	36-37	17.8	12.7	9.0
Series 2	Al	6.5	Ti	36-37	22.9	12.7	12.8
Series 3	Al	12.8	St	59	28.6	12.7	28.0
Series 4	Ti	2.3	Ti	36-37	25.4	12.7	14.05
Series 5	Ti	3.4	Ti	36-37	38.1	12.7	21.28
Series 6	Ti	3.4	St	59	21.8	12.7	21.25
Series 7	Ti	6.5	St	59	22.2	12.7	21.56
Series 8	Ti	13.1	St	63	57.2	19.1	126.3

Table 3. Explanations to notes in tables.

Note label	Description	Note label	Description
(1)	Penetrated	(2)	Contained
(3)	Hole, projectile rebounded	(4)	Projectile lodged, plug ejected
(5)	Contained, plug almost ejected	(6)	Contained, no crack
(7)	Contained, petal	(8)	Contained, crack
(9)	Contained, flap	(10)	Penetrated, flap
(11)	Contained, plug		

4.149.2 *Test results*

Note. Rebound velocity is marked in tables by * when available.

Table 4. Test results for Series 1 of experiments.

v_s (m/s)	θ_s (deg)	v_r (m/s)	Note	v_s (m/s)	θ_s (deg)	v_r (m/s)	Note
225	1.2	12	(1)	220	2.1	54	(1)
215	1.7	21	(1)	205	3.9	0	(2)
238	1.9	117	(1)	196	3.1	0	(2)
191	1	0	(2)	205	1.1	0	(2)
207	3.2	0	(2)	280	1.7	212	(1)
269	1.3	196	(1)	298	6.1	245	(1)
262	1.7	186	(1)	203	3.1	0	(2)
202	2.4	0	(2)				

Table 5. Test results for Series 2 of experiments.

v_s (m/s)	θ_s (deg)	v_r (m/s)	Note	v_s (m/s)	θ_s (deg)	v_r (m/s)	Note
242	5.8	118	(1)	222	3.9	0	(2)
220	2.6	0	(2)	229	4.0	0	(3)
216	2.2	5	(1)	224	5.6	52	(1)
233	3.2	47	(1)	229	7.8	0	(4)
227	2.7	44	(4)	217	4.7	0	(2)
223	3.3	0	(2)	209	3.8	0	(2)
198	1.2	0	(2)	262	0.6	135	(1)
286	0.3	151	(1)				

Table 6. Test results for Series 3 of experiments.

v_s (m/s)	θ_s (deg)	v_r (m/s)	Note	v_s (m/s)	θ_s (deg)	v_r (m/s)	Note
240	1.3	0	(5)	254	1.0	0	(4)
267	1.7	44	(1)	229	1.4	0	(2)
250	0.4	65	(1)	293	9.4	122	(1)
276	2.2	58	(1)	259	0.8	35	(1)
257	0.9	74	(1)	231	1.8	57	(1)
219	1.6	0	(4)				

Table 7. Test results for Series 4 of experiments.

v_s (m/s)	θ_s (deg)	v_r (m/s)	Note	v_s (m/s)	θ_s (deg)	v_r (m/s)	Note
234	3.8	112	(1)	219	4.5	35*	(2)
195	4.4	19*	(6)	218	10.4	20*	(7)

v_s (m/s)	θ_s (deg)	v_r (m/s)	Note	v_s (m/s)	θ_s (deg)	v_r (m/s)	Note
233	6.0	117	(1)	263	2.3	192	(1)
215	4.1	23*	(7)	204	5.1	23*	(7)
215	0.9	17*	(2)	204	3.7	9*	(8)
237	2.4	115	(1)	194	0.1	16*	(2)

Table 8. Test results for Series 5 of experiments.

v_s (m/s)	θ_s (deg)	v_r (m/s)	Note	v_s (m/s)	θ_s (deg)	v_r (m/s)	Note
221	2.7	35*	(8)	212	1.5	19*	(6)
226	2.3	32*	(9)	239	8.0	91	(1)
275	3.7	174	(1)	226	0.7	21*	(2)
239	1.3	80	(1)	236	2.2	11	(1)
276	0.7	191	(1)	262	2.0	177	(1
235	1.6	71	(1)	231	1.4	5*	(2)

Table 9. Test results for Series 7 of experiments.

v_s (m/s)	θ_s (deg)	v_r (m/s)	Note	v_s (m/s)	θ_s (deg)	v_r (m/s)	Note
229	2.2	73	(1)	216	1.7	18*	(10)
218	6.8	12*	(11)	192	5.5	4*	(8)
212	3.0	5*	(9)	198	2.1	6*	(2)
248	1.7	102	(1)	260	1.5	113	(1)
278	5.0	121	(1)	189	–	17*	(2)
232	2.0	84	(1)	230	1.1	86	(1)

Table 10. Test results on BLV.

Series	v_{bl} (m/s)	Series	v_{bl} (m/s)	Series	v_{bl} (m/s)	Series	v_{bl} (m/s)
Series 1	213	Series 2	229	Series 3	244	Series 4	226
Series 5	235	Series 6	198	Series 7	224	Series 8	192

4.150 [Piccione, 1960]

4.150.1 *General data on shields and projectiles*

Shields. Monolithic RHA shields.
Projectiles. 20 mm AP M75 and 20 mm AP M95 projectiles.

4.150.2 *Test results*

Table 1. Test results.

T (mm)	HB	θ_s (deg)	$v_{bl}(m/s)$ 20 mm AP M75	20 mm AP M95
9.52	388	60	457	604
12.7	321	45	504	661
12.7	341	60	691	–
12.7	363	60	–	741
19.1	331–341	20	354	488
19.1	302	45	726	813
19.1	277	60	873	>933
25.4	331–341	0	443	642
25.4	277	20	522	–

4.151 [Piekutowski *et al.*, 1996]

4.151.1 *General data on shields and projectiles*

Shields. Monolithic 6061-T651 AA shields; $c_{l.w} = 304\,mm$, $T = 26.3\,mm$.
Projectiles. 4340 steel (HRC 44) ogive-nose projectiles (Fig. 3.3.1); $D = 12.9\,mm$, $L_{cyl} = 67.5\,mm$, $L_{nose} = 21.4\,mm$, $m \approx 81.2\,g$.

4.151.2 *Test results*

Table 1. Test results for normal impact, $\theta_s = 0°$.

v_s (m/s)	v_r (m/s)	v_s (m/s)	v_r (m/s)	v_s (m/s)	v_r (m/s)	v_s (m/s)	v_r (m/s)
282	0	308	57	341	164	396	266
454	347	508	415	565	482	630	555
633	561	730	665	863	802		

Table 2. Test results for oblique impact, $\theta_s = 30°$.

v_s (m/s)	v_r (m/s)	v_s (m/s)	v_r (m/s)	v_s (m/s)	v_r (m/s)	v_s (m/s)	v_r (m/s)
336	67	341	91	396	215	398	216
401	217	404	217	405	227	446	288
573	468	575	455	580	481	584	453
730	655						

4.152   [Piekutowski *et al.*, 1999]

Shields. Monolithic 6061-T6511 AA shields; $D^{sh} = 254\,mm$.

Projectiles. Ogive-nose projectiles (Fig. 3.3.1; $D = 7.11\,mm$, $L_{cyl} = 59.3\,mm$, $L_{nose} = 11.8\,mm$, $\psi = 3$) made from VAR 4340 steel (HRC 38, $\sigma_y = 1140\,MPa$, $K_{lc} = 130\,MPa \cdot m^{1/2}$, $m = 20.4\,g$) and AerMet 100 steel (HRC 53, $\sigma_y = 1720\,MPa$, $K_{lc} = 126\,MPa \cdot m^{1/2}$, $m = 20.9\,g$).

Note. Experimental results in tables below are given for relatively small velocities whereby projectiles did not deform and penetration channels were straight.

4.152.1   *Test results*

Table 1. Test results for VAR 4340 steel projectiles.

v_s (m/s)	P (mm)	T (mm)	v_s (m/s)	P (mm)	T (mm)	v_s (m/s)	P (mm)	T (mm)
569	58	127	570	55	141	679	72	140
821	102	167	966	140	229	1147	190	248
1237	224	267						

Table 2. Test results for AerMet 100 steel projectiles.

v_s (m/s)	P (mm)	T (mm)	v_s (m/s)	P (mm)	T (mm)	v_s (m/s)	P (mm)	T (mm)
794	103	154	1076	160	230	1255	229	277
1348	254	349	1538	332	406	1654	389	406
1786	452	470						

4.153   [Project THOR, 1961]

See also [Zook, 1977].

4.153.1   *General data on shields and projectiles*

Shields. Monolithic shields made from different materials and face-hardened steel shields.

Table 1. Properties of shield materials.

Material	γ (kg/m^3)	HB
Magnesium alloys	1762;1826	72
Aluminum alloys	2771	120
Titanium alloys	4422; 4550	190
Cast iron	7209	150-220
Face-hardened steel	7770	480-550 (front) 331-375 (rear)
Monolithic steel	7770	Different
Copper	8907	42
Lead	11006	5.5
Tuballoy	18695	235-245

Projectiles. Cylindrical (Fig. 3.1.1) projectiles and FSP (Fig. 3.10.2) steel (SAE 1020) projectiles.

Table 2. Parameters of cylindrical projectiles.

m (g)	D (mm)	L (mm)	m (g)	D (mm)	L (mm)
15.6	14.9	11.4	7.8	12.7	8.0
3.9	10.1	6.2	1.9	7.6	5.4
1.0	5.9	4.6	30.8	17.4	16.6

Table 3. Parameters of FSP projectiles.

m (g)	D (mm)	L_1 (mm)	L_2 (mm)	d (mm)
15.6	14.9	9.0	5.7	10.5
7.8	12.7	5.6	4.3	9.0
3.9	10.1	4.4	3.3	7.2
1.9	7.6	4.2	2.9	5.4
1.0	5.9	3.5	2.4	4.2

Note. Notations in tables below:
*   – parameter was not available; ** – parameter was not specified.

4.153.2 *Test results for Magnesium alloys shields*

Table 4. Test results for shields with $T = 1.3$ *mm*.

m (g)	θ_s (deg)	v_s (m/s)	v_r (m/s)	m_r (g)	m (g)	θ_s (deg)	v_s (m/s)	v_r (m/s)	m_r (g)
1.9	0	610	*	1.9	1.9	45	883	840	1.9
1.9	45	914	*	1.9	1.9	45	1219	*	1.9
1.9	60	672	584	1.9	1.9	60	1549	1446	1.7
1.9	70	883	750	1.9	3.9	45	884	843	3.8
3.9	45	914	*	3.8	3.9	45	914	*	3.8
3.9	45	914	*	3.8	3.9	45	914	*	3.8
3.9	45	914	*	3.8	7.8	60	305	*	7.7
7.8	60	305	*	7.7	7.8	60	305	*	7.7

Table 5. Test results for shields with $T=3.2$ *mm*.

m (g)	θ_s (deg)	v_s (m/s)	v_r (m/s)	m_r (g)	m (g)	θ_s (deg)	v_s (m/s)	v_r (m/s)	m_r (g)
1.9	0	696	487	1.9	1.9	45	1212	1053	1.9
3.9	45	1140	1009	3.8	3.9	45	1219	*	3.8
3.9	45	1219	*	3.8	3.9	45	1219	*	3.8
3.9	45	1219	*	3.8	3.9	45	1219	*	3.8
3.9	60	906	769	3.8	3.9	70	1484	1203	3.8
7.8	60	457	*	7.7	7.8	60	547	422	7.7
7.8	70	867	596.0	7.7	15.6	60	305	*	15.5
15.6	60	305	*	15.5					

Table 6. Test results for shields with $T= 4.7$ *mm*.

m (g)	θ_s (deg)	v_s (m/s)	v_r (m/s)	m_r (g)	m (g)	θ_s (deg)	v_s (m/s)	v_r (m/s)	m_r (g)
1	0	338	0	*	1	45	427	0	*
1	70	771	0	*	3.9	0	214	0	*
3.9	70	579	0	*	15.6	0	130	0	*
15.6	45	187	0	*	15.6	70	260	0	*

Table 7. Test results for shields with $T = 7.6$ *mm*.

m (g)	θ_s (deg)	v_s (m/s)	v_r (m/s)	m_r (g)	m (g)	θ_s (deg)	v_s (m/s)	v_r (m/s)	m_r (g)
1.9	0	919	632	1.9	1.9	45	1512	1110	1.8
3.9	0	860	681	3.8	3.9	45	1476	791	3.4
3.9	60	1554	752	3.1	7.8	0	877	739	7.7
7.8	70	1206	640	*	15.6	70	1219	*	13.7
15.6	70	1219	*	14.9	15.6	70	1219	*	15.1
15.6	70	1219	*	15.5	15.6	70	1248	797	13.5

Table 8. Test results for shields with $T = 7.9$ *mm.*

m (g)	θ_s (deg)	v_s (m/s)	v_r (m/s)	m_r (g)	m (g)	θ_s (deg)	v_s (m/s)	v_r (m/s)	m_r (g)
1	0	521	0	*	1	70	1013	0	*
1.9	60	637	0	*	3.9	70	753	0	*
3.9	80	1679	0	*	15.6	0	203	0	*
15.6	60	376	0	*	15.6	70	521	0	*
15.6	80	956	0	*					

Table 9. Test results for shields with $T = 12.7$ *mm.*

m (g)	θ_s (deg)	v_s (m/s)	v_r (m/s)	m_r (g)	m (g)	θ_s (deg)	v_s (m/s)	v_r (m/s)	m_r (g)
1	70	1737	0	*	1.9	0	1489	784	1.8
1.9	45	1536	594	1.8	3.9	0	1206	842	3.6
3.9	0	1211	860	3.6	3.9	0	1219	*	3.6
3.9	45	3074	2145	0.4	7.8	0	1150	949	*
7.8	60	1304	1011	6.8	7.8	60	1524	*	6.3
7.8	60	1524	*	7.1	7.8	60	1524	*	7.2
7.8	60	1524	*	7.2	7.8	60	1524	*	7.6
7.8	70	1524	*	6.8	7.8	70	1524	*	7.5
7.8	70	2187	151	*	15.6	0	1439	1239	*
15.6	60	1219	*	14.1	15.6	60	1219	*	14.1
15.6	60	1227	781	13.8					

Table 10. Test results for shields with $T = 15.2$ *mm.*

m (g)	θ_s (deg)	v_s (m/s)	v_r (m/s)	m_r (g)	m (g)	θ_s (deg)	v_s (m/s)	v_r (m/s)	m_r (g)
1.0	0	905	0	*	1.0	70	1593	0	*
3.9	0	619	0	*	15.6	0	290	0	*
15.6	70	869	0	*					

Table 11. Test results for shields with $T = 19.1$ *mm.*

m (g)	θ_s (deg)	v_s (m/s)	v_r (m/s)	m_r (g)	m (g)	θ_s (deg)	v_s (m/s)	v_r (m/s)	m_r (g)
3.9	45	1552	695	3.2	7.8	0	1383	1079	5
7.8	0	1524	*	7	7.8	60	1531	603	6.9
7.8	60	305	*	7.4	15.6	0	1530	1202	13.1
15.6	0	1646	*	12.7	15.6	0	1646	*	12.9
15.6	0	1646	*	13	15.6	0	1646	*	13.2
15.6	0	1646	*	13.8	15.6	0	1646	*	14.3
15.6	60	1563	885.0	12.6	15.6	60	1646	*	10.7
15.6	70	1646	*	14.9					

Table 12. Test results for shields with $T = 19.2$ *mm*.

m	θ_s	v_s	v_r	m_r	m	θ_s	v_s	v_r	m_r
(g)	(deg)	(m/s)	(m/s)	(g)	(g)	(deg)	(m/s)	(m/s)	(g)
1.0	0	1042	0	*	1.9	60	1535	0	*
3.9	70	1650	0	*	7.8	70	1216	0	*
15.6	0	434	0	*	15.6	60	695	0	*
15.6	70	956	0	*	15.6	80	2085	0	*

Table 13. Test results for shields with $T = 25.4$ *mm*.

m	θ_s	v_s	v_r	m_r	m	θ_s	v_s	v_r	m_r
(g)	(deg)	(m/s)	(m/s)	(g)	(g)	(deg)	(m/s)	(m/s)	(g)
1.9	0	1373	603	1.6	1.9	0	1828	419	1.5
1.9	0	2498	994	0.7	1.9	0	2891	1165	0.6
3.9	0	1245	447	3.8	3.9	0	1524	*	3.8
7.8	0	1485	821	6.8	7.8	0	1524	*	6.4
7.8	0	1524	*	6.8	7.8	0	1524	*	7.7
7.8	45	1477	547	6.6	7.8	45	1524	*	6.7
7.8	60	1524	*	6.8	7.8	60	1524	*	6.9
7.8	60	2261	124.0	*	15.6	0	1410	981	14.3
15.6	0	1417	538	13.9	15.6	45	1452	832	13.1
15.6	60	1524	*	12	15.6	60	1593	673	*
15.6	70	2889	1109	*	15.6	70	2895	777	*

Table 14. Test results for shields with $T = 50.8$ *mm*.

m	θ_s	v_s	v_r	m_r	m	θ_s	v_s	v_r	m_r
(g)	(deg)	(m/s)	(m/s)	(g)	(g)	(deg)	(m/s)	(m/s)	(g)
7.8	0	1478	199	7	7.8	0	1524	*	6.5
7.8	0	1524	*	6.5	7.8	0	1524	*	6.6
7.8	0	1524	*	6.8	7.8	0	1524	*	6.9
7.8	0	1524	*	7.3	7.8	0	2219	238	*
15.6	0	1438	573	12.8	15.6	0	1524	*	13.8
15.6	0	1524	*	14.2	15.6	45	1676	*	11.3
15.6	45	1676	*	12.3	15.6	45	2938	1782	*

4.153.3 *Test results for aluminum alloys shields*

Table 15. Test results for shields with $T = 0.5$ *mm*.

m	θ_s	v_s	v_r	m_r	m	θ_s	v_s	v_r	m_r
(g)	(deg)	(m/s)	(m/s)	(g)	(g)	(deg)	(m/s)	(m/s)	(g)
0.3	60	643	**	1.2	0.3	70	835	**	1.2
0.3	70	939	**	1.2	0.3	70	1016	**	1.2

m (g)	θ_s (deg)	v_s (m/s)	v_r (m/s)	m_r (g)	m (g)	θ_s (deg)	v_s (m/s)	v_r (m/s)	m_r (g)
1.9	0	2487	**	4.0	1.9	60	2334	**	8.0
1.9	70	2800	**	5.4	1.9	80	183	0	**
3.9	0	2524	**	16.7	7.8	0	2882	**	11.2
15.6	80	85	0	**					

Table 16. Test results for shields with $T = 1.8$ *mm*.

m (g)	θ_s (deg)	v_s (m/s)	v_r (m/s)	m_r (g)	m (g)	θ_s (deg)	v_s (m/s)	v_r (m/s)	m_r (g)
1	70	1744	**	1.6	1.9	0	2500	**	0.6
1.9	0	2512	**	8.1	1.9	60	2816	**	2.3
1.9	70	1128	**	8.8	1.9	70	1667	1304	**
1.9	70	1690	1141	**	1.9	70	2945	**	1.4
1.9	80	910	**	8.8	1.9	80	975	0	**
1.9	80	976	**	8.8	1.9	80	1000	**	8.8
1.9	80	1039	**	8.8	1.9	80	1063	**	8.8
3.9	0	3063	**	3	3.9	60	2493	**	5.2
3.9	70	899	**	16.9	3.9	70	1562	1124	**
3.9	70	1582	1094	**	3.9	70	2590	**	2.9
7.8	70	1804	1373	**	7.8	70	1835	1375	**
15.6	0	3001	**	4.4	15.6	70	1811	1544	**
15.6	70	1814	41	**	15.6	80	259	0	**

Table 17. Test results for shields with $T = 2.3$ *mm*.

m (g)	θ_s (deg)	v_s (m/s)	v_r (m/s)	m_r (g)	m (g)	θ_s (deg)	v_s (m/s)	v_r (m/s)	m_r (g)
1.9	70	762	0	**	1.9	70	764	**	8.8
1.9	70	765	**	8.8	1.9	70	801	**	8.7
1.9	70	827	**	8.7	1.9	70	888	**	8.8
1.9	70	918	**	8.8	3.9	0	224	**	18.0
3.9	0	255	**	18.0	3.9	0	257	**	18.0
3.9	0	300	**	18.0	3.9	0	324	**	18.0
3.9	45	243	**	18.0	3.9	45	244	**	18.0
3.9	45	262	**	18.0	3.9	45	317	**	18.0
3.9	45	329	**	18.0	3.9	45	373	**	18.0
3.9	60	358	**	18.0	3.9	60	379	**	18.0
3.9	60	381	0	**	3.9	60	386	**	18.0
3.9	60	388	**	18.0	3.9	60	401	**	18.0
3.9	60	402	**	18.0	3.9	70	429	**	18.0
3.9	70	490	**	18.0	3.9	70	524	**	18.0
3.9	70	546	**	18.0	3.9	70	548	**	17.9
3.9	70	549	0	**	3.9	70	549	**	17.9
3.9	70	554	**	18.0	3.9	70	555	**	18.0
3.9	70	556	**	18.0	3.9	70	558	**	17.9

| m | θ_s | v_s | v_r | m_r | m | θ_s | v_s | v_r | m_r |
(g)	(deg)	(m/s)	(m/s)	(g)	(g)	(deg)	(m/s)	(m/s)	(g)
3.9	70	561	**	18.0	3.9	70	564	0	**
3.9	70	564	0	**	3.9	70	564	**	18.0
3.9	70	578	**	18.0	3.9	70	584	**	18.0
3.9	70	586	**	18.0	3.9	70	608	**	17.9
3.9	70	609	**	18.0	3.9	70	610	**	18.0
3.9	70	619	**	18.0	3.9	70	622	**	18.0
3.9	70	635	**	18.0	3.9	70	664	**	18.0
3.9	70	683	**	18.0	3.9	70	896	**	18.0
7.8	0	169	**	36.3	7.8	0	194	**	36.3
7.8	0	232	**	36.3	7.8	0	259	**	36.3
7.8	0	276	**	36.3	7.8	70	368	**	36.3
7.8	70	379	**	36.3	7.8	70	385	**	36.3
7.8	70	396	0.0	**	7.8	70	414	**	36.3
7.8	70	421	**	36.3	7.8	70	451	**	36.3
15.6	70	294	0.0	**	15.6	80	351	0	**
15.6	80	1299	**	72.6	15.6	80	1758	**	66.2

Table 18. Test results for shields with $T = 3.2$ *mm*.

| m | θ_s | v_s | v_r | m_r | m | θ_s | v_s | v_r | m_r |
(g)	(deg)	(m/s)	(m/s)	(g)	(g)	(deg)	(m/s)	(m/s)	(g)
0.3	0	1444	**	1.2	1	60	1763	**	2.3
1	70	1841	**	0.4	1.9	0	338	260	**
1.9	0	546	397	**	1.9	0	912	723	**
1.9	0	1116	952	**	1.9	0	2580	**	5.9
1.9	0	2680	**	0.4	1.9	0	2862	**	0.8
3.9	60	2897	**	1.3	3.9	70	2553	**	4.2
3.9	70	2622	**	1.7	3.9	70	2710	**	0.4
3.9	70	2951	**	0.0	7.8	0	557	448	**
7.8	0	918	821	**	7.8	0	1201	1055	**
7.8	70	1156	769	**	7.8	70	1515	868	**
7.8	70	1723	1015	**	7.8	70	1805	1086	**
15.6	70	1815	1283	**	15.6	70	1818	1246	**
15.6	70	1848	1380	**					

Table 19. Test results for shields with $T = 4.8$ *mm*.

| m | θ_s | v_s | v_r | m_r | m | θ_s | v_s | v_r | m_r |
(g)	(deg)	(m/s)	(m/s)	(g)	(g)	(deg)	(m/s)	(m/s)	(g)
1.9	0	546	312	**	1.9	0	872	576	**
1.9	0	881	629	**	1.9	0	882	620	**
1.9	0	1183	921	**	1.9	0	1616	1358	**
1.9	0	1642	1386	**	1.9	45	1636	1220	**
1.9	45	1673	1228	**	1.9	70	914	0	**
7.8	0	580	445	**	7.8	0	925	771.0	**

m (g)	θ_s (deg)	v_s (m/s)	v_r (m/s)	m_r (g)	m (g)	θ_s (deg)	v_s (m/s)	v_r (m/s)	m_r (g)
7.8	0	1198	1033	**	7.8	70	1836	909	**
7.8	70	1842	898	**	15.6	0	1739	1585	**
15.6	0	1776	1575	**	15.6	0	1791	1610	**
15.6	70	1818	1352	**	15.6	70	1824	1184	**
15.6	70	1846	1043	**	15.6	75	884	0	**

Table 20. Test results for shields with $T = 6.4$ *mm*.

m (g)	θ_s (deg)	v_s (m/s)	v_r (m/s)	m_r (g)	m (g)	θ_s (deg)	v_s (m/s)	v_r (m/s)	m_r (g)
0.3	0	1627	**	1.2	1	60	1989	**	0.3
1.9	0	598	289	**	1.9	0	608	296	**
1.9	0	614	290	**	1.9	0	1016	643	**
1.9	0	1023	640	**	1.9	0	1067	654	**
1.9	0	1426	1014	**	1.9	0	1433	1029	**
1.9	0	1447	1052	**	1.9	0	1651	826	**
1.9	0	1670	762	**	1.9	70	1219	0	**
1.9	70	1614	**	2.3	1.9	70	1992	**	0.3
3.9	0	633	378	**	3.9	0	728	458	**
3.9	0	881	565	**	3.9	0	1015	881	**
3.9	0	1229	912	**	3.9	0	1481	1196	**
3.9	60	971	**	18.0	3.9	60	1061	**	18.0
3.9	60	1062	**	14.5	3.9	60	1067	0	**
3.9	60	1083	**	15.4	3.9	60	1083	**	18.0
3.9	60	1091	**	14.4	3.9	60	1100	**	17.8
3.9	60	1128	**	14.4	3.9	60	1133	**	14.3
3.9	60	1832	**	1.8	3.9	70	1614	**	2.3
7.8	0	542	379	**	7.8	0	553	406	**
7.8	0	942	713	**	7.8	0	1037	774	**
7.8	0	1522	1282	**	7.8	0	1569	1333	**
7.8	0	1584	1279	**	7.8	70	1833	**	0.8
15.6	0	603	514	**	15.6	0	635	530	**
15.6	0	1095	920	**	15.6	0	1119	946	**
15.6	0	1489	1296	**	15.6	0	1496	1320	**
15.6	70	770	**	72.6	15.6	70	867	**	68.1
15.6	70	924	**	72.8	15.6	70	932	**	67.8
15.6	70	975	0	**	15.6	70	984	**	55.0
15.6	70	994	**	51.5	15.6	70	1104	**	49.8
15.6	70	1868	931	**	15.6	70	1884	993	**

Table 21. Test results for shields with $T = 9.5$ *mm*.

m (g)	θ_s (deg)	v_s (m/s)	v_r (m/s)	m_r (g)	m (g)	θ_s (deg)	v_s (m/s)	v_r (m/s)	m_r (g)
1.9	60	1311	0	**	1.9	60	1372	0	**

m (g)	θ_s (deg)	v_s (m/s)	v_r (m/s)	m_r (g)	m (g)	θ_s (deg)	v_s (m/s)	v_r (m/s)	m_r (g)
1.9	70	1707	0	**	3.9	70	2405	**	3.5
15.6	70	1003	**	50.4					

Table 22. Test results for shields with $T = 12.7$ *mm*.

m (g)	θ_s (deg)	v_s (m/s)	v_r (m/s)	m_r (g)	m (g)	θ_s (deg)	v_s (m/s)	v_r (m/s)	m_r (g)
1.9	0	1154	330	**	1.9	0	1161	346	**
1.9	0	1442	691	**	1.9	0	1395	627	**
1.9	60	2647	**	0.2	3.9	0	934	283	**
3.9	0	1088	405	**	3.9	0	1479	832	**
3.9	0	1490	869	**	3.9	0	1822	**	8.6
3.9	45	595	**	18.0	3.9	45	1019	**	16.5
3.9	45	1062	**	15.0	3.9	45	1092	**	13.5
3.9	45	1119	**	13.8	3.9	45	1134	**	17.4
3.9	45	1136	**	14.2	3.9	45	1170	**	15.3
3.9	45	1224	**	16.2	3.9	45	1285	**	14.7
3.9	45	1309	**	14.4	3.9	45	1347	**	17.1
3.9	60	1382	**	67.7	3.9	60	1518	**	53.5
3.9	60	1615	0	**	3.9	60	1636	**	11.6
7.8	0	1050	538	**	7.8	0	1070	544	**
7.8	0	1503	983	**	7.8	0	1524	993	**
7.8	60	1806	**	6.2	15.6	0	631	351	**
15.6	0	634	361	**	15.6	0	1060	725	**
15.6	0	1068	714	**	15.6	0	1495	1176	**
15.6	0	1496	1139	**	15.6	45	811	**	70.0
15.6	45	840	**	65.9	15.6	45	843	**	66.9
15.6	45	846	**	72.7	15.6	45	857	**	69.2
15.6	45	866	**	72.3	15.6	60	779	**	53.5
15.6	60	804	**	67.7	15.6	60	940	**	53.3
15.6	60	1000	**	54.6	15.6	60	1006	0	**
15.6	60	1009	**	56.1	15.6	60	1024	**	54.8
15.6	60	1043	**	55.5	15.6	60	1474	**	28.9
15.6	60	1484	**	22.3	15.6	60	1504	**	28.7
15.6	70	1433	0	**					

Table 23. Test results for shields with $T = 19.1$ *mm*.

m (g)	θ_s (deg)	v_s (m/s)	v_r (m/s)	m_r (g)	m (g)	θ_s (deg)	v_s (m/s)	v_r (m/s)	m_r (g)
3.9	0	1477	361	**	3.9	0	1483	433	**
7.8	0	1006	201	**	7.8	0	1061	159	**
7.8	0	1506	658	**	7.8	0	1521	667	**
15.6	0	1043	532	**	15.6	0	1067	493	**
15.6	0	1081	533	**	15.6	0	1497	951	**
15.6	0	1506	941	**	15.6	60	1383	**	48.6

m	θ_s	v_s	v_r	m_r	m	θ_s	v_s	v_r	m_r
(g)	(deg)	(m/s)	(m/s)	(g)	(g)	(deg)	(m/s)	(m/s)	(g)
15.6	60	1478	0	**	15.6	60	1483	**	43.8
15.6	60	1491	**	47.6	15.6	60	1500	**	48.6

Table 24. Test results for shields with $T = 25.4$ *mm*.

m	θ_s	v_s	v_r	m_r	m	θ_s	v_s	v_r	m_r
(g)	(deg)	(m/s)	(m/s)	(g)	(g)	(deg)	(m/s)	(m/s)	(g)
15.6	60	1549	**	46.3	1.9	0	2085	0	**
1.9	0	2666	**	2.7	1.9	45	3658	0	**
3.9	0	2289	**	0.6	7.8	0	1173	0	**
7.8	0	1480	**	35.4	7.8	0	1487	**	18.1
7.8	0	1492	**	20.4	7.8	0	1503	**	32.0
7.8	0	1514	421	**	7.8	0	1531	**	17.4
7.8	0	1550	**	19.4	7.8	0	1699	**	8.7
7.8	0	3045	322	**	7.8	45	1981	0	**
7.8	60	2682	0	**	15.6	0	975	0	**
15.6	0	1062	230	**	15.6	0	1067	246	**
15.6	0	1069	150	**	15.6	0	1093	215	**
15.6	0	1508	667	**	15.6	0	1518	673	**
15.6	45	1311	0	**	15.6	45	1372	0	**
15.6	60	1707	0	**	15.6	70	2438	0	**

Table 25. Test results for shields with $T = 31.8$ *mm*.

m	θ_s	v_s	v_r	m_r	m	θ_s	v_s	v_r	m_r
(g)	(deg)	(m/s)	(m/s)	(g)	(g)	(deg)	(m/s)	(m/s)	(g)
3.9	60	1086	**	16.7	3.9	70	1232	**	4.2
1.9	0	3658	0	**	7.8	0	2377	0	**

Table 26. Test results for shields with T = 50.8 *mm*.

m	θ_s	v_s	v_r	m_r	m	θ_s	v_s	v_r	m_r
(g)	(deg)	(m/s)	(m/s)	(g)	(g)	(deg)	(m/s)	(m/s)	(g)
7.8	0	3658	0	**	15.6	0	2134	0	**
15.6	45	3048	0	**					

4.153.4 *Test results for titanium alloys shields*

Table 27. Test results for shields with $T = 1.3$ *mm*.

m	θ_s	v_s	v_r	m_r	m	θ_s	v_s	v_r	m_r
(g)	(deg)	(m/s)	(m/s)	(g)	(g)	(deg)	(m/s)	(m/s)	(g)
1.9	0	568	521	1.9	1.9	0	610	*	1.9
1.9	0	610	*	1.9	1.9	0	610	*	1.9
1.9	0	610	*	1.9	1.9	0	610	*	1.9

| m | θ_s | v_s | v_r | m_r | m | θ_s | v_s | v_r | m_r |
(g)	(deg)	(m/s)	(m/s)	(g)	(g)	(deg)	(m/s)	(m/s)	(g)
1.9	0	610	*	1.9	1.9	0	1462	1294	0.9
1.9	0	1524	*	1.3	1.9	45	1375	1189	1.8
1.9	60	899	593	1.6	1.9	60	911	632	1.7
1.9	70	905	445	1.4	1.9	70	914	*	1.6
1.9	70	914	*	1.7	3.9	0	799	673	3.8
3.9	0	914	*	3.8	3.9	0	914	*	3.8
3.9	0	914	*	3.8	3.9	0	1033	958	1.7
3.9	0	1676	*	2.9	3.9	0	1676	*	3.2
3.9	0	1676	*	3.4	3.9	0	1676	*	3.6
3.9	60	762	550	3.8	3.9	60	914	*	3.8
3.9	70	731	447	3.8	7.8	0	610	*	7.7
7.8	0	610	*	7.7	7.8	0	610	*	7.7
7.8	0	610	*	7.7	7.8	0	610	*	7.7
7.8	0	610	*	7.7	7.8	0	610	*	7.7
7.8	0	610	*	7.7	7.8	0	641	561	7.7
7.8	0	959	875	7.7	7.8	0	1524	*	2.1
7.8	45	840	716	7.7	7.8	45	914	*	7.7
7.8	60	683	514	7.7	7.8	70	664	397	*
15.6	0	1067	*	14.9	15.6	0	1067	*	15.5

Table 28. Test results for shields with $T = 1.5$ *mm*.

| m | θ_s | v_s | v_r | m_r | m | θ_s | v_s | v_r | m_r |
(g)	(deg)	(m/s)	(m/s)	(g)	(g)	(deg)	(m/s)	(m/s)	(g)
1.9	60	457	0	*	1.9	70	610	0	*
1.9	80	1463	0	*	15.6	60	213	0	*
15.6	70	282	0	*	15.6	80	549	0	*

Table 29. Test results for shields with T=2.5 mm

| m | θ_s | v_s | v_r | m_r | m | θ_s | v_s | v_r | m_r |
(g)	(deg)	(m/s)	(m/s)	(g)	(g)	(deg)	(m/s)	(m/s)	(g)
1.9	0	914	*	1.9	1.9	0	914	*	1.9
1.9	0	914	*	1.9	1.9	0	1524	*	1.7
1.9	60	1524	*	0.9	1.9	60	1524	*	1.1
3.9	60	1219	*	3.5	7.8	0	762	*	7.7
7.8	0	762	*	7.7	7.8	0	1524	*	3.9
7.8	60	914	*	3.3					

Table 30. Test results for shields with $T = 3.2$ *mm*.

| m | θ_s | v_s | v_r | m_r | m | θ_s | v_s | v_r | m_r |
(g)	(deg)	(m/s)	(m/s)	(g)	(g)	(deg)	(m/s)	(m/s)	(g)
1.9	0	880	590	*	1.9	0	914	*	1.6

m	θ_s	v_s	v_r	m_r	m	θ_s	v_s	v_r	m_r
(g)	(deg)	(m/s)	(m/s)	(g)	(g)	(deg)	(m/s)	(m/s)	(g)
1.9	0	914	*	1.9	1.9	0	1355	1084	1.6
1.9	45	1369	962	1.7	1.9	70	3116	*	0.0
3.9	0	620	501	3.8	3.9	0	773	583	3.8
3.9	0	914	*	3.7	3.9	0	914	*	3.8
3.9	0	914	*	3.8	3.9	0	914	*	3.8
3.9	0	1499	1252	2.4	3.9	60	1059	532	3.0
3.9	60	1524	*	2.8	7.8	0	610	*	7.7
7.8	0	610	*	7.7	7.8	0	610	*	7.7
7.8	0	976	785	7.7	7.8	0	1067	*	7.7
7.8	0	1067	*	7.7	7.8	0	1067	*	7.7
7.8	0	1219	*	7.5	7.8	0	1219	*	7.6
7.8	0	1219	*	7.6	7.8	0	1524	*	2.2
7.8	45	1188	887	7.0	7.8	60	914	*	2.0
7.8	60	914	*	6.2	7.8	70	762	*	7.7
7.8	70	1067	*	5.7	15.6	0	610	*	15.5
15.6	0	610	*	15.5	15.6	0	1067	*	15.4
15.6	0	1067	*	15.4	15.6	0	1067	*	15.4
15.6	0	1067	*	15.5	15.6	45	1219	*	14.6
15.6	45	1322	1033	14.2	15.6	60	946	472	2.8
15.6	60	962	761	12.3					

Table 31. Test results for shields with $T = 6.4$ *mm.*

m	θ_s	v_s	v_r	m_r	m	θ_s	v_s	v_r	m_r
(g)	(deg)	(m/s)	(m/s)	(g)	(g)	(deg)	(m/s)	(m/s)	(g)
1.9	0	1491	683	1.7	1.9	0	1986	1127	0.6
1.9	60	3184	803	0.3	3.9	0	1506	774	3.2
3.9	0	1526	979	*	3.9	0	1676	*	2.9
3.9	0	2455	1367	0.1	3.9	60	3037	949	*
7.8	0	1484	995	*	7.8	0	1524	*	4.5
7.8	0	1524	*	5.9	7.8	0	1524	*	6.7
7.8	0	1524	*	6.9	7.8	0	1524	*	6.9
7.8	0	1524	*	7.0	7.8	0	1524	*	7.0
7.8	45	1524	346	4.8	15.6	45	1524	*	12.1
15.6	60	1541	806	4.5					

Table 32. Test results for shields with $T = 12.7$ *mm.*

No.	m	θ_s	v_s	v_r	m_r	No.	m	θ_s	v_s	v_r	m_r
	(g)	(deg)	(m/s)	(m/s)	(g)		(g)	(deg)	(m/s)	(m/s)	(g)
1	1.9	0	2372	382	*	2	1.9	0	3396	*	0.1
3	3.9	0	2552	1166	0.6	4	3.9	60	2997	0	*
5	7.8	0	1524	*	5.4	6	15.6	45	1502	192	*
7	15.6	45	1524	*	7.6						

4.153.5 *Test results for cast iron shields*

Table 33. Test results for shields with $T = 4.8$ *mm*.

m	θ_s	v_s	v_r	m_r	m	θ_s	v_s	v_r	m_r
(g)	(deg)	(m/s)	(m/s)	(g)	(g)	(deg)	(m/s)	(m/s)	(g)
1.0	0	1316	571	0.7	1.9	0	573	193	1.9
1.9	0	1183	668	1.5	1.9	0	1735	1149	0.9
1.9	45	1187	389	1.3	3.9	0	625	379	3.8
3.9	0	1295	894	3.2	3.9	0	1802	1305	2.4
3.9	45	1241	704	0.9	3.9	45	1791	1029	2.0
15.6	0	608	424	15.3	15.6	0	1063	852	13.9
15.6	0	1592	1308	11.9	15.6	45	1292	1006	11.8
15.6	45	1843	1413	10.4					

Table 34. Test results for shields with $T = 9.5$ *mm*.

m	θ_s	v_s	v_r	m_r	m	θ_s	v_s	v_r	m_r
(g)	(deg)	(m/s)	(m/s)	(g)	(g)	(deg)	(m/s)	(m/s)	(g)
1.9	0	1195	148	1.6	1.9	0	1761	552	0.2
3.9	0	1249	357	2.7	3.9	0	1776	755	2.1
15.6	0	1186	702	12.6	15.6	0	1802	1080	11.1
15.6	45	1081	341	10.5	15.6	45	1247	530	lost
15.6	45	1857	888	lost					

Table 35. Test results for shields with $T = 14.3$ *mm*.

m	θ_s	v_s	v_r	m_r	m	θ_s	v_s	v_r	m_r
(g)	(deg)	(m/s)	(m/s)	(g)	(g)	(deg)	(m/s)	(m/s)	(g)
3.9	0	1775	401	1.6	15.6	0	1237	486	11.8
15.6	0	1816	810	7.8	15.6	45	1860	491	2.7

4.153.6 *Test results for face-hardened steel shields*

Table 36. Results for $T = 3.5$ *mm*.

m	θ_s	v_s	v_r	m_r	m	θ_s	v_s	v_r	m_r
(g)	(deg)	(m/s)	(m/s)	(g)	(g)	(deg)	(m/s)	(m/s)	(g)
1.0	70	2679	*	0.0	1.9	0	748	0	*
1.9	0	1017	372	1.6	1.9	0	1082	0	1.6
1.9	0	3373	*	0.1	1.9	45	1536	247	0.0
1.9	60	1729	0	*	1.9	70	2547	*	0.0
3.9	60	1577	0	*	3.9	70	1837	0	*
7.8	60	997	0	*	7.8	60	1042	0	*

m (g)	θ_s (deg)	v_s (m/s)	v_r (m/s)	m_r (g)	m (g)	θ_s (deg)	v_s (m/s)	v_r (m/s)	m_r (g)
7.8	70	1631	0	*	7.8	70	1671	0	*
15.6	70	721	0	*	15.6	70	1088	0	*
15.6	70	1797	493	0.0					

Table 37. Test results for shields with $T = 3.6\ mm$.

m (g)	θ_s (deg)	v_s (m/s)	v_r (m/s)	m_r (g)	m (g)	θ_s (deg)	v_s (m/s)	v_r (m/s)	m_r (g)
1.9	30	1589	533	1.0	1.9	45	1490	335	0.7
15.6	45	1190	664	3.0	15.6	45	1220	773	11.0

Table 38. Test results for shields with $T = 6.4\ mm$.

m (g)	θ_s (deg)	v_s (m/s)	v_r (m/s)	m_r (g)	m (g)	θ_s (deg)	v_s (m/s)	v_r (m/s)	m_r (g)
0.3	0	2896	0	*	1.0	0	2208	567	0.1
1.0	0	3136	*	0.0	1.9	0	1073	*	0.5
1.9	0	1283	*	0.5	1.9	0	1306	*	1.0
1.9	0	1372	*	0.6	1.9	0	1765	551	0.8
1.9	0	1820	717	0.4	1.9	0	2882	1116	0.1
1.9	0	2936	1263	0.2	1.9	30	1584	386	*
3.9	0	931	*	0.6	3.9	0	967	*	0.7
3.9	0	999	*	0.7	3.9	0	1067	0	*
3.9	0	1164	331	2.0	3.9	0	1786	767	0.3
3.9	0	2151	901	0.7	3.9	0	2799	1515	0.3
3.9	60	2700	*	0.0	7.8	0	764	*	1.0
7.8	0	871	*	1.5	7.8	0	1151	484	6.4
7.8	0	1766	871	1.4	7.8	0	1897	960	0.6
7.8	0	2985	2043	0.2	7.8	45	1225	252	*
7.8	60	2111	0	*	7.8	60	3069	524	0.0
15.6	0	1191	699	9.5	15.6	0	1636	1039	1.5
15.6	45	1554	610	6.7	15.6	45	1558	651	6.8
15.6	60	1807	*	0.0	15.6	70	2757	876	0.1
15.6	70	2834	1847	*					

Table 39. Test results for shields with $T = 12.7\ mm$.

m (g)	θ_s (deg)	v_s (m/s)	v_r (m/s)	m_r (g)	m (g)	θ_s (deg)	v_s (m/s)	v_r (m/s)	m_r (g)
1.9	0	3251	*	0.1	3.9	0	1791	0	*
3.9	0	2276	305	1.1	3.9	0	2694	694	0.1
7.8	0	1802	369	6.6	7.8	0	2086	525	0.7
7.8	0	2225	766	1.0	7.8	0	2881	978	0.4
7.8	30	1839	205	0.8	15.6	0	1200	290	12.3
15.6	0	1840	710	7.5	15.6	0	2753	1351	0.5

m (g)	θ_s (deg)	v_s (m/s)	v_r (m/s)	m_r (g)	m (g)	θ_s (deg)	v_s (m/s)	v_r (m/s)	m_r (g)
15.6	30	1738	213	4.5	15.6	45	2454	*	0.0
15.6	60	2581	557	0.1					

4.153.7 *Test results for monolithic steel shields*

Table 40. Test results for monolithic steel shields having different hardness.

HB	T (mm)	m (g)	θ_s (deg)	v_s (m/s)	v_r (m/s)	m_r (g)
90	0.7	1.9	0	183	0	*
90	1.5	1.9	0	274	0	*
92	6.4	1.9	0	975	0	*
100	0.7	1.9	60	1623	1323	*
100	0.7	1.9	60	1649	1425	*
100	0.7	1.9	70	1628	1060	*
100	0.7	1.9	70	1661	785	*
100	0.7	1.9	70	1698	1170	*
100	1.5	3.9	60	1553	1000	*
100	1.5	3.9	60	1565	1000	*
100	1.5	7.8	60	1756	1153	*
100	1.5	7.8	60	1831	1212	*
100	1.5	7.8	70	1796	1031	*
100	1.5	7.8	70	1822	730	*
100	1.5	7.8	70	1831	909	*
100	1.5	15.6	60	1844	1482	*
100	1.5	15.6	60	1849	1483	*
100	1.5	15.6	70	1834	1354	*
100	1.5	15.6	70	1853	1267	*
100	3.4	7.8	60	1812	777	*
100	3.4	7.8	60	1814	792	*
100	3.4	15.6	60	1848	1206	*
100	3.4	15.6	60	1862	1242	*
100	3.4	15.6	70	1799	965	*
100	3.4	15.6	70	1843	1003	*
105	3.4	1.9	0	503	0	*
105	19.1	3.9	0	3050	*	0.2
107	6.3	1.9	0	991	0	*
110	0.8	0.3	0	1086	*	0.3
110	0.8	1.0	0	1275	*	0.3
110	0.8	1.0	60	1302	*	0.6
110	0.8	1.0	70	1693	*	0.1
110	0.8	1.9	70	1470	*	0.8
110	0.8	3.9	70	1494	*	1.4
110	0.8	7.8	70	1133	*	6.1

HB	T (mm)	m (g)	θ_s (deg)	v_s (m/s)	v_r (m/s)	m_r (g)
110	0.8	15.6	70	1285	*	12.1
110	1.7	15.6	70	335	0	*
110	25.4	7.8	0	2829	*	0.0
110	25.4	15.6	0	2801	*	1.6
115	2.5	1.0	0	1699	*	0.0
115	2.5	1.0	0	2563	*	0.0
115	2.5	1.9	0	1625	*	0.1
115	2.5	1.9	60	1646	*	0.1
115	2.5	1.9	70	2679	*	0.0
115	2.5	3.9	60	1580	*	0.6
115	2.5	3.9	70	2074	*	0.0
115	2.5	7.8	70	1845	*	0.0
115	2.5	15.6	70	1508	*	1.7
120	6.4	1.9	0	1550	*	0.4
120	6.4	1.9	60	3039	*	0.0
120	6.4	3.9	0	1585	*	0.2
120	6.4	3.9	60	2830	*	0.0
120	6.4	7.8	45	1825	*	1.2
120	6.4	15.6	70	2667	*	0.0
120	12.7	3.9	0	2601	*	0.7
120	12.7	7.8	0	1825	*	0.5
120	12.7	15.6	0	1756	*	3.7
125	9.5	1.9	0	1936	*	0.9
125	9.5	1.9	0	2402	*	0.1
128	25.4	1.9	0	3658	0	*
128	25.4	3.9	0	3200	0	*
128	25.4	7.8	0	2743	0	*
128	25.4	15.6	0	1951	0	*
128	25.4	15.6	45	2743	0	*
135	0.5	1.9	0	888	849	*
135	0.5	1.9	70	894	732	*
135	0.5	3.9	0	302	277	*
135	0.5	3.9	70	636	434	*
135	1.5	1.9	0	1212	1015	*
135	1.5	1.9	60	1089	492	*
135	1.5	3.9	0	393	256	*
135	1.5	3.9	60	1199	817	*
135	1.5	7.8	60	805	558	*
135	1.5	7.8	70	1195	747	*
135	3.2	3.9	0	884	543	*
135	3.2	7.8	60	1492	858	*
141	38.1	15.6	0	2896	0	*
150	1.4	1.9	60	3217	*	0.3
150	1.4	1.9	70	2977	*	0.1
150	2.5	1.9	0	3232	*	0.1
150	6.4	3.9	0	2644	*	0.2

HB	T (mm)	m (g)	θ_s (deg)	v_s (m/s)	v_r (m/s)	m_r (g)
150	6.4	3.9	0	2967	*	0.2
200	12.7	1.9	0	2743	0	*
200	12.7	7.8	70	2865	0	*
200	12.7	15.6	60	2155	0	*
300	3.2	1.9	0	1521	1196	*
300	3.2	3.9	0	880	584	*
300	3.2	3.9	0	1466	1164	*
300	3.2	7.8	0	910	690	*
300	3.2	15.6	0	917	754	*
300	3.2	15.6	0	1426	1027	*
300	3.2	15.6	60	1507	1049	*
300	6.4	3.9	0	1466	687	*
300	6.4	15.6	0	1433	1038	*
300	6.4	15.6	60	1431	450	*
300	6.6	1.1	0	1492	0	*
300	6.6	1.1	30	1798	0	*
300	6.9	2.9	0	1097	0	*
300	6.9	2.9	30	1298	0	*
300	6.9	2.9	45	1510	0	*
300	6.9	13.4	0	687	0	*
300	6.9	13.4	0	694	0	*
300	6.9	13.4	30	777	0	*
300	6.9	13.4	30	798	0	*
300	6.9	13.4	45	919	0	*
300	6.9	13.4	45	919	0	*
300	6.9	13.4	60	1136	0	*
300	6.9	13.4	60	1232	0	*
300	6.9	53.5	0	471	0	*
300	6.9	53.5	30	517	0	*
300	6.9	53.5	45	640	0	*
300	6.9	53.5	60	764	0	*
300	7.1	1.1	0	1555	0	*
300	7.1	1.1	30	1812	0	*
300	13.2	13.4	0	1148	0	*
300	13.2	13.4	30	1439	0	*
300	13.2	13.4	45	1806	0	*
300	13.2	53.5	0	788	0	*
300	13.2	53.5	30	906	0	*
300	13.2	53.5	45	1048	0	*
305	6.4	1.9	0	1394	460	*
305	6.4	15.6	0	1556	1109	*
332	12.7	15.6	0	1661	760	*
350	2.0	1.9	60	1006	0	*
350	2.0	7.8	60	579	0	*
350	2.0	15.6	70	533	0	*
350	3.2	1.9	60	1463	0	*

HB	T (mm)	m (g)	θ_s (deg)	v_s (m/s)	v_r (m/s)	m_r (g)
350	3.2	7.8	60	1097	0	*
350	3.2	7.8	70	2134	0	*
350	3.2	15.6	70	914	0	*
350	6.4	7.8	60	1981	0	*
350	6.4	15.6	70	1524	0	*
370	2.0	1.9	0	996	*	1.7
370	2.0	1.9	0	3457	*	0.0
370	2.0	1.9	60	1552	*	0.1
370	2.0	1.9	70	2472	*	0.1
370	2.0	3.9	60	1551	*	0.5
370	2.0	3.9	70	2077	*	0.3
370	2.0	7.8	70	1815	*	2.2
370	2.0	7.8	70	1856	*	0.7
370	3.2	15.6	60	1297	867	*
371	3.2	1.9	0	1611	*	0.1
371	3.2	1.9	60	2400	*	0.0
371	3.2	3.9	45	1566	*	2.1
371	3.2	3.9	70	2430	*	0.1
371	3.2	7.8	45	1866	*	4.0
371	3.2	7.8	60	1804	*	1.4
371	3.2	15.6	45	1479	*	11.2
371	3.2	15.6	60	1469	*	4.3
371	3.2	15.6	70	1554	*	10.1
380	2.0	1.9	70	3211	*	0.1
380	3.2	3.9	60	2918	*	0.4
380	3.2	3.9	60	3011	*	0.2
380	6.4	1.9	0	3130	*	0.2
390	6.4	1.9	0	1840	*	0.4
390	6.4	3.9	0	1511	*	0.6
390	6.4	7.8	0	1573	*	1.4
390	6.4	7.8	70	2736	*	0.1
390	6.4	15.6	0	1377	*	2.3
390	6.4	15.6	60	1559	*	0.0
390	6.4	15.6	70	2561	*	0.1
390	12.7	3.9	0	2571	*	0.1
390	12.7	7.8	0	1840	*	1.2
390	12.7	15.6	0	2671	*	0.1
390	12.7	15.6	45	2469	*	0.1
393	1.5	1.9	0	610	368	*
393	1.5	7.8	70	1272	767	*
393	1.5	15.6	70	1548	1143	*
400	6.9	1.1	0	1465	0	*
400	6.9	1.1	30	1621	0	*
400	6.9	2.9	0	1006	0	*
400	6.9	2.9	30	1181	0	*
400	6.9	2.9	45	1394	0	*

HB	T (mm)	m (g)	θ_s (deg)	v_s (m/s)	v_r (m/s)	m_r (g)
400	6.9	13.4	0	636	0	*
400	6.9	13.4	30	767	0	*
400	6.9	13.4	45	879	0	*
400	6.9	13.4	60	1223	0	*
400	6.9	53.5	0	425	0	*
400	6.9	53.5	30	506	0	*
400	6.9	53.5	45	485	0	*
400	12.8	13.4	0	1055	0	*
400	12.8	13.4	30	1222	0	*
400	13.0	53.5	0	715	0	*
400	13.0	53.5	30	731	0	*
410	15.9	1.9	0	3505	0	*

4.153.8 *Test results for copper shields*

Table 41. Test results for shields with $T = 1.5$ *mm*.

m (g)	θ_s (deg)	v_s (m/s)	v_r (m/s)	m_r (g)	m (g)	θ_s (deg)	v_s (m/s)	v_r (m/s)	m_r (g)
1.0	0	411	*	0.9	1.0	0	411	*	0.9
1.0	0	414	240	*	1.0	0	457	*	0.9
1.0	0	457	*	0.9	1.0	0	862	611	0.9
1.0	0	914	*	0.9	1.0	0	1097	*	0.9
1.0	0	1433	*	0.6	1.0	0	1446	1107	*
1.0	60	1045	487	*	1.0	60	1461	579	0.3
1.9	0	305	*	1.9	1.9	60	853	515	1.9
1.9	60	914	*	1.7	1.9	60	2158	**	0.5
1.9	70	1817	**	0.3	1.9	70	2636	**	0.1
3.9	0	369	265	3.8	3.9	0	790	623	3.5
3.9	60	792	525	*	3.9	70	1287	*	0.9
3.9	70	1287	890	1.4	3.9	70	2499	**	0.0
3.9	70	2948	**	0.1	7.8	0	305	*	7.7
7.8	0	349	261	7.7	15.6	0	305	*	15.5
15.6	0	305	*	15.5	15.6	0	305	*	15.5
15.6	0	305	*	15.5	15.6	0	396	*	15.5
15.6	0	402	331	15.5	15.6	0	402	331	15.5

Table 42. Test results for shields with $T = 3.2$ *mm*.

m (g)	θ_s (deg)	v_s (m/s)	v_r (m/s)	m_r (g)	m (g)	θ_s (deg)	v_s (m/s)	v_r (m/s)	m_r (g)
1.0	0	849	411	0.9	1.0	0	853	*	0.8
1.0	0	914	*	0.9	1.0	0	1439	906	0.6

m	θ_s	v_s	v_r	m_r	m	θ_s	v_s	v_r	m_r
(g)	(deg)	(m/s)	(m/s)	(g)	(g)	(deg)	(m/s)	(m/s)	(g)
1.0	0	1448	*	0.5	1.0	0	1524	*	0.1
1.0	0	1524	*	0.9	1.9	0	762	*	1.7
1.9	0	775	440	1.8	1.9	60	1490	625	0.5
1.9	60	2739	**	0.1	1.9	70	3464	**	0.0
3.9	0	1132	805	3.5	3.9	60	1270	400	2.1
3.9	70	2474	**	0.0	3.9	70	3011	**	0.0
7.8	0	610	*	7.6	7.8	0	746	502	7.5
7.8	60	910	336	5.9	7.8	60	910	*	6.0
15.6	0	464	326	*	15.6	0	610	*	15.5
15.6	0	610	*	15.5	15.6	70	1297	492	5.4
15.6	70	3028	**	0.1					

Table 43. Test results for shields with $T = 6.4$ *mm*.

m	θ_s	v_s	v_r	m_r	m	θ_s	v_s	v_r	m_r
(g)	(deg)	(m/s)	(m/s)	(g)	(g)	(deg)	(m/s)	(m/s)	(g)
1.0	0	3228	963	*	1.9	0	1170	666	1.9
1.9	0	3480	1325	*	3.9	45	2495	553	0.4
3.9	45	2973	841	*	3.9	45	3076	1305	*
3.9	60	3201	710	*	7.8	0	912	*	6.2
7.8	0	912	393	7.7	7.8	0	914	*	7.2
7.8	60	2641	315	*	15.6	0	914	*	11.6
15.6	0	914	*	12.9	15.6	0	914	*	15.5
15.6	0	1020	706	12.7	15.6	60	1297	523	4.6
15.6	60	1433	*	9.1	15.6	60	1463	*	9.2
15.6	60	1472	*	4.6	15.6	60	1472	477	9.2
15.6	60	1524	*	8.0	15.6	70	2385	152	*

Table 44. Test results for shields with $T = 12.7$ *mm*.

m	θ_s	v_s	v_r	m_r	m	θ_s	v_s	v_r	m_r
(g)	(deg)	(m/s)	(m/s)	(g)	(g)	(deg)	(m/s)	(m/s)	(g)
1.9	0	2675	386	0.2	1.9	0	3209	468	0.1
3.9	0	1463	*	3.8	3.9	0	1509	279	3.8
3.9	0	3312	800	0.2	7.8	0	1524	*	7.2
7.8	0	1524	377	0.3	7.8	0	2635	861	*
7.8	45	3153	274	*	15.6	0	1403	608	15.5
15.6	0	1438	559	*	15.6	45	1403	359	10.1
15.6	70	2726	1806	0.4					

Table 45. Test results for shields with $T = 25.4$ *mm*.

m	θ_s	v_s	v_r	m_r	m	θ_s	v_s	v_r	m_r
(g)	(deg)	(m/s)	(m/s)	(g)	(g)	(deg)	(m/s)	(m/s)	(g)
15.6	0	1565	224	*	15.6	0	1576	226	*

4.153.9　*Test results for lead shields*

Table 46. Test results for shields with $T = 1.7$ *mm*.

m (g)	θ_s (deg)	v_s (m/s)	v_r (m/s)	m_r (g)	m (g)	θ_s (deg)	v_s (m/s)	v_r (m/s)	m_r (g)
1.0	60	892	244	0.4	1.0	60	893	366	0.9
1.0	60	926	244	0.3					

Table 47. Test results for shields with $T = 3.2$ *mm*.

m (g)	θ_s (deg)	v_s (m/s)	v_r (m/s)	m_r (g)	m (g)	θ_s (deg)	v_s (m/s)	v_r (m/s)	m_r (g)
1.9	0	2401	1067	0.1	1.9	0	2440	914	0.1
1.9	60	2499	*	0.0	1.9	60	2721	*	0.0
1.9	60	3140	1034	0.1	3.9	0	2322	*	0.3
3.9	60	2911	*	0.2	7.8	0	1811	1102	0.8

Table 48. Test results for shields with $T = 3.3$ *mm*.

m (g)	θ_s (deg)	v_s (m/s)	v_r (m/s)	m_r (g)	m (g)	θ_s (deg)	v_s (m/s)	v_r (m/s)	m_r (g)
1.0	0	871	474	**	1.0	0	1002	516	**
1.9	0	1219	**	1.4	1.9	0	1219	**	1.5
1.9	0	1235	869	1.4	1.9	60	401	23	**
1.9	60	848	232	**	1.9	60	849	212	**
1.9	60	905	181	**	1.9	60	1151	*	0.0
7.8	60	363	177	**	15.6	0	1199	1006	13.3
15.6	0	1219	**	13.2	15.6	0	1219	**	13.4
15.6	60	326	268	**	15.6	70	1219	**	3.1
15.6	70	1219	**	3.6					

Table 49. Test results for shields with $T = 3.5$ *mm*.

m (g)	θ_s (deg)	v_s (m/s)	v_r (m/s)	m_r (g)	m (g)	θ_s (deg)	v_s (m/s)	v_r (m/s)	m_r (g)
1.0	0	724	518	0.9	1.0	0	1455	762	0.1
1.0	0	1466	610	0.0	1.9	0	958	457	1.8
1.9	0	1046	610	0.7	1.9	0	1722	762	0.2

Table 50. Test results for shields with $T = 6.4$ *mm*.

m (g)	θ_s (deg)	v_s (m/s)	v_r (m/s)	m_r (g)	m (g)	θ_s (deg)	v_s (m/s)	v_r (m/s)	m_r (g)
1.9	0	2789	*	0.0	1.9	60	3019	*	0.0
3.9	0	799	366	1.3	3.9	60	2654	*	0.0
15.6	0	2609	1006	0.3	15.6	60	2605	518	0.1

Table 51. Test results for shields with $T = 6.6$ *mm*.

m (g)	θ_s (deg)	v_s (m/s)	v_r (m/s)	m_r (g)	m (g)	θ_s (deg)	v_s (m/s)	v_r (m/s)	m_r (g)
1.0	0	386	0	**	1.0	60	1522	**	0.1
1.9	0	1189	**	0.8	1.9	0	1189	**	1.4
1.9	0	1753	*	0.0	1.9	60	963	**	0.1
1.9	70	2223	92	**	3.9	60	624	0	**
3.9	60	2720	1195	**	3.9	60	2777	838	**
3.9	70	2747	210	**	7.8	0	720	367	**
7.8	0	733	373	**	7.8	0	1218	640	6.4
7.8	0	1236	701	**	7.8	70	1985	707	**
15.6	0	168	0	**	15.6	0	1892	*	3.5
15.6	60	604	43	**	15.6	70	1219	**	5.0

Table 52. Test results for shields with $T = 12.7$ *mm*.

m (g)	θ_s (deg)	v_s (m/s)	v_r (m/s)	m_r (g)	m (g)	θ_s (deg)	v_s (m/s)	v_r (m/s)	m_r (g)
1.9	0	1962	260	**	1.9	45	3168	365	**
3.9	0	874	212	**	3.9	0	884	175	**
3.9	45	2852	435	**	7.8	0	937	297	1.3
7.8	0	1211	368	**	7.8	0	1222	336	**
7.8	0	1267	575	2.9	7.8	45	1540	*	2.6
7.8	60	2077	0	**	7.8	60	2317	0	**
15.6	0	893	371	7.3	15.6	0	897	377	7.1
15.6	0	942	386	15.4	15.6	0	1730	610	1.4
15.6	60	1216	0	**	15.6	60	1803	396	0.6
15.6	60	1896	*	0.1	15.6	70	2677	541	**

Table 53. Test results for shields with $T = 25.4$ *mm*.

m (g)	θ_s (deg)	v_s (m/s)	v_r (m/s)	m_r (g)	m (g)	θ_s (deg)	v_s (m/s)	v_r (m/s)	m_r (g)
1.9	0	2537	553	**	3.9	0	1511	0	**
7.8	0	1186	122	**	15.6	0	938	197	11.2
15.6	0	973	169	8.3	15.6	0	1264	762	11.1
15.6	0	1463	**	6.4					

4.153.10 *Test results for tuballoy shields*

Table 54. Test results for shields with $T = 2.5$ *mm*.

m (g)	θ_s (deg)	v_s (m/s)	v_r (m/s)	m_r (g)	m (g)	θ_s (deg)	v_s (m/s)	v_r (m/s)	m_r (g)
1.9	0	1091	435	1.5	1.9	0	1447	549	1.0
1.9	0	2868	1372	0.1	1.9	30	1359	414	0.8

m	θ_s	v_s	v_r	m_r	m	θ_s	v_s	v_r	m_r
(g)	(deg)	(m/s)	(m/s)	(g)	(g)	(deg)	(m/s)	(m/s)	(g)
1.9	45	1662	*	0.0	3.9	0	926	321	0.9
3.9	0	1472	823	0.3	3.9	30	1139	408	2.8
3.9	30	1356	571	2.3	3.9	60	1703	*	0.0
7.8	0	495	174	7.6	7.8	0	686	387	5.5
7.8	30	1207	667	5.6	7.8	60	1728	*	0.0
7.8	60	1798	695	*	7.8	60	2850	914	0.1
15.6	0	732	443	14.3	15.6	30	960	551	13.0
15.6	45	2939	1036	1.6	15.6	60	1817	549	0.1
29.9	45	1618	762	13.9	30.1	0	975	701	10.7
30.1	60	2239	884	1.4					

Table 55. Test results for shields with $T = 3.2\ mm$.

m	θ_s	v_s	v_r	m_r	m	θ_s	v_s	v_r	m_r
(g)	(deg)	(m/s)	(m/s)	(g)	(g)	(deg)	(m/s)	(m/s)	(g)
1.9	0	1735	488	0.2	1.9	45	1672	*	0.0
1.9	45	1813	914	0.1	3.9	45	1715	244	0.1
15.6	0	1418	762	14.6	15.6	45	1476	396	5.7
15.6	45	1512	610	0.1	15.6	45	1697	610	0.1
15.6	60	1761	549	0.1					

Table 56. Test results for shields with $T = 3.8\ mm$.

m	θ_s	v_s	v_r	m_r	m	θ_s	v_s	v_r	m_r
(g)	(deg)	(m/s)	(m/s)	(g)	(g)	(deg)	(m/s)	(m/s)	(g)
1.9	0	1619	457	0.3	3.9	0	1485	415	0.5
7.8	0	1394	914	0.2	7.8	45	1546	274	0.1
15.6	0	1788	792	0.3	15.6	45	1551	762	5.8
30.1	30	1311	518	7.1	30.3	45	1548	701	7.8
30.3	30	1638	701	10.0	30.5	30	1036	701	23.0

Table 57. Test results for shields with $T = 5.1\ mm$.

m	θ_s	v_s	v_r	m_r	m	θ_s	v_s	v_r	m_r
(g)	(deg)	(m/s)	(m/s)	(g)	(g)	(deg)	(m/s)	(m/s)	(g)
1.9	0	2840	762	0.1	1.9	45	3076	*	0.0
1.9	60	2785	*	0.0	3.9	0	1718	610	0.1
3.9	0	2535	792	0.2	3.9	45	1780	*	0.0
7.8	0	1700	1219	0.1	7.8	0	1843	962	0.1
7.8	45	1866	274	0.1	15.6	0	1779	732	0.8
15.6	0	1859	1219	0.1	15.6	45	1512	610	0.1
30.1	45	2336	945	0.3	30.1	0	2171	640	1.4
30.2	45	1531	457	3.8	30.3	45	1573	671	5.2
30.3	0	1561	808	5.4					

4.154 [Radin and Goldsmith, 1988]

4.154.1 *General data on shields and projectiles*

Shields. 2024-0 Al (HB 42-52) monolithic and layered (plates in-contact and with air gaps) shields; $\widetilde{D}^{sh} = 114\,mm$.

Projectiles. Steel (HRC 56-59) projectiles with $D = 12.57\,mm$ and $L = 38.1\,mm$: cylindrical projectiles (Fig. 3.1.1; $m = 35\,g$) and cone-nose projectiles (Fig. 3.2.1; $\beta = 30°$, $m = 29\,g$). Projectiles did not deform in all tests.

4.154.2 *Test results*

Table 1. Test results for cylindrical projectiles.

T_{sum} (mm)	Shield structure	v_{bl} (m/s)	T_{sum} (mm)	Shield structure	v_{bl} (m/s)
1.6	1×[1.6]	61.9	3.2	1×[3.2]	93.0
3.2	2×[1.6]	90.0	4.8	1×[4.8]	135
4.8	3×[1.6]	113.6	6.4	1×[6.4]	142.4
6.4	4×[1.6]	137.0	6.4	2×[3.2]	160.4

Table 2. Test results for cone-nose projectiles.

T_{sum} (mm)	Shield structure	v_{bl} (m/s)	T_{sum} (mm)	Shield structure	v_{bl} (m/s)
1.6	1×[1.6]	52.8	3.2	1×[3.2]	95.2
3.2	2×[1.6]	93.2	3.2	1×[1.6]\6.4	90.6
4.8	1×[4.8]	144.0	4.8	3×[1.6]	124.0
4.8	[1.6]+[3.2]	129.9	6.4	1×[6.4]	184.4
6.4	4×[1.6]	157.7	6.4	2×[3.2]	160.4
6.4	2×[3.2]\6.4	153.4			

4.155 [Raguraman *et al.*, 2010]

See also [Raguraman *et al.*, 2009].

4.155.1 *General data on shields and projectiles*

Shields. Monolithic 1100 H14 AA shields; $\widetilde{D}^{sh} = 205\,mm$, $T = 1\,mm$.

Projectiles. Mild steel (HRC 50-52) cone-nose (Fig. 3.2.1; $L = 59.7\,mm$, $L_{cyl} = 30.0\,mm$), ogive-nose (Fig. 3.3.1; $L = 49.6\,mm$, $L_{cyl} = 30.0\,mm$), hemispherical-nose (Fig. 3.4.1; $L_{cyl} = 34.9\,mm$) and cylindrical (Fig. 3.1.1; $L = 39.9\,mm$) projectiles with $D = 15mm$ and $m = 55 \pm 0.1g$.

4.155.2 *Test results*

Table 1. Test results.

Projectile shape	v_s (m/s)	θ_s (deg)	v_r (m/s)
Cone-nose projectile	32.0	5.7	25.7
Ogive-nose projectile	30.7	15.9	24.1
Hemispherical-nose projectile	33.9	14.5	25.7
Cylindrical projectile	34.8	4.8	23.1

4.156 [Recht and Ipson, 1963]

See also [Recht and Ipson, 1962].

4.156.1 *General data on shields and projectiles*

Shields. Monolithic 190 HB mild steel and aluminum shields.
Projectiles. Steel (HRC 30) cylindrical projectiles (Fig. 3.1.1).

4.156.2 *Test results for 190 HB mild steel shield*

4.156.2.1 *Test results for normal impact,* $\theta_s = 0°$

Table 1#. Test results for $T/L = 0.15$.

v_s (m/s)	v_r (m/s)	v_s (m/s)	v_r (m/s)	v_s (m/s)	v_r (m/s)	v_s (m/s)	v_r (m/s)
145	2	193	115	204	131	302	235
342	271	403	327	584	481	670	550
966	778	977	785				

Table 2#. Test results for $T/L = 0.29$.

v_s (m/s)	v_r (m/s)	v_s (m/s)	v_r (m/s)	v_s (m/s)	v_r (m/s)
296	0	321	145	383	196
413	214	515	298	664	394
849	525	961	578	965	590

Table 3#. Test results for $T/L = 0.44$.

v_s (m/s)	v_r (m/s)	v_s (m/s)	v_r (m/s)	v_s (m/s)	v_r (m/s)
475	0	515	140	591	207
713	293	805	330	934	440

4.156.2.2 *Test results for oblique impact, $\theta_s = 45°$*

Table 4#. Test results for $T/L = 0.15$.

v_s (m/s)	v_r (m/s)	v_s (m/s)	v_r (m/s)	v_s (m/s)	v_r (m/s)	v_s (m/s)	v_r (m/s)
226	0	299	193	306	210	386	306
398	285	456	317	473	332	612	443
611	457	639	481	775	581	1004	738

Table 5#. Test results for $T/L = 0.29$.

v_s (m/s)	v_r (m/s)	v_s (m/s)	v_r (m/s)	v_s (m/s)	v_r (m/s)	v_s (m/s)	v_r (m/s)
461	3	476	47	483	109	506	127
506	207	522	184	528	113	598	246
615	226	617	352	624	268	627	348
668	270	727	331	730	301	751	385
766	354	770	402	794	356	898	432
899	454	902	418	909	404	936	512
949	524	967	423				

Table 6#. Test results for $T/L = 0.44$.

v_s (m/s)	v_r (m/s)	v_s (m/s)	v_r (m/s)	v_s (m/s)	v_r (m/s)	v_s (m/s)	v_r (m/s)
699	4	718	61	727	51	729	75
735	62	732	233	898	281	923	281
936	289	939	273	1066	405		

4.156.3 *Test results for aluminum shield*

Table 7#. Test results for normal impact, $\theta_s = 0°$, and $T/L = 1.45$.

v_s (m/s)	v_r (m/s)	v_s (m/s)	v_r (m/s)	v_s (m/s)	v_r (m/s)	v_s (m/s)	v_r (m/s)
602	5	609	123	636	143	673	199
871	371	940	453	958	464		

4.157   [Rodríguez-Martínez *et al.*, 2013]

See also [Kpenyigba *et al.*, 2013], [Kpenyigba *et al.*, 2015].

4.157.1   *General data on shields and projectiles*

Shields. Monolithic AISI 304 austenitic stainless steel shields; $\widetilde{c}_{l.w} = 100\,mm$.

Projectiles. Maraging steel ($\sigma_y = 2000\,MPa$) projectiles with $D = 13mm$, $L_{cyl} = 25mm$ and $m = 30\,g$: cone-nose projectiles (Fig. 3.2.1; $L_{nose} = 8.95mm$, $\beta = 36°$ and hemispherical-nose projectiles (Fig. 3.4.1).

Note. Yield stress of projectile material that underwent heat treatment was higher than that of AISI 304 steel.

4.157.2   *Test results*

Table 1#. Test results for cone-nose projectiles.

T (mm)	v_s (m/s)	v_r (m/s)	v_s (m/s)	v_r (m/s)	v_s (m/s)	v_r (m/s)	v_s (m/s)	v_r (m/s)
0.5	36	0	44	18	54	42	72	61
0.5	88	78	146	138				
1.0	73	0	88	48	122	98	139	110

Table 2#. Test results for hemispherical-nose projectiles.

T (mm)	v_s (m/s)	v_r (m/s)	v_s (m/s)	v_r (m/s)	v_s (m/s)	v_r (m/s)	v_s (m/s)	v_r (m/s)
0.5	95	0	112	62	121	80	138	107
0.5	176	153						
1.0	130	0	143	60	151	80	159	106
1.0	181	135						

4.158   [Rodríguez-Millán *et al.*, 2014]

4.158.1   *General data on shields and projectiles*

Shields. Monolithic 5754-H111 and 6082-T6 aluminum alloys plates, $\widetilde{c}_{l.w} = 100\,mm$, $T = 4\,mm$.

Projectiles. Cylindrical projectiles (Fig. 3.1.1, *Projectile-1*), hemispherical-nose projectiles (Fig. 3.4.1, *Projectile-2*) and cone-nose projectiles (Fig. 3.2.1, $\beta = 36^\circ$, *Projectile-3*). Projectile material was maraging steel after heat treatment to $\sigma_y \approx 2\,GPa$. For all projectiles $D = 13\,mm$, $L = 25\,mm$ and $m = 30\,g$.

4.158.2 *Test results for 5754-H111 AA shields*

Table 1#. Test results for Projectile-1.

v_s (m/s)	v_r (m/s)	v_s (m/s)	v_r (m/s)	v_s (m/s)	v_r (m/s)
120	0	130	30.4	137	62.9
149	95.1	164	120.1	179	139.9

$v_{bl} = 120\,m/s$; $a = 1$, $p = 2.09$.

Table 2#. Test results for Projectile-2.

v_s (m/s)	v_r (m/s)	v_s (m/s)	v_r (m/s)	v_s (m/s)	v_r (m/s)
160	0	165	0	167	0
171	43.9	179	75.5		

$v_{bl} = 166\,m/s$; $a = 1$, $p = 2.17$.

Table 3#. Test results for Projectile-3.

v_s (m/s)	v_r (m/s)	v_s (m/s)	v_r (m/s)	v_s (m/s)	v_r (m/s)	v_s (m/s)	v_r (m/s)
145	0	147	0	149	0	151	31.2
162	61.6	172	89.0	176	94.9	180	103.1

$v_{bl} = 147\,m/s$; $a = 1$, $p = 1.95$.

4.158.3 *Test results for 6082-T6 AA shields*

Table 4#. Test results for Projectile-1.

v_s (m/s)	v_r (m/s)	v_s (m/s)	v_r (m/s)	v_s (m/s)	v_r (m/s)
129	0	137	52.5	150	107.9
160	134.0	178	152.5		

$v_{bl} = 129\,m/s$; $a = 1$, $p = 2.77$.

Table 5#. Test results for Projectile-2.

v_s (m/s)	v_r (m/s)	v_s (m/s)	v_r (m/s)	v_s (m/s)	v_r (m/s)
137	0	144	38.1	148	55.4
150	60.4	171	109.7	180	123.9

$v_{bl} = 137\,m/s; a = 1, p = 2.08.$

Table 6#. Test results for Projectile-3.

v_s (m/s)	v_r (m/s)	v_s (m/s)	v_r (m/s)	v_s (m/s)	v_r (m/s)
144	0	150	34.7	156	54.9
161	63.4	172	90.4	180	105.4

$v_{bl} = 143\,m/s; a = 1, p = 1.88.$

4.159 [Roisman *et al.*, 1999]

4.159.1 *General data on shields and projectiles*

Shields. Layered (plates-in-contact and with air gaps) shields.

Table 1. Material properties of shield layers.

Notation	Material	T (mm)	σ_u (MPa)	σ_y (MPa)	δ (%)	HV*
Al-1	Al-6061-T651	25	328	301	11.8	104
Al-1	Al-6061-T651	30	339	301	13.1	99.5
Al-1	Al-6061-T651	40	330	284	12.9	104
Al-2	Al-F36	85.4	36	220	7	–
St	Steel	50	510	–	–	170

* At 20 kg=196 N load.

Projectiles. Tungsten alloy (HV 502 at 20 kg = 196 N load, $\sigma_u = 1435\,MPa$, $\sigma_y = 1430\,MPa$, $\delta = 7.8\%$) ogive-nose projectiles (Fig. 3.3.1); $D = 11.3\,mm$, $L = 56.5\,mm$, $\rho = 22.6\,mm$, $\psi = 2.0$.

4.159.2 *Test results*

Table 2. Test results for the cases when projectile was embedded in Al-1 plates.

T_{sum} (mm)	Shield structure	m (g)	v_s (m/s)	θ_s (deg)	P (mm)
120.0	3×[40.0]	88.6	651	0	BLV
85.0	[25.0]+ 2×[30.0]	87.9	581	45	47.2*

T_{sum} (mm)	Shield structure	m (g)	v_s (m/s)	θ_s (deg)	P (mm)
105.0	2×[40.0]+[25.0]	88.3	522	30	65.0
105.0	[25.0]+ 2×[40.0]	88.2	593	30	84.2
85.0	[25.0]+ 2×[30.0]	88.6	578	45	51.6
120.0	3×[40.0]	87.9	638	0	116.5
105.0	[25.0]+ 2×[40.0]	88.2	662	30	BLV

* Projectile was broken.

Table 3. Test results for the cases when projectile perforated aluminum plates.

Shield structure	m (g)	v_s (m/s)	θ_s (deg)	v_r * (m/s)	P ** (mm)
2×(Al-1)[40.0]\200.0\2×(St)[50.0]	88.4	570	0	–	23.7
[(Al-2) 85.4]\200.0\2×(St)[50.0]	88.9	751	45	–	–
(Al-1){[25.0]+2×[40.0]}\200.0\2×(St)[50.0]	86.5	706	30	–	18.5
3×(Al-1)[40.0]\200.0\2×(St)[50.0]	86.7	845	0	537	49.0
(Al-1){[25.0]+2×[40.0]}\200.0\2×(St)[50.0]	87.7	810	30	462	14.4
3×(Al-1)[40.0]\200.0\2×(St)[50.0]	88.4	1108	0	887	73.0
(Al-1){[25.0]+2×[40.0]}\200.0\2×(St)[50.0]	88.0	1175	30	965	63.5
(Al-1){[25.0]+2×[30.0]}\200.0\2×(St)[50.0]	88.0	1146	45	927	73.9

* After aluminum plates; ** Into steel plates.

4.160   [Rosenberg and Forrestal, 1988]

See also [Anderson *et al.*, 1992].

4.160.1   *General data on shields and projectiles*

Shields. Monolithic 6061-T6 Al ($\sigma_y = 300\,MPa$); $c_{l,w} = 305\,mm$, $T = 25.4\,mm$.

Projectiles. Maraging T-200 steel ($\sigma_y = 1379\,MPa$) and maraging C-300 steel ($\sigma_y = 2069\,MPa$; one test) slightly spherically blunted cone-nose projectiles (Figs. 3.2.2a and 3.2.2b).

Table 1. Geometrical parameters of projectiles.

Designation	Shape	D (mm)	D_{cyl} (mm)	L (mm)	L_{nose} (mm)	ρ_1 (mm)	ρ_0 (mm)
Projectile 1	Fig. 3.2.2a	7.1	–	81.8	10.7	0.51	–
Projectile 2	Fig. 3.2.2b	7.1	6.6	93.2	10.7	0.76	3.2

4.160.2 *Test results*

Table 2. Test results for Projectile 1.

v_s (m/s)	m (g)	v_r (m/s)	v_s (m/s)	m (g)	v_r (m/s)	v_s (m/s)	m (g)	v_r (m/s)
314	25.1	0	315	25.1	0	327	25.1	0
383	25.1	175	419	25.1	260	515	25.1	399
886	25.1	827	1394	24.0	1334	1442	24.0	1397
1516	24.0	1445	1575*	25.3	1509			

* Projectile material is maraging C-300 steel.

Table 3. Test results for Projectile 2.

v_s (m/s)	m (g)	v_r (m/s)	v_s (m/s)	m (g)	v_r (m/s)	v_s (m/s)	m (g)	v_r (m/s)
301	23.4	0	315	23.4	040	360	23.6	175
421	23.5	282	532	23.7	433			

4.161 [Ryan and Cimpoeru, 2015]

4.161.1 *General data on shields and projectiles*

Shields. Monolithic shields made from different materials (see Table 1).

Table 1. Properties of shield material.

Material	E (GPa)	σ_y (MPa)	Material	E (GPa)	σ_y (MPa)
2024-T351	73.1	324	5083-H112	70.3	200
5083-H131	70.3	276	6061-T651	69	262
7075-T651	71.1	520			

Projectiles. 7.62 mm APM2 projectiles (Fig. 3.11.1, Table 3.11.1).

4.161.2 *Test results*

Table 2. Test results.

Material	T (mm)	v_{bl} (m/s)	T (mm)	v_{bl} (m/s)	T (mm)	v_{bl} (m/s)
2024-T351	26	652	30	715	34	774
	38	831	–	–	–	–
5083-H112	20	474	30	589	40	698

Material	T (mm)	v_{bl} (m/s)	T (mm)	v_{bl} (m/s)	T (mm)	v_{bl} (m/s)
5083-H131	38	738	–	–	–	–
6061-T651	25	578	50	896	–	–
7075-T651	26	718	30	784	32	817
	38	909	–	–	–	–

4.162 [Saburi *et al.*, 2008]

4.162.1 *General data on shields and projectiles*

Shields. Monolithic 5052S AA shields, $c_{l.w} = 200\,mm$.
Projectiles. Cylindrical projectiles (Fig. 3.1.1); $D = 10mm$, $L = 10mm$, $m = 6\,g$. Projectile material: SNCM 439 (nickel–chromium–molybdenum steel of the Japanese Industrial Standard).

4.162.2 *Test results*

Table 1#. Test results.

T (mm)	v_s (m/s)	v_r (m/s)	v_s (m/s)	v_r (m/s)	v_s (m/s)	v_r (m/s)	v_s (m/s)	v_r (m/s)
5.0	743	553	908	633				
10.0	307	212	380	0	385	294	402	0
10.0	724	383						
15.0	573	0	741	123	904	232		

4.163 [Scheffler and Magness, 1998]

4.163.1 *General data on shields and projectiles*

Shields. Monolithic 7039 aluminum plates, $T = 76.2\,mm$.
Projectiles. Tungsten alloy penetrators; hemispherical-nose projectiles (Fig. 3.4.1, *Projectile-1*) and ogive-nose projectiles (Fig. 3.3.1, $\rho = 20.27mm$, *Projectile-2*). For both projectiles, $D = 6.76mm$, $L = 101.34mm$.

Table 1. Types of projectile conditions after penetration (TPC).

Notation	Description	Notation	Description
(1)	Eroded	(2)	Fractured
(3)	Bulged, fractured	(4)	Rigid, fractured
(5)	Slight bulge	(6)	Eroded, bent

4.163.2 *Test results*

Table 2. Test results for Projectile-1.

m (g)	v_s (m/s)	v_r (m/s)	TPC	m (g)	v_s (m/s)	v_r (m/s)	TPC
65.2	1198	1020	(1)	65.2	1038	964	(2)
65.5	1093	961	(3)				

Table 3. Test results for Projectile-2.

m (g)	v_s (m/s)	v_r (m/s)	TPC	m (g)	v_s (m/s)	v_r (m/s)	TPC
63.3	1474	1414	(4)	63.5	1595	1528	(4)
~63.5	1755	1677	(5)	~63.5	1768	1652	(6)

4.164 [Scheffler, 1997]

Note. The author has noted that test results used in his report are unpublished data by L.S. Magness, Jr. (U.S. Army Research Laboratory, Aberdeen Proving Ground, MD). See also [Scheffler, 1996].

4.164.1 *General data on shields and projectiles*

Shields. Monolithic 7039 Al ($\gamma = 2770\,kg/m^3$, $\sigma_y = 500\,MPa$) and 5083 Al ($\gamma = 2660\,kg/m^3$, $\sigma_y = 450\,MPa$) shields.

Projectiles. Tungsten alloy ($\gamma = 18160\,kg/m^3$, $\sigma_y = 3.5\,GPa$) projectiles with $D = 6.76\,mm$ and $m = 63\,g$: hemispherical-nose projectiles (Fig. 3.4.1; $L = 97.79\,mm$) and ogive-nose projectiles (Fig. 3.3.1; $L = 101.35\,mm$, $\psi = 3.0$).

4.164.2 *Test results*

Table 1. Test results for hemispherical-nose projectiles.

Shield material	v_s (m/s)	P (mm)	v_s (m/s)	P (mm)	v_s (m/s)	P (mm)
7039 Al	1165	114*	1038	283	1248	130
5083 Al	1086	448	1200	200	1296	216

* Estimation.

Table 2. Test results for ogive-nose projectiles.

Shield material	v_s (m/s)	P (mm)	v_s (m/s)	P (mm)	v_s (m/s)	P (mm)
7039 Al	1156	377	1291	>533	1075	354
5083 Al	923	417	1070	399	1227	>533

4.165 [Schwer *et al.*, 2006]

4.165.1 *General data on shields and projectiles*

Projectiles. Cylindrical projectiles (Fig. 3.1.1), $D = 25.4mm$, $L = 76.2mm$, $m = 302\,g$.

Shields. A36 steel plates; $\widetilde{c}_{l,w} = 203mm$.

Note. The data indicated as *METS* data and *Battelle* data were obtained in two independent facilities.

4.165.2 *Test results*

Table 1#. Test results (*Battelle* data).

T (mm)	v_s (m/s)	v_r (m/s)	T (mm)	v_s (m/s)	v_r (m/s)	T (mm)	v_s (m/s)	v_r (m/s)
3.175	142	0	3.175	304	272	6.35	82	0
6.35	198	190	12.7	305	203			

Table 2#. Test results (*METS* data).

T (mm)	v_s (m/s)	v_r (m/s)	T (mm)	v_s (m/s)	v_r (m/s)	T (mm)	v_s (m/s)	v_r (m/s)
6.35	210	110	6.35	210	153	3.175	207	183
3.175	209	173	12.7	199	50	12.7	200	41
12.7	202	53						

Table 3. Recht-Ipson approximation ($p = 2$).

T (mm)	a	v_{bl} (m/s)	T (mm)	a	v_{bl} (m/s)	T (mm)	a	v_{bl} (m/s)
3.175	0.98	369	6.35	1.0	475	12.7	0.86	630

4.166 [Segletes *et al.*, 2000]

See also [Segletes *et al.*, 2001].

4.166.1 *General data on shields and projectiles*

Shields. Monolithic and layered shields with layers manufactured from different materials including shields with acrylic interplay described below as a layer.

Table 1. Materials of layers.

Notation	Material description	γ (kg/m^3)	HB
Al	5083 Al	2700	103
MS	Mild steel	7850	93
HHA	High-hard armor	7850	500
Ac	Acrylic	1190	–

Projectiles. 14.5 mm caliber B32 AP projectiles (Fig. 3.14.2, Table 3.14.2) consisting of hardened steel core (HRC 65, $\gamma = 7850\,kg/m^3$, $\sigma_y = 4.46\,GPa$) enclosed in brass jacket.

4.166.2 *Test results*

Table 2. Test results for semi-infinite Al shields.

v_s (m/s)	P (mm)	v_s (m/s)	P (mm)	v_s (m/s)	P (mm)	v_s (m/s)	P (mm)
679	61.1	795	77.2	835	87.9	955	109.2

Table 3. Test results for Al shields with $T = 23\ mm$.

v_s (m/s)	v_r (m/s)	v_s (m/s)	v_r (m/s)	v_s (m/s)	v_r (m/s)
620	499	682	574	751	663
865	777	997	930		

Table 4. Test results for [(Al) 23.0]+[(MS) 0.8] shields.

v_s (m/s)	v_r (m/s)	v_s (m/s)	v_r (m/s)	v_s (m/s)	v_r (m/s)
588	454	693	576	774	689
898	822	1000	931		

Table 5. Test results for [(Al) 23.0]+[(HHA) 9.6] shields.

v_s (m/s)	v_r (m/s)	v_s (m/s)	v_r (m/s)	v_s (m/s)	v_r (m/s)	v_s (m/s)	v_r (m/s)
609	233	608	278	688	397	705	415
794	550	805	557	897	690	903	690
987	782	990	794				

Table 6. Test results for [(Al) 23.0]+[(Ac) 2.8]+[(HHA) 9.6] shields.

v_s (m/s)	v_r (m/s)	v_s (m/s)	v_r (m/s)	v_s (m/s)	v_r (m/s)	v_s (m/s)	v_r (m/s)
591	186	598	227	710	425	714	404
806	515	812	567	904	644	908	673
995	789	999	792				

4.167 [Segletes, 2006]

See also [Segletes, 2004].

4.167.1 *General data on shields and projectiles*

Shields. Monolithic aluminum plates: *Shield-1* (HB 110, $T = 1.75\,mm$) and *Shield-2* (HB 190, $T = 50.8\,mm$).

Projectiles. 14.5 mm caliber B32 penetrators (Fig. 3.14.2, Table 3.14.2), consisting of a hardened steel core (HRC 65) enclosed in brass jacket (*Projectile-1*) and tungsten alloy hemispherical-nose projectiles (Fig. 3.4.1, $D = 6.81\,mm$, $L = 102.3\,mm$, $m = 65\,g$, HRC 35.5, *Projectile-2*).

Table 1. Results of experiment (RE) versions.

Notation	Description
(1)	Perforation
(2)	v_r not obtained
(3)	Partial penetration
(4)	Ricochet
(5)	Projectile core fractured, v_r obtained via momentum average of residual-fragment velocities
(6)	Ricochet after rod penetrated into shield and reemerged later from the frontal surface
(7)	Projectile core was fractured, v_r denotes the largest velocity of residual fragments

4.167.2 *Test results*

Table 2. Test results for Projectile-1 against Shield-1.

θ_s (deg)	v_s (m/s)	v_r (m/s)	RE	θ_s (deg)	v_s (m/s)	v_r (m/s)	RE
0	1019	–	(1),(2)	0	999	815	(1)
30	991	726	(1)	30	538	–	(3)
45	583	–	(4)	45	1004	543	(1), (5)
50	1003	561	(1),(5)	50	596	–	(4)
55	1005	–	(4)	55	619	–	(4)
60	595	–	(4)	60	999	–	(4)

Table 3. Test results for Projectile-2 against Shield-2.

θ_s (deg)	v_s (m/s)	v_r (m/s)	RE	θ_s (deg)	v_s (m/s)	v_r (m/s)	RE
15	686	–	(3)	20	836	–	(1)
25	628	–	(3)	35	691	–	(3),(4),(6)
40	907	255	(1),(7)	50	1198	986	(1),(7)
50	937	–	(4)	55	1079	–	(4), (6)
65	1295	812	(1),(7)	65	1088	–	(4)
70	1293	–	(3)	70	1402	893	(1), (7)

4.168 [Senthil *et al.*, 2015]

4.168.1 *General data on shields and projectiles*

Shields. Monolithic Armox 500T steel shields, $T = 8\,mm$, $c_{l,w} = 200\,mm$.

Projectiles. 7.62 mm API projectiles.

Impact velocity. $v_s \approx 850\,m/s$.

4.168.2 *Test results*

Table 1. Test results.

θ_s (deg)	v_r (m/s)	θ_s (deg)	v_r (m/s)	θ_s (deg)	v_r (m/s)
0	335	15	291	20	276
25	97	30	0		

4.169 [Shadbolt *et al.*, 1983]

4.169.1 *General data on shields and projectiles*

Shields. Monolithic mild steel (hereafter *MS*; $\sigma_y = 375\,MPa$, $\sigma_u = 400\,MPa$), stainless steel (hereafter *SS*; $\sigma_y = 370\,MPa$, $\sigma_u = 713\,MPa$), and aluminum alloy (hereafter AA; $\sigma_y = 460\,MPa$, $\sigma_u = 480\,MPa$) shields.

Projectiles. Steel (HV 850 at 20 kg=196 N load) cylindrical projectiles (Fig. 3.1.1); $D = 12.5\,mm$, $m = 34.6\,g$.

4.169.2 *Test results*

Table 1#. Test results on BLV.

Shield material	T (mm)	v_{bl} (m/s)	T (mm)	v_{bl} (m/s)	T (mm)	v_{bl} (m/s)
MS	1.3	67	2.0	90	3.0	145
MS	5.0	184	6.4	186		
SS	1.3	104	2.0	147	2.7	170
SS	3.4	180	4.1	168	6.5	200
AA	1.3	45	2.6	96	4.0	134
AA	6.4	145				

4.170 [Showalter *et al.*, 2008a]

4.170.1 *General data on shields and projectiles*

Shields. Monolithic shields made from 5059 AA that was subjected to different treatment (details are given in tables in section Test results). The temper number in tables indicates treatment procedure during material production. H131 indicates that material was strengthened by mechanical strain hardening. Temper number H136 indicates that material was only stretched, and not cold rolled. Other characteristics of treatment procedure are not presented.

Projectiles. 0.30 Caliber AP M2 projectiles (Fig. 3.11.1, Table 3.11.1), 0.50 caliber AP M2 projectiles (Fig. 3.13.1, Table 3.13.1), 0.50 caliber FSP and 20 mm FSP (Fig. 3.10.1, Table 3.10.1).

4.170.2 *Test results*

Table 1. Test results for 0.30 caliber AP M2 projectiles and $\theta_s = 0°$.

Temper	T (mm)	HB	v_{bl} (m/s)	Temper	T (mm)	HB	v_{bl} (m/s)
H131	25.1	118	588	H131	25.4	126	590
H131	26.3	126	610	H131	30.4	116	661
H131	34.6	114	704	H131	37.9	116	763
H131	38.9	121	773	H131	47.8	107	859
H131	51.2	126	906	H131	51.4	121	912
H136	25.6	109	583	H136	38.7	107	738
H136	51.0	105	873				

Table 2. Test results for 0.30 caliber AP M2 projectiles and $\theta_s = 30°$.

Temper	T (mm)	HB	v_{bl} (m/s)	Temper	T (mm)	HB	v_{bl} (m/s)
H131	12.8	118	435	H131	13.3	121	451
H131	13.4	116	450	H131	14.9	121	490
H131	18.8	121	559	H131	19.8	118	586
H131	19.9	124	584	H136	12.4	107	421
H136	19.1	112	555				

Table 3. Test results for 0.50 caliber AP M2 projectiles and $\theta_s = 0°$.

Temper	T (mm)	HB	v_{bl} (m/s)	Temper	T (mm)	HB	v_{bl} (m/s)
H131	47.8	107	661	H131	51.2	126	680
H131	51.4	121	670	H131	61.6	121	756
H131	64.2	114	753	H131	64.5	114	769
H131	76.4	99	834	H131	76.6	105	820
H136	51.0	105	647	H136	64.3	103	742
H136	77.1	109	830				

Table 4. Test results for 0.50 caliber FSPs and $\theta_s = 0°$.

Temper	T (mm)	HB	v_{bl} (m/s)	Temper	T (mm)	HB	v_{bl} (m/s)
H131	18.8	121	558	H131	19.8	118	631
H131	19.9	124	607	H131	20.1	121	607
H131	20.4	126	648	H131	21.5	121	691
H131	23.6	121	772	H131	25.1	121	856

Temper	T (mm)	HB	v_{bl} (m/s)	Temper	T (mm)	HB	v_{bl} (m/s)
H131	25.4	126	908	H131	26.3	126	976
H136	19.1	112	559	H136	25.6	109	868

Table 5. Test results for 20 mm FSPs and $\theta_s = 0°$.

Temper	T (mm)	HB	v_{bl} (m/s)	Temper	T (mm)	HB	v_{bl} (m/s)
H131	25.1	118	588	H131	25.1	118	419
H131	25.1	121	426	H131	25.5	116	456
H131	25.4	126	448	H131	26.3	126	471
H131	30.4	116	579	H131	34.6	114	680
H131	37.9	116	801	H131	38.9	121	849
H131	47.8	107	1212	H136	25.6	109	433
H136	38.7	107	779	H136	50.9	105	1242

4.171 [Showalter *et al.*, 2008b]

4.171.1 *General data on shields and projectiles*

Shields. Monolithic ARMOX 600T (HB 570-640, $\tilde{\sigma}_y = 1500\,MPa$, $\sigma_u = 2000\,MPa$, $\delta = 7\%$; hereafter *Mat-1*) and ARMOX ADVANCE (HRC 58-63, $\tilde{\sigma}_y = 1600\,MPa$, $\sigma_u = 2250\,MPa$, $\delta = 9\%$; hereafter *Mat-2*) shields.

Projectiles. 0.30 Caliber AP M2 projectiles (Fig. 3.11.1, Table 3.11.1) and 0.50 caliber AP M2 projectiles (Fig. 3.13.1, Table 3.13.1).

Angle of obliquity. $\theta_s = 30°$.

4.171.2 *Test results*

Table 1. Test results for 0.30 caliber AP M2 projectiles.

Shield material	T (mm)	v_{bl} (m/s)	T (mm)	v_{bl} (m/s)	T (mm)	v_{bl} (m/s)
Mat-1	4.52	723.0	5.46	760.1	6.32	781.8
Mat-1	7.42	863.2	8.36	891.5		
Mat-2	4.52	666.6	5.38	819.0	6.40	859.5
Mat-2	7.34	905.5	8.33	892.4		

Table 2. Test results for 0.50 caliber AP M2 projectiles.

Shield material	T (mm)	v_{bl} (m/s)	T (mm)	v_{bl} (m/s)	T (mm)	v_{bl} (m/s)
Mat-1	8.36	691.9	10.41	754.3	12.25	826.0
Mat-2	8.33	725.7	10.29	788.2	12.24	824.7

4.172   [Showalter *et al.*, 2009]

4.172.1   *General data on shields and projectiles*

Shields. Monolithic ATI 500-MIL (HB 477-534, $\sigma_y = 1034\,MPa$, $\sigma_u = 1792\,MPa$, $\delta = 13\%$) shields.
Projectiles. 0.30 caliber AP M2 projectiles (Fig. 3.11.1, Table 3.11.1), 0.50 caliber AP M2 projectiles (Fig. 3.13.1, Table 3.13.1), 14.5 mm BS41 (Fig. 3.14.1, Table 3.14.1) and 14.5 mm B32 (Fig. 3.14.2, Table 3.14.2).
Angle of obliquity. $\theta_s = 30°$.

4.172.2   *Test results*

Table 1. Test results on BLV.

Projectile type	T (mm)	v_{bl} (m/s)	T (mm)	v_{bl} (m/s)	T (mm)	v_{bl} (m/s)
0.30 Caliber AP M2	5.1	663	6.9	819	7.7	814
0.50 Caliber AP M2	7.7	627	9.7	723	3.1	787
14.5 mm B32	15.6	730	15.4	739		
14.5 mm BS41	18.8	841	24.5	869		

4.173   [Siriphala *et al.*, 2012]

4.173.1   *General data on shields and projectiles*

Shields. Monolithic Armox 600T steel plates, $c_{l.w} = 300\,mm$, $T = 7\,mm$.
Projectiles. 5.56 cartridge M193 bullets (copper jacket, lead core; $L = 19.3\,mm$, $m = 56\,g$) [TM 43-0001-27, 1994].

4.173.2 *Test results*

Table 1. Test results on DOP.

v_s (m/s)	P (mm)	v_s (m/s)	P (mm)	v_s (m/s)	P (mm)
967.6	0.90	969.4	1.13	967.5	0.90

4.174 [Squillacioti, 1994]

4.174.1 *General data on shields and projectiles*

Shields. Shields comply with requirements in [MIL-A-46100D, 1988].
Projectiles. Tungsten-carbide core projectiles: 14.5 mm API BS41 (*Projectile-1*) and 20 mm API-T M602 (*Projectile-2*) [MIL-A-46100D, 1988] and 1/2 scale 30 mm (Fig. 3.7.1e) (*Projectile-3*).

4.174.2 *Test results (Series 1)*

Table 1. Test results for *Projectile-1 and* $\theta_s = 30°$.

T (mm)	v_{bl} (m/s)	T (mm)	v_{bl} (m/s)	T (mm)	v_{bl} (m/s)	T (mm)	v_{bl} (m/s)
21.9	766	22.2	798	22.4	788	22.5	790
22.5	798	25.2	888	25.5	892	25.7	892

Table 2. Test results for *Projectile-2 and* $\theta_s = 30°$ (data from [Allison, 1992]).

T (mm)	v_{bl} (m/s)	T (mm)	v_{bl} (m/s)	T (mm)	v_{bl} (m/s)	T (mm)	v_{bl} (m/s)
21.9	648	22.2	563	22.4	615	22.5	573
22.5	615	25.2	697	25.5	678	25.7	676
28.3	794	28.4	721	28.4	823	28.4	798
28.4	756	28.5	764	28.5	736	28.6	745
28.6	768	28.6	736	28.8	723	29.0	727
29.0	747	29.1	776	30.0	838	31.5	898
31.7	900	31.7	891	31.8	882	31.8	859
32.0	848	32.2	924	32.3	940	34.6	865
34.7	915	34.9	993	35.2	997	35.3	1011
35.3	968	40.6	1157				

Table 3. Test results for *Projectile-2 and* $\theta_s = 0°$.

T (mm)	v_{bl} (m/s)	T (mm)	v_{bl} (m/s)	T (mm)	v_{bl} (m/s)	T (mm)	v_{bl} (m/s)
28.3	596	28.4	615	28.4	600	28.4	640
28.4	600	28.5	545	28.5	609	28.6	558
28.6	626	28.6	624	28.8	589	29.0	567
29.0	626	29.1	620	30.0	673	31.5	655
31.7	678	31.7	666	31.8	659	31.8	659
32.0	635	32.2	721	32.3	693	34.6	694
34.7	697	34.9	737	35.2	750	35.3	721
35.3	692	40.6	840				

4.174.3　*Test results (Series 2)*

Table 4. Test results for *Projectile-2 and* $\theta_s = 0°$.

T (mm)	HRC	v_{bl} (m/s)	T (mm)	HRC	v_{bl} (m/s)
25.5	48	491	28.1	48	607
28.3	48	626	28.3	48	605
28.4	48	620	28.4	48	604
28.4	48	607	28.5	48	616
28.5	48	634	28.6	52	616
28.7	48	613	28.8	48	602
28.8	48	612	28.8	48	645
28.8	48	620	28.9	48	613
29.0	48	616	30.9	49.2	654
32.8	48	709	35.5	52	743
39.0	50.2	778			

Table 5. Test results for *Projectile-3 and* $\theta_s = 0°$.

T (mm)	HRC	v_{bl} (m/s)	T (mm)	HRC	v_{bl} (m/s)
28.3	48	1118	28.4	52	1122
28.5	52	1110	28.5	52	1123
28.6	48	1124	28.7	48	1111
28.7	48	1119	28.8	48	1134
28.8	48	1144	28.9	48	1134
28.9	48	1153	29.0	48	1143
29.2	48	1151			

4.175 [Stargel, 2005]

4.175.1 *General data on shields and projectiles*

Shields. Monolithic 2024-T3 Al plates with varying curvatures. Plates are convex upwards or plane, radius of curvature is denoted by R_{cur}; $c_{l,w} = 203\,mm$, $T = 2\,mm$.

Projectiles. Steel spherical projectiles (Fig. 3.5.1); $D = 12.7\,mm$, $m = 8.57 \pm 0.05\,g$.

4.175.2 *Test results*

Table 1. Test results for shields with $R_{cur} = 111.8\ mm$.

v_s (m/s)	v_r (m/s)	v_s (m/s)	v_r (m/s)	v_s (m/s)	v_r (m/s)	v_s (m/s)	v_r (m/s)
123-147	Ricochet	149	23	157	57	160	57
180	105	189	116	271	223		

Table 2. Test results for shields with $R_{cur} = 203.2\ mm$.

v_s (m/s)	v_r (m/s)	v_s (m/s)	v_r (m/s)	v_s (m/s)	v_r (m/s)	v_s (m/s)	v_r (m/s)
120-153	Ricochet	156	41	159	56	165	74
257	206	274	228				

Table 3. Test results for shields with $R_{cur} = 304.8\ mm$.

v_s (m/s)	v_r (m/s)	m^{pl} (g)	v_s (m/s)	v_r (m/s)	m^{pl} (g)	v_s (m/s)	v_r (m/s)	m^{pl} (g)
104-151	Ricochet	–	152	52	–	154	45	–
154	31	0.2276	158	48	0.2295	188	123	–
209	151	–	220	165	–	266	195	–

Table 4. Test results for plane shield.

v_s (m/s)	v_r (m/s)	m^{pl} (g)	v_s (m/s)	v_r (m/s)	m^{pl} (g)	v_s (m/s)	v_r (m/s)	m^{pl} (g)
128-144	Ricochet	–	147	27	0.2118	147	31	0.2093
154	63	0.2181	242	192	0.3304			

4.176   [Stepanov and Zubov, 1998]

4.176.1   *General data on shields and projectiles*

Shields. Monolithic shields made from Al-Zn-Mg hard AA and AMg6 soft AA (labels of materials correspond to SU standard classification).

Projectiles. Steel cylindrical projectiles (Fig. 3.1.1); $D = 14.5mm$, $L = 50mm$.

4.176.2   *Test results*

Table 1. Test results on DOP.

Shield material	T (mm)	v_s (m/s)	P (mm)
Al-Zn-Mg hard AA	30	220	7.0
AMg6 soft AA	30	380	20.0
AMg6 soft AA	60	460	18.2
AMg6 soft AA	60	575	30.0

4.177   [Stock and Thompson, 1970]

4.177.1   *General data on shields and projectiles*

Shields. Monolithic 1200 AA, 2014 AA (HV 75) and 2014-T6 AA (HV 145) shields; HV at 20 kg = 196 N load.

Projectiles. Hardened steel spherical projectiles (Fig. 3.5.1).

4.177.2   *Test results*

Table 1#. Test results for 1200 AA shields.

D (mm)	v_s (m/s)	P (mm)	v_s (m/s)	P (mm)	v_s (m/s)	P (mm)	v_s (m/s)	P (mm)
6.35	229	2.6	398	5.6	629	11.2	687	13.2
6.35	843	18.5						
9.52	269	4.9	355	7.1	380	8.1	381	8.4
9.52	382	8.6	505	12.7	564	15.8	620	17.4

Table 2#. Test results for 2014 AA shields.

D (mm)	v_s (m/s)	P (mm)	v_s (m/s)	P (mm)	v_s (m/s)	P (mm)	v_s (m/s)	P (mm)
6.35	252	2.8	399	4.2	447	4.8	626	7.9
6.35	830	12.1						
9.52	273	4.2	429	7.1	587	10.3	625	11.4

Table 3#. Test results for 2014-T6 AA shields.

D (mm)	v_s (m/s)	P (mm)	v_s (m/s)	P (mm)	v_s (m/s)	P (mm)	v_s (m/s)	P (mm)
6.35	254	2.2	399	3.2	685	5.0	686	5.3
6.35	841	7.6	897	9.4				
9.52	272	3.5	354	4.7	384	5.1	506	7.6
9.52	565	7.4	566	7.9				

4.178 [Sullivan, 1942]

4.178.1 *General data on shields and projectiles*

Shields. Monolithic rolled face hardened armor (*RFHA*) and *RHA* shields; data on material hardness is given in tables below.

Projectiles. 0.50 Caliber AP M2 and 37 mm APC M51 projectiles.

4.178.2 *Test results on BLV for 0.50 caliber AP M2 projectiles*

Table 1. Test results for shields with $T = 9.52$ *mm*.

Shield		v_{bl} (m/s)				
Type	HB	$\theta_s = 0°$	$\theta_s = 20°$	$\theta_s = 30°$	$\theta_s = 40°$	$\theta_s = 45°$
RHA	241	446	500	608	639	733
RHA	245	429	502	573	733	788
RHA	329	464	–	651	–	801
RHA	331	449	665	648	801	828
RHA	341	461	579	618	806	854
RHA	415	457	660	689	804	884
RFHA	601/375	640	692	710	703	829

Table 2. Test results for shields with $T = 12.7$ *mm*.

Shield		v_{bl} (*m/s*)				
Type	HB	$\theta_s = 0°$	$\theta_s = 20°$	$\theta_s = 30°$	$\theta_s = 40°$	$\theta_s = 45°$
RHA	261	533	–	668	731	749
RHA	282	539	–	–	–	–
RHA	302	561	–	685	697	831
RHA	321	579	–	672	732	852
RHA	415	464	–	580	803	879
RFHA	653/363	349	687	730	817	905

Table 3. Test results for shields with $T = 15.9$ *mm*.

Shield		v_{bl} (*m/s*)			
Type	HB	$\theta_s = 0°$	$\theta_s = 20°$	$\theta_s = 30°$	$\theta_s = 40°$
RHA	255	578	765	849	–
RHA	302	608	645	744	840
RHA	359	664	659	819	–
RHA	409	617	718	853	–
RHA	415	609	751	878	–
RFHA	–	704	799	–	–

Table 4. Test results for shields with $T = 19.1$ *mm*.

Shield		v_{bl} (*m/s*)			
Type	HB	$\theta_s = 0°$	$\theta_s = 10°$	$\theta_s = 20°$	$\theta_s = 30°$
RHA	269	666	663	743	766
RHA	271	649	660	702	784
RHA	302	683	699	778	873
RHA	304	686	690	829	875
RHA	363	706	706	804	894
RHA	378	693	735	869	936
RHA	388	707	714	851	–
RHA	388	690	730	798	–
RFHA	597/435	694	801	871	922

Table 5. Test results for shields with $T = 25.4$ *mm*.

Shield		v_{bl} (*m/s*)		
Type	HB	$\theta_s = 0°$	$\theta_s = 10°$	$\theta_s = 20°$
RHA	263	753	764	–
RHA	272	754	833	807
RHA	279	745	–	–

Shield		v_{bl} (m/s)		
Type	HB	$\theta_s = 0°$	$\theta_s = 10°$	$\theta_s = 20°$
RHA	304	777	799	840
RHA	361	824	836	901
RHA	363	822	822	862
RHA	368	827	875	886
RHA	370	826	819	888
RHA	387	831	845	894
RFHA	601/363	908	–	–

4.178.3 *Test results on BLV for 37 mm APC M51*

Table 6. Test results for shields with $T = 25.4$ *mm*.

Shield		v_{bl} (m/s)		
Type	HB	$\theta_s = 0°$	$\theta_s = 20°$	$\theta_s = 30°$
RHA	244	432	–	483
RHA	263	422	551	–
RHA	279		463	469
RHA	304	421	444	–
RHA	361	393	410	–
RHA	368	307	375	–
RFHA	555/384	421	514	–

4.179 [Sullivan, 1945]

Note. Systematization of experimental data from reports of Watertown Arsenal Laboratory.

4.179.1 *General data on shields and projectiles*

Shields. Plates made from various steels and subjected to different types of metallurgical treatment.

Projectiles. Caliber 0.22 FSPs (Fig. 3.10.1, Table 3.10.1; hereafter *FSP*); Caliber 0.45 steel jacketed ball projectiles (Fig. 3.5.1 hereafter *SJBP*); $D = 11.4 mm$, $m = 14.9 \, g$.

4.179.2　*Test results*

Table 1. Test results for some types of steels.

Type of steel	T (mm)	HRC	v_{bl} (m/s) FSP	SJBP
Full hard stainless	1.12	45	341	201
1/2 Hard stainless	1.22	33	358	197
SAE 4330	1.22	36	473	213
SAE 4330	1.27	34	471	203
Hadfield manganese (average)	1.27	~90*	533	305
1/4 Hard stainless	1.30	27	511	278

* HRB.

Table 2. Test results for Mn-Mo steel.

T (mm)	HRC	v_{bl} (m/s) FSP	SJBP	T (mm)	HRC	v_{bl} (m/s) FSP	SJBP
0.58	35	268	–	0.76	26	239	–
0.76	18	229	101	0.79	29	236	–
0.97	34	320	125	0.97	33	322	< 140
0.99	41	419	213	1.02	35	337	151
1.22	29	540	266	1.24	35	583	313
1.27	37	585	318	1.37	31	574	158
1.02*	–	488	274				

* Hadfield manganese steel (average).

Table 3. Test results for 0.70% carbon Amola steel.

T (mm)	HRC	v_{bl} (m/s) FSP	SJBP	T (mm)	HRC	v_{bl} (m/s) FSP	SJBP
0.99	41	321	–	0.99	42	–	157
0.99	51	–	187	1.02	49	322	
1.04	53	311	–	1.04	54	0	252
1.12	50	–	253	1.17	49	355	–

Table 4. Test results for Hadfield manganese steel.

T (mm)	HRC	v_{bl} (m/s) FSP	SJBP	T (mm)	HRC	v_{bl} (m/s) FSP	SJBP
1.02	38	376	–	1.02	39	–	187
1.02	37	–	191	1.02	38	352	–
1.12	89*	–	289	1.12	88*	–	–
1.12	87*	479	–				

* HRB.

Table 5. Test results for modified SAE 4340 steel.

T (mm)	HRC	v_{bl} (m/s)		T (mm)	HRC	v_{bl} (m/s)	
		FSP	*SJBP*			*FSP*	*SJBP*
0.99	40	–	150	0.99	47	–	191
0.99	52	344	197	1.02	29	336	–
1.02	39	318	–	1.04	31	–	180

Table 6. Test results for Si-Ni-Cr-Mo steel and Cr-Mo-V steel.

T (mm)	HRC	v_{bl} (m/s)		T (mm)	HRC	v_{bl} (m/s)	
		FSP	*SJBP*			*FSP*	*SJBP*
1.04	23	332	162	1.04	37	345	155
1.04	51	403	237	1.07	46	355	171
1.07	46	369	187	1.07	49	–	208

Table 7. Test results for austenitic steel.

T (mm)	HB	v_{bl} (m/s)		T (mm)	HB	v_{bl} (m/s)	
		FSP	*SJBP*			*FSP*	*SJBP*
1.07	255	–	222	1.07	255	357	–
1.07	282	–	191	1.07	88*	497	280
1.12	282	352	–	1.14	270	–	210
1.14	270	373	–	1.14	301	–	209
1.17	295	377	–				

* HRB, Hadfield manganese steel.

Table 8. Test results for light gauge Mn-Mo steel.

T (mm)	HRC	v_{bl} (m/s)		T (mm)	HRC	v_{bl} (m/s)	
		FSP	*SJBP*			*FSP*	*SJBP*
0.99	51-53	354	< 198	1.04	40-42	327	115
1.04	41-43	306	128	1.04	47-50	404	173
1.30	39-41	535	296	1.30	45-46	523	294
1.30	47-49	531	293	1.30	50-52	533	*

* Plate split.

Table 9. Test results for Mn-Mo steel.

T (mm)	HRC	v_{bl} (m/s)		T (mm)	HRC	v_{bl} (m/s)	
		FSP	*SJBP*			*FSP*	*SJBP*
1.02	40	387	133	1.02	45	331	180
1.04	45	376	166	1.07	44	370	128
1.07	47	333	166	1.07	47	333	166

Table 10. Test results for silico-manganese spring steel.

T (mm)	HRC	v_{bl} (m/s)		T (mm)	HRC	v_{bl} (m/s)	
		FSP	*SJBP*			*FSP*	*SJBP*
0.99	51	–	207	1.07	40	–	208
1.07	49	536	–	1.09	43	357	–

Table 11. Test results for *SJBP* and austempered Mn-Mo steel.

T (mm)	HRC	v_{bl} (m/s)	T (mm)	HRC	v_{bl} (m/s)
1.02	46	169	1.07	38	163
1.07	44	177	1.09	42	149
1.09	45	193			

Table 12. Test results for X4130 steel and 8630 steel.

T (mm)	HRC	v_{bl} (m/s)		T (mm)	HRC	v_{bl} (m/s)	
		FSP	*SJBP*			*FSP*	*SJBP*
1.07	49	424	268	1.09	24	–	115
1.22	22	–	188	1.22	51	513	280

Table 13. Test results for *SJBP* and 18-8 stainless steel.

T (mm)	HRC	v_{bl} (m/s)	T (mm)	HRC	v_{bl} (m/s)
1.02	37	115	1.04	31	151
1.04	34	119	1.04	39	141
1.04	42	175	1.07	27	188
1.07	27	194	1.07	30	132
1.07	34	136	1.07	35	190
1.07	41	189	1.07	42	200

Table 14. Test results for *SJBP* and N.E. 8620 steel.

T (mm)	HRC	v_{bl} (m/s)	T (mm)	HRC	v_{bl} (m/s)
1.12	25	134	1.12	29	130
1.12	43	160	1.12	43	167
1.14	19	138	1.14	31	148

Table 15. Test results for Ni-Mo and Si-Cr-Mo-Zr steels.

T (mm)	HRC	v_{bl} (m/s)		T (mm)	HRC	v_{bl} (m/s)	
		FSP	*SJBP*			*FSP*	*SJBP*
1.14	44	514	–	1.17	35	334	–
1.19	23	322	–	1.19	37	319	–
1.19	39	314	–	1.19	40	352	–
1.19	41	398	–	1.19	44	506	–
1.22	17	–	173	1.22	17	331	–
1.22	26	–	136	1.22	28	290	–
1.22	28	418	–	1.22	35	–	149
1.22	46	–	164	1.24	20	–	152
1.24	20	272	–	1.24	20	–	164
1.24	29	–	148	1.24	31	285	–
1.24	34	–	157	1.24	36	–	149
1.24	43	–	176	1.24	45	–	210
1.27	29	–	193	1.27	33	–	187
1.27	35	–	134	1.27	38	–	208
1.32	43	364	–				

Table 16. Test results for Hadfield manganese steel.

T (mm)	HRB	v_{bl} (m/s)		T (mm)	HRB	v_{bl} (m/s)	
		FSP	*SJBP*			*FSP*	*SJBP*
0.81	83	370	–	0.81	85	–	215
1.04	83	–	288	1.04	85	488	–
1.32	90	–	335	1.32	90	543	–
1.35	90	–	334	1.60	89	–	386

Table 17. Test results for *FSP* and Hadfield manganese steel in form of M1 helmet.

T (mm)	v_{bl} (m/s)	T (mm)	v_{bl} (m/s)	T (mm)	v_{bl} (m/s)
0.91	294	0.94	300	0.97	305
0.99	309	1.02	317	1.04	326
1.07	338	1.09	352	1.12	369
1.14	389				

4.180 [Sun and Pei, 1995]

4.180.1 *General data on shields and projectiles*

Shields. Layered, spaced plates $(20\times[2.0]\backslash30.0)$ manufactured from LY12CZ Al $c_{l.w} = 60\,mm$.

Projectiles. Tungsten alloy spherical projectiles (Fig. 3.5.1); $D \approx 4.7\,mm$, $m = 1\,\mathrm{g}$.

4.180.2 *Test results*

Table 1. Test results on number of perforated layers.

θ_s (deg)	v_s (m/s)	N_{perf}	v_s (m/s)	N_{perf}	v_s (m/s)	N_{perf}	v_s (m/s)	N_{perf}
0	707	7	867	9	906	11	1097	13
0	1216	14	1400	12	1492	11		
15	738	7	883	10	910	11	1109	13
15	1215	14	1400	14	1550	11		
30	736	7	864	10	914	10	1115	13
30	1237	14	1420	14	1493	15		
45	744	6	853	7	967	9	1115	10
45	1208	11	1420	14	1493	14		
60	743	3	829	5	953	6	1109	8
60	1201	8	1410	11	1480	11		

Table 2. Test results on projectile deformation for $\theta_s = 0^\circ$.

v_s (m/s)	800	900	1000	1100	1200	1300	1400
D (m/s)	4.70	4.64	4.50	4.44	4.42	4.26	4.10
D_r (m/s)	4.70	4.72	4.74	4.76	4.78	4.82	5.00

4.181 [Sundararajan and Dikshit, 2009]

See also [Dikshit and Sundararajan, 1992b].

4.181.1 *General data on shields and projectiles*

Shields. Monolithic RHA steel (HV 350, $\sigma_y = 1080\,MPa$, $\sigma_u = 1160\,MPa$, $\varepsilon_u = 5\%$ $\varepsilon_f = 15\%$) shields; $c_{l,w} = 1m$, $T = 20\,mm$.
Projectiles. Steel (HV 750) ogive-nose projectiles (Fig. 3.3.1); $D = 6.2\,mm$, $m = 5.2\,g$.
Note. Vickers hardness (VH) at 30 kg=294 N load.

4.181.2 *Test results*

Table 1#. Test results on DOP.

v_s (m/s)	P (mm)	v_s (m/s)	P (mm)	v_s (m/s)	P (mm)	v_s (m/s)	P (mm)
309	6.2	404	7.0	451	7.5	513	8.8
533	9.0	598	10.0	645	11.0		

4.182 [Teng *et al.*, 2008]

Note. Data from [Liu and Shu, 1999].

4.182.1 *General data on shields and projectiles*

Shields. Monolithic steel ($E = 205\,GPa$, $\gamma = 7800\,kg/m^3$, $v = 0.3$, $\sigma_y = 330\,MPa$) shields; $T = 3\,mm$.

Projectiles. Steel ($E = 205\,GPa$, $\gamma = 7800\,kg/m^3$, $v = 0.3$, $\sigma_y = 625\,MPa$) spherical projectiles (Fig. 3.5.1); $D = 8\,mm$.

4.182.2 *Test results*

Table 1. Test results on v_r .

v_s (m/s)	v_r (m/s)	v_s (m/s)	v_r (m/s)	v_s (m/s)	v_r (m/s)
928	531	852	459	835	431
717	300	623	163		

4.183 [Van Valkenburg *et al.*, 1956]

4.183.1 *General data on shields and projectiles*

Shields. Monolithic shields made from different materials.
Projectiles. Spherical projectiles (Fig. 3.5.1); $D = 3.175\,mm$.

4.183.2 *Test results*

Table 1#. Test results for identical materials of shield and projectile.

Material	$\tilde{v}_s$	P (mm)	$\tilde{v}_s$	P (mm)	$\tilde{v}_s$	P (mm)	$\tilde{v}_s$	P (mm)
Aluminum	0.31	1.2	0.37	2.1	0.38	2.3	0.41	2.0
	0.47	2.8	0.51	4.0	0.52	2.5	0.53	3.6
	0.58	3.3	0.64	3.8	0.71	3.4	0.72	3.7
	0.86	3.2						
Magnesium	0.39	1.9	0.46	2.7	0.49	3.1	0.56	3.2
	0.59	3.6	0.62	3.1	0.62	3.0	0.62	2.7
	0.64	3.4	0.77	3.9	0.82	2.8		
Brass	0.34	0.9	0.35	0.9	0.53	1.8	0.59	2.0
	0.65	3.2	0.91	3.4	1.02	3.2		
Steel	0.21	1.1	0.23	1.0	0.33	1.2		
Zinc	0.46	1.9	0.47	2.1	0.48	2.5		

Table 2. Test results for penetration into lead shield with v_s =2500 *m/s*.

Impactor material	P (mm)	Impactor material	P (mm)	Impactor material	P (mm)
Magnesium	3.60	Aluminum	4.60	Brass	6.73

4.184 [Verolme *et al.*, 1999]

See also [Cimpoeru, 2002].

4.184.1 *General data on shields and projectiles*

Shields. Layered shields. Witness packs WP-1 and WP-2 with plates separated by polystyrene foam (25.4 mm).

Table 1. Structure of witness pack WP-1.

Plate number (i)	Material	T_i (mm)	Plate number (i)	Material	T_i (mm)
1	1100 H14 Al	1.0	2	1100 H14 Al	1.0
3	1100 H14 Al	3.0	4	AISI 1020 steel	1.5
5	AISI 1020 steel	1.5	6	AISI 1020 steel	1.5
7	AISI 1020 steel	1.5			

Table 2. Structure of witness pack WP-2.

Plate number (i)	Material	T_i (mm)	Plate number (i)	Material	T_i (mm)
1	AISI 1010 steel	0.8	2	AISI 1020 steel	1.5
3	AISI 1020 steel	1.5	4	AISI 1020 steel	3.0
5	AISI 1020 steel	3.0	6	AISI 1020 steel	6.0

Projectiles. Different fragment simulating projectiles (Fig. 3.10.1).

Table 3. Characteristics of FSPs [Cimpoeru, 2002].

FSP notation $\Rightarrow$	FSP-1	FSP-2	FSP-3	FSP-4	FSP-5	FSP-6
m (g)	0.162	0.486	1.1	2.85	5.3	13.4
D (mm)	2.65	4.05	5.38	7.70	9.10	12.9

4.184.2 *Test results*

Table 4. BLV of plates in witness pack WP-1, v_{bl} (m/s) .

FSP $\Rightarrow$ Plate number $\Downarrow$	FSP-1	FSP-2	FSP-3	FSP-4	FSP-5	FSP-6
1	187	128	111	–	–	–
2	314	219	174	143	115	–
3	–	468	348	267	223	–
4	–	713	511	382	343	235
5	–	–	716	520	399	287
6	–	–	–	686	512	419
7	–	–	–	–	619	474

Table 5. BLV of plates in witness pack WP-2, v_{bl} (m/s) .

FSP $\Rightarrow$ Plate number $\Downarrow$	FSP-2	FSP-3	FSP-4	FSP-5	FSP-6
1	200	170	–	–	–
2	554	392	270	236	–
3	–	648	430	340	273
4	–	–	–	754	509

4.185   [Vijayan *et al.*, 2013]

4.185.1   *General data on shields and projectiles*

Shields. Monolithic aluminum shields; $D^{sh} = 206\,mm$, $T = 2.5\,mm$.
Projectiles. Truncated cone-nose aluminum projectiles (Fig. 3.6.1) with different d; $D = 15.6\,mm$, $\beta = 17°$, $m = 29\,g$.

4.185.2   *Test results*

Table 1#. Test results for BLV.

$\dfrac{d}{D}$	v_{bl} (m/s)	$\dfrac{d}{D}$	v_{bl} (m/s)	$\dfrac{d}{D}$	v_{bl} (m/s)	$\dfrac{d}{D}$	v_{bl} (m/s)
0	120	0.25	111	0.5	107	0.75	106

4.186   [Virostek and Goldsmith, 1987]

4.186.1   *General data on shields and projectiles*

Shields. Monolithic 2024-0 Al ($E = 71\,GPa$, $\gamma = 2710\,kg/m^3$, $v = 0.33$, $\sigma_y = 75.8\,MPa$, $\sigma_u = 186\,MPa$)   and   1010   steel   ($E = 207\,GPa$, $\gamma = 2810\,kg/m^3$, $v = 0.29$, $\sigma_y = 179\,MPa$, $\sigma_u = 324\,MPa$) shields.
Projectiles. Steel (HRC 58-62) cone-nose (Fig. 3.2.1; $L = 38.1\,mm$, $\beta = 30°$, $m = 29.7\,g$) and hemispherical-nose (Fig. 3.4.1; $m = 29.5\,g$) projectiles with $D = 12.7\,mm$.
Notes. Values of v_r in tables below were calculated based on the area of recorded force curve, while * indicates that v_r was verified by camera data; R instead of v_r indicates ricochet.

4.186.2   *Test results for cone-nose projectile*

4.186.2.1   *Test results for 2024-0 Al plates*

Table 1. Test results for shields with $T = 1.27\ mm$.

θ_s (deg)	v_s (m/s)	v_r (m/s)	θ_s (deg)	v_s (m/s)	v_r (m/s)	θ_s (deg)	v_s (m/s)	v_r (m/s)
0	46.3	31.1	15	46.6	28.7	30	51.2	21.3
40	46.0	0*	45	50.3	16.2*	45	64.3	43.3

Table 2. Test results for shields with $T = 3.18$ *mm*.

θ_s (deg)	v_s (m/s)	v_r (m/s)	θ_s (deg)	v_s (m/s)	v_r (m/s)	θ_s (deg)	v_s (m/s)	v_r (m/s)
0	98.8	R	0	101	16.5	0	105	28.0
15	96.3	0	30	113	39.9	45	111	0.9

Table 3. Test results for shields with $T = 4.76$ *mm*.

θ_s (deg)	v_s (m/s)	v_r (m/s)	θ_s (deg)	v_s (m/s)	v_r (m/s)	θ_s (deg)	v_s (m/s)	v_r (m/s)
0	144	0	0	151	60.0	15	157	71.0
30	116	R	30	172	51.5	45	170	48.5

4.186.2.2 *Test results 1010 steel plates*

Table 4. Test results for shields with $T = 1.59$ *mm*.

θ_s (deg)	v_s (m/s)	v_r (m/s)	θ_s (deg)	v_s (m/s)	v_r (m/s)	θ_s (deg)	v_s (m/s)	v_r (m/s)
0	82.0	R	0	93.3	R	0	111	62.2
0	125	82.0	15	101	38.7*	15	114	65.2
30	127	78.9	45	131	69.5			

4.186.3 *Test results for hemispherical-nose projectile*

4.186.3.1 *Test results for 2024-0 Al plates*

Table 5. Test results for shields with $T = 1.27$ *mm*.

θ_s (deg)	v_s (m/s)	v_r (m/s)	θ_s (deg)	v_s (m/s)	v_r (m/s)	θ_s (deg)	v_s (m/s)	v_r (m/s)
0	49.4	R	0	61.6	R	0	70.1	37.8*
0	103	84.7	15	64.0	R	15	68.6	41.5
30	87.8	55.8	45	99.4	70.7			

Table 6. Test results for shields with $T = 3.18$ *mm*.

θ_s (deg)	v_s (m/s)	v_r (m/s)	θ_s (deg)	v_s (m/s)	v_r (m/s)	θ_s (deg)	v_s (m/s)	v_r (m/s)
0	97.5	R	15	104	0*	15	107	35.1
30	112	R	30	121	40.8	45	135	35.1

Table 7. Test results for shields with $T = 4.76\ mm$.

θ_s (deg)	v_s (m/s)	v_r (m/s)	θ_s (deg)	v_s (m/s)	v_r (m/s)	θ_s (deg)	v_s (m/s)	v_r (m/s)
0	153	83.5	0	159	96.0	15	148	78.6
15	152	100	30	167	66.4	45	160	R

4.186.3.2 *Test results 1010 steel plate*

Table 8. Test results for shields with $T = 1.59\ mm$.

θ_s (deg)	v_s (m/s)	v_r (m/s)	θ_s (deg)	v_s (m/s)	v_r (m/s)	θ_s (deg)	v_s (m/s)	v_r (m/s)
0	110	0	0	114	47.5	15	119	54.9
30	129	64.3	45	143	R	45	159	57.9

4.187 [Warren and Poormon, 2001]

4.187.1 *General data on shields and projectiles*

Shields. Monolithic 6061-T6511 Al, $\gamma = 2710\,kg/m^3$, $\sigma_y = 276\,MPa$ shields; $D^{sh} = 254\,mm$.

Projectiles. VAR 4340 steel (HRC 44.5) ogive-nose projectiles (Fig. 3.3.1); $D = 7.11\,mm$, $L_{cyl} = 59.3\,mm$, $L_{nose} = 11.8\,mm$, $\psi = 3$, $m = 20.5\,g$.

4.187.2 *Test results*

Table 1. Test results for $\theta_s = 15°$.

v_s (m/s)	$\widetilde{P}$ (mm)	$\widetilde{P}_x$ (mm)	$\widetilde{P}'_x$ (mm)	$\widetilde{P}_y$ (mm)	$\widetilde{P}'_y$ (mm)
1209	213	−70.5	−201.2	−45.7	−136.0
985	136	−39.1	−130.7	−17.6	−63.7
759	92	−34.4	−85.0	−7.1	−20.0
590	60	−26.1	−54.7	6.8	8.3

Table 2. Test results for $\theta_s = 30°$.

v_s (m/s)	$\widetilde{P}$ (mm)	$\widetilde{P}_x$ (mm)	$\widetilde{P}'_x$ (mm)	$\widetilde{P}_y$ (mm)	$\widetilde{P}'_y$ (mm)
1156	188	−113.2	−149.5	−66.4	−96.8

v_s (m/s)	$\widetilde{P}$ (mm)	$\widetilde{P}_x$ (mm)	$\widetilde{P}_x'$ (mm)	$\widetilde{P}_y$ (mm)	$\widetilde{P}_y'$ (mm)
853	108	−68.7	−83.4	−21.5	−32.3
753	91	−71.4	−55.2	−14.3	−15.0
577	62	−50.6	−35.4	5.7	5.0

Table 3. Test results for $\theta_s = 45°$.

v_s (m/s)	$\widetilde{P}$ (mm)	$\widetilde{P}_x$ (mm)	$\widetilde{P}_x'$ (mm)	$\widetilde{P}_y$ (mm)	$\widetilde{P}_y'$ (mm)	Penetration mode
1184	217	−203.3	−7.25	−141.3	−35.0	−
963	168	−149.9	8.63	−90.1	−24.5	(1)
802	109	−	−	−	−	(2)
553	37	−	−	−	−	(3)

Table 4. Description of penetration modes in Table 3.

Penetration mode	Description
(1)	Projectile nose exited from the frontal face of shield 145 mm below the entrance hole measured along the shield face. Exit angle was 95–100 degrees with respect to the shot-line axis. 12 mm of projectile nose protruded from shield
(2)	Projectile exited from the frontal face of shield 99 mm below the entrance hole measured along the shield face. Exit angle was 90–95 degrees with respect to the shot-line axis and exit velocity was estimated as 80 m/s. Bent rod was recovered
(3)	Projectile exited from the frontal face of shield 40 mm below the entrance hole measured along the shield face. Exit angle was 85–90 degrees with respect to the shot-line axis and exit velocity was estimated as 300 m/s. Bent rod was recovered

4.188 [Weidemaier *et al.*, 1993]

4.188.1 *General data on shields and projectiles*

Shields. Layered (plates-in-contact) steel (HB 280-322) shields; $T_{sum} = 43\,mm$.

Projectiles. Tungsten alloy (HRC 40-42, $\gamma = 1810\,kg/m^3$, $\widetilde{\sigma}_y = 1180\,MPa$, $\delta = 4-5\%$) hemispherical-nose projectiles (Fig. 3.4.1); $D = 17.0\,mm$, $L = 55.0\,mm$, $m = 214\,g$.

4.188.2 *Test results*

Table 1#. Test results on v_r.

Shield structure	v_s (m/s)	v_r (m/s)	v_s (m/s)	v_r (m/s)	v_s (m/s)	v_r (m/s)
[31.0]+2×[6.0]	839	0	952	400	1131	674
	1241	842				
[6.0]+[31.0] +[6.0]	996	315	1122	627	1237	809
2×[6.0]+[31.0]	982	259	1127	543	1240	788
3×[14.3]	1016	321	1015	360		

4.189 [Woodward and Cimpoeru, 1998]

4.189.1 *General data on shields and projectiles*

Shields. Monolithic and layered (plates-in-contact) 2024-T351 AA shields; $T_{sum} = 9.53\,mm$.

Projectiles. Hardened steel cone-nose (Fig. 3.2.1; $\beta = 45°$, $m = 3.80\,g$) and cylindrical (Fig. 3.1.1; $m = 3.83\,g$) projectiles with $D = 6.35\,mm$.

4.189.2 *Test results*

Table 1. Test results on BLV, v_{bl} (m/s).

Shield structure	Cone-nose projectiles	Cylindrical projectiles	Shield structure	Cone-nose projectiles	Cylindrical projectiles
1×[9.53]	462	392	[3.18]+[6.35]	421	404
2×[4.76]	484	445	[6.35]+[3.18]	474	421
3×[3.18]	473	433	6×[1.59]	454	401

4.190 [Woodward and De Morton, 1976]

4.190.1 *General data on shields and projectiles*

Shields. Monolithic shields; $c_{l,w} = 50\,mm$.

Table 1. Data on shield material.

Shield label	Description of material	T (mm)	HV*
Shield-1	SAE 4130 steel oil hardened and tempered at 640°C	1.55	275
Shield-2	SAE 4130 steel oil hardened and tempered at 640°C	3.23	275

Shield label	Description of material	T (mm)	HV*
Shield-3	Mild steel	4.78	136
Shield-4	Mild steel	1.60	117
Shield-5	7039 aluminum armor	12.7	155
Shield-6	Commercial aluminum (cold rolled)	6.35	75
Shield-7	Commercial aluminum (annealed)	6.35	38

* At 10 kg = 98 N load.

Projectiles. Silver hardened steel (HV 750 at 10 kg=98 N load) cylindrical rigid projectiles (Fig. 3.1.1); $D = 6.23\,mm$, $L = 25.4\,mm$.

4.190.2 *Test results*

Table 2. Test results for all shields on BLV and plug thickness.

Shield	v_{bl} (m/s)	T^{pl} (mm)	Shield	v_{bl} (m/s)	T^{pl} (mm)
Shield-1	114 ± 1	1.37	Shield-2	257 ± 3	3.00
Shield-3	298 ± 12	4.37	Shield-4	95 ± 12	1.30
Shield-5	732 ± 16	8.08	Shield-6	152 ± 6	5.51
Shield-7	146 ± 6	4.95			

Table 3#. Test results for Shield-6.

v_s (m/s)	v^{pl} (m/s)	T^{pl} (mm)	v_s (m/s)	v^{pl} (m/s)	T^{pl} (mm)	v_s (m/s)	v^{pl} (m/s)	T^{pl} (mm)
110	–	5.05	155	0	–	182	–	4.70
225	164	4.91	262	174	–	332	–	4.92
354	194	–	368	209	–	436	234	–
458	–	4.74	504	302	–	504	275	4.01
549	265	–	638	–	4.41	679	–	4.24
693	315	–	700	326	–	710	333	4.19
715	366	3.64	746	–	4.65	753	347	3.95

Table 4#. Test results for Shield-3.

v_s (m/s)	v^{pl} (m/s)	T^{pl} (mm)	v_s (m/s)	v^{pl} (m/s)	T^{pl} (mm)	v_s (m/s)	v^{pl} (m/s)	T^{pl} (mm)
290	0	4.35	510	134	4.20	529	–	4.36
590	169	4.26	625	268	–	640	–	4.21
650	219	4.14	660	233	–	690	198	4.25
700	201	–	720	–	4.11	725	295	3.93

4.191   [Woodward, 1978a]

4.191.1   *General data on shields and projectiles*

Shields. Monolithic shields.

Table 1. Data on shield materials.

Label	Description of material	T (mm)	HV*
Sh.1	SAE 4130 steel oil hardened and tempered at 705°C	3.20	285
Sh.2	SAE 4130 steel oil hardened and tempered at 675°C	1.60	290
Sh.3	SAE 4130 steel oil hardened and tempered at 450°C	1.60	410
Sh.4	Mild steel	4.78	136
Sh.5	Mild steel	1.60	117
Sh.6	Commercial aluminum	6.35	75
Sh.7	5083 aluminum	12.7	105
Sh.8	Hadfield manganese steel	6.35	194

* At 5 kg=49 N load.

Projectiles. Hardened steel cone-nose projectiles (Fig. 3.2.1; $\beta = 22.5°$): *Projectile-1* with $D = 4.76\,mm$, $L = 25.4\,mm$, $m = 2.83\,g$ and *Projectile-2* with $D = 6.35\,mm$, $L = 17.5\,mm$, $m = 3.03\,g$. Formally, projectiles belong to a class of truncated cone-nose projectiles (Fig. 3.6.1) with $d \leq 0.015\,mm$. The projectiles, except for one case, did not deform during the experiments.

Table 2. Shield damage types (SDTs).

Label	Description
(1)	Ductile hole formation
(2)	Dishing
(3)	Petaling (treated as dishing)

4.191.2   *Test results*

Table 3. Results for Projectile-1.

Shield	v_{bl} (m/s)	SDT	Shield	v_{bl} (m/s)	SDT	Shield	v_{bl} (m/s)	SDT
Sh.1	330	(1)	Sh.4	375	(1)	Sh.6	225	(1)
Sh.7	454	(1)	Sh.5	202	(2)	Sh.2	243	(2)
Sh.3	240	(3)						

Table 4. Results for Projectile-2.

Shield	v_{bl} (m/s)	SDT	Shield	v_{bl} (m/s)	SDT
Sh.4	442	(1)	Sh.6	293	(1)
Sh.8	603	(1)	Sh.7	536	(1)

4.192 [Woodward, 1978b]

4.192.1 *General data on shields and projectiles*

Shields. Monolithic shields; $c_{l.w} = 50\,mm$.

Table 1. Data on shield materials.

Notation	Material	Condition	HV
Shield-1	5083 AA	Cold rolled	105
Shield-2	IMI titanium 125	Rolled and annealed	160
Shield-3	IMI titanium 318 (AMS 4911B)	Forged and annealed	315
Shield-4	Titanium 811 Alloy (AMS 4916B)	Hot rolled and duplex annealed	315

Projectiles. Cone-nose projectiles (Fig. 3.2.1), $D = 4.76mm$. Formally, projectiles belong to a class of truncated cone-nose projectiles (Fig. 3.6.1) with $d \leq 0.015mm$.

Table 2. Parameters of projectiles.

Notation	β (deg)	m (g)	Notation	β (deg)	m (g)
Projectile-1	22.5	2.85	Projectile-2	30	3.08
Projectile-3	37.5	3.15	Projectile-4	45	3.21
Projectile-5*	90	3.34			

* Cylinder.

Table 3. Shield damage types (SDTs).

Notation	Description
(1)	Ductile hole formation
(2)	Adiabatic shear failure
(3)	Plug was formed with a hole in the center through which the projectile tip had penetrated
(4)	Conventional plug was formed

Note. Parameter "critical energy" was converted to BLV.

4.192.2 *Test results*

Table 4#. Results for Projectile-1 and Shield-1.

T (mm)	v_{bl} (m/s)	SDT	T (mm)	v_{bl} (m/s)	SDT
4.2	235	(1)	6.35	305	(1)
8.5	356	(1)	12.8	436	(1)

Table 5#. Results for Projectile-1 and Shield-2.

T (mm)	v_{bl} (m/s)	SDT	T (mm)	v_{bl} (m/s)	SDT	T (mm)	v_{bl} (m/s)	SDT
3.1	275	(1)	4.7	347	(1)	6.35	418	(1)

Table 6#. Results for Projectile-1 and Shield-3.

T (mm)	v_{bl} (m/s)	SDT	T (mm)	v_{bl} (m/s)	SDT
3.1	186	(2)	4.8	294	(2)
6.35	358	(2)	9.6	491	(2)

Table 7. Results for Shield-4 with $T = 6.35$ *mm*.

Projectile	v_{bl} (m/s)	SDT	Projectile	v_{bl} (m/s)	SDT
Projectile-1	460	(2)	Projectile-3	420	(2)

Table 8. Results for Shield-2 with $T = 6.35$ *mm*.

Projectile	v_{bl} (m/s)	SDT	Projectile	v_{bl} (m/s)	SDT	Projectile	v_{bl} (m/s)	SDT
Projectile-1	418	(1)	Projectile-2	384	(3)	Projectile-3	326	(4)
Projectile-4	300	(4)	Projectile-5	295	(4)			

4.193 [Woodward, 1982]

4.193.1 *General data on shields and projectiles*

Shields. Monolithic 2024 T351 Al (HV 135, hereafter aluminum), copper (HV 80), and mild steel (HV 188) shields; $D^{sh} = 50\,mm$. Most shields had generally the thickness $T = 25\,mm$, although some shields with the thickness $T = 50\,mm$ were used for deeper penetration.

Projectiles. Cone-nose projectiles (Fig. 3.2.1; $\beta = 45°$) and cylindrical (Fig. 3.1.1) projectiles with $D = 4.8mm$ made from copper (HV 114) and sintered tungsten alloy "Kennertium W2" (HV 305, hereafter tungsten): cone-nose copper projectiles ($m = 2.14\,g$), cone-nose tungsten projectiles ($m = 4.48\,g$), cylindrical copper projectiles ($m = 2.0\,g$), and cylindrical tungsten projectiles ($m = 3.3\,g$). Formally, projectiles belong to a class of slightly rounded cone-nose projectiles (Fig. 3.2.2a) with $\rho_1 \leq 0.0075mm$ for the tungsten and $\rho_1 \leq 0.03mm$ for the copper projectiles.

Additional notation. D^{hole} is the residual diameter of the hole at the front surface of shield.

Notes.

- All Vickers hardness values (HVs) at 5 kg = 49 N load.
- Because of uncertainties in graphical presentation of experimental results and different parameters in the figures in the original study, in some instances it is not feasible to determine unambiguously whether a particular parameter corresponds to one or several experiments.

4.193.2 *Test results for cone-nose projectiles*

Table 1#. Test results for copper projectiles and aluminum shields.

v_s (m/s)	D_r (mm)	D^{hole} (mm)	h^{lip} (mm)	P (mm)
250	9.9	2.9	–	0.3
270	9.9	2.8	–	0.4
340	13.2	4.4	0.0	0.8
385	–	–	0.1	–
425	11.8	7.7	0.3	1.3
455	11.0	8.3	0.5	1.6
515	10.3	9.7	0.9	2.1
545	9.6	9.0	1.2	–
570	8.4	9.0	–	2.7
575	–	9.5	1.1	3.1
590	9.8	10.0	1.7	2.8
600	10.0	10.3	2.5	3.0
630	9.4	10.0	3.1	4.2
650	–	9.0	2.3	4.3
680	8.3	10.0	2.7	4.5

v_s (m/s)	D_r (mm)	D^{hole} (mm)	h^{lip} (mm)	P (mm)
705	8.3	9.0	2.5	4.8
735	8.2	9.3	2.5	5.1
790	–	9.0	2.4	7.8
850	–	9.1	2.5	8.5
975	7.7	9.6	2.7	10.2

Table 2#. Test results for tungsten projectiles and mild steel shields.

v_s (m/s)	D_r (mm)	D^{hole} (mm)	h^{lip} (mm)	P (mm)
160	4.9	6.0	0.9	2.2
200	5.2	6.6	1.3	2.7
300	5.5	8.0	1.5	3.7
350	5.7	8.0	1.6	4.2
395	6.0	8.5	1.7	5.2
500	–	9.2	2.0	5.5
515	6.8	9.2	2.3	5.9
545	6.9	9.1	2.2	6.2
625	7.0	10.0	2.5	6.6
700	7.3	10.0	2.4	7.1
740	6.8	10.0	2.7	7.9
770	–	10.5	2.5	7.7
870	–	10.8	2.6	9.3
920	–	11.3	2.7	11.0

Table 3#. Test results for tungsten projectiles and copper shields.

v_s (m/s)	D_r (mm)	D^{hole} (mm)	h^{lip} (mm)	P (mm)
175	4.7	5.5	0.6	4.2
300	4.8	5.6	1.4	9.2
360	4.8	6.9	1.6	9.9
420	4.8	8.2	1.8	14.6
500	4.8	9.1	1.7	19.1
550	4.9	10.0	1.6	18.3
565	5.0	9.5	2.2	–
585	5.1	9.5	2.1	22.7
660	5.2	10.5	2.1	22.0
775	5.5	11.9	3.0	–
790	–	11.8	2.1	20.1
940	–	12.9	3.3	25.9

4.193.3 *Test results for cylindrical projectiles*

Table 4#. Test results for copper projectiles and aluminum shields.

v_s	D_r	D^{hole}	h^{lip}	P
(m/s)	(mm)	(mm)	(mm)	(mm)
240	9.8	5.0	0.0	0.4
430	11.8	7.1	0.5	1.3
480	10.7	8.1	0.9	1.5
490	10.5	8.5	0.6	1.5
500	10.5	9.0	0.6	1.7
525	9.9	–	1.1	1.9
540	9.0	10.1	1.8	2.2
555	9.5	9.9	1.7	2.3
580	8.7	9.6	2.4	2.4
580	9.2	9.0	2.1	2.6
580	8.7	9.5	1.9	3.1
595	8.8	9.0	–	2.7
650	8.7	8.5	1.9	3.8
660	8.3	8.6	2.4	4.1
675	8.3	8.5	2.6	4.5
680	7.7	8.1	2.2	3.4
700	8.3	8.5	1.9	4.3
710	8.2	8.5	–	4.8
740	8.2	8.5	2.2	5.7

Table 5#. Test results for tungsten projectiles and mild steel shields.

v_s	D_r	D^{hole}	h^{lip}	P
(m/s)	(mm)	(mm)	(mm)	(mm)
200	5.6	5.8	0.3	0.6
330	6.5	6.1	0.9	1.6
400	7.4	7.1	0.7	2.0
420	7.3	–	1.1	–
435	7.5	6.5	0.9	1.7
440	–	–	–	2.1
450	7.6	6.8	1.1	–
470	7.8	6.5	1.3	–
480	7.4	6.8	1.7	2.5
485	–	–	–	2.9
495	7.6	7.0	1.4	3.4
615	7.5	7.2	1.0	–
625	6.1	7.1	1.1	4.1
630	–	–	–	3.9
630	–	–	–	5.3

v_s (m/s)	D_r (mm)	D^{hole} (mm)	h^{lip} (mm)	P (mm)
645	8.0	–	1.4	3.6
680	7.5	8.8	2.1	–
695	7.9	–	2.1	6.4
725	5.5	9.2	2.2	7.0
775	–	9.2	2.7	7.8
820	–	9.3	2.7	8.4
855	–	9.0	2.5	8.8
1010	–	9.8	2.9	–

4.194 [Wu *et al.*, 1994]

4.194.1 *General data on shields and projectiles*

Shields. Monolithic 6061-T6 Al shields; $D^{sh} = 120\,mm$, $T = 1\,mm$.

Projectiles. Hard steel (HRC 60) hemispherical-nose projectiles (Fig. 3.4.1); $D = 12.7\,mm$, $L = 38.5 \pm 0.2\,mm$ $m = 35.6 \pm 0.1\,g$.

4.194.2 *Test results*

Table 1. Test results.

v_s (m/s)	v_r (m/s)	v_s (m/s)	v_r (m/s)	v_s (m/s)	v_r (m/s)	v_s (m/s)	v_r (m/s)
25.6-48.9	Ricochet	53.0	0	54.3	0	59.5	26.1
63.3	34.3	69.8	47.9	70.5	49.9	82.8	64.7
88.1	68.9	93.5	81.1				

$v_{bl} = 54\,m/s$.

4.195 [Xue *et al.*, 2010]

4.195.1 *General data on shields and projectiles*

Shields. Monolithic DH-36 steel shields; $D^{sh} = 152\,mm$, $T = 4.76\,mm$.

Projectiles. 4340 steel (HRC 38-40) projectiles with $D = 36.2\,mm$, $L_{nose} = 18.1\,mm$, $\beta = 45°$ and $m = 145\,g$: cone-nose projectiles (Fig. 3.2.1; $L_{cyl} = 11.6\,mm$) and truncated cone-nose projectile (Fig. 3.6.1; $L_{cyl} = 14.5\,mm$, $d = 27.2\,mm$). Some geometrical parameters of projectiles were estimated using pictures in the original paper.

Table 1#. Test results cone-nose projectiles.

$\overline{v}_s$	$\overline{v}_r$	$\overline{v}_s$	$\overline{v}_r$	$\overline{v}_s$	$\overline{v}_r$	$\overline{v}_s$	$\overline{v}_r$
0.87	0.00	1.00	0.00	0.99	0.36	1.00	0.40
1.11	0.42	1.26	0.38	1.27	0.69	1.48	0.93
1.84	1.45	2.04	1.64				

$a = 0.97$, $p = 1.8$.

Table 2#. Test results truncated cone-nose projectiles.

$\overline{v}_s$	$\overline{v}_r$	$\overline{v}_s$	$\overline{v}_r$	$\overline{v}_s$	$\overline{v}_r$	$\overline{v}_s$	$\overline{v}_r$
0.91	0.00	0.96	0.00	1.03	0.00	1.09	0.54
1.18	0.72	1.27	0.81	1.58	1.17	1.93	1.57

$a = 0.86$, $p = 2.8$.

4.196 [Yong *et al.*, 2010]

4.196.1 *General data on shields and projectiles*

Shields. Monolithic 2014-T4 Al ($T = 1.22\,mm$, $c_{l,w} = 50\,mm$), 6082-T6 Al ($T = 1.63\,mm$), and mild cold rolled steel ($T = 0.87\,mm$) shields.
Projectiles. Stainless steel spherical projectiles (Fig. 3.5.1); $D = 6\,mm$.

4.196.2 *Test results*

Table 1#. Test results for Al2014-T4 shields.

v_s (m/s)	v_r (m/s)	v_s (m/s)	v_r (m/s)	v_s (m/s)	v_r (m/s)	v_s (m/s)	v_r (m/s)
185	0	199	64	215	91	215	111
242	153	257	177	287	220	305	229
307	223	355	271	371	283		

Table 2#. Test results for Al6082-T6 shields.

v_s (m/s)	v_r (m/s)	v_s (m/s)	v_r (m/s)	v_s (m/s)	v_r (m/s)	v_s (m/s)	v_r (m/s)
157	0	187	123	189	107	210	148
228	175	238	178	255	202	302	226
304	246	308	260	318	251	352	303
372	306						

Table 3#. Test results for mild cold rolled steel shields.

v_s (m/s)	v_r (m/s)	v_s (m/s)	v_r (m/s)	v_s (m/s)	v_r (m/s)	v_s (m/s)	v_r (m/s)
221	0	222	32	223	41	232	79
247	113	250	125	254	124	255	106
259	143	264	107	263	149	272	164
283	163	289	192	310	243	314	183
315	208	322	203	342	229	355	242

4.197 [Zaid and Travis, 1974]

4.197.1 *General data on shields and projectiles*

Shields. Monolithic and layered (plates-in-contact) mild steel BS 1449 E1A shields; $D^{sh} = 298\,mm$.

Projectiles. Steel (HV 870 at 50 kg=490 N load) cylindrical projectiles (Fig. 3.1.1); $D = 9.52\,mm$, $L = 35.56\,mm$, $m = 22.1\,g$.

4.197.2 *Test results*

Table 1#. Test results on v_{bl} (m/s) for monolithic and two-layer shields.

T_{sum} (mm)	$1\times\left[T_{sum}\right]$	$2\times\left[0.5\cdot T_{sum}\right]$	T_{sum} (mm)	$1\times\left[T_{sum}\right]$	$2\times\left[0.5\cdot T_{sum}\right]$
1.2	49	34	1.7	97	87
2.1	112	99	2.5	146	130
3.2	172	197	4.7	282	338
6.5	420	541	9.5	842	1094

4.198 [Zener and Peterson, 1943]

4.198.1 *General data on shields and projectiles*

Shields. Monolithic armor shields.

Projectiles. Cylindrical projectiles (Fig. 3.1.1); $D = 7.62\,mm$, $m = 10.8\,g$.

4.198.2 *Test results*

Table 1. Test results on BLV.

T (mm)	HB	v_{bl} (m/s)	T (mm)	HB	v_{bl} (m/s)	T (mm)	HB	v_{bl} (m/s)
4.76	205	216	4.76	275	207	4.76	388	184
6.35	302	296	7.62	321	358			

4.199 [Zhang *et al.*, 2011]

4.199.1 *General data on shields and projectiles*

Shields. Monolithic Weldox 460 E SA shields; $D^{sh} = 280\,mm$, $T = 12\,mm$.
Projectiles. Ame tool steel cylindrical projectiles (Fig 3.1.1); $D = 20\,mm$, $L = 80\,mm$.

4.199.2 *Test results*

Table 1. Test results.

v_s (m/s)	v_r (m/s)	v^{pl} (m/s)	v_s (m/s)	v_r (m/s)	v^{pl} (m/s)	v_s (m/s)	v_r (m/s)	v^{pl} (m/s)
182	0	0	190	42	62	200	71	104
225	114	169	244	133	188	285	181	225
304	200	242						

4.200 [Zhang *et al.*, 2012b]

4.200.1 *General data on shields and projectiles*

Shields. Monolithic and layered (plates in-contact and with air gaps) Q235 low carbon steel ($\sigma_y = 229\,MPa$) shields.

Projectiles. Hardened steel 38CrSi (HRC 55) cylindrical projectiles (Fig. 3.1.1); $D = 12.7\,mm$, $m = 34.5\,g$.

4.200.2 *Test results for shields with* $T_{sum} = 2\,mm$

Table 1. Test results for 1×[2.0] shields.

v_s (m/s)	v_r (m/s)	v_s (m/s)	v_r (m/s)	v_s (m/s)	v_r (m/s)	v_s (m/s)	v_r (m/s)
122.3	0	125.1	26.5	126.7	37.9	128.6	50.9
129.5	53.4	132.9	64.2	136.7	74.7	137.0	77.4
144.0	87.5	150.0	95.5	155.1	101.9		

$v_{bl} = 124.2\,m/s$; $a = 0.94$, $p = 2.58$.

Table 2. Test results for 2×[1.0] shields.

v_s (m/s)	v_r (m/s)	v_s (m/s)	v_r (m/s)	v_s (m/s)	v_r (m/s)
105.3	0.0	111.0	0.0	117.3	25.0

v_s (m/s)	v_r (m/s)	v_s (m/s)	v_r (m/s)	v_s (m/s)	v_r (m/s)
118.8	39.0	126.1	53.9	129.3	64.6
135.5	76.4	146.6	97.7	152.0	102.2

$v_{bl} = 115.9\,m/s; a = 0.95, p = 2.39.$

Table 3. Test results for [0.5]+[1.5] shields.

v_s (m/s)	v_r (m/s)	v_s (m/s)	v_r (m/s)	v_s (m/s)	v_r (m/s)	v_s (m/s)	v_r (m/s)
101.6	0	106.9	0	111.5	25.6	111.8	20.6
112.3	28.0	113.2	22.3	117.0	43.2	119.8	54.7
123.5	46.3	124.2	63.9	127.1	74.0	139.8	92.0

$v_{bl} = 110.3\,m/s; a = 0.97, p = 2.31.$

Table 4. Test results for [1.5]+[0.5] shields.

v_s (m/s)	v_r (m/s)	v_s (m/s)	v_r (m/s)	v_s (m/s)	v_r (m/s)
114.9	0	116.2	0	119.5	17.9
121.1	35.5	122.5	45.6	126.3	57.1
134.5	74.4	136.2	81.0	146.4	93.0

$v_{bl} = 118.8\,m/s; a = 0.94, p = 2.6.$

Table 5. Test results for 4×[0.5] shields.

v_s (m/s)	v_r (m/s)	v_s (m/s)	v_r (m/s)	v_s (m/s)	v_r (m/s)
102.5	0	107.8	28.0	108.7	30.0
110.1	38.9	119.0	58.3	128.2	75.7
138.9	90.6	149.6	104.2	163.6	117.5

$v_{bl} = 104.8\,m/s; a = 0.9, p = 2.31.$

Table 6. Test results for 2×[1.0]\100.0 shields.

v_s (m/s)	v_r (m/s)	v_s (m/s)	v_r (m/s)	v_s (m/s)	v_r (m/s)	v_s (m/s)	v_r (m/s)
105.1	0	105.8	0	107.6	0	107.9	0
110.6	49.2	110.6	42.7	110.6	21.0	113.5	39.8
119.1	32.6	121.7	65.6	121.7	53.6	125.6	55.5
126.4	55.3	128.5	53.6	130.1	61.5	131.6	85.6
135.3	77.0	141.6	91.1	142.2	99.6	147.5	94.8
151.1	104.8	153.9	110.6				

Table 7. Test results for 2×[1.0]\6.0 shields.

v_s (m/s)	v_r (m/s)	v_s (m/s)	v_r (m/s)	v_s (m/s)	v_r (m/s)	v_s (m/s)	v_r (m/s)
99.3	0	102.7	30.1	104.2	27.7	116.1	63.9
133.7	89.1	145.3	106.4	167.6	137.9		

$v_{bl} = 101.5\,m/s; a = 0.97, p = 2.31.$

4.200.3 *Test results for shields with* $T_{sum} = 6\,mm$

Table 8. Test results for 1×[6.0] shield.

v_s (m/s)	v_r (m/s)	v_s (m/s)	v_r (m/s)	v_s (m/s)	v_r (m/s)	v_s (m/s)	v_r (m/s)
181.8	0	215.3	0	224.6	64.9	263.9	146.3
287.1	181.1	303.0	204.6	305.5	198.4	311.7	202.0
315.8	209.8	428.6	327.3	449.2	340.9		

$v_{bl} = 220\,m/s; a = 0.82, p = 2.5.$

Table 9. Test results for 3×[2.0]\2.0 shields.

v_s (m/s)	v_r (m/s)	v_s (m/s)	v_r (m/s)	v_s (m/s)	v_r (m/s)	v_s (m/s)	v_r (m/s)
198.4	0	227.3	107.3	250.0	133.0	265.2	166.7
272.7	163.6	273.7	170.5	287.1	181.8	287.9	181.8
314.7	218.2	320.9	217.4	327.3	200.0	332.0	227.3
337.7	230.8	339.8	251.8	355.7	259.7	422.1	311.0
446.3	327.3						

$v_{bl} = 203\,m/s; a = 0.8, p = 2.45.$

4.201 [Zhou and Stronge, 2008]

4.201.1 *General data on shields and projectiles*

Shields. Monolithic and layered (plates-in-contact) 316L stainless steel shields.

Projectiles. Cylindrical (Fig. 3.1.1) and hemispherical-nose (Fig. 3.4.1) projectiles with $D = 12.68\,mm$. Additional data is given in tables below.

4.201.2 *Test results for cylindrical projectiles*

Table 1. Test results on BLV.

L (mm)	m (g)	T_{sum} (mm)	Shield structure	θ_s (deg)	v_{bl} (m/s)
18.7	20.5	0.5	1×[0.5]	0	96
18.7	20.5	0.5	1×[0.5]	30	57
18.7	20.5	0.5	1×[0.5]	45	66
18.7	20.5	0.4	2×[0.2]	0	83
18.7	20.5	0.4	2×[0.2]	30	66
18.7	20.5	0.5	2×[0.25]	0	89
18.7	20.5	0.5	2×[0.25]	30	69
28.1	30.8	0.5	1×[0.5]	0	85
28.1	30.8	0.5	1×[0.5]	30	54
37.4	41.0	0.5	1×[0.5]	0	72
37.4	41.0	0.5	1×[0.5]	30	48
46.8	51.3	0.5	1×[0.5]	30	40

Table 2. Test results on v_r for monolithic shields with $T = 0.5\,mm$, and projectiles with $L = 18.7\,mm$, $m = 20.5\,g$, and $\theta_s = 30°$.

v_s (m/s)	v_r (m/s)	v_s (m/s)	v_r (m/s)	v_s (m/s)	v_r (m/s)
54	13	55	27	59	33
62	30	70	43		

4.201.3 *Test results for hemispherical-nose projectiles and monolithic shields with $T = 0.5\,mm$*

Table 3. Test results on BLV.

L (mm)	m (g)	θ_s (deg)	v_{bl} (m/s)	L (mm)	m (g)	θ_s (deg)	v_{bl} (m/s)
20.3	19.9	0	125	20.3	19.9	30	86
20.3	19.9	45	103	30.5	29.9	30	77
50.8	49.8	30	63				

Table 4. Test results on v_r for shields with $T = 0.5\,mm$, and projectiles with $L = 30.5\,mm$, $m = 29.9\,g$ and $\theta_s = 30°$.

v_s (m/s)	v_r (m/s)	v_s (m/s)	v_r (m/s)	v_s (m/s)	v_r (m/s)	v_s (m/s)	v_r (m/s)
86	34	91	35	101	51	111	72

4.202 [Zhou *et al.*, 2012]

4.202.1 *General data on shields and projectiles*

Shields. Layered (explosively welded layered plates) shields made from 304L steel (hereafter St) and LY12 Al (hereafter Al) layers and $T_{sum} = 5mm$.

Projectiles. Steel spherical projectiles (Fig. 3.5.1); $D = 6mm$.

4.202.2 *Test results*

Table 1#. Test results for two-layer shields.

Shield structure	v_{bl} (m/s)	v_s (m/s)	v_r (m/s)	v_s (m/s)	v_r (m/s)	v_s (m/s)	v_r (m/s)
[(St) 1.0]+ [(Al) 4.0]	410	349	0	389	0	478	162
		696	480				
[(St) 2.0]+ [(Al) 3.0]	460	279	0	360	0	875	614
[(St) 3.0]+ [(Al) 2.0]	480	400	0	619	310		
[(St) 4.0]+ [(Al) 1.0]	501	449	0	600	216	687	378

Table 2#. Test results for three-layer shields.

Shield structure	v_{bl} (m/s)	v_s (m/s)	v_r (m/s)	v_s (m/s)	v_r (m/s)
[(St) 1.0]+ [(Al) 3.0]+[(St) 1.0]	507	476	0	494	0
		532	125	640	251
[(St) 1.5]+ [(Al) 2.0]+[(St) 1.5]	585	489	0	542	0
		560	103	650	219
[(St) 2.0]+ [(Al) 1.0]+[(St) 2.0]	590	483	0	561	0
		638	137	684	210
[(Al) 1.0]+ [(St) 3.0]+[(Al) 1.0]	490	385	0	496	68
		514	98		
[(Al) 1.5]+ [(St) 2.0]+[(Al) 1.5]	445	379	0	449	28
		490	138	507	162
		514	172		
[(Al) 2.0]+ [(St) 1.0]+[(Al) 2.0]	495	362	0	409	0
		420	0	550	213

4.203 [Zook and Frank, 1985]

Data from [Grace, 1995].

4.203.1 *General data on shields and projectiles*

Shields. Monolithic RHA shields, $\gamma = 7850 \, kg/m^3$.

Table 1. Shield parameters.

T (mm)	σ_u (MPa)	σ_{ss} (MPa)	T (mm)	σ_u (MPa)	σ_{ss} (MPa)	T (mm)	σ_u (MPa)	σ_{ss} (MPa)
25.4	1300	751	50.8	1200	693	76.2	1100	535

Projectiles. Tungsten alloy ($\gamma = 17300\,kg/m^3$, $\sigma_u = 1510\,MPa$) cylindrical (Fig. 3.1.1) and hemispherical-nose (Fig. 3.4.1) projectiles with $D = 7.87\,mm$ and $L = 78.7\,mm$.

4.203.2   *Test results*

Table 2#. Test results for cylindrical projectiles.

T (mm)	v_s (m/s)	v_r (m/s)	v_s (m/s)	v_r (m/s)	v_s (m/s)	v_r (m/s)
25.4	836	369	848	0	857	354
	1008	722	1098	896	1130	918
50.8	1074	0	1081	438	1114	487
	1175	594	1233	802	1455	1207
76.2	1433	89	1473	4	1476	293
	1495	492	1514	504	1590	703

Table 3#. Test results for hemispherical-nose projectiles.

T (mm)	v_s (m/s)	v_r (m/s)	v_s (m/s)	v_r (m/s)	v_s (m/s)	v_r (m/s)
25.4	803	0	835	415	836	148
	978	754	1093	943	1223	1095
50.8	991	0	1016	191	1053	0
	1053	89	1072	166	1072	208
	1144	567	1263	879	1462	1098
76.2	1439	0	1449	319	1495	646
	1526	753	1536	930		

4.204   [Zook *et al.*, 1983]

4.204.1   *General data on shields and projectiles*

Shields. Monolithic Al 2024-T351 and Al 24S-T shields; $c_l = 76\,mm$ or $c_l = 100\,mm$, $c_w = 50\,mm$.

Table 1. Shield parameters.

Notation	Material	T (mm)	HB	Notation	Material	T (mm)	HB
Shield-1	Al 2024-T351	6.35	143	Shield-2	Al 2024-T351	50.8	153
Shield-3	Al 24S-T	50.8	163	Shield-4	Al 24S-T	76.2	163

Projectiles. Steel (HRC 65) spherical projectiles (Fig. 3.5.1): *Projectile-1* ($D = 6.35mm$, $m = 1.044\,g$) and *Projectile-2* ($D = 11.91mm$, $m = 6.888\,g$).

Table 2. Explanation to notations in tables below.

Notation	Description
CP	Complete penetration
*	Not applicable
**	Penetrator was broken up
∧	Ball was embedded but extracted
∧∧	Added mass due to stuck aluminum

4.204.2 *Test results for* $\theta_s = 0°$

Table 3. Test results for Projectile-1 and Shield-1.

v_s (m/s)	v_r (m/s)	θ_r (deg)	P (mm)	m_r (g)
341	5	171.0	2.1	–
388	6	179.7	2.5	1.044
533	–	–	4.3	1.045
537	–	–	4.5	1.044
576	0	*	4.7	–
587	0	*	5.4	–
631	0	*	7.0	–
650	32	0.5	CP	–
658	98	5.2	CP	1.044
757	340	0.3	CP	1.044
836	457	0.1	CP	1.046
1097	777	0.2	CP	1.046
1144	825	0.3	CP	1.047
1338	992	0.5	CP	**

Table 4. Test results for Projectile-1 and Shield-2.

v_s (m/s)	v_r (m/s)	θ_r (deg)	P (mm)	m_r (g)
206	36	178.8	1.4	–
214	38	178.6	1.4	1.044

v_s (m/s)	v_r (m/s)	θ_r (deg)	P (mm)	m_r (g)
225	37	179.5	1.5	1.043
228	37	179.8	1.6	–
264	40	179.4	1.9	–
272	40	180.0	2.1	1.043
434	46	181.0	2.4	1.044
437	–	–	3.1	1.044
469	47	181.0	3.0	1.044
536	45	179.5	3.6	1.044
635	16	179.6	4.8	1.044
956	0	*	11.0	–
967	0	*	10.2	–
993	0	*	11.2	–
1003	0	*	11.3	–
1009	0	*	11.7	–
1016	0	*	11.5	–
1020	0	*	13.1	–
1031	0	*	13.4	–
1032	0	*	15.4	–
1038	–	–	CP	–
1045	137	-1.4	CP	1.045
1050	157	0.4	CP	–
1052	–	–	CP	–
1062	112	-0.1	CP	–
1065	97	0.4	CP	–
1071	159	3.6	CP	1.044
1072	–	–	CP	–
1183	423	-0.4	CP	–
1205	485	-0.2	CP	1.044
1230	522	0.7	CP	–
1301	–	–	CP	–
1323	651	0.5	CP	–
1326	681	0.8	CP	1.044
1340	672	0.5	CP	–
1417	766	0.3	CP	–
1430	782	0.2	CP	–
1486	834	0.5	CP	–
1492	834	0.4	CP	–
1519	863	0.7	CP	–
1572	909	0.5	CP	**
1600	927	0.5	CP	**
1606	927	0.3	CP	**
1607	934	0.2	CP	**
1615	942	0.6	CP	**
1625	939	0.5	CP	**

v_s	v_r	θ_r	P	m_r
(m/s)	(m/s)	(deg)	(mm)	(g)
1626	948	0.8	CP	**
1629	–	–	CP	**
1634	962	0.5	CP	**

Table 5. Test results for Projectile-1 and Shield-3.

v_s	v_r	θ_r	P	m_r
(m/s)	(m/s)	(deg)	(mm)	(g)
212	43	176.4	0.9	0.634
281	47	179.4	1.3	0.635
359	52	179.6	2.0	0.634
384	53	180.3	2.0	0.634
406	53	179.8	2.3	0.634
412	–	–	2.5	–
472	–	–	3.2	0.636
511	59	181.1	3.3	–
587	66	182.3	4.1	0.634
625	65	184.7	4.4	0.635
696	69	179.7	5.1	0.635
771	70	180.6	6.3	–
786	68	180.6	6.3	–
872	57	180.9	7.2	0.636
891	38	186.7	7.3	0.638
940	0	*	6.0	–
975	0	*	6.8	–
1305	0	*	13.0	–
1707	0	*	16.6	–

Table 6. Test results for Projectile-2 and Shield-1

v_s	v_r	θ_r	P	m_r
(m/s)	(m/s)	(deg)	(mm)	(g)
143	18	184.6	1.3	6.891
267	24	182.7	2.7	6.888
412	12	180.2	6.4	6.888
415	–	–	6.4	6.888
419	20	182.9	6.4	6.888
425	–	–	CP	6.888
439	122	–7.1	CP	6.888
441	138	5.0	CP	6.888
447	156	–2.7	CP	6.888
473	–	–	CP	6.888

v_s (m/s)	v_r (m/s)	θ_r (deg)	P (mm)	m_r (g)
475	227	1.4	CP	6.888
578	368	0.4	CP	6.888
744	566	0.4	CP	6.888
927	765	0.5	CP	6.888
1025	866	0.5	CP	6.885
1039	880	1.0	CP	**

Table 7. Test results for Projectile-2 and Shield-2.

v_s (m/s)	v_r (m/s)	θ_r (deg)	P (mm)	m_r (g)
329	5	178.5	4.2	6.888
427	6	179.2	5.6	6.890
476	6	186.0	7.0	6.888
500	12	179.6	7.2	6.888
547	–	–	8.4	6.888
591	56	-2.6	CP	6.888
637	154	-0.4	CP	6.889
637	158	1.1	CP	6.889
746	325	0.4	CP	6.890
749	–	–	CP	6.892
949	601	0.0	CP	6.887
1048	720	0.3	CP	6.886^
1266	951	-0.3	CP	**
1394	1076	0.1	CP	**
1489	1158	0.2	CP	**

Table 8. Test results for Projectile-2 and Shield-4.

v_s (m/s)	v_r (m/s)	θ_r (deg)	P (mm)	m_r (g)
189	40	179.8	1.9	6.888
360	51	180.2	3.9	6.889
371	52	181.0	4.2	6.888
430	55	180.9	5.0	6.888
586	66	180.8	8.1	6.889
722	66	178.6	10.9	6.889
733	66	177.3	10.9	6.889
835	53	180.4	13.0	6.888
908	34	182.1	14.8	6.890
938	0	*	11.7	–
992	0	*	15.3	–
1361	0	*	25.5	–

4.204.3 *Test results for* $\theta_s = 45°$

Table 9. Test results for Projectile-1 and Shield-1.

v_s (m/s)	v_r (m/s)	θ_r (deg)	P (mm)	m_r (g)
109	66	117.8	2.0	0.2
187	99	120.7	2.3	0.6
211	113	114.0	3.0	0.6
220	–	–	2.5	0.7
244	129	115.5	3.5	0.9
323	144	120.2	4.0	1.1
344	145	123.0	3.5	1.4
368	147	126.8	4.1	1.6
395	–	–	5.1	1.5
445	149	127.7	5.5	2.2
544	108	145.2	6.5	2.9
615	52	188.3	7.8	3.8
654	48	195.1	8.9	4.2
673	27	232.6	9.7	4.4
687	26	229.2	9.2	4.2
716	9	201.7	10.3	5.5
732	13	219.4	10.1	5.8
754	–	–	CP	CP
755	0	*	12.8	7.1
775	100	9.9	CP	CP
775	77	-0.9	CP	CP
831	209	17.9	CP	CP
857	276	27.6	CP	CP
989	–	–	CP	CP
1007	506	37.8	CP	CP
1036	542	38.3	CP	CP
1036	538	38.7	CP	CP
1038	544	38.8	CP	CP
1213	762	42.8	CP	CP
1286	834	43.4	CP	CP
1405	963	44.6	CP	CP

Table 10. Test results for Projectile-1 and Shield-2.

v_s (m/s)	v_r (m/s)	θ_r (deg)	P (mm)	m_r (g)
64	–	–	1.4	0.1
170	99	120.5	2.8	0.5
192	107	118.3	2.8	0.4
197	112	120.0	2.9	0.6
218	119	118.9	3.8	0.8
308	152	115.9	4.0	1.1

v_s (m/s)	v_r (m/s)	θ_r (deg)	P (mm)	m_r (g)
325	150	127.7	4.0	1.4
334	156	126.5	4.0	1.4
362	156	130.5	4.8	1.6
392	168	125.6	4.4	1.7
492	166	127.0	6.2	2.3
534	163	133.4	5.9	2.6
647	141	136.6	6.9	3.0
786	137	160.0	9.5	4.4
815	122	158.8	10.0	4.4
875	85	160.4	11.9	5.1
947	92	172.6	12.1	6.0
949	50	203.5	12.3	6.1
1021	58	201.3	13.7	7.0
1032	43	197.3	14.2	7.1
1058	38	194.6	15.0	6.9
1093	31	227.5	15.6	7.8
1136	23	223.3	16.4	–
1136	0	*	–	–
1187	0	*	17.2	8.5
1235	0	*	18.6	–
1244	0	*	19.6	–
1269	0	*	22.0	14.1
1273	28	-0.8	CP	CP
1313	177	8.0	CP	CP
1317	156	6.3	CP	CP
1369	291	21.5	CP	CP
1425	355	27.0	CP	CP
1546	528	36.7	CP	CP

Table 11. Test results for Projectile-1 and Shield-3.

v_s (m/s)	v_r (m/s)	θ_r (deg)	P (mm)	m_r (g)
255	138	114.9	0.9	1.044
303	154	118.3	1.2	1.044
408	181	123.3	1.5	1.044
451	183	128.4	1.8	1.044
454	179	126.9	1.9	1.044
552	180	137.0	2.4	1.044
624	166	135.7	2.9	1.045
655	168	138.1	3.2	1.044
798	172	148.8	4.1	1.044
857	144	154.9	4.3	1.044
918	–	–	5.0	1.044
927	–	–	5.0	1.044
951	124	167.9	5.4	1.045

v_s (m/s)	v_r (m/s)	θ_r (deg)	P (mm)	m_r (g)
989	107	181.9	5.6	1.044
1075	57	198.1	6.7	1.045
1165	46	198.3	7.4	1.044
1222	52	187.6	8.1	1.044
1233	48	205.6	8.2	1.044
1247	54	207.1	8.2	1.044
1267	0	*	7.8	–
1277	41	225.4	8.6	–
1293	39	200.2	9.2	1.044
1293	24	231.2	8.8	1.045
1317	0	*	–	–
1346	11	*	9.5	1.044
1413	0	*	–	–

Table 12. Test results for Projectile-2 and Shield-1.

v_s (m/s)	v_r (m/s)	θ_r (deg)	P (mm)	m_r (g)
155	80	109.7	1.4	6.888
190	90	110.4	1.8	6.888
249	103	114.3	2.7	6.888
299	100	113.2	3.4	6.887
332	98	116.2	4.0	6.888
358	92	118.6	4.1	6.888
407	76	126.5	5.2	6.887
408	65	131.1	5.1	6.887
444	33	155.2	6.7	6.887
451	13	182.0	5.9	6.887
455	28	150.7	6.2	6.888
460	15	210.2	4.9	6.888
481	18	181.9	10.5	6.886
497	20	177.1	–	6.887
497	16	185.5	–	6.887
502	26	3.9	CP	6.886
516	85	−0.5	CP	6.886
519	88	−4.5	CP	6.888
522	89	7.8	CP	6.887
530	115	4.3	CP	6.888
616	294	34.1	CP	6.887
708	419	38.4	CP	6.888
794	530	41.6	CP	6.888
930	687	43.5	CP	6.887
1044	810	44.2	CP	6.887
1239	1016	45.1	CP	**
1275	1051	45.0	CP	**

Table 13. Test results for Projectile-2 and Shield-2.

v_s (m/s)	v_r (m/s)	θ_r (deg)	P (mm)	m_r (g)
174	99	117.5	1.4	6.886
229	124	111.4	1.9	6.885
303	153	116.8	2.5	6.886
361	–	–	2.9	6.886
366	157	124.1	3.0	6.886
421	166	128.2	4.1	6.886
482	131	131.3	4.4	6.886
482	143	126.2	4.4	6.887
494	144	139.3	5.1	6.888
587	102	146.6	6.2	6.887
644	64	181.0	7.4	6.887
648	72	137.8	7.3	6.887
694	38	202.5	8.7	6.888
750	17	216.3	10.8	6.888
777	0	*	–	6.888^
790	75	-1.7	CP	6.888
814	141	7.3	CP	6.899^^
956	374	33.3	CP	6.888
1079	558	39.3	CP	6.889
1083	–	–	CP	6.889
1197	720	42.8	CP	6.888
1323	855	43.9	CP	**
416	952	44.4	CP	**
1477	1013	43.8	CP	**

Table 14. Test results for Projectile-2 and Shield-3.

v_s (m/s)	v_r (m/s)	θ_r (deg)	P (mm)	m_r (g)
237	137	115.3	1.7	6.888
373	177	122.9	3.4	6.889
412	186	124.7	3.2	6.889
508	187	135.3	4.5	6.889
543	171	128.7	5.6	6.890
580	170	130.7	5.4	6.888
607	175	129.9	5.7	6.889
697	182	133.4	7.1	6.889
810	168	141.9	8.7	6.889
857	132	139.6	9.1	6.890
1033	117	171.7	13.2	6.889
1039	109	165.0	13.2	6.891
1110	–	–	14.9	6.890
1114	71	157.8	15.0	6.891

v_s (m/s)	v_r (m/s)	θ_r (deg)	P (mm)	m_r (g)
1118	28	218.1	14.7	6.889
1131	83	183.1	15.7	6.891
1211	38	192.1	16.7	6.889
1216	56	181.6	17.1	6.890
1264	41	203.4	17.7	6.890
1339	30	211.5	19.6	6.890
1374	0	*	19.8	6.890
1440	0	*	20.5	–
1542	49	176.7	16.7	**

4.204.4 *Test results for* $\theta_s = 60°$

Table 15. Test results for Projectile-1 and Shield-1.

v_s (m/s)	v_r (m/s)	θ_r (deg)	P (mm)	m_r (g)
118	93	110.0	0.3	1.044
123	94	110.6	0.2	1.044
168	125	110.0	0.4	1.044
250	182	110.0	0.7	1.045
339	230	108.9	0.8	1.044
409	259	108.7	1.0	1.045
440	270	110.0	1.3	1.044
524	294	113.2	1.4	1.044
569	302	115.2	1.8	1.045
573	306	115.4	1.9	1.044
634	303	116.0	2.2	1.044
753	279	122.6	2.9	1.044
964	205	134.8	3.9	1.044
946	116	158.8	5.7	1.044
953	111	171.7	5.0	1.044
960	77	163.7	5.0	1.044
997	43	181.1	5.9	1.044
1016	2	163.1	6.7	1.044
1025	4	188.8	7.0	1.044
1040	0	*	7.1	–
1047	–	–	CP	1.044
1068	78	-7.6	CP	1.044
1073	85	6.2	CP	1.045
1082	–	–	CP	1.044
1142	264	34.5	CP	1.045
1239	459	48.3	CP	1.044
1335	599	53.3	CP	1.044
1341	595	53.7	CP	1.044
1635	926	59.8	CP	**

Table 16. Test results for Projectile-1 and Shield-2.

v_s (m/s)	v_r (m/s)	θ_r (deg)	P (mm)	m_r (g)
89	–	–	0.1	1.044
126	106	96.4	0.2	1.044
190	–	–	0.3	1.044
226	165	110.1	0.6	1.044
234	–	–	0.8	1.044
254	–	–	0.8	1.044
313	219	110.8	0.8	1.044
315	219	110.5	0.8	1.045
347	232	108.6	0.9	1.046
391	256	110.2	1.0	1.046
408	262	110.8	0.9	1.044
463	–	–	1.4	1.044
477	290	113.0	1.3	1.046
662	335	116.6	2.0	1.045
754	339	118.3	2.4	1.045
830	332	122.3	2.8	1.044
989	307	129.9	3.6	1.045
1103	257	133.1	4.5	1.045
1387	94	175.5	6.5	1.045
1453	45	186.4	7.5	1.045
1544	47	214.2	7.9	1.045
1567	31	230.9	8.2	1.044
1591	–	–	9.2	1.044
1625	30	234.2	–	1.041
1636	24	237.0	–	1.043
1649	–	–	8.2	**
1651	13	223.3	–	1.037
1696	28	196.2	8.9	**
1709	–	–	7.9	**
1728	54	180.3	8.1	**
1748	68	181.1	8.7	**
1749	–	–	8.0	**
1754	–	–	8.0	**
1757	75	179.5	8.3	**
1766	–	–	8.0	**
1787	–	–	8.5	**
1801	–	–	8.6	**
1805	–	–	8.9	**

Table 17. Test results for Projectile-1 and Shield-3.

v_s (m/s)	v_r (m/s)	θ_r (deg)	P (mm)	m_r (g)
257	184	112.2	0.6	1.042
389	254	112.6	1.1	1.044
517	304	114.1	1.2	1.043

Table 18. Test results for Projectile-2 and Shield-1.

v_s (m/s)	v_r (m/s)	θ_r (deg)	P (mm)	m_r (g)
133	104	102.4	0.6	6.886
249	175	102.3	1.7	6.886
269	184	104.1	1.8	6.886
315	–	–	2.4	–
333	216	104.5	2.8	6.885
388	210	110.3	4.0	6.885
466	210	111.5	5.3	6.885
591	199	114.8	7.8	6.885
632	158	119.7	6.35	6.885
673	67	135.7	6.35	6.886
682	44	156.1	6.35	6.886
686	44	160.6	6.35	6.885
695	17	162.1	6.35	6.885
697	–	–	CP	6.886
711	67	-6.2	CP	6.885
719	138	25.1	CP	–
746	–	–	CP	6.886
774	279	46.6	CP	6.886
816	342	49.6	CP	6.886
818	–	–	CP	6.886
859	–	–	CP	6.885
931	522	55.2	CP	6.885
1095	727	58.1	CP	6.885
1299	952	59.5	CP	6.886
1417	1069	60.1	CP	6.885

Table 19. Test results for Projectile-2 and Shield-2.

v_s (m/s)	v_r (m/s)	θ_r (deg)	P (mm)	m_r (g)
109	86	109.8	0.3	6.886
124	98	110.1	0.5	6.886
138	–	–	0.4	6.885
190	144	109.5	0.8	6.885

v_s (m/s)	v_r (m/s)	θ_r (deg)	P (mm)	m_r (g)
292	202	108.0	1.1	6.886
387	255	111.2	1.9	6.885
477	281	113.2	2.5	6.885
534	301	115.3	3.2	6.885
590	311	112.4	3.7	6.886
718	305	118.7	5.0	6.886
805	276	122.6	5.9	6.885
839	251	126.3	6.5	6.885
874	215	133.6	7.3	6.886
916	172	143.3	8.3	6.885
958	150	143.4	9.1	6.885
971	141	141.8	9.6	6.886
1004	87	158.6	10.0	6.885
1011	16	221.2	9.9	6.885
1029	19	220.6	10.2	6.885
1040	28	-9.6	CP	6.886
1052	–	–	CP	6.886
1062	52	-27.2	CP	6.885
1081	105	5.6	CP	6.886
1083	94	-3.3	CP	6.885
1087	102	9.7	CP	6.886
1095	–	–	CP	6.886
1141	226	32.2	CP	6.886
1177	297	37.9	CP	6.885
1199	–	–	CP	6.886
1246	441	47.2	CP	6.885
1339	565	51.8	CP	6.885
1399	–	–	CP	6.885
1405	661	54.3	CP	6.885

Table 20. Test results for Projectile-2 and Shield-3.

v_s (m/s)	v_r (m/s)	θ_r (deg)	P (mm)	m_r (g)
308	208	109.9	1.7	6.887
513	300	113.9	3.0	6.888

Chapter 5

Penetration into concrete and reinforced concrete

5.1 Cumulative indexes of experiments

Classification of experiments according to the types of strikers is presented in Tables 1-7. In the column "Structure of shield" M and L correspond to monolithic and layers-in-contact shields, respectively; column "Meshes" includes information about a type of the reinforcement (R – rebars, W – wires, minus – absence of both); in the column "Fibers" – plus (+) and minus (–) indicate the presence or the absence of fiber-reinforcement, respectively; in the column "Liners" plus (+) and minus (–) indicate the presence or the absence of liners at frontal or rear faces of the shield, respectively. In the column containing values of ranges of parameter f'_c, the circle and square indicate that the tests were conducted with the specimens having cylindrical or cubic shapes, respectfully, while the absence of these symbols indicates the lack of this information.

When both variants are realized, two symbols in the column are separated by a slash.

The note in the beginning of Section 4.1 holds with respect to bullets and shells (Table 7).

Table 1. Experiments on normal penetration of cylindrical projectiles (Fig. 3.1.1).

Section	D (mm)	Structure of shield	Meshes	Fibers	Liners	f'_c (MPa)
5.8	40	M	R	–	+/–	○30-50
5.10	100-300	M	R/–	–	–	33-50
5.11	100-300	M	R/–	–	–	15-79

Section	D (mm)	Structure of shield	Meshes	Fibers	Liners	f_c' (MPa)
5.12	100-250	M	R/–	–	–	32-60
5.81	35	M, L	R/–	–	+/–	25
5.85	12.7	M	–	–	–	27
5.89	12	M	–	–	–	35
5.90	12	M	–	–	–	35
5.91	20-203	M	–	–	–	24-42
5.95	101	M	R	–	–	24
	300	M	R	–	–	24, 35
5.98	25	M	–	+	–	25, 162-193
5.100	12	M	–	–	–	35

Table 2. Experiments on penetration of cone-nose projectiles (Fig. 3.2.1).

Section	D (mm)	Structure of shield	Meshes	Fibers	Liners	f_c' (MPa)	θ_s
5.12	100	M	R/–	–	–	39-48	0
5.23	25	M	R/W	+/–	–	□24, 34, 104, 110	0
5.24	35	M	W	–	–	○25, 35, 39, 112	0
5.25	49	M	R	–	–	□30	0
5.26	49	M	R	+/–	–	□40, 93, 101-119	0
5.27	50	M, L	R	+/–	–	□90-117	0
5.28	25	M	R/W	–	–	□○34-39	0
5.29	25	M	R/W	–	–	□34-39, 112	0
5.55	60	M	–	–	–	□34-51	
5.69	37	M	–	+	–	□107, 116	>0
5.79	5.0	M	–	+/–	–	39-66	0
5.104	12.7	M	–	–	+	30-41	0

Table 3. Experiments on penetration of ogive-nose projectiles (Fig. 3.3.1).

Section	D (mm)	Structure of shield	Meshes	Fibers	Liners	f_c' (MPa)	θ_s
5.13	12.7	M	–	–	–	○26-31	0
5.17	7.62	M	–	–	–	35	0
5.18	50.8	M	–	–	–	37-40	0
5.19	26.9	M	–	+	–	85, 157	0
5.20	26.9	M	–	–	–	○32-40	0
5.21	25	M	–	–	–	□45	≥0
5.30	25	M	–	–	–	140, 200, 600	0
5.31	57	M	–	+	–	30, 186	0
	80	M	–	+/–	–	30, 186	0
5.34	22.0, 88.6	M	–	–	–	○49	0

Section	D (mm)	Structure of shield	Meshes	Fibers	Liners	f'_c (MPa)	θ_s
5.35	52	M	–	–	–	30	0
5.36	26.9	M	–	–	–	○32-40, 90-108	0
5.37	20.3	M	–	–	–	○56-70	0
	30.5	M	–	–	–	○48-55	0
5.38	76.2	M	–	–	–	23, 39	0
5.39	20.3, 30.5	M	–	–	–	○58	0
5.40	76.2	M	–	–	–	○23	0
5.42	31	M	–	–	–	–	0
5.43	30	M	R/–	+/–	–	34	0
5.44	30	M	–	–	–	31	0
5.47	6.35	M	–	–	–	38	0
5.48	50.8	M	–	–	–	○43	0
5.49	60	M	–	–	–	□40-45	0
5.50	12	M	–	–	–	42	0
5.51	25.4	M	R	–	–	48, 140	0
5.52	152	M	–	–	–	○92	0
5.53	50	M	R/–	–	–	○37-55, 97, 135, 146	0
5.58	64	M	–	–	–	34	0
5.59	12.7	M	–	–	–	56	0
5.65	20	M	–	–	–	77	≥0
5.66	62	M	–	–	–	30	0
5.70	12.6	M	–	+	–	□70, ○55	0
5.72	152	M	R/–	+/–	–	39, 103, 120, 149-158	0
5.77	5.7	M	W	–	–	□55	0
5.78	12	M	–	–	–	□51	0
5.80	26.9	M	–	+	–	○155–162	0
5.82	13.35	M,L	–	–	–	○84-111	0
5.87	15	L	–	–	–	32-40	0
5.88	62	L	–	–	–	30, 35, 45	0
5.89	11-307	M	–	–	–	30–39, 46, 90, 140, 180–200	0
5.90	12	M	–	–	–	35	0
5.92	12.6	M	–	+	–	67	0
5.96	152	M	–	–	–	○103	0
5.99	75	M, L	–	+/–	–	30, 38, 90, 140, 153, 180, 200	0
5.101	20	M	–	–	–	20	0
5.103	75	M	R/–	–	–	○153	0
5.105	13.35	M	–	+	–	○82–91	0
5.107	157	M	–	–	–	27	0
5.108	25.3	M	–	+/–	–	35, 67, 87-99, 114-142	0

Section	D (mm)	Structure of shield	Meshes	Fibers	Liners	f'_c (MPa)	θ_s
5.109	25.3	M, L	R	–	+	41	0
5.110	25.3	M	–	+	–	111-129	0
5.112	12.6	M	–	+/–	–	□46, 58, 87, 93, 112-151, 183-237	0
5.113	12.6	M	–	+/–	–	□40-46, 60-76, 85-93, 106-130	0
5.115	144	M	–	–	–	30-35	0

Table 4. Experiments on normal penetration of hemispherical-nose projectiles (Fig. 3.4.1).

Section	D (mm)	Structure of shield	Meshes	Fibers	Liners	f'_c (MPa)
5.5	40	M	R	+/–	–	○65-75
5.6	40	M	R	+/–	–	○49
5.12	100	M	R	–	–	50
5.13	12.7	M	–	–	–	○28
5.45	12.7-20.0	M	R/–	–	+/–	37
5.63	60	M, L	R/–	–	+/–	27

Table 5. Experiments on normal penetration of spherical projectiles (Fig. 3.5.1) into monolithic shields.

Section	D (mm)	Meshes	Fibers	Liners	f'_c (MPa)
5.13	12.7	–	–	–	○29
5.22	7.9, 11.1, 12.7	–	+	–	○190-244*

* Data from external sources.

Table 6. Experiments on penetration into monolithic shields by projectiles having complex shapes.

Section	Projectile's shape	D (mm)	Structure of shield	Meshes	Fibers	Liners	f'_c (MPa)	θ_s
5.2	Fig. 3.8.1	23	M	R/–	–	+/–	□26	0
5.4	Fig. 3.7.1a	40	M	R	+/–		○29-44, ○54-71	0
5.9	Fig. 3.4.2	25	M	–	–	–	25	0
5.12	*		M	R	–	–	31, 41	0
	*		M	R	–	–	33, 41, 46	0
5.13	*	12.7	M	–	–	–	○26	0

Section	Projectile's shape	D (mm)	Structure of shield	Meshes	Fibers	Liners	f_c' (MPa)	θ_s
5.14	Fig. 3.6.1	20	M	–	+	–	○79, 129, 180 □88, 140, 197	0
5.15	Fig. 3.6.1	20	M	–	+	–	○14, 41-197	
5.16	Fig. 3.10.1	5.5	M	–	–	–	43	0
5.32	Fig. 3.7.1a	37	M	–	+	+	75; 88	0
5.33	*	305	M	R			40	
5.41	*	31	M	–	–	–	30	0
5.46	*	200-300	M	R	–	–	42-51	0
5.54	Fig. 3.9.1	45	M	R/–	–	+/–	29	0
5.56	Fig. 3.1.2b	12.7	M	–	–	–	27	0
	Fig. 3.1.2d	12.7	M	–	–	–	27	0
5.60	Fig. 3.8.1	23	L	R/–	–	–	35	0
5.61	*	40	M	R/–	–	–	48	0
5.62	Fig. 3.8.1	150	M	–	–	+	34	0
5.67	Fig. 3.6.2	25.3	M	–	–	-	30	0
		65	M	–	–	–	32	0
5.68	*	7.85	M	R/–	+	–	□40-45 ○32-37	0
5.71	*	7.92	M	–	+/–	–	○132, 148-152	0
5.75	Fig. 3.4.2	25	M	–	–	–	25	0
5.76	Fig. 3.8.1	23	L	W/–	–	–	35	0
5.83	*	See text	M	R	–	–	19-34	0
5.84	Fig. 3.1.2b-d	12.7	M	–	–	–	27	0
5.85	Fig. 3.1.2b	12.7	M	–	–	–	27	0
5.86	Fig. 3.1.2c	12.7	M	W	–	–	27	≥0
5.93	*	7.92	M	–	+/-	–	○38, 69, 132-164	0
5.94	Fig.3.1.2a-b	12.7	M	–	–	–	40**	0
5.97	*	85.1	M	–	–	–	35	>0
5.99	*	152	M, L	–	+/–	–	30-41, 83, 90, 108-250	0
5.102	*	39.5	M	R	–	+/–	24	0
5.111	Fig.3.6.4b	37	M	R/–	+/–	+/–	○46-101	0
5.114	*	12.6	M	–	+/–	–	□49-72	0
5.115	Fig. 3.3.2	48	M	–	–	–	30-35	0

* See corresponding section.
** Data from external sources.

Table 7. Experiments on penetration of bullets and shells.

Section	Caliber or D (mm)	Structure of shield	Meshes	Fibers	Liners	f'_c (MPa)	θ_s
5.3	5.56	M	–	+/–		○40	0
	7.62	M	–	+/–		○40	0
	12.7	M	–	+/–		○40	0
5.7	7.62	M	–	+/–		–	0
5.57	5.56	M	–	–	–	21-58	0
	7.62	M	–	–	–	23-58	0
5.64	14.8	M	–	+/–	–	124-251	0
5.72	57	M, L	R/–	+	–	30, 39, 149, 158;180;220	0
	152	M, L	R/–	+/–	–	38, 83, 108, 141, 153, 180, 200	0
	75	M, L	R/–	+/–	–	34-41, 83, 90, 108-141, 180-220	0
5.73	5.56	M	–	+	–	2.1-3.5	0
5.74	14.5	M	–	–	–	60	>0
5.106	–	M	–	–	–	15-42	0

5.2 [Abdel-Kader and Fouda, 2014]

5.2.1 *General data on shields and projectiles*

Shields. RC slabs with one mesh without liners and slabs with liner (liners) without meshes; $c_{l,w} = 500\,mm$, $T = 100\,mm$.

Concrete. $\gamma = 2400\,kg/m^3$, $f'_c = 26\,MPa$ (150 mm edge cubic specimens were tested), $f_r = 4.6\,MPa$ ($100 \times 100 \times 500$ mm beam specimens were tested), $\sigma_u = 2.5\,MPa$, $f_{sl} = 85\,mm$.

Meshes. Steel rebars ($\sigma_y = 240\,MPa$, $\varepsilon_u = 20\%$). *Shield-F, Shield-I* and *Shield-R* have identical meshes with the following parameters: $d_{l,w} = 6\,mm$, $g_{l,w} = 80\,mm$, $\rho_{l,w} = 0.4\%$.

Shields with one mesh. Location of the meshes: frontal face ($t^F = 15\,mm$) for Shield-F, rear face ($t^R = 15\,mm$) for Shield-R, and intermediate position ($t^I = 50\,mm$) for Shield-I.

Liners. Steel, $\gamma = 7850\,kg/m^3$, HB102, $\sigma_y = 240\,MPa$, $\sigma_u = 360\,MPa$, $\varepsilon_f = 20\%$. Data on liners' locations are given in Table 1.

Table 1. Shields with liners only.

Shield label	T_{lin}^{F} (mm)	T_{lin}^{R} (mm)	Shield label	T_{lin}^{F} (mm)	T_{lin}^{R} (mm)
Shield-0-0	–	–	Shield-3-0	3	–
Shield-0-2	–	2	Shield-0-3	–	3
Shield-1-1	1	1	Shield-1-2	1	2
Shield-2-1	2	1			

Table 2. Spherical-segment-nose projectiles (Fig. 3.8.1).

D (mm)	L (mm)	HB	σ_y (MPa)	σ_u (MPa)	ε_f (%)	m (g)
23	64	475	1726	1900	7	175

Table 3. Shield damage types (SDTs).

Notations	Description
(1)	No visible damage to visible cracks without scabbing
(2)	Scabbing without perforation (even with rear steel plate lining)
(3)	Perforation without complete penetration of shield by projectile — projectile stuck inside the shield or was found near the shield
(4)	Projectile perforated the shield and exited with a residual velocity

5.2.2 *Test results*

Table 4. SDT versus shield structure and impact velocity.

Shield	v_s (m/s)	SDT	Shield	v_s (m/s)	SDT
Shield-0-0	201	(1)	Shield-0-0	240	(2)
Shield-0-0	270	(3)	Shield-0-0	299	(4)
Shield-0-0	354	(4)	Shield-F	302	(4)
Shield-I	307	(4)	Shield-R	313	(3)
Shield-3-0	316	(3)	Shield-3-0	334	(4)
Shield-3-0	339	(4)	Shield-0-2	336	(2)
Shield-0-2	359	(2)	Shield-0-2	367	(2)
Shield-0-3	270	(2)	Shield-0-3	282	(2)
Shield-0-3	317	(2)	Shield-1-1	316	(2)
Shield-1-1	338	(2)	Shield-1-1	349	(4)
Shield-1-2	320	(2)	Shield-1-2	324	(2)
Shield-1-2	331	(2)	Shield-2-1	335	(2)
Shield-2-1	350	(2)	Shield-2-1	431	(4)

5.3 [Almansa and Cánovas, 1999]

See also [Almansa and Cánovas, 1996].

5.3.1 *General data on shields and projectiles*

Shields. Single PC and steel fiber RC slabs; $c_{l,w} = 600\,mm$; $f_c' = 40\,MPa$ for concrete without fibers (Ø150×300 mm cylinders were tested to determine f_c').

Fibers. Cold-drawn low-carbon steel fibers of hooked-end type; parameter μ below refers to fibers; $L^{fib} = 50\,mm$, $d^{fib} = 0.5\,mm$.

Table 1. Number of tested shields for different values of T and μ.

$\mu\ (kg/m^3) \rightarrow$ T (cm) $\downarrow$	0	40	80	120
4	4	0	4	1
6	10	2	10	2
8	8	0	8	2
10	10	0	10	2
12	8	0	8	0
14	0	2	2	2
18	0	2	0	2
20	0	0	0	2

Table 2. Types of projectiles.

Label	Name	m (g)	m^{core} (g)
Projectile-1	5.56 × 45 mm (SS 109 NATO)	4.13	0.495
Projectile-2	7.62 × 51 mm AP	9.70	4.60
Projectile-3	12.7 × 99 mm AP	42.0	22.8

5.3.2 *Test results*

Table 3. Values of $\eta = (w_s - w_r)/T$ for different types of projectiles (averaged over experiments with shields having different T).

Projectile	μ (kg/m^3)	Average η (J/cm)	Standard deviation (J/cm)	Number of data points
Projectile-1	0	320	21	8

Projectile	μ (kg/m^3)	Average η (J/cm)	Standard deviation (J/cm)	Number of data points
Projectile-1	80	345	20	8
Projectile-2	0	394	25	4
Projectile-2	80	403	27	22
Projectile-3	0	1022	84	11
Projectile-3	40	1035	44	6
Projectile-3	80	1048	86	37
Projectile-3	120	1130	85	20

5.4 [Almusallam *et al.*, 2013]

See also [Abadel *et al.*, 2014].

5.4.1 *General data on shields and projectiles*

Shields. Slabes made of normal strength concrete (NSC) ($f'_c \approx 30\,MPa$) and high strength concrete (HSC) ($f'_c \approx 60\,MPa$) reinforced by rebars and combination of steel and plastic fibers; single PC slab; $c_{l,w} = 600\,mm$, $T = 90\,mm$; nominal $q^{agr} = 10\,mm$, Ø150×300 mm cylinders were tested to determine f'_c.

Rebars. Steel grade 420 ($\sigma_y = 420\,MPa$); $t^R = 15\,mm$, $d^R = 8\,mm$, $g^R = 100\,mm$.

Table 1. Characteristics of fibers.

Material	L^{fib} (mm)	d^{fib} (mm)	σ_u (MPa)	E (GPa)
Crimped steel	13	0.20	2600	200
Polyolyphene	50	0.015	30	4

Table 2. Double-tapered projectile (Fig. 3.7.1a).

D (mm)	L_{cyl} (mm)	L_1 (mm)	L_2 (mm)	Material	m (kg)
40	75	10	30	Hardened steel	0.8

Table 3. Shield damage types (SDTs).

Notation	Description
(1)	Projectile perforated the shield and exited with a residual velocity.
(2)	Visible cracks without scabbing.

Notation	Description
(3)	Heavy cracking, pattern of rear crater indicating formation of a rear shear plug
(4)	Rear face damage (scabbing and spalling or shear plug formation) without perforation
(5)	Penetration without complete perforation of the shield

Table 4. Parameters of shields.

Steel fiber		Polyolyphene fiber		NSC		HSC	
ρ_V (%)	ρ_m (%)	ρ_V (%)	ρ_m (%)	Shield label	f'_c (MPa)	Shield label	f'_c (MPa)
0.0	0.00	0.0	0.00	Sh-1N	40.5	Sh-1H	59.5
0.6	2.05	0.0	0.00	Sh-2N	44.0	Sh-2H	62.0
0.4	1.37	0.2	0.08	Sh-3N	37.5	Sh-3H	61.5
0.2	0.68	0.4	0.16	Sh-4N	28.8	Sh-4H	65.0
0.0	0.00	0.6	0.23	Sh-5N	32.0	Sh-5H	68.0
0.9	3.07	0.0	0.00	Sh-6N	43.0	Sh-6H	71.0
0.6	2.05	0.3	0.12	Sh-7N	43.6	Sh-7H	61.5
0.3	1.02	0.6	0.23	Sh-8N	42.3	Sh-8H	54.0
0.0	0.0	0.9	0.35	Sh-9N	41.0	Sh-9H	55.0

5.4.2 *Test results*

Note. Asterisk in column for P indicates perforation.

Table 5. Test results for NSC shields.

Shield	Shot 1			Shot 2			Shot 3		
	v_s (m/s)	P (mm)	SDT	v_s (m/s)	P (mm)	SDT	v_s (m/s)	P (mm)	SDT
Sh-1N	125.5	*	(1)	107.9	*	(5)	91.1	33.2	(2)
Sh-2N	107.9	40.8	(3)	125.4	*	(5)	90.9	34.8	(2)
Sh-3N	108.4	*	(5)	118.2	*	(1)	90.6	31.7	(3)
Sh-4N	107.6	*	(5)	115.3	*	(1)	91.0	33.1	(2)
Sh-5N	107.9	*	(5)	90.6	34.8	(2)	107.6	*	(5)
Sh-6N	107.5	40.6	(4)	125.0	*	(1)	90.9	34.4	(2)
Sh-7N	107.7	39.8	(3)	125.0	*	(1)	118.1	*	(5)
Sh-8N	108.0	*	(5)	115.4	*	(1)	91.3	35.3	(2)
Sh-9N	108.4	*	(5)	90.7	32.7	(2)	107.8	*	(5)

Table 6. Test results for HSC shields.

	Shot 1			Shot 2			Shot 3		
Shield	v_s (m/s)	P (mm)	SDT	v_s (m/s)	P (mm)	SDT	v_s (m/s)	P (mm)	SDT
Sh-1H	124.9	*	(5)	107.6	35.0	(4)	91.0	30.0	(2)
Sh-2H	124.8	*	(1)	131.9	*	(1)	107.7	*	(5)
Sh-3H	125.1	*	(5)	128.8	*	(1)	108.0	36.2	(2)
Sh-4H	125.0	42.7	(4)	135.2	*	(5)	108.0	34.4	(2)
Sh-5H	125.4	42.0	(3)	128.5	*	(5)	107.8	33.1	(2)
Sh-6H	124.6	40.6	(3)	135.0	*	(5)	108.5	33.3	(2)
Sh-7H	124.9	42.6	(3)	131.6	*	(5)	107.9	33.3	(2)
Sh-8H	125.0	*	(5)	107.8	33.8	(2)	125.2	*	(5)
Sh-9H	125.2	*	(5)	108.0	36.1	(2)	125.2	*	(5)

5.5 [Almusallam *et al.*, 2015a]

Note. This paper includes also results obtained in other studies which are presented in [Almusallam *et al.*, 2013], [Soe *et al.*, 2013], [Dancygier *et al.*, 2014], [Luo *et al.*, 2001], [Zhang *et al.*, 2005] (see corresponding sections).

5.5.1 *General data on shields and projectiles*

Shields. Single high strength concrete (HSC) ($f_c' \approx 60\,MPa$) slabs reinforced by rebars and combination of steel and plastic fibers; $c_{l,w} = 600\,mm$, $T = 90\,mm$; nominal $q^{agg} = 5-10\,mm$, Ø150×300 mm cylinders were tested to determine f_c' .

Rebars. Steel ($\sigma_y = 510\,MPa$); $t^R = 15\,mm$, $d^R = 8\,mm$, $g^R = 100\,mm$.

Table 1. Fiber properties.

Designation	Material	Shape	Cross-section dimensions (mm)	L^{fib} (mm)	σ_u (MPa)	E (GPa)
SF	Steel fibers	Hooked ends	Ø0.75	60	1225	200
PF	Polypropylene fibers	Crimped	1.0×0.6	50	550	4
KF	Kevlar fibers	Plain	Ø0.50	45	3220	131

Projectiles.   Steel   hemispherical-nose   projectiles   (Fig. 3.4.1); $D = 40\,mm$, $L = 100\,mm$, $m = 800\,g$.

Table 2. Parameters of shields.

Shield label	f_c' (MPa)	ρ_V (%)		
		SF	PF	KF
Sh-1	64.5	0	0	0
Sh-2	73.5	1.2	0	0
Sh-3	70.0	1.0	0.2	0
Sh-4	74.4	1.4	0	0
Sh-5	71.9	1.2	0.2	0
Sh-6	65.4	0.9	0	0.3
Sh-7	66.0	1.1	0	0.3
Sh-8	66.7	0.9	0.2	0.3
Sh-9	65.6	0.7	0.2	0.3

5.5.2 *Test results*

Table 3. Test results on DOP.

Shield	Shot 1		Shot 2		Shot 3	
	v_s (m/s)	P (mm)	v_s (m/s)	P (mm)	v_s (m/s)	P (mm)
Sh-1	135.2	*	125.1	32.5	108.1	31.2
Sh-2	135.1	30.9	160.2	*	147.2	39.4
Sh-3	135.1	31.1	160.2	*	147.2	38.0
Sh-4	135.1	30.0	178.5	*	147.2	33.2
Sh-5	135.1	29.4	168.2	*	147.2	32.2
Sh-6	135.1	33.0	125.0	31.1	147.1	*
Sh-7	135.0	30.7	160.2	*	147.2	33.1
Sh-8	135.2	30.5	158.1	*	147.3	35.9
Sh-9	135.3	30.2	153.7	*	147.3	36.0

* Perforation.

5.6 [Almusallam *et al.*, 2015b]

5.6.1 *General data on shields and projectiles*

Shields. RC slabs ($c_{l,w} = 600\,mm$, $T = 90\,mm$) and the same RC slabs strengthened using one layer of unidirectional carbon fiber reinforced polymer (CFRP) sheet. The sheet was attached to RC slabs so that the fibers were directed along the span of the slab.

Concrete. $q_{max}^{agg} = 10\,mm$, $f_{sl} = 18\,mm$; $f_c' = 49\,MPa$ (cylinders Ø150 mm × 300 mm were tested).

Rebars. $d_{l,w}^R = 8\,mm$, $g_{l,w}^R = 100\,mm$, $t^R = 15\,mm$, $\sigma_y = 575\,MPa$.

Projectiles. Hard hemispherical-nose projectiles (Fig. 3.4.1); $D = 40\,mm$, $m = 800$ g.

5.6.2 Test results

Table 1. Test results for RC slabs without strengthening.

v_s (m/s)	P (mm)	SDT
108.0	27.4	Formation of shear plug without dislodging; rebar separation from concrete in some parts of the shield
125.0	–	Total perforation of concrete slabs with almost zero residual velocity
92.0	20.0	Visible scabbing without visible shear plug at the rear face

Table 2. Test results for RC slabs with strengthening.

v_s (m/s)	P (mm)	SDT
125.0	33.0	Bulging of CFRP sheet with concrete stuck to CFRP sheet
147.5	–	CFRP sheet fails; total perforation of concrete slab with almost zero residual velocity
158.0	–	CFRP sheet fails; perforation of concrete slab with non-zero residual velocity

5.7 [Anderson *et al.*, 1984]

5.7.1 General data on shields and projectiles

Shields. FRC specimens prepared from mixtures having different composition were tested. Ordinary Portland cement was used throughout; water-cement ratio, aggregate-cement ratio and fine to coarse aggregates ratio were 0.5, 4.0 and 2:1, respectively. Three coarse aggregate types (limestone, river gravel and basalt; $q_{max}^{agg} = 10\,mm$) and three types of steel fibers: melt extract carbon steel, brass coated circular steel and brass coated indented circular steel ($d^{fib} = 0.25\,mm$, $L^{fib} = 25\,mm$) were used in different combinations; $\widetilde{D}^{sh} = 450\,mm$.

Projectiles. Copper-sheathed hardened steel projectiles, $D = 7.62\,mm$, $m = 9.6 - 9.9\,g$.

Impact velocity. $v_s \approx 800\,m/s$.

5.7.2 *Test results*

Table 1. Test results for melt extract carbon steel fibers.

Aggregate type	Fiber ρ_m (%)	P (mm)	Aggregate type	Fiber ρ_m (%)	P (mm)
Limestone	0	76	Limestone	2.5	92
Limestone	5.0	80	Limestone	7.5	80
Limestone	10.0	66	River gravel	0	56
River gravel	2.5	50	River gravel	5.0	86
River gravel	7.5	80	River gravel	10.0	68
Basalt	0	87	Basalt	2.5	62
Basalt	5.0	102	Basalt	7.5	91
Basalt	10.0	101			

Table 2. Test results for brass coated circular steel fibers.

Aggregate type	Fiber ρ_m (%)	P (mm)	Aggregate type	Fiber ρ_m (%)	P (mm)
Limestone	0	83	Limestone	2.5	94
Limestone	5.0	91	Limestone	7.5	57
Limestone	10.0	74	River gravel	0	74
River gravel	2.5	67	River gravel	5.0	93
River gravel	7.5	75	Basalt	0	89
Basalt	2.5	84	Basalt	5.0	81
Basalt	7.5	84	Basalt	10.0	91

Table 3. Test results for brass coated indented circular steel fibers.

Aggregate type	Fiber ρ_m (%)	P (mm)	Aggregate type	Fiber ρ_m (%)	P (mm)
Limestone	0	68	Limestone	2.5	87
Limestone	5.0	83	Limestone	7.5	60
Limestone	10.0	68	River gravel	0	77
River gravel	2.5	93	River gravel	5.0	67
River gravel	7.5	71	River gravel	10.0	68
Basalt	0	93	Basalt	2.5	85

Aggregate type	Fiber ρ_m (%)	P (mm)	Aggregate type	Fiber ρ_m (%)	P (mm)
Basalt	5.0	102	Basalt	7.5	86
Basalt	10.0	79			

5.8 [Barr *et al.*, 1983]

5.8.1 *General data on shields and projectiles*

Shields. Monolithic circular panels with integral circumferential reinforcement; with liners and without liners; $D^{sh} = 767\,mm$, $T = 82\,mm$, $f'_c \approx 40\,MPa$ (cylindrical specimens were tested), $q_{max}^{agg} = 2\,mm$. Detailed information regarding preparation of shields to the test can be found in the original study.

Steel liners. $v_{bl} = 40.7\,m/s$ and $v_{bl} = 81.8\,m/s$ for plates having the thicknesses $1\,mm$ and $3\,mm$, correspondingly.

Projectiles. Steel cylindrical projectiles (Fig. 3.1.1); $D = 40\,mm$, $m = 1\,kg$.

Table 1. Methods of attaching the rear plate.

Notation	Methods of attaching	Notation	Methods of attaching	Notation	Methods of attaching
E	Attaching by edge clamp	S	Attaching by Spitz bolts	T	Attaching by through bolts

5.8.2 *Test results*

Table 2. Test results for unreinforced concrete shields.

f'_c (MPa)	Rear liner T^R_{lin} (mm)	Attaching	v_s (m/s)	v_r (m/s)
39.5	–	–	80	26
40.0	–	–	81	17
36.4	–	–	74	*
29.9	1	E	170	*
33.4	1	S	187	*
37.3	1	T	178	60

f_c' (MPa)	Rear liner		v_s (m/s)	v_r (m/s)
	T_{lin}^R (mm)	Attaching		
40.1	1	T	174	*
39.9	1	T	188	68

* Non-perforated.

Table 3. Test results for circular reinforced concrete shields with liners attached by through bolts.

f_c' (MPa)	Liners		v_s (m/s)	v_r (m/s)
	T_{lin}^F (mm)	T_{lin}^R (mm)		
36.6	–	–	100	*
36.2	–	–	115	39
35.7	–	–	111	36
42.1	–	1	146	*
39.8	–	1	172	38
37.4	–	1	166	42
48.9	1	–	123	28
49.1	1	–	126	38
41.2	1	1	188	54
43.3	1	1	192	68
45.7	–	3	235	*
40.0	–	3	256	0
43.7	-	3	266	665
45.1	3	3	284	50
49.3	3	3	282	51

* Non-perforated.

5.9 [Beppu *et al.*, 2008]

5.9.1 *General data on shields and projectiles*

Shields. Single PC slabs; $c_{l,w} = 500\,mm$, $f_c' = 25\,MPa$.
Projectiles. Steel mushroom-shaped projectile (Fig. 3.4.2), $D = 25\,mm$, $m = 50\,g$.

Table 1. Shield damage types (SDTs).

Label	Description	Label	Description	Label	Description
(1)	Cratering	(2)	Spalling	(3)	Perforation

5.9.2   *Test results*

Table 2#. Test results for "semi-infinite" shield.

v_s (m/s)	P (mm)	SDT	v_s (m/s)	P (mm)	SDT	v_s (m/s)	P (mm)	SDT
196	15	(2)	213	17	(1)	308	28	(2)
308	21	(1)	309	24	(2)	312	21	(1)
412	30	(2)	415	25	(2)	415	31	(2)
486	36	(1)	488	36	(1)			

Table 3#. Test results for shield having finite thickness.

T (mm)	v_s (m/s)	SDT	T (mm)	v_s (m/s)	SDT	T (mm)	v_s (m/s)	SDT
30	177	(3)	30	213	(3)	60	196	(2)
70	307	(3)	70	310	(3)	80	214	(1)
80	310	(2)	80	414	(2)	80	417	(3)
90	415	(2)	100	311	(1)	100	489	(3)
100	491	(3)	120	417	(1)	120	417	(1)
130	486	(1)	130	488	(1)	130	488	(1)

5.10   [Berriaud *et al.*, 1978]

See also [Berriaud *et al.*, 1977], [Sliter, 1980].

5.10.1   *General data on shields and projectiles*

Shields. Single PC and RC slabs with rebars reinforced by different steel meshes; μ is the reinforcement quality of meshes in the shield; RLD is linear density of rebars (number of rebars per one meter of a square mesh).

Projectiles. Cylindrical projectiles (Fig. 3.1.1).

Table 1. Types of shields for Set-1 of experiments ($c_{l,w} = 1.46\,m$).

Shield label	T (mm)	μ (kg/m^3)	$d_{l,w}$ (mm)	RLD	Location of meshes
D_0	260	260	10	12	4 equidistant meshes
D_1	260	160	8	12	4 equidistant meshes
D_2	260	0	—	—	—
D_3	260	160	8	12	1 mesh at distance of 20 mm from frontal face and 3 meshes-in-contact at distance of 20 mm from rear face

Shield label	T (mm)	μ (kg/m^3)	$d_{l,w}$ (mm)	RLD	Location of meshes
D_4	208	260	8	15	4 equidistant meshes
D_5	416	260	16	7.5	4 equidistant meshes
D_6	156	260	6	20	4 equidistant meshes
D_7	104	260	4	30	4 equidistant meshes
D_8	260	260	10	12	2 meshes-in-contact at frontal and rear faces

Table 2. Types of shields for Set-2 of experiments (2 meshes-in-contact at frontal and rear faces, $c_{l,w} = 5m$).

Shield label	T (mm)	μ (kg/m^3)	$d_{l,w}$ (mm)	RLD
G_1	400	265	16	8
G_2	500	220	16	8
G_3	600	200	16	8

5.10.2 *Test results*

Table 3. Test results for Set-1 of experiments.

Shield	T (mm)	f'_c (MPa)	m (kg)	D (mm)	v_s (m/s)	v_r (m/s)	v_{bl} (m/s)	P (mm)
D_0	260	41	30	200	113	NP	–	–
D_0	260	41	50	200	112	NP	–	–
D_0	260	43.5	30	250	~ 150	NP	–	80
D_0	260	41.5	34	200	157	21	154	–
D_0	260	50.5	50	250	166	48	156	–
D_0	260	49	99.5	250	125	59	98	–
D_0	260	45	50	200	145	47	129	–
D_0	260	49.5	70	300	156	–	–	–
D_0	260	43.5	119	250	93	38	80	–
D_0	260	38	17.5	200	238	NP	–	70
D_0	260	42.5	17.5	200	269	NP	–	110
D_0	260	36.5	25.1	250	252	NP	–	125
D_0	260	39	25.1	250	309	NP	–	250
D_1	260	44.5	50	200	138	8	138	–
D_1	260	36.5	30	200	127	NP	–	80
D_1	260	43.5	30	250	216	44	200	–
D_1	260	41.5	50	250	156	34	147	–
D_1	260	44	100	250	104	≤53	≥88	–
D_1	260	44.5	70	300	152	44	135	–
D_1	260	34	30	250	187	AP	187	260

Shield	T (mm)	f'_c (MPa)	m (kg)	D (mm)	v_s (m/s)	v_r (m/s)	v_{bl} (m/s)	P (mm)
D_2	260	45	51	200	114	42	98	–
D_3	260	43	70	300	128	35	116	–
D_4	208	39	34	278	148	NP	–	150
D_4	208	41.5	34	278	151	17	148	–
D_4	208	43	37.2	278	155	31	146	–
D_4	208	38	34	278	102	NP	–	35
D_4	208	40	52	278	99	NP	–	93
D_4	208	41.5	57	278	102	0	102	–
D_4	208	44	51.6	300	129	29	120	–
D_4	208	43.5	32.8	300	153	NP	–	70
D_4	208	40	32.8	300	187	20	183	–
D_4	208	43.5	34.5	278	186	38	174	–
D_5	416	38.5	31	300	445	NP	–	–
D_5	416	50	303	100	49	12	47.4	–
D_6	156	40.5	300	300	30.3	14.5	25.5	–
D_7	104	38	300	300	21.9	14.9	15.1	–
D_8	260	40.5	295	100	29.7	12.9	26.6	–
D_8	260	40	297	300	52.4	15.5	49.3	–
D_8	260	37.5	147	300	91	33.8	79.8	–
D_8	260	39.5	70	300	164	55.2	140.6	–
D_8	260	40.5	35	300	220	NP	–	150

P – Perforation; AP – Almost perforation; NP – Non-perforation.

Table 4. Test results for Set-2 of experiments.

Shield	T (mm)	f'_c (MPa)	m (kg)	D (mm)	v_s (m/s)	v_r (m/s)	v_{bl} (m/s)	P (mm)
G_1	400	~40	227	305	110	0	110	P
G_1	400	~40	227	305	90	NP	–	170
G_1	400	~40	227	305	90	NP	–	200
G_1	400	~40	227	305	106	0	106	P
G_1	400	~40	160	305	110	NP	–	125
G_1	400	~40	227	305	97	NP	–	185
G_1	400	~40	160	305	108	NP	–	105
G_1	400	33.5	160	305	133	NP	–	400
G_1	400	33.5	191	305	94	NP	–	100
G_1	400	33.5	160	305	110	NP	–	125
G_1	400	33.5	192	305	123	23	119	P
G_1	400	33.5	192	305	110	NP	–	240
G_1	400	36	240	200	86	32	75	P
G_1	400	36	240	200	72	NP	–	430
G_1	400	36	300	305	89	37	165	P
G_1	400	36	160	305	86	NP	–	65

Shield	T (mm)	f'_c (MPa)	m (kg)	D (mm)	v_s (m/s)	v_r (m/s)	v_{bl} (m/s)	P (mm)
G_1	400	36	300	305	65	NP	–	110
G_2	500	33.5	227	305	117	NP	–	225
G_2	500	33.5	227	305	107	NP	–	150
G_2	500	33.5	160	305	148	NP	–	175
G_2	500	33.5	160	305	77	NP	–	20
G_2	500	33.5	192	305	134	NP	–	165
G_2	500	38.5	227	305	127	NP	–	450
G_2	500	38.5	160	305	160	NP	–	290
G_2	500	38.5	192	305	144	0	144	P
G_2	500	38.5	192	305	79	NP	–	30
G_2	500	38.5	192	305	120	NP	–	135
G_3	600	36	300	305	143	32	136	P
G_3	600	36	160	305	173	NP	–	230
G_3	600	36	240	305	73	NP	–	100
G_3	600	36	300	305	100	NP	–	–
G_3	600	36	160	305	143	NP	–	80

P – Perforation; AP – Almost perforation; NP – Non-perforation.

5.11 [Berriaud *et al.*, 1979]

5.11.1 *General data on shields and projectiles*

Shields. Single PC and RC slabs with rebars reinforced by different steel meshes; μ is the reinforcement quality of meshes in the shield without stirrups; RLD is linear density of rebars (number of rebars per one meter of a square mesh); $c_{l,w} = 1460\,mm$.

Projectiles. Cylindrical projectiles (Fig. 3.1.1).

Table 1. Types of shields.

Shield label	T (mm)	μ (kg/m^3)	$d_{l,w}$ (mm)	RLD	Location of meshes
Sh-1	260	139	8	12	4 equidistant meshes
Sh-2	260	–	–	–	–
Sh-3	300	69	8	12	1 mesh at distance of 20 mm from frontal and rear faces
Sh-4	260	108	10	12	1 mesh at distance of 20 mm from frontal and rear faces
Sh-5	260	108	10	12	2 meshes in-contact at distance of 20 mm from frontal face
Sh-6	260	108	10	12	2 meshes in-contact at distance of 20 mm from rear face

Shield label	T (mm)	μ (kg/m^3)	$d_{l,w}$ (mm)	RLD	Location of meshes
Sh-7	260	34.5	4	12	4 equidistant meshes
Sh-8	416	217	16	7.5	4 equidistant meshes

5.11.2 *Test results*

Table 2. Values of v_s, v_r and v_{bl} or P for different projectiles and shields.

Shield label	f'_c (MPa)	m (kg)	D (mm)	v_s (m/s)	v_r (m/s)	v_{bl} (m/s)	P (mm)
Sh-1	78.5	74	200	100	32	91.4	–
Sh-1	68.5	74	200	92	14	90.3	–
Sh-1	55	74	200	113	42	99.7	–
Sh-1	31.5	74	200	84	16	81.5	–
Sh-1	29	74	200	84	26	77.3	–
Sh-1	15	72	200	78.5	NCP	–	285
Sh-1	43.3	72.8	200	117	51.8	96.6	–
Sh-1	47.5	73.4	200	114	46.9	97.2	–
Sh-1	52.4	73.3	200	97.7	21.1	94	–
Sh-1	61.3	72.8	200	94.6	14.5	92.8	–
Sh-2	44	310	100	18.8	10.9	15.1	–
Sh-2	40	72	200	78.5	29	69.3	–
Sh-3	42.5	70	200	123	44	110	–
Sh-3	41	310	100	24.9	8.9	23.2	–
Sh-4	41	72	300	142	34	132	–
Sh-4	39	295	100	26.2	17	19.7	–
Sh-4	42	53	200	138	54.5	116	–
Sh-5	42	72	300	144	40	139	–
Sh-5	40.5	295	100	25.1	9.45	23.2	–
Sh-5	42	53	200	141	45	127	–
Sh-6	45.5	70	200	96	NP	–	–
Sh-6	41	315	100	27.3	12.4	24.2	–
Sh-7	35	72	200	101	46	82.3	–
Sh-8	–	103	300	187	NCP	–	160

NCP – non-complete penetration; NP – nearly perforation.

5.12 [Berriaud *et al.*, 1982]

5.12.1 *General data on shields and projectiles*

Shields: Single PC and RC slabs with rebars reinforced by different steel meshes; μ^* and μ^{**} are reinforcement qualities of meshes in the shield, with or without stirrups, correspondingly; RLD is linear density of rebars (number of rebars per one meter of a square mesh); $c_{l,w} = 1460\,mm$.

Projectiles. Projectiles having different shapes (including projectiles with non-circular cross-sections).

Table 1. Types of shields.

Shield label	T (mm)	μ^{*} (kg/m^3)	μ^{**} (kg/m^3)	$d_{l,w}$ (mm)	RLD	Location of meshes
Sh-1	260	153	139	8	12	4 equidistant meshes
Sh-2	260	–	–	–	–	–
Sh-3	208	238	217	8	15	4 equidistant meshes
Sh-4	156	255	217	6	20	4 equidistant meshes
Sh-5	260	230	217	10	12	2 meshes-in-contact at distance of 20 mm from frontal and rear faces
Sh-6	260	37	34.5	4	12	Ibid
Sh-7	260	34.5	34.5	4	12	Ibid
Sh-8	260	153	139	8	12	Ibid
Sh-9	260	139	139	8	12	Ibid
Sh-10	156	–	–	–	–	–

5.12.2 *Test results*

Additional notation. NCP – non-complete penetration.

Table 2. Test results for projectiles shaped as regular triangular prisms with the length of side of the triangle in the cross-section equal to 224 mm.

Shield type	f_c' (MPa)	m (kg)	v_s (m/s)	v_r (m/s)	v_{bl} (m/s)	P (mm)
Sh-3	31.3	43.3	90.7	–	–	260
Sh-3	41.5	42	98	NCP	–	50

Table 3. Test results for projectiles shaped as parallelepipeds with the lengths of sides of the rectangle in the cross-section a_w and a_l.

Shield type	f_c' (MPa)	m (kg)	a_w (mm)	a_l (mm)	v_s (m/s)	v_r (m/s)	v_{bl} (m/s)
Sh-4	32.6	160	125	63	19.8	9	17.6
Sh-4	41.3	220	177	44	19.7	12.6	15
Sh-4	45.5	220	196	40	20	10.5	16.9

Table 4. Test results for cylindrical projectiles (Fig. 3.1.1).

Shield type	f'_c (MPa)	m (kg)	D (mm)	v_s (m/s)	v_r (m/s)	v_{bl} (m/s)	P (mm)
Sh-1	45.5	72	200	120	< 55	>98.6	–
Sh-1	52	80	200	126	64	97.8	–
Sh-1	46.7	79.6	200	113	47	97.2	–
Sh-1	46.2	74.3	200	105	43	90.8	–
Sh-1	60.2	33.3	128	144	42	135	–
Sh-2	55.6	35.7	128	101	25	96.4	–
Sh-4	44	167	100	19.8	10.5	16.6	–
Sh-4	46.2	160	100	19	11.2	15.2	–
Sh-5	43.3	29.9	200	226	60	206	–
Sh-5	41.8	25	200	258	54	243	–
Sh-5	39.2	24.7	250	316	57	297	–
Sh-6	50.3	33.1	128	143	52	128	–
Sh-6	46.3	33.4	128	138	46	126	–
Sh-6	40.2	31.9	128	144	62.5	122	–
Sh-7	32.3	31.5	130	133	48	119	–
Sh-8	47.5	74.9	200	102	41	88.5	–
Sh-8	48.3	73.5	200	95.2	30.6	87.3	–
Sh-8	48.2	74	200	115	50	97	–
Sh-8	48.3	80.6	200	105	41	92.5	–
Sh-8	49.3	74.6	200	108	39	97	–
Sh-8	45.5	20	200	205	NCP	–	–
Sh-8	46	16.4	200	333	47	321	–
Sh-8	42.3	7	100	315	40	308	–
Sh-8	43	25.4	250	220	NCP	–	80
Sh-8	40.5	19.9	200	288	68	263	–
Sh-8	38.8	3.55	50	291	NCP	–	–
Sh-8	38.8	7.05	50	216	74	198	–
Sh-9	41.3	32.1	128	145	49	132	–
Sh-10	31.6	130	100	12.5	7.65	9.9	–

Table 5. Test results for cone-nose projectiles (Fig. 3.2.1), $D = 100$ *mm*.

Shield type	f'_c (MPa)	m (kg)	θ_s (deg)	v_s (m/s)	v_r (m/s)	v_{bl} (m/s)	P (mm)
Sh-4	45	133	60	18.7	NCP	–	30
Sh-4	45	187	60	18.9	5	18.3	–
Sh-4	38.7	230	45	19.3	9.55	16.7	–
Sh-4	45	230	30	19	NCP	–	–
Sh-4	47.2	247	30	18.9	NCP	15	–
Sh-4	48	252	30	19	9.1	16.7	–

Table 6. Test results for hemispherical-nose projectile (Fig. 3.4.1), $D = 100\ mm$.

Shield type	f_c' (MPa)	m (kg)	v_s (m/s)	v_r (m/s)	v_{bl} (m/s)
Sh-4	50.4	210.7	19.5	0	19.5

5.13   [Beth, 1945]

5.13.1   *General data on shields and projectiles*

Shields. Single PC slabs; quartz aggregates with $q_{max}^{agg} = 9.5\,mm$; $\varnothing 100 \times 200$ mm cylinders were tested to determine f_c'.

Table 1. Parameters of projectiles ($D = 12.7\,mm$).

Label	Shape	Figure	Special name	ψ
Projectile-1	Spherical projectiles	Fig. 3.5.1	–	–
Projectile-2	Cylindrical projectiles	Fig. 3.1.1	–	–
Projectile-3	Hemispherical-nose projectiles	Fig. 3.4.1	–	0.5
Projectile-4H	Ogive-nose projectiles	Fig. 3.3.1	Hollow projectile E-6	1.5
Projectile-4S	Ogive-nose projectiles	Fig. 3.3.1	Standard projectile E-6	1.5
Projectile-4T	Ogive-nose projectiles	Fig. 3.3.1	Tungsten carbide projectile E-6	1.5
Projectile-5	Ogive-nose projectiles	Fig. 3.3.1	Long nose	3.1
Projectile-6	Shape is shown in the original paper		Service AP	–

5.13.2   *Test results*

Notes.
- Symbol ± indicates that the measurement is considered to be less accurate than other similar measurements.
- Symbol * indicates that the bullet stuck and the penetration depth is measured from the center of the base of the bullet plus the length of the bullet unless otherwise noted in the "Remarks" column.
- As a rule flat-nosed projectiles did not leave a definite nose imprint.

Table 2. Test results for Projectile-1, $f_c' = 28.8\,MPa, m = 8.2\,g$.

v_s (m/s)	P (mm)	P^{cr} (mm)	v_s (m/s)	P (mm)	P^{cr} (mm)
193.9	6.9	6.9	250.9	8.4	10.7
400.8	16.5	17.0	485.9	24.1	21.3
482.5	21.3	21.6	567.8	27.4	25.9
681.5	35.6	35.6			

Table 3. Test results for Projectile-2 (Series 1), $f_c' = 30.6\,MPa, m = 28.4\,g$.

v_s (m/s)	P (mm)	P^{cr} (mm)	Remarks
224.0	21.6	20.3	–
250.9	24.1	24.1	–
267.0	27.2	26.7	–
334.1	33.0±	33.0	Flat aggregates imprint at 30.7 mm.
403.9	42.2	41.9	Partial flat imprint.
488.6	55.9*	35.6	–
536.1	61.7*	39.4	–
559.9	62.0*	39.4	–
620.9	75.9*	40.6	–
684.0	92.7*	35.6	Band off. Stuck behind bullet.

Table 4. Test results for Projectile-2 (Series 2), $f_c' = 28.0\,MPa, m = 28.3\,g$.

v_s (m/s)	P (mm)	P^{cr} (mm)	Remarks
264.6	26.9	26.9	–
352.3	38.4	28.4	–
481.3	53.3*	36.8	Bullet removed for measuring.
488.9	55.9*	34.0	Bullet at 10°. Removed.
526.7	71.4*	38.1	–
591.6	79.2*	40.1	Bullet removed for measuring.

Table 5. Test results for Projectile-3, $f_c' = 28.0\,MPa, m = 28.1\,g$.

v_s (m/s)	P (mm)	P^{cr} (mm)	Remarks
251.5	27.7	22.9	–
399.9	50.8*	25.4	Bullet removed for measuring.
403.3	54.1*	30.0	Bullet at 10°. Removed.
427.0	54.9*	30.5	Bullet at 10°. Removed.
436.2	59.9*	31.8	Bullet at 5°. Removed.
496.8	66.5*	40.4	Bullet removed for measuring.
605.0	103.9*	32.5	–

Table 6. Test results for Projectile-4S (Series 1), $f_c' = 30.6\,MPa, m = 29.3\,g$.

v_s (m/s)	P (mm)	P^{cr} (mm)	Remarks
175.0	20.6±	17.8	Yaw 45°?
226.5	25.9	25.4	–
288.0	35.6	29.2	–
296.0	36.1	27.9	–
346.9	47.5	30.5	–
482.5	72.6*	27.9	Bullet at 20°.
520.0	76.5*	30.5	Bullet at 7°.
555.7	87.9*	30.5	–

Table 7. Test results for Projectile-4S (Series 2), $f_c' = 29.2\,MPa, m = 29.1\,g$.

v_s (m/s)	P (mm)	P^{cr} (mm)	Remarks
367.9	50.0*	34.3	Bullet at slight angle. Removed.
397.2	59.7*	30.5	Bullet removed for measuring.
400.2	55.6±*	29.2	Bullet at 30°. Removed.
488.6	75.2*	30.5	–
566.3	98.0*	27.9	–

Table 8. Test results for Projectile-4S (Series 3), $f_c' = 28.0\,MPa, m = 29.1\,g$.

v_s (m/s)	P (mm)	P^{cr} (mm)	Remarks
236.2	31.0	24.1	–
296.6	38.9	31.8	–
352.0	50.5*	28.4	Bullet at slight angle. Removed.
432.2	57.9*	33.0	Bullet at 15°. Removed.
437.7	64.5*	30.5	Bullet at small angle.
503.5	82.6*	33.5	Bullet at slight angle.
663.9	131.1*	31.8	Bullet at 10°.

Table 9. Test results for Projectile-4S (Series 4), $f_c' = 28.8\,MPa, m = 25.9\,g$.

v_s (m/s)	P (mm)	P^{cr} (mm)	Remarks
204.8	22.9	18.3	Imprint at 30°.
260.0	32.0	24.1	Imprint at 10°.
342.0	39.6	28.2	Imprint at 40° .
422.1	55.4	39.6	Imprint at 10°.
510.5	71.9	30.5	Stuck at 25°.
568.5	84.3	43.4	–

Table 10. Test results for Projectile-4S (Series 5), $f_c' = 26.0\,MPa$, $m = 29.0\,g$.

v_s (m/s)	P (mm)	P^{cr} (mm)	Remarks
192.0	20.8±	17.8	Imprint at 30°.
272.2	30.5±	25.4	Imprint at 25°.
366.7	48.3	34.3	Imprint at 20°.
479.8	70.1	38.6	–
710.2	124.5±	–	Cube demolished.

Table 11. Test results for Projectile-4H, $f_c' = 29.2\,MPa$, $m = 20.3\,g$.

v_s (m/s)	P (mm)	P^{cr} (mm)	Remarks
454.8	49.5*	30.0	Bullet at 15°. Removed.
466.3	53.3	40.6	–
522.7	62.0*	25.1	–
543.5	65.8*	27.9	–

Table 12. Test results for Projectile-4T, $f_c' = 29.2\,MPa$, $m = 58.7\,g$.

v_s (m/s)	P (mm)	P^{cr} (mm)	Remarks
242.0	47.0	39.4	Yaw 20°?
280.1	63.8*	22.9	Bullet removed for measuring.
308.5	69.1*	26.7	Base broken. Removed.
331.3	82.6±*	30.5	Bullet base broken.
346.6	81.8±*	26.7	Bullet at 15°. Base broken.
399.0	119.4±*	32.5	Bullet base shattered.
445.9	142.2±*	33.0	Bullet base chipped.
453.8	140.2*	22.6	–

Table 13. Test results for Projectile-5, $f_c' = 28.8\,MPa$, $m = 30.5\,g$.

v_s (m/s)	P (mm)	P^{cr} (mm)	Remarks
132.3	17.8	13.0	Imprint at 45°.
231.3	32.3	23.4	–
339.5	45.7	34.3	Imprint at 15°.
421.5	62.2	28.4	–
508.7	84.1	38.6	Bullet at 20°.
576.1	103.1	38.1	Bullet at 45°.
637.9	116.8	35.6	–

Table 14. Test results for Projectile-6, $f_c' = 26.0\,MPa, m = 45.6\,g$.

v_s (m/s)	P (mm)	P^{cr} (mm)	Remarks
204.8	24.4±	19.8	Imprint at 45°.
224.0	32.8±	21.3	Imprint at 25°.
237.7	32.8	22.9	Imprint at 20°.
306.9	46.7	33.0	–
321.9	50.0	33.5	Imprint at 10°.
385.0	65.3±*	31.2	Bullet at 35°.
485.5	87.1±*	66.0	Bullet at 10°.
510.8	99.1*	40.1	Path at 5°.
696.5	170.2	–	Cube demolished.

5.14 [Børvik *et al.*, 2002b]

5.14.1 *General data on shields and projectiles*

Shields. Single high and ultra-high performance fiber reinforced concrete slabs, $\widetilde{D}^{sh} = 500\,mm, T = 100\,mm$.

Fibers. Steel fibers (parameter μ in tables below is reinforcement quality of fibers), $L^{fib} = 12.5\,mm$, $d^{fib} = 0.4\,mm$.

Table 1. Averages (over several tests) concrete mix properties.

Label	μ (kg/m^3)	100 mm cubes f_c' (MPa)	Ø150×300 mm cylinders f_c' (MPa)	E (GPa)
C75	79.4	88	79	36
C150	73.2	140	129	51
C200	71.3	197	180	76

Table 2. Truncated cone-nose projectiles (Fig. 3.6.1).

D (mm)	d (mm)	L_{cyl} (mm)	L_{nose} (mm)	Material	HRC
20	2	68	30	Arne oil hardened steel	53

Table 3. Shield damage types (SDTs).

Notation	Description	Notation	Description	Notation	Description
(1)	Perforation	(2)	Embedment	(3)	Piercing

5.14.2 *Test results*

Table 4. Test results for C75 shield.

m (g)	m^{sh} (kg)	v_s (m/s)	m_r (g)	Δm^{sh} (kg)	v_r (m/s)	SDT
195.6	92.6	367.0	194.3	1.6	41.6	(1)
195.7	93.4	347.5	194.5	3.0	36.3	(1)
195.7	92.7	319.4	194.7	1.3	25.4	(1)
195.6	93.7	316.1	194.6	1.5	26.9	(1)
195.5	94.6	308.4	194.5	1.4	–	(1)
195.7	92.8	296.8	–	2.5	–	(2)

$v_{bl} = 302.6\,m/s; a = 0.14, p = 3.62.$

Table 5. Test results for C150 shield.

m (g)	m^{sh} (kg)	v_s (m/s)	m_r (g)	Δm^{sh} (kg)	v_r (m/s)	SDT
195.7	100.0	344.2	194.3	1.7	37.1	(1)
195.6	101.0	321.2	194.5	2.3	–	(1)
195.4	99.0	317.5	194.5	4.2	–	(3)
195.2	101.8	307.4	194.2	0.6	–	(2)
195.6	100.0	306.9	194.6	1.0	–	(2)

$v_{bl} = 318.5\,m/s.$

Table 6. Test results for C200 shield.

m (g)	m^{sh} (kg)	v_s (m/s)	m_r (g)	Δm^{sh} (kg)	SDT
195.6	106.8	366.2	–	1.1	(3)
195.2	1.5.7	359.9	193.7	2.1	(2)
195.2	107.8	328.8	193.9	0.6	(2)

$v_{bl} = 366.2\,m/s.$

5.15 [Børvik *et al.*, 2007]

5.15.1 *General data on shields and projectiles*

Shields. Single PC and steel FRC slabs, $\widetilde{D}^{sh} = 600\,mm, T = 100\,mm$; cube-shaped specimens with the length of the edge 100 mm were tested to determine f_c', while Ø100×200 mm cylinders were tested to determine E. **Fibers.** Steel fibers (parameter μ in tables below is reinforcement quality of fibers), $L^{fib} = 12.5\,mm$, $d^{fib} = 0.4\,mm$.

Table 1. Properties of concrete mixes in shields.

Notation	μ (kg/m^3)	f_c' (MPa)	γ (kg/m^3)	E (GPa)
C35-1	–	41.4	2500	22.2
C75-1	–	77.5	2500	27.2
C110-1	–	116.7	2500	35.0
C75-2	79	88.4	2500	35.6
C150-2	73	14.01	2500	51.1
C200-2	71	196.6	2500	75.8
C75-3a	–	101.5	3137	39.7
C75-3b	–	89.8	3828	40.6

Table 2. Truncated cone-nose projectiles (Fig. 3.6.1).

D (mm)	d (mm)	L_{cyl} (mm)	L_{nose} (mm)	Material	HRC
20	2	68	30	Arne oil hardened steel	53

5.15.2 *Test results*

Table 3#. Residual velocity versus initial velocity for the concrete mix C75-2.

v_s (m/s)	v_r (m/s)	v_s (m/s)	v_r (m/s)	v_s (m/s)	v_r (m/s)
297	0	316	27	320	25
348	36	367	42		

Table 4#. Test results on BLV.

Shield	v_{bl} (m/s)	Shield	v_{bl} (m/s)
C35-1	245	C75-1	280
C110-1	292	C75-2	303
C150-2	317	C200-2	366
C75-3a	269	C75-3b	299

5.16 [Buchar *et al.*, 2002b]

See also [Buchar *et al.*, 2001].

5.16.1 *General data on shields and projectiles*

Shields. Single PC slabs; $c_{l,w} = T = 100\,mm$, $f_c' = 43.1\,MPa$.

Projectiles. FSP Caliber 0.22, Type 1 (Fig. 3.10.1, Table 3.10.1).

5.16.2 *Test results*

Table 1#. Test results on DOP.

v_s (m/s)	P (mm)	v_s (m/s)	P (mm)	v_s (m/s)	P (mm)	v_s (m/s)	P (mm)
358	2.0	401	5.6	482	7.5	565	7.7
576	11.5	649	11.7	650	7.5	737	9.4
799	12.2	902	11.0	968	11.8	993	17.1
1099	15.0	1198	16.3	1318	18.1	1395	25.8

5.17 [Canfield and Clator, 1966]

Note. Data from [Forrestal *et al.*, 1994]. See also data from [Bernard, 1977].

5.17.1 *General data on shields and projectiles*

Shields. Single PC slabs.

Table 1. Ogive-nose projectiles (Fig. 3.3.1).

Projectile label	D (mm)	ψ	m (kg)	Projectile label	D (mm)	ψ	m (kg)
Projectile-1	76.2	1.5	5.90	Projectile-2	7.62	1.5	0.0059

5.17.2 *Test results*

Table 2#. Test results for Projectile-1 and shields with $\gamma = 2310\,kg/m^3$, $f_c' = 35.1\,MPa$.

v_s (m/s)	P (mm)	v_s (m/s)	P (mm)	v_s (m/s)	P (mm)	v_s (m/s)	P (mm)
307	199	312	229	381	248	452	368
541	419	602	598	616	498	709	657
717	606	742	698	773	738	809	747

Table 3#. Test results for Projectile-2 and shields with $\gamma = 2240\,kg/m^3$, $f_c' = 34.6\,MPa$.

v_s (m/s)	P (mm)	v_s (m/s)	P (mm)	v_s (m/s)	P (mm)	v_s (m/s)	P (mm)
328	19	337	21	346	19	408	23
419	27	552	45	564	44	588	49
608	49	609	50	616	52	709	66

v_s (m/s)	P (mm)	v_s (m/s)	P (mm)	v_s (m/s)	P (mm)	v_s (m/s)	P (mm)
712	68	728	68	759	68	767	74
774	74	809	83	823	77	828	74
828	83						

5.18 [Cargile, 1993]

5.18.1 *General data on shields and projectiles*

Shields. Single PC slabs, $D^{sh} = 1.52\,m$, $q_{max}^{agg} = 9.5\,mm$.
Projectiles. Heat treated 4340 steel (HRC 43-45) ogive-nose projectiles (Fig. 3.3.1); $D = 50.8\,mm$, $L_{nose} = 84.25\,mm$ $L = 355.6\,mm$, $\psi = 3$.

5.18.2 *Test results*

Table 1. v_r versus v_s for different shields.

m (kg)	T (mm)	f_c' (MPa)	v_s (m/s)	v_r (m/s)
2.319	284.4	39.9	309	0
2.330	284.4	39.9	317	0
2.330	254.0	36.5	318	41
2.314	254.0	36.5	309	46
2.319	254.0	36.5	309	32
2.268	215.9	39.9	325	104
2.322	215.9	39.9	311	116
2.330	127.0	36.5	319	233
2.355	127.0	36.5	306	221
2.354	127.0	36.5	306	218
2.341	254.0	36.5	379	138
2.365	254.0	39.9	379	141
2.330	254.0	39.9	473	258
2.325	254.0	39.9	470	257

5.19 [Cargile *et al.*, 2002]

5.19.1 *General data on shields and projectiles*

Shields. Single high strength, steel fiber reinforced concrete (HSSFRC) slabs ($D^{sh} = 1370\,mm$, $f_c' = 85\,MPa$; fibers with $\mu = 158\,kg/m^3$), and very high strength concrete (VHSC) slabs ($D^{sh} = 762\,mm$, $f_c' = 157\,MPa$; fibers with $\rho_V = 3\%$).

Fibers. Hooked-ended steel fibers, $L^{fib} = 30\,mm$, $d^{fib} = 0.5\,mm$.

Projectiles. Heat treated 4340 steel (HRC 43-45) ogive-nose projectiles (Fig. 3.3.1); $D = 26.9\,mm$, $L_{nose} = 35.6\,mm$, $L = 242.4\,mm$, $\rho = 53.6\,mm$, $m = 906\,g$.

5.19.2 *Test results*

Table 1#. Test results for HSSFRC shields (results of experiments by J.D. Gargile, 1998).

v_s (m/s)	P (mm)	v_s (m/s)	P (mm)	v_s (m/s)	P (mm)	v_s (m/s)	P (mm)
561	361	585	379	763	672	780	687

Table 2#. Test results for VHSC shields.

v_s (m/s)	P (mm)	v_s (m/s)	P (mm)	v_s (m/s)	P (mm)	v_s (m/s)	P (mm)
239	78	285	124	391	161	404	182
570	281	588	281	765	429	775	461

5.20 [Cargile, 1999]

See also [Cargile *et al.*, 1993].

5.20.1 *General data on shields and projectiles*

Shields. Single conventional strength Portland cement slabs, $\varnothing152\times305$ mm cylinders were tested to determine f'_c.

Projectiles. Heat treated 4340 steel (HRC 43-45) ogive-nose projectiles (Fig. 3.3.1); $D = 26.9\,mm$, $L_{nose} = 35.6\,mm$ $L = 242.4\,mm$.

5.20.2 *Test results*

Table 1. Test results for values on DOP.

m (g)	D^{sh} (mm)	T (mm)	f'_c (MPa)	v_s (m/s)	P (mm)
906	1370	760	35.2	277	173
910	1370	910	37.8	410	310
907	1370	910	38.1	431	411
912	1370	760	33.5	499	480
910	1370	1070	38.4	567	525

m	D^{sh}	T	f'_c	v_s	P
(g)	(mm)	(mm)	(MPa)	(m/s)	(mm)
901	1220	1830	40.1	591	513
903	1220	1830	35.4	631	607
905	1220	1830	34.7	642	620
901	1220	1830	36.0	773	866
904	1220	1830	32.4	800	958

5.21   [Chen *et al.*, 2010]

5.21.1   *General data on shields and projectiles*

Shields. Single PC cylinders, $D^{sh} = 640\,mm$, $T = 800\,mm$; $q^{agg} = 3 - 8\,mm$, $f'_c = 45\,MPa$ (150 mm edge cube-shaped specimens were tested).

Projectiles. Ogive-nose projectile (Fig. 3.3.1) made of D6A high-strength alloy steel ($\sigma_u = 1570\,MPa$) with macromolecule composite backfill ($\gamma = 1650\,kg/m^3$); the tail attachment is made of duralumin 2AL2 and is attached to the projectile body by a screw-thread connection; $D = 25\,mm$, $\psi = 3$.

Table 1. Types of projectile deformation (TPDs).

Notation	Description	Notation	Description
(1)	Integrity	(2)	Gentle bending
(3)	Bending	(4)	Bending & fracture

5.21.2   *Test results*

Table 2. Test results for $\theta_s = 0°$.

L	T	m	v_s	P	m_r	TPD
(mm)	(mm)	(g)	(m/s)	(mm)	(g)	
150	2.50	312.6	775	395.25	304.1	(1)
150	3.75	362.8	765	395.25	349.8	(1)
200	2.50	415.6	721	444.75	403.7	(1)
200	3.75	488.2	674	–	475.7	(1)
250	2.50	516.3	656	425.00	501.6	(1)
250	3.75	608.3	615	454.50	593.1	(1)

Table 3. Test results for $\theta_s \approx 20°$.

L (mm)	T (mm)	m (g)	v_s (m/s)	θ_s (deg)	P (mm)	m_r (g)	θ_r (deg)	TPD
150	2.50	313.3	815	19.6	504.00	300.4	<21.6	(1)
150	3.75	363.2	767	20.0	494.00	351.5	28.5	(1)
200	2.50	414.7	723	24.0	464.50	–	36.9	(2)
200	3.75	482.5	681	17.0	444.75	469.2	24.0	(1)
250	2.50	516.4	657	19.0	405.25	502.9	28.9	(2)
250	3.75	602.7	613	18.0	504.00	593.3	26.0	(2)

Table 4. Test results for $\theta_s \approx 30°$.

L (mm)	T (mm)	m (g)	v_s (m/s)	θ_s (deg)	P (mm)	m_r (g)	θ_r (deg)	TPD
150	2.50	310.5	815	27.9	504.00	299.2	33.6	(2)
150	3.75	360.8	769	30.6	415.00	347.9	51.9	(1)
200	2.50	415.5	721	29.4	405.25	402.7	56.2	(3)
200	3.75	488.9	676	30.1	454.50	475.6	48.2	(2)
250	2.50	519.4	656	30.1	395.25	–	52.0	(3)
250	3.75	593.0	620	31.6	395.25	577.8	52.9	(4)

5.22 [Chen *et al.*, 2011]

5.22.1 *General data on shields and projectiles*

Shields. Single FRC panels ($c_{l,w} = 305\,mm$) produced using Cor-Tuf ultra-high-strength concrete ([Williams *et al.*, 2009]).

Fibers. Hooked-end Bekaert Dramix® ZP305 steel fibers; $L^{fib} \approx 30\,mm$.

Projectiles. Spherical steel ($\sigma_y = 2000\,MPa$, $\sigma_u = 2150\,MPa$, $\delta = 7\%$, $G = 80\,GPa$, $B = 140\,GPa$) projectiles (Fig. 3.5.1): *Projectile-1* ($D = 7.9\,mm$, $m = 2.07\,g$); *Projectile-2* ($D = 11.1\,mm$, $m = 5.57\,g$) and *Projectile-3* ($D = 12.7\,mm$, $m = 8.36\,g$).

5.22.2 *Test results*

Table 1. Test results for Projectile-1.

T (mm)	v_s (m/s)	v_r (m/s)	m_r (g)	m^{fr} (g)	T (mm)	v_s (m/s)	v_r (m/s)	m_r (g)	m^{fr} (g)
39.7	2048	54	0.7	185.5	39.7	605	0	2.0	2.7

T (mm)	v_s (m/s)	v_r (m/s)	m_r (g)	m^{fr} (g)	T (mm)	v_s (m/s)	v_r (m/s)	m_r (g)	m^{fr} (g)
27.0	1988	391	0.8	86.2	27.0	919	24	–	43.5
17.5	874	297	2.0	20.0	13.5	239	0	–	–

Table 2. Test results for Projectile-2.

T (mm)	v_s (m/s)	v_r (m/s)	m_r (g)	m^{fr} (g)	T (mm)	v_s (m/s)	v_r (m/s)	m_r (g)	m^{fr} (g)
36.5	797	15	–	–	39.7	1708	685	5.5	190.5
25.4	1115	545	5.5	77.1	25.4	539	0	5.6	49.4
25.4	269	0	5.6	4.5	27.0	577	0	5.6	72.6
14.3	262	0	5.6	25.4					

Table 3. Test results for Projectile-3.

T (mm)	v_s (m/s)	v_r (m/s)	m_r (g)	m^{fr} (g)	T (mm)	v_s (m/s)	v_r (m/s)	m_r (g)	m^{fr} (g)
41.3	1718	792	8.2	194.6	39.7	408	0	8.3	5.4
25.4	482	23	8.3	59.4	14.3	462	199	8.3	27.2
15.9	115	0	8.3	2.3					

5.23   [Dancygier and Yankelevsky, 1996]

See also [Dancygier end Yankelevsky, 1994].

5.23.1   *General data on shields and projectiles*

Shields. Single NSC slabs ($f_c' = 34 - 35\,MPa$ measured in tests with cube-shaped specimens having the lenth of the edge 100 mm) and HSC slabs ($f_c' = 95 - 110\,MPa$) with different reinforcement; $c_{l,w} = 400\,mm$.

Table 1. Reinforcement types.

Notation	Description
R-1	Single mesh with deformed steel bars; $t^R = 10\,mm$, $d_{l,w}^R = 5\,mm$, $g_{l,w}^R = 100\,mm$. Reinforcement ratios are calculated with respect to the effective depth that in these experiments was T =10 mm
R-2	Steel fibers, $\mu = 60\,kg/m^3$
R-3	Two wire meshes ($d_{l,w} = 0.5\,mm$, $g_{l,w} = 7\,mm$) at distances 10 mm and 20 mm from the rear face of the shield

Notation	Description
R-4	Concrete shields with relatively low reinforcement ratios, or with relatively wide spacing between the reinforcing bars. The type of the steel fibers was Dramix ZP 30/0.50 (hooked steel fibers with $L^{fib} = 30\,mm$, $d^{fib} = 0.5\,mm$)

Projectiles. Hardened steel cone-nose projectiles (Fig. 3.2.1); $D = 25\,mm$, $L_{cyl} = 40\,mm$, $L_{nose} = 35\,mm$ $m = 120\,g$.

Table 2. Shield damage types (SDTs).

Notation	Description	Notation	Description
P	Perforation	B	Projectile perforated the plate but bounced and did not fully penetrate
S	Split specimen		

5.23.2 *Test results*

Table 3. Test results for reinforcement R-1.

T (mm)	f_c' (MPa)	$\rho_{l,w}^{R}$ (%)	μ (kg/m^3)	v_s (m/s)	P (mm) or SDT
40	34	0.65	77	85	21
40	34	0.65	77	144	P
40	104	0.65	77	116	B
40	104	0.65	77	143	P
62	24	0.39	51	142	40
60	104	0.39	51	142	33

Table 4. Test results for reinforcement R-2.

T (mm)	f_c' (MPa)	v_s (m/s)	P (mm) or SDT	T (mm)	f_c' (MPa)	v_s (m/s)	P (mm) or SDT
40	34	84	18	40	34	144	P
40	104	143	P	50	95	141	B
60	34	130	40	60	34	138	34
60	104	144	30				

Table 5. Test results for reinforcement R-3.

T (mm)	f_c' (MPa)	$\rho_{l,w}^{R}$ (%)	μ (kg/m^3)	v_s (m/s)	P (mm) or SDT
40	34	0.22	22	145	P
40	104	0.22	22	114	B

T (mm)	f_c' (MPa)	$\rho_{l,w}^R$ (%)	μ (kg/m^3)	v_s (m/s)	P (mm) or SDT
40	104	0.22	22	145	B
50	35	0.16	18	158	P
50	110	0.16	18	145	21
50	110	0.16	18	161	B
50	110	0.16	18	206	S
60	34	0.12	15	145	44
60	34	0.12	15	185	34
60	34	0.12	15	228	S
60	104	0.12	15	145	35

Table 6. Test results for reinforcement type R-4.

T (mm)	f_c' (MPa)	v_s (m/s)	SDT	T (mm)	f_c' (MPa)	v_s (m/s)	SDT
40	34	145	P	40	104	142	P
50	35	140	P	50	110	143	B
60	34	145	S	60	104	144	S

5.24 [Dancygier *et al.*, 1995]

5.24.1 *General data on shields and projectiles*

Shields. Single NSC slabs ($f_c' = 35 - 39\,MPa$ measured in tests with cube-shaped specimens having the length of the edge of 70 mm; and $\sigma_u = 3\,MPa$ measured in tests with $\varnothing 100 \times 250$ mm cylinders) and HSC slabs (average $f_c' = 112\,MPa$ and $\sigma_u = 6\,MPa$) with different reinforcement; $c_{l,w} = 400\,mm$, $T = 50\,mm$.

Table 1. Mesh types.

Label	Description
W-1	Woven smooth wires in two layers, 10 mm apart
W-2	Woven fence mesh
W-3	Non-woven, non-welded, cold-worked bars
W-4	Same as W-3 but deformed bars

Projectiles. Hardened steel cone-nose projectiles (Fig. 3.2.1); $D = 25\,mm$, $L_{nose} = 35\,mm$, $m = 120 - 125\,g$.

5.24.2 *Test results*

Table 2. Test results for shields having different types of mesh.

Concrete			Reinforcement				
f_c' (MPa)	Mesh type	$d_{l,w}$ (mm)	$g_{l,w}$ (mm)	$\rho_{l,w}$ (%)	σ_y (MPa)	v_s (m/s)	SDT
35	W1	0.5	7	0.16	183	158	PP
112	W1	0.5	7	0.16	183	206	PP
25	W2	2.5	34	0.36	382	165	PS
112	W2	2.5	34	0.36	382	215	PS
39	W2	2	16	0.49	316	160	–
112	W2	2	16	0.49	316	208	–
35	W2	3	25	0.71	382	204	PP
39	W2	4	34	0.92	316	171	–
112	W2	4	34	0.92	316	170	–
39	W3	5	27	1.82	534	221	PP
112	W3	5	27	1.82	534	246	–
35	W4	5	27	1.82	473	191	PP
112	W4	5	27	1.82	473	225	PS

PP – Projectile perforated shield, PS – Projectile was stuck in shield.

5.25 [Dancygier *et al.*, 1999]

5.25.1 *General data on shields and projectiles*

Shields. Single RC with rebars, $f_c' = 30\,MPa$ (measured in tests with cube-shaped specimens with the length of the edge of 100 mm), 127 mm slump, $q_{max}^{agg} = 19\,mm$; $c_{l,w} = 800\,mm$, $T = 250\,mm$.

Table 1. Mesh parameters.

$t^{F,R}$ (mm)	$d_{l,w}^{F,R}$ (mm)	$g_{l,w}^{F}$ (mm)	$\rho_{l,w}^{F}$ (%)	$g_{l,w}^{R}$ (mm)	$\rho_{l,w}^{R}$ (%)
15	8	200	0.14	100	0.28

Rebars. Steel, $\sigma_y = 400\,MPa$, $\varepsilon_u = 12\%$.

Projectiles. Hardened steel cone-nose projectiles (Fig. 3.2.1); $D = 49\,mm$, $L_{nose} = 74$, $L_{cyl} = 125$, $m = 1.5\,kg$.

Table 2. Types of shields responses.

Notation	Description
(1)	Cracks without scabbing
(2)	Heavy cracking at scabbing limit
(3)	Heavy cracking without perforation, rear face crater was formed

Notation	Description
(4)	Partial perforation, formation of rear face crater, projectile did not fully penetrate the specimen

5.25.2 *Test results*

Table 3. DOP and shield response.

v_s (m/s)	P (mm)	Rear face damage	v_s (m/s)	P (mm)	Rear face damage
299	–	(4)	300	–	(4)
263	160	(2)	281	192	(3)
264	178	(2)	273	146	(1) – (2)
273	173	(3)			

5.26   [Dancygier *et al.*, 2007]

5.26.1   *General data on shields and projectiles*

Shields. Single RC barriers having different compositions and reinforcement ; $c_{l,w} = 800\,mm$, $T = 200\,mm$.

Table 1. Types of meshes.

Mesh	$t^{F,R}$ (mm)	$d_{l,w}^{F,R}$ (mm)	$g_{l,w}^{F}$ (mm)	$\rho_{l,w}^{F}$ (%)	$g_{l,w}^{R}$ (mm)	$\rho_{l,w}^{R}$ (%)
Mesh-1	15	8	200	0.14	100	0.28
Mesh-2	15	8	100	0.28	100	0.28
Mesh-3	15	8	100	0.28	50	0.56

Rebars. Steel, $\sigma_y = 400\,MPa$, $\varepsilon_u = 12\%$.

Fibers. Hooked-end steel fibers: long fibers (LF) with $L^{fib} = 60\,mm$, $d^{fib} = 0.9\,mm$ and short fibers (SF) with $L^{fib} = 30\,mm$, $d^{fib} = 0.5\,mm$.

Table 2. Types of shields.

Shield label	Mesh type	Micro-silica (Y/N)	Fibers	Aggregates		f'_c (MPa)
				Material	q^{agg} (mm)	
Sh-1	Mesh-1	N	–	Dolomite	22	40
Sh-2	Mesh-1	N	LF	Dolomite	22	108
Sh-3	Mesh-1	N	–	Dolomite	22	119
Sh-4	Mesh-1	Y	LF	Dolomite	22	102

| Shield label | Mesh type | Micro-silica (Y/N) | Fibers | Aggregates | | f_c' (MPa) |
				Material	q^{agg} (mm)	
Sh-5	Mesh-1	Y	–	Dolomite	22	117
Sh-6	Mesh-1	Y	LF	Dolomite	50	104
Sh-7	Mesh-1	Y	SF	Dolomite	22	113
Sh-8	Mesh-1	Y	LF	Basalt	22	109
Sh-9	Mesh-1	Y	LF	Flint	12	106
Sh-10	Mesh-1	N	LF	Dolomite	50	101
Sh-11	Mesh-1	Y	LF	Basalt	50	93
Sh-12	Mesh-1	Y	LF	Flint	50	94
Sh-13	Mesh-2	Y	LF	Dolomite	22	114
Sh-14	Mesh-3	Y	LF	Dolomite	22	102
Sh-15	Mesh-2	Y	LF	Dolomite	22	110

Note. f_c' was measured using cube-shaped specimens with the length of the edge 100 mm for mixes with $q^{agg} = 22\,mm$, and using cube-shaped specimens with the length of the edge 120 mm and/or 150 mm for mixes with $q^{agg} = 50\,mm$.

Projectiles. Hardened steel cone-nose projectiles (Fig. 3.2.1); $D = 49\,mm$, $L_{nose} = 74$, $L_{cyl} = 125$, $m = 1.5\,kg$.

Table 3. Types of shields responses.

Designation	Description
(1)	No visible damage to hairline cracks at the rear face
(2)	Visible cracks without scabbing
(3)	Heavy cracking, pattern of rear crater indicates either scabbing limit or the development of a rear shear plug
(4)	Rear face damage (scabbing and spalling or shear plug formation) without perforation
(5)	Perforation without complete penetration through the shield
(6)	Projectile perforated the shield and exited with a residual velocity
(7)	Projectile was stuck in the shield

Note. Additional symbols were used to refine the description for intermediate cases.

5.26.2 *Test results*

Table 4. Shield response.

Shield	v_s (m/s)	P (mm)	Response	Shield	v_s (m/s)	P (mm)	Response
Sh-1	199	102	(2)	Sh-1	199	110	(2)–(3)
Sh-1	203	135	(3)–(4)	Sh-1	204	127	(3)+

Shield	v_s (m/s)	P (mm)	Response	Shield	v_s (m/s)	P (mm)	Response
Sh-1	204	120	(2)	Sh-1	211	87	(2)
Sh-1	216	145	(4)	Sh-1	245	*	(5)
Sh-1	248	*	(5)	Sh-1	250	*	(6)
Sh-1	250	*	(6)	Sh-1	289	*	(6)
Sh-1	296	*	(6)	Sh-2	273	142	(3)
Sh-2	276	*	(5)	Sh-2	282	152	(4)
Sh-2	285	*	(6)	Sh-3	254	*	(5)
Sh-3	270	*	(7)	Sh-4	243	90	(1)–(2)
Sh-4	276	152	(3)+	Sh-4	281	180	(4)
Sh-4	289	*	(6)	Sh-4	309	*	(6)
Sh-5	278	*	(6)	Sh-5	282	*	(7)
Sh-6	287	180	(4)	Sh-6	291	*	(6)
Sh-7	262	128	(2)	Sh-7	289	*	(6)
Sh-8	279	150	(3)	Sh-8	289	145	(2)–(3)
Sh-9	291	135	(3)-	Sh-9	307	*	(7)
Sh-10	286	147	(3)	Sh-10	292	*	(7)
Sh-11	292	168	(4)	Sh-11	313	*	(7)
Sh-12	313	*	(5)	Sh-12	314	163	(4)
Sh-13	270	140	(3)	Sh-13	286	165	(3)+
Sh-14	287	170	(4)	Sh-14	289	*	(7)
Sh-15	284	*	(7)	Sh-15	286	*	(6)

* Complete or partial perforation.

5.27 [Dancygier *et al.*, 2014]

5.27.1 *General data on shields and projectiles*

Shields. Single and double-layer (manufactured as a single shield) HRC barriers with rebars, with and without fibers (parameter μ in tables below is the reinforcement quality of fibers); concrete contains dolomite aggregates; f_c' is measured using cube-shaped specimens having the length of the edge 150 mm; $c_{l,w} = 800\,mm$.

Rebars. Steel, $\sigma_y = 240\,MPa$.

Meshes. $t^{F,R} = 15\,mm$, $d_{l,w}^{F,R} = 8\,mm$, $g_{l,w}^{F} = 200\,mm$, $g_{l,w}^{R} = 100\,mm$.

Fibers. Hooked-end steel fibers: *F-1* ($L^{fib} = 35\,mm$, $d^{fib} = 0.55\,mm$, $\sigma_u = 1345\,MPa$) and *F-2* ($L^{fib} = 60\,mm$, $d^{fib} = 0.90\,mm$, $\sigma_u = 1000\,MPa$).

Projectiles. Hardened steel cone-nose projectiles (Fig. 3.2.1); $D = 50\,mm$, $L_{nose} = 75\,mm$, $L_{cyl} = 20\,mm$, $m = 1.75\,kg$.

Note. The BLV was determined either from the direct results when projectile was stuck in the shield or by taking the average of the highest and lowest velocities in the cases when the shield was not perforated and perforated, respectively.

5.27.2 *Test results*

Table 1. Test results for single-layer shields ($T = 200\,mm$).

f'_c (MPa)	q^{agg}_{max} (mm)	Fibers type	μ (kg/m^3)	v_{bl} (m/s)
89.5	19	–	0	239
115.4	19	F-1	60	258
108.1	19	F-1	60	280
112.3	37	–	0	242-264
111.4	37	F-2	60	289

Table 2. Test results for double-layer shields ($T_1 = T_2 = 100\,mm$).

Frontal layer				Rear layer				v_{bl} (m/s)
f'_c (MPa)	q^{agg}_{max} (mm)	Fibers type	μ (kg/m^3)	f'_c (MPa)	q^{agg}_{max} (mm)	Fibers type	μ (kg/m^3)	
107.4	37	–	0	96.1	19	–	0	258
97.9	37	–	0	103.5	19	F-1	60	275
116.1	37	F-2	60	116.9	19	F-1	60	289
100.0	19	–	0	113.8	19	F-1	60	279
113.2	19	F-1	60	102.4	37	–	0	271

Table 3. Test results for double-layer shields ($T_1 = 140\,mm$, $T_2 = 60\,mm$).

Frontal layer				Rear layer				v_{bl} (m/s)
f'_c (MPa)	q^{agg}_{max} (mm)	Fibers type	μ (kg/m^3)	f'_c (MPa)	q^{agg}_{max} (mm)	Fibers type	μ (kg/m^3)	
101.5	37	–	0	100.3	19	–	0	261-277
99.6	37	–	0	109.9	19	F-1	60	292
103.7	37	F-1	60	107.8	19	F-1	60	305
105.5	37	–	0	101.6	19	F-1	80	289-295

5.28 [Dancygier, 1997]

5.28.1 *General data on shields and projectiles*

Shields. Series-A of experiments. Single NSC slabs reinforced by woven fence meshes and non-woven rebar meshes; $f'_c = 35-39\,MPa$

measured in tests with cube-shaped specimens with the length of the edge 70 mm, and $\sigma_u = 3\,MPa$ measured in tests of $\varnothing 100 \times 250$ mm cylinders; $c_{l,w} = 400\,mm$.

Shields. Series-B of experiments. Reinforcement consisted of smooth rebar meshes; $f_c' = 34 - 35\,MPa$ measured in tests with cube-shaped specimens having the length of the edge 100 mm, and $\sigma_u = 3.4\,MPa$ measured in tests of $\varnothing 100 \times 250$ mm cylinders; $c_{l,w} = 400\,mm$, $T = 50\,mm$.

Table 1. Mesh types.

Mesh label	Description
W-1	Woven smooth wires in two layers, 10 mm apart
W-2	Woven fence mesh
W-3	Non-woven, non-welded, cold-worked bars
W-4	Same as W-3 but deformed bars

Projectiles. Hardened steel cone-nose projectiles (Fig. 3.2.1); $D = 25\,mm$, $L_{cyl} = 40\,mm$, $L_{nose} = 35\,mm$, $m = 120\,g$.

5.28.2 *Test results*

Table 2. Test results for Series-A of experiments with $T = 50\,mm$.

Concrete		Reinforcement				Results
f_c' (MPa)	Mesh type	$d_{l,w}^R$ (mm)	$g_{l,w}^R$ (mm)	$\rho_{l,w}^R$ (%)	σ_y (MPa)	v_{bl} (m/s)
35	W-1	0.5	7	0.16	382	140-158
35	W-2	2.5	34	0.36	382	165
39	W-2	2.0	16	0.49	316	>160
35	W-2	3.0	25	0.71	183	222
39	W-2	4.0	34	0.92	316	>171
39	W-3	5.0	27	1.82	534	202-221
35	W-4	5.0	27	1.82	483	<191

Table 3. Test results for Series-A of experiments with $T = 60\,mm$.

Concrete		Reinforcement				Results
f_c' (MPa)	Mesh type	$d_{l,w}^R$ (mm)	$g_{l,w}^R$ (mm)	$\rho_{l,w}^R$ (%)	σ_y (MPa)	v_{bl} (m/s)
39	W-1	0.5	7	0.12	382	>228
35	W-2	2.5	35	0.29	382	208-224

Concrete	Reinforcement					Results
f'_c (MPa)	Mesh type	$d^R_{l,w}$ (mm)	$g^R_{l,w}$ (mm)	$\rho^R_{l,w}$ (%)	σ_y (MPa)	v_{bl} (m/s)
35	W-2	3.0	25	0.57	183	232
39	W-3	5.0	27	1.46	534	<279
35	W-4	5.0	27	1.46	473	236

Table 4. Test results for Series-B of experiments.

Concrete	Reinforcement				Results
f'_c (MPa)	$d^R_{l,w}$ (mm)	$g^R_{l,w}$ (mm)	$\rho^R_{l,w}$ (%)	σ_y (MPa)	v_{bl} (m/s)
35	2.5	61	0.20	650	<147
34	3.25	104	0.20	600	126-141
35	5.0	245	0.20	450	146-219
34	3.25	21	0.99	600	210
34	2.5	80	0.15	650	157-158
34	3.25	80	0.26	600	148-159
35	5.0	80	0.61	450	158-165
34	2.5	20	0.61	650	162

5.29 [Dancygier, 1998]

See also [Dancygier and Yankelevsky, 1999].

5.29.1 *General data on shields and projectiles*

Shields. Single normal strength concrete (NSC) and high strength concrete (HSC) slabs; $c_{l,w} = 400\,mm$, $T = 50\,mm$; f'_c was measured in tests with cube-shaped specimens having a lenth of the edge 70 mm and 100 mm in *Series-A* and *Series-B* experiments, respectively.

Table 1. Mesh types.

Mesh label	Description
W-2	Woven fence mesh
W-1	Woven smooth wires in two layers, 10 mm apart
HTB	Smooth heat treated steel bars
DB	Deformed steel bars

Table 2. Shield damage types (SDTs).

Notation	Description
(1)	Perforation (complete penetration of the projectile)
(2)	Projectile was stuck in the shield
(3)	Specimen was punched but projectile bounced
(4)	Scabbing
(5)	Scabbing limit
(6)	Very fine radial cracks
(7)	Heavy radial cracking through
(8)	Not damaged

Projectiles. Hardened steel cone-nose projectiles (Fig. 3.2.1); $D = 25\,mm$, $L_{cyl} = 40\,mm$, $L_{nose} = 35\,mm$, $m = 125\,g$.

Note. The BLV was determined either from the direct results when projectile was stuck in the shield or by taking the average of the highest and lowest velocities in the cases when the shield was not perforated and perforated, respectively.

5.29.2 *Parameters of shields and results for NSC*

Table 3. Shields details and BLVs.

Shield label	Series	Concrete f'_c (MPa)	Mesh type	$d^{R}_{l,w}$ (mm)	$g^{R}_{l,w}$ (mm)	$\rho^{R}_{l,w}$ (%)	σ_y (MPa)	v_{bl} (m/s)
Sh-1	A	35	W-2	3.0	25	0.71	180	222
Sh-2	A	35	W-2	2.5	34	0.36	350	165
Sh-3	A	35	W2	0.5	7	0.16	250	149
Sh-4	A	39	HTB	5.0	27	1.82	450	211.5
Sh-5	A	35	DB	5.0	27	1.82	500	~185
Sh-6	A	39	W-2	4.0	34	0.92	250	~185
Sh-7	A	39	W-2	2.0	16	0.49	200	~185
Sh-8	B	35	HTB	2.5	61	0.20	650	~140
Sh-9	B	34	HTB	3.25	104	0.20	600	133.5
Sh-10	B	35	HTB	5.0	245	0.20	450	182.5
Sh-11	B	34	HTB	3.25	21	0.99	600	210
Sh-12	B	34	HTB	2.5	80	0.15	650	155.5
Sh-13	B	34	HTB	3.25	80	0.26	600	153.5
Sh-14	B	35	HTB	5.0	80	0.61	450	161.5
Sh-15	B	34	HTB	2.5	20	0.61	650	161.5

Table 4. Shield responses.

Shield	v_s (m/s)	SDT	Shield	v_s (m/s)	SDT	Shield	v_s (m/s)	SDT
Sh-1	204	(1)	Sh-1	209	(4)	Sh-1	222	(2)
Sh-2	158	(3)	Sh-2	165	(2)	Sh-2	159	(4)
Sh-3	165	(1)	Sh-3	158	(1)	Sh-3	140	(8)
Sh-4	202	(4)	Sh-4	221	(1)	Sh-5	210	(1)
Sh-5	191	(1)	Sh-6	171	(3)	Sh-7	160	(4)
Sh-8	166	(1)	Sh-8	147	(1)	Sh-9	156	(1)
Sh-9	113	(6)	Sh-9	126	(6)	Sh-9	141	(1)
Sh-9	145	(1)	Sh-9	117	(6)	Sh-10	115	(6)
Sh-10	146	(3)	Sh-10	143	(3)	Sh-10	219	(1)
Sh-11	181	(3)	Sh-11	175	(3)	Sh-11	198	(4)
Sh-11	251	(1)	Sh-11	210	(2)	Sh-11	159	(3)
Sh-12	141	(4)	Sh-12	138	(5)	Sh-12	154	(1)
Sh-12	158	(1)	Sh-12	157	(3)	Sh-13	136	(5)
Sh-13	159	(1)	Sh-13	148	(3)	Sh-14	158	(5)
Sh-14	165	(1)	Sh-14	158	(4)	Sh-14	153	(3)
Sh-14	174	(1)	Sh-14	240	(1),(7)	Sh-15	159	(2)
Sh-15	150	(3)	Sh-15	162	(2)	Sh-15	157	(5)
Sh-15	253	(1)						

5.29.3 *Parameters of shields and results for HSC*

Table 5. Parameters of shields and BLVs.

Shield label	Series	Concrete f'_c (MPa)	Reinforcement Mesh type	$d^R_{l,w}$ (mm)	$g^R_{l,w}$ (mm)	$\rho^R_{l,w}$ (%)	σ_y (MPa)	Results v_{bl} (m/s)
Sh-16	A	112	W-2	2.5	34	0.36	350	215
Sh-17	A	112	W-1	0.5	7	0.16	250	183.5
Sh-18	A	112	HTB	5.0	27	1.82	450	~250
Sh-19	A	112	DB	5.0	27	1.82	500	225
Sh-20	A	112	W-2	4.0	34	0.92	250	~200
Sh-21	A	112	W-2	2.0	16	0.49	200	208
Sh-22	B	111	HTB	3.25	104	0.20	600	189.5

Table 6. Shield responses.

Shield	v_s (m/s)	SDT	Shield	v_s (m/s)	SDT	Shield	v_s (m/s)	SDT
Sh-16	161	(3)	Sh-16	215	(2)	Sh-16	215	(3), (7)
Sh-17	161	(3)	Sh-17	145	(4)	Sh-17	206	(1)
Sh-18	162	(4)	Sh-18	209	(4)	Sh-18	246	(4)
Sh-19	159	(4)	Sh-19	210	(4)	Sh-19	225	(2)

Shield	v_s (m/s)	SDT	Shield	v_s (m/s)	SDT	Shield	v_s (m/s)	SDT
Sh-20	158	(3)	Sh-20	170	(3)	Sh-20	142	(4)
Sh-21	137	(6)	Sh-21	165	(4)	Sh-21	208	(2)
Sh-22	174	(3)	Sh-22	176	(3)	Sh-22	203	(1)

5.30   [Darrigade and Buzaud, 1999]

Note. Data from [Sjøl and Teland, 2001].

5.30.1   *General data on shields and projectiles*

Shields. Single high strength concrete and very high strength concrete slabs.

Projectiles. Ogive-nose projectiles (Fig. 3.3.1); $D = 25\,mm$, $\rho = 75\,mm$, $m = 500\,g$.

5.30.2   *Test results*

Table 1. Test results for v_r.

f_c' (MPa)	T (mm)	v_s (m/s)	v_r (m/s)	f_c' (MPa)	T (mm)	v_s (m/s)	v_r (m/s)
140	1030	445	234	140	1050	545	350
140	1560	547	20	200	500	135	10
200	550	250	168	200	560	335	278
200	1070	445	305	200	1100	550	425
600	500	352	249	600	520	250	151
600	530	140	0	600	1040	440	255
600	1050	560	351				

5.31   [Fan *et al.*, 2013]

5.31.1   *General data on shields and projectiles*

Shields. Single steel fiber reactive powder concrete (SFRPC) slabs ($\gamma = 2850\,kg/m^3$, $f_c' = 186\,MPa$, $K_{Ic} = 10.54\,MPa \cdot m^{1/2}$), and ordinary C30 concrete slabs.

Fibers. Shield contained 5% cold drawn low-carbon steel fibers, $L^{fib} = 13\,mm$, $d^{fib} = 0.16\,mm$, $\sigma_u = 2000\,MPa$.

Table 1. Ogive-like nosed steel (HRC 49) projectiles (Fig. 3.3.1).

Label	D (mm)	L (mm)	m (kg)	Label	D (mm)	L (mm)	m (kg)
Projectile-1	57	138	2.32	Projectile-2	80	750	24.0

5.31.2 *Test results*

Table 2. Test results for Projectile-1 against SFRPC shields ($D^{sh} = 1.2\,m$).

T (mm)	v_s (m/s)	P (mm)	T (mm)	v_s (m/s)	P (mm)	T (mm)	v_s (m/s)	P (mm)
200	308	106	200	328	125	200	364	140
400	342	135	400	360	145	400	364	142
400	380	147	400	410	178	400	550	197
400	582	210						

Table 3. Test results for Projectile-2 against SFRPC shields ($T = 2.0\,m$, $D^{sh} = 1.6\,m$).

v_s (m/s)	P (m)	v_s (m/s)	P (m)	v_s (m/s)	P (m)
847	1.325	852	1.330	887	1.445

Table 4. Test results for Projectile-2 against C30 shields ($T = 3.6\,m$, $D^{sh} = 1.6\,m$).

v_s (m/s)	P (m)	v_s (m/s)	P (m)	v_s (m/s)	P (m)
808	3.600	836	Run through	880	Run through

5.32 [Fang *et al.*, 2006]

5.32.1 *General data on shields and projectiles*

Shields. Single polyethylene FRC slabs ($D^{sh} = 200\,mm$, $T = 200\,mm$) having different fiber volume content and the steel liner with $T_{lin}^{R} = 3\,mm$. **Projectiles.** 35CrMnSi alloy steel projectiles. The shape of the projectile is shown in Fig. 3.7.1a where all dimensions, except for $D = 37\,mm$, are approximate since they are taken from the sketch of the projectile.

5.32.2 *Test results*

Table 1. Additional data and test results.

ρ_V (%)	f'_c (MPa)	m (g)	v_s (m/s)	P (mm)
0	74.8	694	671	200
0.2	88.2	695	436	130
0.2	88.2	694	509	160
0.3	88.0	697	419	103
0.3	88.0	695	501	210

5.33 [Fiquet and Dacquet, 1977]

5.33.1 *General data on shields and projectiles*

Shields. Single MRC slabs; $c_{l,w} = 5m$, $f'_c = 40\,MPa$, $\sigma_u = 3.1\,MPa$.

Meshes. Four identical meshes with the following characteristics: $d_{l,w} = 16mm$ and $g_{l,w} = 125mm$, $\sigma_y = 400\,MPa$ for rebar. Two meshes at distances of 58 mm and 114 mm from the frontal face of the slab, and two meshes at the distances of 58 mm and 114 mm from the rear face of the slab.

Projectiles. Projectile can be assumed to have a cylindrical shape with $D = 305\,mm$.

5.33.2 *Test results*

Table 1. Test results for shields with $T = 400\,mm$.

m (kg)	v_s (m/s)	P (mm)	Note
227	86.9	200	–
227	90	170	Speed determined inaccurately
227	97.2	185	–
227	106	550	Nose distorted reinforcement on back face
227	106.4	–	Projectile hung by the tail on shield back face
192	93.6	100	–
192	110.6	240	–
192	125.6	–	Projectile penetrated the shield, $v_r \approx 25\,m/s$
160	98.5	85	–

m (kg)	v_s (m/s)	P (mm)	Note
160	106	105	–
160	110	125	–
160	132.6	40	Concrete plug between projectile nose and back distorted irons

Table 2. Test results for shields with $T = 500\,mm$.

m (kg)	v_s (m/s)	P (mm)	Note
227	107	150	–
227	117.3	225	–
227	126.6	450	Concrete plug between nose grid very distorted irons
192	79.1	30	–
192	120.3	135	–
192	134.5	165	–
192	144.1	–	Projectile penetrated the shield, $v_r \approx 5m/s$
50	77.5	20	–
50	148.5	176	–
50	160.4	290	–

5.34 [Folsom, 1987]

5.34.1 *General data on shields and projectiles*

Shields. Predrilled PC shields (shields with an inline cylindrical hole) and monolithic PC shields.

Table 1. Properties of concrete shields (dimensions include aluminum case).

Shield label	D^{sh} (mm)	T (mm)	γ (kg/m^3)	f_c' (MPa)	q_{max}^{agg} (mm)
Shield-1	406	1537	2370	48.5/37.0	25
Shield-2	102	–	2390	49.0/44.4	10
Shield-3	203	–	2390	49.0/44.4	10

Note. Ø152×305 mm cylinders were tested to determine f_c'. In numerator and denominator in the column for f_c', volumes are given at the time of penetration tests and after 28 days, correspondingly.

Table 2. Hardened 4330 steel (HRC 32-36) ogive-nose projectiles (Fig. 3.3.1).

Projectile label	D (mm)	L_{cyl} (mm)	L_{nose} (mm)	ρ (mm)	m (g)
Projectile-1	88.6	177	88.6	111	5930
Projectile-2	22.0	44.0	22.0	27.6	88.7

5.34.2 *Test results*

Note. Asterisk in column P indicates that projectile fell out of the hole after penetration.

Table 3. Test results for Shield-1 and Projectile-1.

$\dfrac{\hat{D}}{D}$	v_s (m/s)	P (mm)	$\dfrac{\hat{D}}{D}$	v_s (m/s)	P (mm)	$\dfrac{\hat{D}}{D}$	v_s (m/s)	P (mm)
0	206	140*	0.431	209	164*	0.718	206	215*
0.860	198	330	0.930	214	486	0.934	107	232
0.934	355	1098	0.969	204	933			

Table 4. Test results for Shield-2 and Projectile-2.

$\dfrac{\hat{D}}{D}$	v_s (m/s)	P (mm)	$\dfrac{\hat{D}}{D}$	v_s (m/s)	P (mm)	$\dfrac{\hat{D}}{D}$	v_s (m/s)	P (mm)
0	227	24*	0	391	46*	0.868	319	113
0.869	94	28*	0.869	207	63	0.898	216	79
0.900	401	197	0.908	91	36	0.927	221	91
0.929	309	162	0.929	395	212			

Table 5. Test results for Shield-3 and Projectile-2.

$\dfrac{\hat{D}}{D}$	v_s (m/s)	P (mm)	$\dfrac{\hat{D}}{D}$	v_s (m/s)	P (mm)	$\dfrac{\hat{D}}{D}$	v_s (m/s)	P (mm)
0.868	316	106	0.869	183	54	0.869	199	58

5.35 [Forquin *et al.*, 2015]

5.35.1 *General data on shields and projectiles*

Shields. Monolithic concrete slabs made of R30A7 concrete; $f_c' = 30\,MPa$, $q_{\max}^{agg} = 8\,mm$, $D^{sh} = 800\,mm$.

Projectiles. 35NCD16 steel ogive-nose projectiles (Fig. 3.3.1); $D = 52\,mm$, $L = 300\,mm$, $\psi = 5.77$, $m = 2.4\,kg$.

5.35.2 *Test results*

Table 1. Test results for shields with $T = 800$ *mm*.

w_s (kJ)	P (mm)	w_s (kJ)	P (mm)	w_s (kJ)	P (mm)	w_s (kJ)	P (mm)
62#	195#	145	269	149#	274#	154#	260#
158#	311#	160#	272#	220#	345#		

Test results for $T = 300\,mm$: $v_s = 333\,m/s$, $v_r = 71\,m/s$.

5.36 [Forrestal *et al.*, 1994]

5.36.1 *General data on shields and projectiles*

Shields. Single PC slabs; Ø152×305 mm cylinders were tested to determine f_c'.

Projectiles. HT 4340 steel (HRC 43-45) ogive-nose projectile (Fig. 3.3.1); $D = 26.9\,mm$, $L_{cyl} = 206.8\,mm$, $L_{nose} = 35.6\,mm$, $\psi = 2$.

5.36.2 *Test results*

Table 1. Test results for shields with $\gamma = 2370\,kg/m^3$, $f_c' \approx 32 - 40\,MPa$.

T (m)	D^{sh} (m)	f_c' (MPa)	m (g)	v_s (m/s)	P (mm)
0.76	1.37	35.2	906	277	173
0.91	1.37	37.8	910	410	310
0.91	1.37	38.1	907	431	411
0.76	1.37	33.5	912	499	480
1.07	1.37	38.4	910	567	525
1.83	1.22	36.9	905	590	729
1.83	1.22	40.1	901	591	513
1.83	1.22	35.4	903	631	607
1.83	1.22	34.7	905	642	620
1.83	1.22	36.0	901	773	866
1.83	1.22	32.4	904	800	958

Table 2. Test results for shields with $\gamma = 2340\,kg/m^3$, $D^{sh} = 1.37\,m$, $T = 1.22\,m$, $f_c' \approx 90 - 108\,MPa$.

f_c' (MPa)	m (g)	v_s (m/s)	P (mm)	f_c' (MPa)	m (g)	v_s (m/s)	P (mm)
90.5	907	561	353	91.0	898	584	384
95.0	908	608	422	101.4	905	622	437
94.0	907	750	630	108.3	900	793	605

5.37   [Forrestal *et al.*, 1996]

5.37.1   *General data on shields and projectiles*

Shields. Single PC slabs; $\varnothing 51 \times 102$ mm cylinders were tested to determine f_c'.

Table 1. 4330 Steel ogive-nose projectiles (Fig. 3.3.1), $\psi = 3$.

Projectile label	D (mm)	L_{cyl} (mm)	L_{nose} (mm)	Projectile label	D (mm)	L_{cyl} (mm)	L_{nose} (mm)
Projectile-1	20.3	169.5	33.7	Projectile-2	30.5	254.3	50.5

5.37.2   *Test results*

Table 2. Test results for Projectile-1 and shields with $D^{sh} = 0.51\,m$, $f_c' = 56.0 - 70.2\,MPa$; $m = 0.478\,kg$.

T (mm)	v_s (m/s)	P (mm)	$\Delta m/m$ (%)	T (mm)	v_s (m/s)	P (mm)	$\Delta m/m$ (%)
910	450	300	1.5	910	612	480	2.7
1220	821	760	4.5	1520	926	950	5.5
1520	987	920	6.6	1520	1024	940	6.2

Table 3. Test results for Projectile-2 and shields with $D^{sh} = 0.91\,m$, $f_c' = 48.3 - 55.2\,MPa$.

T (m)	m (kg)	v_s (m/s)	P (m)	$\Delta m/m$ (%)
1.83	1.608	405	0.37	1.2
1.83	1.600	446	0.42	1.5
2.13	1.600	545	0.56	2.0
2.13	1.603	651	0.78	3.1
2.13	1.604	804	1.05	4.7
2.13	1.600	821	1.23	4.4
2.44	1.603	900	1.41	5.4

T	m	v_s	P	$\Delta m / m$
(m)	(kg)	(m/s)	(m)	(%)
2.44	1.599	1009	1.75	6.4
2.74	1.609	1069	1.96	7.0
2.74	1.603	1201	2.03	6.8

5.38 [Forrestal *et al.*, 2003]

5.38.1 *General data on shields and projectiles*

Shields. Single PC slabs: $D^{sh} = 1.83m$, $q_{max}^{agg} = 9.5mm$; quartz aggregate, $f_c' = 23\,MPa$ for *Shield-1*, and limestone aggregate, $f_c' = 39\,MPa$ for *Shield-2*.

Projectiles. Ogive-nose projectiles (Fig. 3.3.1, nose is spherically rounded by a spherical surface having a radius 3.18 mm); $D = 76.2\,mm$, 4340 steel (HRC 45); $\psi = 3$, $L = 530.73\,mm$ for *Projectile-1*, and $\psi = 6$, $L = 528.47\,mm$ for *Projectile-2*.

5.38.2 *Test results*

Table 1. Test results for Projectile-1 and Shield-1.

m	T	v_s	P	m	T	v_s	P
(kg)	(m)	(m/s)	(m)	(kg)	(m)	(m/s)	(m)
13.043	1.83	139.3	0.24	13.037	1.83	200.0	0.42
13.085	1.22	250.0	0.62	13.158	1.22	283.7	0.76
13.080	1.83	336.6	0.93	13.119	1.83	378.6	1.18

Table 2. Test results for Projectile-2 and Shield-1.

m	T	v_s	P	m	T	v_s	P
(kg)	(m)	(m/s)	(m)	(kg)	(m)	(m/s)	(m)
13.061	1.83	238.4	0.58	13.064	1.83	378.6	1.25

Table 3. Test results for Projectile-1 and Shield-2.

m	T	v_s	P	m	T	v_s	P
(kg)	(m)	(m/s)	(m)	(kg)	(m)	(m/s)	(m)
12.923	1.22	238.1	0.30	12.900	1.83	275.7	0.38
12.910	1.22	314.0	0.45	12.914	1.22	369.5	0.53
12.957	1.83	456.4	0.94				

Table 4. Test results for Projectile-2 and Shield-2.

m	T	v_s	P	m	T	v_s	P
(kg)	(m)	(m/s)	(m)	(kg)	(m)	(m/s)	(m)
12.873	1.83	312.5	0.61	12.909	1.83	448.5	0.99

5.39 [Frew *et al.*, 1998]

5.39.1 *General data on shields and projectiles*

Shields. Single PC slabs; $\gamma = 2320\,kg/m^3$, $f_c' = 58.4\,MPa$ ($\varnothing 51 \times 102$ mm cylinders were tested to determine f_c'); limestone aggregate ($q^{agg} = 9.5\,mm$, Mohs hardness scale of 3.0).

Table 1. Ogive-nose projectiles (Fig. 3.3.1 $\psi = 3.0$).

Projectile label	D (mm)	L_{nose} (mm)	Material	HRC	m (kg)
Projectile-1	20.3	33.7	4340 Steel	45	0.478
Projectile-2	20.3	33.7	AerMet 100	53	0.478
Projectile-3	30.5	50.5	4340 Steel	45	1.62
Projectile-4	30.5	50.5	AerMet 100	53	1.62

5.39.2 *Test results*

Table 2. Test results for Projectile-1, $D^{sh} = 0.51m$.

L (mm)	T (m)	v_s (m/s)	P (m)	$\Delta m/m$ (%)
203.2	0.94	442	0.287	0.81
203.2	0.94	610	0.491	1.55
203.2	1.07	815	0.840	2.69
203.2	1.52	1009	1.300	3.52
203.2	1.93	1162	1.590	4.12
220.1	2.03	797	1.010	2.85
270.8	2.26	803	1.226	3.02

Table 3. Test results for Projectile-2 with $L = 203.2\,mm$; $D^{sh} = 0.51m$.

T (m)	v_s (m/s)	P (m)	$\Delta m/m$ (%)	T (m)	v_s (m/s)	P (m)	$\Delta m/m$ (%)
2.28	791	0.73	2.58	1.98	994	1.16	3.76
2.28	1165	1.46	4.64				

Table 4. Test results for Projectile-3; $L = 304.7\,mm$, $D^{sh} = 0.91\,m$.

| T | v_s | P | $\Delta m/m$ | T | v_s | P | $\Delta m/m$ |
(m)	(m/s)	(m)	(%)	(m)	(m/s)	(m)	(%)
1.07	445	0.46	0.7	1.07	584	0.79	1.5
1.68	796	1.23	2.5	2.44	980	1.95	3.5
2.44	992	1.96	3.4	3.05	1176	2.67	3.8

Table 5. Test results for Projectile-4; $L = 304.7\,mm$, $D^{sh} = 0.91\,m$.

| T | v_s | P | $\Delta m/m$ | T | v_s | P | $\Delta m/m$ |
(m)	(m/s)	(m)	(%)	(m)	(m/s)	(m)	(%)
2.74	972	1.96	3.1	3.05	1225	2.83	3.6

5.40 [Frew *et al.*, 2006]

5.40.1 *General data on shields and projectiles*

Shields. Single PC slabs, quartz aggregate ($q_{max}^{agg} = 9.5\,mm$), $f_c' = 23\,MPa$ (Ø50.8×114 mm cylinders were tested to determine f_c'), $\gamma = 2040\,kg/m^3$. **Projectiles.** Ogive-nose projectiles (Fig. 3.3.1, nose is spherically rounded by a spherical surface having a radius 3.18 mm); $D = 76.2\,mm$, 4340 steel (HRC 45), $\psi = 3$, $L = 530.73\,mm$.

5.40.2 *Test results*

Table 1. Test results for P.

| m | D^{sh} | T | v_s | P |
(kg)	(m)	(m)	(m/s)	(m)
13.08	1.83	1.37	335	0.94
13.04	1.37	1.52	332	0.96
13.05	0.91	1.52	337	1.02
12.89	1.83	1.83	280	0.73
13.06	1.37	1.22	279	0.69
12.92	0.91	0.91	201	0.45
13.00	1.37	0.91	164	0.31

5.41 [Gao *et al.*, 2006]

Note. Data from [Gao *et al.*, 1995], [Gao *et al.*, 2003].

5.41.1 *General data on shields and projectiles*

Shields. Single PC slabs; $\gamma = 2400\ kg/m^3$, $f_c' = 30\ MPa$.

Projectiles. Steel parabolic nose projectiles (bodies of revolution). Equation of the generatrix reads: $\tilde{x}/L_{nose} = (\tilde{\rho}/D)^2$ in the cylindrical coordinate system $\tilde{\rho}O\tilde{x}$ with the origin at the top of the paraboloid, axis $O\tilde{x}$ coincides with the axis of the paraboloid and is directed in the opposite direction to the impact velocity; $D = 31\ mm$.

5.41.2 *Test results*

Table 1. Parameters of projectile and test results on DOP.

m	L_{nose}	L_{cyl}	v_s	P
(kg)	(mm)	(mm)	(m/s)	(mm)
3.777	100	55	763	830
3.034	70	57	577	340
3.747	70	57	666	560
3.022	70	57	538	370
3.154	70	57	630	460

5.42 [Goel *et al.*, 1986]

5.42.1 *General data on shields and projectiles*

Shields. Single PC slabs, $c_{l,w} = 750\ mm$, $T = 250\ mm$; fabricated using cement, sand and coarse aggregate ($q^{agg} = 12.5\ mm$) in the ratio of 1:1:2.

Projectiles. Ogive-nose projectiles (Fig. 3.3.1) with $D = 31\ mm$, $\psi = 2.5$ and different lengths.

5.42.2 *Test results*

Table 1. θ_{ric} as a function of v_s for the impactor with $L/D = 3$.

v_s	θ_{ric}	v_s	θ_{ric}	v_s	θ_{ric}
(m/s)	(deg)	(m/s)	(deg)	(m/s)	(deg)
163 ± 6	15	221 ± 5	17	278 ± 10	18

Table 2#. Values of θ_{ric} for impactors with different L for $v_s = 250 \pm 8 \, m/s$.

L/D	θ_{ric} (deg)	L/D	θ_{ric} (deg)	L/D	θ_{ric} (deg)
3	18.4	4.5	26.8	6.4	28.9
7.8	30.2	8.9	31.2		

5.43 [Goel *et al.*, 1990]

5.43.1 *General data on shields and projectiles*

Shields. Single slabs manufactured from PC, steel fiber reinforced concrete (SFRC), and steel fiber and steel rods reinforced concrete (SFRRC); $c_{l,w} = 750 \, mm$, $T = 175 \, mm$, $f_c' = 34 \, MPa$.

Fibers. Steel fibers, $L^{fib} = 80 \, mm$, $d^{fib} = 0.45 \, mm$, $\widetilde{\rho}_V = 1.25\%$.

Rebars. $t^{F,R} = 25 \, mm$, $d_{l,w}^{F,R} = 10 \, mm$, $g_{l,w}^{F,R} = 110 \, mm$.

Projectiles. EN24 steel ogive-nose projectiles (Fig. 3.3.1); $D = 30 \, mm$, $\psi = 1$, $L = 150 \, mm$, $m = 472 \pm 4 \, g$.

5.43.2 *Test results*

Table 1. Test results for PC shields.

v_s (m/s)	P (mm)	v_s (m/s)	P (mm)	v_s (m/s)	P (mm)	v_s (m/s)	P (mm)
45	9	72	15	79	16	92	19
105	26	118	28	128	30	141	34
150	36	166	37	215	55	224	64
225	63	256	65	300	86	320	80
337	92	348	91	366	86		

Table 2. Test results for SFRC shields.

v_s (m/s)	P (mm)	v_s (m/s)	P (mm)	v_s (m/s)	P (mm)	v_s (m/s)	P (mm)
73	16	86	19	110	26	160	37
231	42	269	59	321	78	374	90
437	80						

Table 3. Test results for SFRRC shields.

v_s (m/s)	P (mm)	v_s (m/s)	P (mm)	v_s (m/s)	P (mm)	v_s (m/s)	P (mm)	v_s (m/s)	P (mm)
60	15	88	22	115	28	165	32	223	49
234	52	285	60	339	80	424	90	463	114

5.44   [Goel *et al.*, 2000]

5.44.1   *General data on shields and projectiles*

Shields. Single PC slabs; $c_{l,w} = 750\,mm$, $T = 250\,mm$, $f_c' = 31\,MPa$.
Projectiles. Ogive-nose projectiles (Fig. 3.3.1) with $D = 30\,mm$, $\psi = 1$ and different length.

5.44.2   *Test results*

Table 1. Values of θ_{ric} versus L/D for $v_s = 250 \pm 10\,m/s$.

L/D	θ_{ric} (deg)	L/D	θ_{ric} (deg)	L/D	θ_{ric} (deg)
3	22	4.5	25	6.4	26
8	28	9	28		

5.45   [Gold *et al.*, 1996]

See also [Gold *et al.*, 1993], [Gold, 1996].

5.45.1   *General data on shields and projectiles*

Shield-1. Single PC slab; $D^{sh} = 910\,mm$, $T = 910\,mm$, $q_{max}^{agg} = 19\,mm$, $f_c' = 37.4\,MPa$.
Shield-2. was manufactured on the basis of Shield-1. It included a 1.6-cm thick and 30.5-cm square steel plate imitating a reinforcing rebar, which was embedded at the distance of 3.8 cm from the frontal face of the shield and welded to the frontal grid of the transverse reinforcing bars; $\sigma_y = 410\,MPa$ for steel.
Projectiles. Copper and tantalum hemispherical-nose projectiles (Fig. 3.4.1).

5.45.2   *Test results*

Table 1. Test results for Shield-1.

Projectile material	m (g)	D (mm)	L (mm)	v_s (m/s)	P (mm)
Copper	224	13.0	190.0	1836	363
Tantalum	215	12.7	103.2	1872	409

Table 2. Test results for Shield-2.

Projectile material	m (g)	D (mm)	L (mm)	v_s (m/s)	P (mm)
Copper	224	13.0	190.0	1626	325
Copper	164	13.0	140.0	1875	250
Tantalum	216	12.7	103.2	1717	354
Copper	381	20.0	140.0	1663	335
Tantalum	158	13.0	76.6	1475	250
Tantalum	397	20.0	79.0	1770	338
Copper	164	13.0	140.0	1756	315
Copper	164	13.0	140.0	1750	325

5.46 [Goldstein *et al.*, 1977]

5.46.1 *General data on shields and projectiles*

Shields. Single MRC slabs ($c_{l,w} = 1450\,mm$, $T = 260\,mm$) of two types (*Sields-1* and *Shields-2*) with four identical meshes: two meshes at the distance of 10 mm and 90 mm from the frontal face of the slab, and two meshes at the distance of 10 mm and 90 mm from the rear face of the slab.

Additional characteristics of reinforcement. $\mu = 260\,kg/m^3$, $d_{l,w} = 10\,mm$ and $g_{l,w} = 80\,mm$ (for *Shields-1*), and $\mu = 160\,kg/m^3$, $d_{l,w} = 8\,mm$ and $g_{l,w} = 80\,mm$ (for *Shields-2*).

Projectiles. Flat nosed rigid projectiles.

5.46.2 *Test conditions and results*

Notes.
- Values of v_s in numerator and denominator are determined by different methods.
- Asterisk indicates "no perforation".

Table 1. Test results for Shields-1.

E (GPa)	f'_c (MPa)	σ_u (MPa)	m (kg)	D (mm)	v_s (m/s)	v_r (m/s)
40.9	41.6	3.8	30	200	106 / 109	*
40.9	41.6	3.8	50	200	106 / 112	*

E (GPa)	f_c' (MPa)	σ_u (MPa)	m (kg)	D (mm)	v_s (m/s)	v_r (m/s)
40.8	41.5	4.1	34	200	152 / 157	*
39.5	50.5	3.1	50	250	166 / 175	50
38.9	49.0	3.4	99.5	250	124 / 127	62
40.1	45.0	3.6	50	200	133 / 140	50
23.7	43.5	3.3	119	250	89 / 92	40
32.1	45.0	2.7	51	200	– / 115	80

Table 2. Test results for Shields-2.

E (GPa)	f_c' (MPa)	σ_u (MPa)	m (kg)	D (mm)	v_s (m/s)	v_r (m/s)
24.8	44.5	4.4	50	200	127 / 138	35
22.9	36.5	3.5	30	200	142 / 150	*
22.9	43.5	3.0	30	250	– / 215	45
27.5	41.5	3.5	50	250	153 / 164	38
26.6	44.0	4.6	100	250	105 / 106	45
35.7	44.5	2.9	70	300	146 / 151	55
30.0	34.0	2.9	30	250	187 / 187	*

5.47 [Gomez and Shukla, 2001]

See also [Shukla and Sadd, 2000].

5.47.1 *General data on shields and projectiles*

Shields. Single Air Force G mix PC slabs, $c_{l,w} = T = 406\,mm$, limestone aggregate ($q^{agg} = 9.5\,mm$), $\gamma = 2337\,kg/m^3$, $f_c' = 38.15\,MPa$, $\sigma_{tsp} = 3.14\,MPa$.

Projectiles. Steel ogive-nose projectiles (Fig. 3.3.1); $D = 6.35\,mm$, $L_{cyl} = 50.79\,mm$, $L = 64.34\,mm$, $\rho = 30.48\,mm$.

Table 1. Material properties of projectile.

E (GPa)	γ (kg/m^3)	σ_y (MPa)	σ_u (MPa)	Charpy (10 mm) (J)	HRC
207	8118	1999	2068	20	53

Note. Several series of experiments on multiple impact. The projectile was removed after every shot.

5.47.2 *Test results*

Table 2. Test results for Series 1 of tests (single shot).

m (g)	v_s (m/s)	P (mm)
15.3	616	135.1

Table 3. Test results for Series 2 of tests (5 shots).

Shot No.	m (g)	v_s (m/s)	P (mm)	Shot No.	m (g)	v_s (m/s)	P (mm)
1	15.1	298	40.0	2	14.4	291	46.3
3	14.9	295	63.5	4	14.7	291	80.1
5	15.1	310	99.2				

Table 4. Test results for Series 3 of tests (3 shots).

Shot No.	m (g)	v_s (m/s)	P (mm)	Shot No.	m (g)	v_s (m/s)	P (mm)
1	14.9	375	58.1	2	14.8	362	80.2
3	14.9	359	118.2				

Table 5. Test results for Series 4 of tests (5 shots), $m = 15.0\,g$.

Shot No.	v_s (m/s)	P (mm)	Shot No.	v_s (m/s)	P (mm)	Shot No.	v_s (m/s)	P (mm)
1	186	24.3	2	183	32.3	3	207	42.0
4	203	54.7	5	212	69.0			

5.48 [Gran and Frew, 1997]

5.48.1 *General data on shields and projectiles*

Shields. Single PC slabs; $D^{sh} = 1.37\,m$, $T = 1.22\,m$; $f_c' = 43\,MPa$ (determined by testing cylindrical specimens).

Projectiles. Hardened 4340 steel ogive-nose projectile (Fig. 3.3.1); $D = 50.80\,mm$, $L = 355.6\,mm$, $L_{nose} = 83.82\,mm$, $\rho = 152.40\,mm$, $m = 2.3\,kg$.

5.48.2 *Test results*

Table 1. Test results for DOP.

v_s (m/s)	P (mm)	v_s (m/s)	P (mm)	v_s (m/s)	P (mm)
320	185	310	175	316	157

5.49 [Guo *et al.*, 2014a]

5.49.1 *General data on shields and projectiles*

Shields. Single PC cylinders, $D^{sh} = 1800\,mm$, limestone aggregate ($q^{agg} = 3-8\,mm$), $f_c' = 40-45\,MPa$ (cube-shaped specimens having the length of the edge 150 mm were tested), $\gamma = 2400\,kg/m^3$.

Projectiles. 30CrMnSiNi2A steel ogive-nose projectiles (Fig. 3.3.1); $D = 60\,mm$, $\psi = 3$, $L = 300\,mm$, $\sigma_u = 1800\,MPa$.

5.49.2 *Test results*

Table 1. Test results.

m_s (kg)	v_s (m/s)	m_r (kg)	L_r (mm)	P (m)
3.813	853	3.738	298.85	1.670
3.883	1401	3.611	288.76	3.200

5.50 [Guo *et al.*, 2014b]

5.50.1 *General data on shields and projectiles*

Shields. Single PC cylinders, $D^{sh} = 400\,mm$, limestone aggregate ($q^{agg}_{max} = 5\,mm$), $f_c' = 42\,MPa$, $\gamma = 2400\,kg/m^3$.

Projectiles. 35CrMnSiA steel ogive-nose projectiles (Fig. 3.3.1); $D = 12\,mm$, $\psi = 3$, $L = 96\,mm$, $\rho = 36\,mm$.

5.50.2 *Test results*

Table 1. Test results.

m_s	v_s	m_r	L_r	P
(g)	(m/s)	(g)	(mm)	(m)
68.5	839	66.9	91.2	360
68.5	901	65.1	90.7	407
69.0	967	65.8	87.1	394
68.8	1046	65.6	87.4	448
68.4	1149	64.2	85.9	462
68.8	1213	52.2	75.0	350
69.2	1353	52.4	66.0	328

5.51 [Hanchak, 1992]

5.51.1 *General data on MRC shields and projectiles*

Shields. Single MRC slabs with three identical meshes located near the frontal and rear faces, and in the middle of the slab; $c_{l,w} = 610\,mm$, $T = 178\,mm$.

Meshes. $t^{F,R} = 12.7\,mm$, $d_{l,w} = 5.69\,mm$, $g_{l,w} = 76.2\,mm$.

Table 1. Ogive-nose projectiles (Fig. 3.3.1).

D	L_{cyl}	L_{nose}	ψ	Material	σ_y	γ	m
(mm)	(mm)	(mm)			(MPa)	(kg/m^3)	(kg)
25.4	101.6	42.1	3	T-250 maraging steel	1720	8020	0.5

5.51.2 *Test results*

Table 2. Test results for shields with $f_c' = 48\,MPa$ and $\sigma_u \approx 4\,MPa$.

v_s	v_r	v_s	v_r	v_s	v_r	v_s	v_r
(m/s)	(m/s)	(m/s)	(m/s)	(m/s)	(m/s)	(m/s)	(m/s)
301	0	360	67	381	136	434	214
606	449	746	605	749	615	1058	947

Table 3. Test results for shields with $f_c' = 140\,MPa$ and $\sigma_u \approx 5\,MPa$.

v_s	v_r	v_s	v_r	v_s	v_r	v_s	v_r
(m/s)	(m/s)	(m/s)	(m/s)	(m/s)	(m/s)	(m/s)	(m/s)
376	0	382	0	443	171	522	265
587	368	743	544	998	842		

5.52 [Hansson, 2001]

Note. Data are taken from [Hansson, 2004] where the actual study is referenced as a source.

See also [Hansson and Skoglund, 2002], [Hansson, 2002].

5.52.1 *General data on shields and projectiles*

Shields. Single PC slabs ($D^{sh} = 2.4m$, $T = 0.75m$); $f'_c = 92 \pm 2\,MPa$ and $E = 44.5 \pm 0.9\,GPa$ (Ø150×300 mm cylinders were tested), $\sigma_{tsp} = 6.5 \pm 0.2\,MPa$ (cube-shaped specimens having the length of the edge 150 mm were tested).

Projectiles. Steel ogive-nose projectiles (Fig. 3.3.1); $D = 152\,mm$, $\rho = 380\,mm$, $L = 552\,mm$, $m = 46.2 \pm 0.1\,g$.

5.52.2 *Test results*

Table 1. Test results on v_r.

v_s (m/s)	v_r (m/s)	v_s (m/s)	v_r (m/s)
460.0±0.5	183±6	455.5±0.2	204±4
458.8±0.2	181±4	458.1±0.2	190±14

5.53 [Hansson, 2005a]

See also [Hansson, 2011], [Hansson, 2005b].

5.53.1 *General data on shields and projectiles (FOI Tests)*

Shields. Single MRC and non-reinforced normal strength concrete (NSC) and high performance concrete (HPC) slabs; Ø100×200 mm cylinders were tested to determine f'_c.

Reinforcement. (Series-5 of experiments). Five identical meshes ($d_{l,w} = 14\,mm$ and $g_{l,w} = 60\,mm$); two meshes are located near the frontal and rear faces ($t^{F,R} = 30\,mm$) and the rest three meshes are located at equal distances from the first two meshes. The meshes with ordinal numbers 2 and 4 are shifted laterally by the distance 28 mm.

Table 1. Material properties of shields.

Material	γ (kg/m^3)	f'_c (MPa)	σ_{tsp} (MPa)	E (GPa)
NSC K45	2300	45–55	3–4	30–35
HPC K100	2400	85–100	6.0-6.5	40
HPC K140	2500	135–145	7	45

Table 2. 34CrNiMo6 ogive-nose projectiles (Fig. 3.3.1, $D = 50\,mm$, $L = 450\,mm$).

Projectile label	ρ (mm)	ψ	HRC	HV	m (kg)
Projectile-1	400	8	50.2–50.6	560–620	3.65
Projectile-2	150	3	–	500	4.5
Projectile-3	400	8	–	500	4.5
Projectile-4	600	12	–	500	4.5

Table 3. Parameters that were kept constant and varied in series of tests.

Test series	Shield material	Reinforcement	D^{sh} (m)	c_l (m)	Projectile	θ_s (deg)
Series-1	Varied	No	Varied	–	Projectile-1	0
Series-2	NSC K45	No	1.2	–	Varied	0
Series-3	HPC 140*	No	1.2	–	Varied	0
Series-4	Varied	No	1.5	–	Projectile-3	30.5
Series-5	NSC K45	Yes	–	1.2	Varied	Varied

* $f'_c \approx 135\,MPa$.

Note. Values of v_s in numerator and denominator in the tables below are determined by different methods.

5.53.2 *Test results for non-reinforced shields (FOI Tests)*

Table 4. Results for Series-1 tests.

Shield material	f'_c (MPa)	D^{sh} (m)	T (m)	v_s (m/s)	v_r (m/s)	P (m)
NSC K45	48	1.25	1.50	420	–	0.49
HPC K100	97	1.20	1.50	411	–	0.50
HPC K100	97	1.20	1.50	417	–	0.47
HPC K100	97	1.20	1.50	462	–	0.46
HPC K140	146	1.20	1.20	407	–	1.38

Shield material	f_c' (MPa)	D^{sh} (m)	T (m)	v_s (m/s)	v_r (m/s)	P (m)
HPC K140	146	1.20	1.20	410	–	0.41
HPC K140	146	1.20	0.45	424	129	–
HPC K140	146	1.20	0.45	417	64	–

Table 5. Results for Series-2 tests.

T (m)	Projectile	v_s (m/s)	v_r (m/s)	P (m)	P^{cr} (m)
0.90	Projectile-2	415/415	–	0.545	0.10
0.90	Projectile-2	419/415	–	0.57	0.10
0.90	Projectile-3	409/404	–	0.62	0.125
1.2	Projectile-3	463/465	–	0.69	0.150
0.90	Projectile-4	422/424	–	0.640	0.155
0.60	Projectile-3	425/425	139	–	0.110

Table 6. Results for Series-3 tests.

T (m)	Projectile	v_s (m/s)	v_r (m/s)	P (m)	P^{cr} (m)
0.51	Projectile-3	421/419	136	–	0.150
0.75	Projectile-3	422/423	–	0.480	0.210
0.65	Projectile-3	426/423	–	0.500	0.250
0.90	Projectile-2	422/422	–	0.410	0.210
0.90	Projectile-4	417/414	–	0.495	0.165
0.90	Projectile-3	457/452	–	0.535	0.165
0.90	Projectile-4	460/455	–	0.560	0.190
0.90	Projectile-2	460/456	–	0.475	0.155
0.66	Projectile-3	456/448	–	0.540	0.165
0.66	Projectile-4	455/448	–	0.550	0.250

Table 7. Results for Series-4 tests.

Shield material	T (m)	f_c' (MPa)	v_s (m/s)	v_r (m/s)	P (m)	θ_r (deg)	P^{cr} (m)
NSC K45	0.54	55	424/419	16	–	–	0.24
NSC K45	0.54	55	422/422	–	0.50	50	0.23
HPC K140	0.45	135	421/418	–	0.25	–	0.25
HPC K140	0.45	135	456/459	–	0.25	–	0.25

5.53.3 *Test results for reinforced shields (FOI tests)*

Table 8. Results for Series-5 tests.

c_w (m)	T (m)	ρ_V (%)	μ (kg/m^3)	Projectile	v_s (m/s)	θ_s (deg)	v_r (m/s)	P (m)	θ_r (deg)	P^{cr} (m)
1.2	0.60	4.71	291	Projectile-2	422/416	0	–	0.485	–	0.045
1.2	0.60	4.71	291	Projectile-3	424/416	0	–	0.530	–	0.045
1.5	0.54	5.19	325	Projectile-2	424/422	30.5	–	0.350	51	0.105
1.5	0.54	5.19	325	Projectile-2	423/425	30.5	–	0.340	47	0.030
1.5	0.54	5.19	325	Projectile-3	421/420	30.5	–	0.390	44.5	0.080
1.5	0.54	5.19	325	Projectile-3	420/417	30.5	–	0.345	61	0.085

5.53.4 *Some penetration data from the literature*

Table 9. Test conditions and results for ogive-nose projectiles (Fig. 3.3.1).

D (mm)	ψ	L (mm)	m (kg)	f_c' (MPa)	v_s (m/s)	P (m)	Reference
25.0	3.00	151	0.50	38.4	292	123	[Buzaud *et al.*, 1999a] [Buzaud *et al.*, 1999ab]
25.0	3.00	151	0.50	38.4	132	55	Ibid
363	2.07	1200	485	37	259	1350	[Riedel *et al.*, 1999]
363	2.07	1200	433	39	261	960	Ibid
90.8	2.03	300	6.63	40.5	238	185	Ibid
90.8	2.03	300	6.64	42.4	254	210	Ibid
25.0	4.00	–	0.595	50	389	215	[Hiermaier and Thoma, 1997]
50.8	3.00	387	4.37	40.5	516	694	[Gran *et al.*, 1999]
50.8	3.00	387	4.38	41.6	510	666	Ibid
50.8	3.00	387	4.38	42.4	505	689	Ibid
105	3.00	632	30.4	34.5	273	463	[Baty *et al.*, 2003]

5.54 [Hashimoto *et al.*, 2005]

5.54.1 *General data on shields and projectiles*

Shields. Single MRC shields with liners and without liners; $c_{l,w} = 750\,mm$; $f_c' = 28.7\,MPa$, $\sigma_{tsp} = 2.53\,MPa$.

Liners. Steel plates, $\sigma_u = 300-330\,MPa$.

Table 1. Rebars and liners.

Shield label	Reinforcement characteristics
Shield-1	$t^{F,R}=15mm$, $d^{F,R}_{l,w}=6.35mm$, $g^{F,R}_{l,w}=100mm$
Shield-2	$t^{F}=15mm$, $d^{F}_{l,w}=6.35mm$, $g^{F}_{l,w}=100mm$, T^{R}_{lin}=variable
Shield-3	$T^{F}_{lin}=T^{R}_{lin}$= variable; without rebars

Table 2. Steel non-deformable double cylinder projectile with hemispherical-nose (Fig. 3.9.1).

D (mm)	D_1 (mm)	L_1 (mm)	L_2 (mm)	ρ (mm)	m (kg)
45	40	60	35	20	0.5

Table 3. Shield damage types (SDTs).

Notation	Description	Notation	Description
(1)	Perforation	(2)	Scabbing
(3)	Splitting	(4)	Bulging

5.54.2 *Test results*

Table 4. Test results for Shield-1.

T (mm)	v_s (m/s)	SDT	T (mm)	v_s (m/s)	SDT	T (mm)	v_s (m/s)	SDT
60	175	(1)	80	175	(1)	100	175	(2)
100	215	(1)	100	215	(1)	120	215	(2)
120	250	(1)	150	250	(1)			

Table 5. Test results for Shield-2, $T^{R}_{lin}=0.5mm$.

T (mm)	v_s (m/s)	SDT	T (mm)	v_s (m/s)	SDT	T (mm)	v_s (m/s)	SDT
60	140	(3)	80	140	(4)	60	175	(1)
80	175	(4)	80	215	(1)	100	250	(4)

Table 6. Test results for Shield-2, $T^{R}_{lin}=0.8mm$.

T (mm)	v_s (m/s)	SDT	T (mm)	v_s (m/s)	SDT	T (mm)	v_s (m/s)	SDT
60	175	(4)	60	175	(4)	80	175	(4)
60	215	(1)	80	215	(4)	80	250	(1)
100	250	(4)						

Table 7. Test results for Shield-2; $T = 80\,mm$, $T_{lin}^R = 1.2\,mm$.

v_s (m/s)	SDT	v_s (m/s)	SDT	v_s (m/s)	SDT
175	(4)	215	(4)	250	(3)

Table 8. Test results for Shield-3.

T_{lin}^F, T_{lin}^R (mm)	T (mm)	v_s (m/s)	SDT	T_{lin}^F, T_{lin}^R (mm)	T (mm)	v_s (m/s)	SDT
0.8	60	175	(4)	0.8	60	175	(4)
0.8	80	250	(3)	1.2	60	175	(4)

5.55 [Heider and Günter, 1998]

5.55.1 *General data on shields and projectiles*

Shields. Single PC slabs, $D^{sh} = T = 1\,m$; $q_{max}^{agg} = 16\,mm$.
Projectiles. High strength steel cone-nose projectiles (Fig. 3.2.1); $D = 60\,mm$, $L_{nose} = 180\,mm$, $L = 510\,mm$.

5.55.2 *Test results*

Table 1. Test results on DOP.

m (kg)	E (GPa)	f_c' (MPa)	v_s (m/s)	P (m)
6.013	29.5	34.8	556	1.34
6.039	29.5	34.8	509	1.145
5.980	33.5	46.3	563	1.11
6.023	33.5	46.3	563	1.105
6.045	29.5	34.3	560	1.28
5.986	35.3	51.3	562	1.08

5.56 [Heimdahl and Schulz, 1982]

5.56.1 *General data on shields and projectiles*

Shields. Single slab made of Thorite (® Standard Dry Wall Products), a fast-setting concrete, $f_c' = 27.2\,MPa$.

Projectiles. Projectiles (Figs. 3.1.2b and 3.1.2d) were hollow cylinders ($D = 12.70\,mm$, $L = 50.80\,mm$, $d = 11.68\,mm$, $L_1 = 6.35\,mm$)   machined from heat-treated 4340 steel (HRC 38-40). Generally, eight longitudinal grooves, equally spaced around the circumference, were machined into the surface of the cavity (Fig. 3.1.2d), $\chi = 0.5(d_1 - d) = 0.10\,mm$. The cavities of some of the projectiles were filled with plasticine.

Table 1. Geometric parameters of projectiles.

Projectile label	Shape	L_2 (mm)
Projectile-1	Fig. 3.1.2b	–
Projectile-2	Fig. 3.1.2d	11.68
Projectile-3	Fig. 3.1.2d	18.03

Table 2. Explanation of the numbers in brackets in the column PD (projectile deformation) in tables below.

Number	Explanation	Number	Explanation
(1)	Survived	(2)	Failed
(3)	Small cracks in bulge	(4)	Bulged
(5)	Sheared off at bulge	(6)	Sheared off nose, peeled case
(7)	Sheared off at bulge	(8)	Canted nose, small cracks on bulge
(9)	Slightly canted nose	(10)	Sheared off at nose
(11)	Rounded bulge	(12)	Large rounded bulge
(13)	Overweight	(14)	Split case, nose attached
(15)	Split case, nose detached	(16)	Moderate rounded bulge
(17)	Split case		

5.56.2 *Test results*

Additional notation. d_{bul}^{cav} is residual cavity bulge diameter of projectile.

Table 3. Test results for projectiles with unfilled cavity.

Projectile	m (g)	v_s (m/s)	P (mm)	D_r (mm)	d_{bul}^{cav} (mm)	L_r (mm)	PD
Projectile-1	19.71	728	78.7	13.1	15.6	48.2	(1), (3)
Projectile-1	19.70	750	86.4	13.2	15.2	48.3	(1), (4)

Projectile	m (g)	v_s (m/s)	P (mm)	D_r (mm)	d_{bul}^{cav} (mm)	L_r (mm)	PD
Projectile-1	19.72	750	68.6	13.3	–	–	(2), (5)
Projectile-2	19.60	722	76.2	13.1	15.2	48.6	(1), (4)
Projectile-2	19.88	732	78.7	13.2	15.3	48.5	(1), (4)
Projectile-2	20.01	750	63.5	13.3	–	–	(2), (5)
Projectile-2	19.64	756	61.0	13.3	–	–	(2), (6)
Projectile-2	20.12	756	86.4	13.3	–	–	(2), (7)
Projectile-3	19.73	738	83.8	13.1	15.7	47.9	(1), (8)
Projectile-3	19.70	756	88.9	13.1	15.6	48.0	(1), (9)
Projectile-3	19.87	765	91.4	13.2	15.4	48.1	(1), (4)
Projectile-3	19.59	768	94.0	13.1	15.6	48.0	(1), (4)
Projectile-3	19.71	768	73.7	13.2	–	–	(2), (10)

Table 4. Test results for projectiles with filled cavity.

Projectile	m (g)	Empty mass (g)	v_s (m/s)	P (mm)	D_r (mm)	d_{bul}^{cav} (mm)	L_r (mm)	PD
Projectile-1	26.44	19.90	693	119.4	13.0	14.9	49.0	(1), (11)
Projectile-1	26.55	19.98	722	106.7	13.1	16.3	47.9	(1), (12)
Projectile-1	28.12	22.10	725	101.6	13.1	16.1	47.8	(1), (13)
Projectile-1	26.40	20.01	725	124.5	13.2	16.0	48.4	(1), (11)
Projectile-1	26.48	19.87	728	81.3	13.2	–	–	(2), (14)
Projectile-1	26.44	19.85	759	88.9	13.2	–	–	(2), (14)
Projectile-1	26.52	19.92	792	81.3	13.3	–	–	(2), (15)
Projectile-1	26.58	19.89	844	83.8	13.5	–	–	(2), (15)
Projectile-2	26.58	19.94	637	101.6	12.9	14.0	49.7	(1), (16)
Projectile-2	26.58	20.04	664	83.8	13.0	14.6	49.3	(1), (11)
Projectile-2	26.45	19.81	689	76.2	13.2	–	–	(2), (14)
Projectile-2	26.51	19.94	695	81.3	13.2	–	–	(2), (17)
Projectile-2	26.48	19.91	698	91.4	13.1	–	–	(2), (17)
Projectile-2	26.55	19.94	719	91.4	13.1	–	–	(2), (17)
Projectile-3	26.50	19.85	672	91.4	13.0	–	–	(2), (17)
Projectile-3	26.44	19.84	677	101.6	13.0	14.8	49.1	(1), (11)
Projectile-3	26.50	19.86	692	116.8	12.9	14.6	49.3	(1), (11)
Projectile-3	26.21	19.73	707	86.4	13.1	–	–	(2), (14)
Projectile-3	26.17	19.72	719	68.6	13.1	–	–	(2), (15)

5.57 [Iremonger *et al.*, 1989]

5.57.1 *General data on shields and projectiles*

Shields. Single concrete panels ($c_{l,w} = 400\,mm$, $T = 30-150\,mm$) manufactured of ordinary Portland cement; 5 mm down fine limestone aggregate were used. Three types of coarse aggregate were used to provide different concrete grades: 20 mm–5 mm river valley limestone rock gravel (*Mix-1*); 20 mm–5mm angular flint based gravel (*Mix-2*), and 10 mm down river valley limestone rock gravel (*Mix-3*).

Projectile-1 is NATO SS109 5.56 mm × 45 mm bullet, $m = 4.00\,g$. Its core comprises an air gap in front of a flat-tipped hardened steel core ($m^{core} = 0.495\,g$, minimum HV 600). The rear part of the core is a lead/antimony alloy. The core is encased in a gilding metal (copper/zinc) envelope. $v_s = 924 \pm 10\,m/s$.

Projectile-2 is NATO SS77 7.62×51 mm bullet, $m = 9.33\,g$. The core is made of lead/antimony, and the jacket is made of gilding metal. $v_s = 799 \pm 7\,m/s$.

5.57.2 *Test results*

Table 1#. Test results for Mix-1 and Projectile-1.

f_c' (MPa)	P (mm)	f_c' (MPa)	P (mm)	f_c' (MPa)	P (mm)	f_c' (MPa)	P (mm)
23.3	42.9	40.2	39.5	54.1	29.7	58.4	30.9

Table 2#. Test results for Mix-1 and Projectile-2.

f_c' (MPa)	P (mm)	f_c' (MPa)	P (mm)	f_c' (MPa)	P (mm)	f_c' (MPa)	P (mm)
23.3	43.0	40.3	41.0	54.1	32.8	58.3	31.8

Table 3#. Test results for $T = 150\,mm$, Mix-1 and Projectile-1.

f_c' (MPa)	P (mm)	f_c' (MPa)	P (mm)	f_c' (MPa)	P (mm)	f_c' (MPa)	P (mm)
22.9	43.0	39.9	40.3	53.8	30.0	57.8	31.3

Table 4#. Test results for $T = 150\,mm$, Mix-2 and Projectile-1.

f_c' (MPa)	P (mm)	f_c' (MPa)	P (mm)	f_c' (MPa)	P (mm)	f_c' (MPa)	P (mm)
20.9	44.0	35.6	34.1	46.2	31.9	49.8	23.8

Table 5#. Test results for $T = 150\,mm$, Mix-3 and Projectile-1.

f_c' (MPa)	P (mm)	f_c' (MPa)	P (mm)	f_c' (MPa)	P (mm)	f_c' (MPa)	P (mm)
28.9	45.9	47.5	37.1	49.1	40.0	54.4	34.6

5.58 [Jinzhu *et al.*, 2013]

5.58.1 *General data on shields and projectiles*

Shields. Single C30 PC slabs ($D^{sh} = 1200\,mm$); quartz aggregate ($q^{agg} = 5-15mm$), $\gamma = 2420\,kg/m^3$, $f_c' = 34.3\,MPa$.

Projectiles. Ogive-nose 35CrMnSiA steel projectiles (Fig. 3.3.1); $D = 64\,mm$, $\psi = 3$, $L_{cyl} = 214\,mm$, $L_{nose} = 106\,mm$.

5.58.2 *Test results*

Table 1. Test results (value of v_r in nominator and denominator in the table below are determined by different methods).

m (kg)	T (mm)	v_s (m/s)	v_r (m/s)
5.040	300	405.7	249.8/264.2
5.052	400	400.8	220.6/213.4
5.304	500	415.4	149.5
5.036	600	405.1	30
4.986	700	411.9	0

5.59 [Jones *et al.*, 2001]

See also [Jerome *et al.*, 2000], [Davis, 2003], [Jones *et al.*, 2003].

Note. Data on projectile mass loss and some values of v_s were taken from [Davis *et al.*, 2004].

5.59.1 *General data on shields and projectiles*

Shields. Single PC slabs; $f_c' = 56.3\,MPa$, $\gamma = 2300$ (kg/m^3).

Projectiles. Ogive-nose projectiles (Fig. 3.3.1) manufactured of different types of high strength steel alloys; $D = 12.7\,mm$, $\psi = 3$, $L = 88.9\,mm$, $\sigma_y = 1240 - 1760\,MPa$, $m = 65\,g$.

5.59.2 *Test results*

Table 1#. Test results for Steel 2 projectiles.

v_s (m/s)	P (mm)	Δm (g)	v_s (m/s)	P (mm)	Δm (g)	v_s (m/s)	P (mm)	Δm (g)
708	221	0.003	718	287	–	911	333	0.02
1049	512	0.10	1097	535	–	1199	671	0.36
1277	737	–	1403	890	0.50	1617	1077	–

Table 2#. Test results for AerMet 100 projectiles.

v_s (m/s)	P (mm)	Δm (g)	v_s (m/s)	P (mm)	Δm (g)	v_s (m/s)	P (mm)	Δm (g)
805	257	0.41	1081	544	0.88	1220	506	1.04
1397	674	1.22	1471	892	1.34			

Table 3#. Test results for AF 1410 projectiles.

v_s (m/s)	P (mm)	Δm (g)	v_s (m/s)	P (mm)	Δm (g)	v_s (m/s)	P (mm)	Δm (g)
846	241	0.66	894	278	1.1	1000	342	1.30
1219	525	1.31	1437	670	1.59			

Table 4#. Test results for Steel 1 projectiles.

v_s (m/s)	P (mm)	v_s (m/s)	P (mm)	v_s (m/s)	P (mm)
740	206	1073	310	1071	412
1123	458	1504	910		

Table 5#. Test results for HY 180 Steel projectiles.

v_s (m/s)	P (mm)	v_s (m/s)	P (mm)	v_s (m/s)	P (mm)	v_s (m/s)	P (mm)
731	225	927	424	1268	619	1476	858

Table 6#. Test results for 4340 steel projectiles.

v_s (m/s)	P (mm)	v_s (m/s)	P (mm)	v_s (m/s)	P (mm)	v_s (m/s)	P (mm)
947	336	984	318	989	346	1071	398
1109	329	1229	521	1256	550	1280	580

5.60 [Kamal and Eltehewy, 2012]

5.60.1 *General data on shields and projectiles*

Shields. Each shield consisted of two identical blocks. The block ($\widetilde{c}_{l,w} = 500\,mm$, $T = 200\,mm$) did not include meshes (*Shield-0*), or included single (*Shield-1*), two (*Shield-2*) or three (*Shield-3*) meshes in-contact not far from the frontal and rear faces of the block.

Concrete mix.   Dolomite   aggregate, $q_{max}^{agg} = 19\,mm$; $f_c' = 35\,MPa$ $E = 29\,GPa$, $\gamma = 2350\,kg/m^3$, $\sigma_u = 3.1\,MPa$.

Mesh. Galvanized woven wire mesh, $d_{l,w} = 2\,mm$, $g_{l,w} = 50\,mm$; for packages of meshes, $t^{F,R} = 20\,mm$. Material properties: 1006 steel, $\gamma = 7850\,kg/m^3$, $\sigma_y = 250\,MPa$, $\sigma_u = 360\,MPa$, $E = 210\,GPa$.

Table 1. Spherical-segment-nose projectiles (Fig. 3.8.1).

D (mm)	L (mm)	ρ (mm)	Material	HB	σ_y (MPa)	σ_u (MPa)	ε_f (%)	m (g)
23	64	17	Steel	475	1726	1900	7	175

5.60.2 *Test results*

Table 2. Test results on DOP.

Shield	v_s (m/s)	P (mm)	Shield	v_s (m/s)	P (mm)
Shield-0	976	400	Shield-1	994	290
Shield-2	996	287	Shield-3	978	280

5.61 [Kim *et al.*, 2008]

5.61.1 *General data on shields and projectiles*

Shields. Single PC and MRC barriers; $f_c' = 48\,MPa$, $T = 60\,mm$.
Rebars (MRC shield). Steel; $t^{F,R} = 20\,mm$, $d_{l,w}^{F,R} = 1\,mm$, $g_{l,w}^{F,R} = 10\,mm$.
Projectiles. Ogive-like nose projectiles; $D = 40\,mm$, $L = 120\,mm$, $m = 470\,g$.

5.61.2 *Test results*

Table 1. Test results for $v_s = 280\,m/s$.

Shield	v_r (m/s)	Shield	v_r (m/s)
PC shield	167	MRC shield	143

5.62 [Kim *et al.*, 2015]

5.62.1 *General data on shields and projectiles*

Shields. Monolithic PC slabs ($f_c' = 34.5\,MPa$) with ASTM A36 steel liners at rear face; $c_{l,w} = 2\,m$.

Table 1. Shields parameters.

Shield	T (mm)	T_{lin}^R (mm)	Shield	T (mm)	T_{lin}^R (mm)
Shield 1	175	4	Shield 2	250	4
Shield 3	175	6	Shield 4	250	6

Projectiles. ASTM A36 steel projectiles having spherical-segment nose (Fig. 3.8.1); $D = 150\,mm$, $L_{cyl} = 280\,mm$, $L = 300\,mm$, $\rho = 131\,mm$, $m \approx 40\,g$ (including sabot).

5.62.2 *Test results*

Table 2. Test results.

Shield	v_s (m/s)	v_r (m/s)	SDT	Shield	v_s (m/s)	v_r (m/s)	SDT
Shield 1	147.5	85.0	*	Shield 2	95.6	0	*
Shield 3	152.4	33.9	*	Shield 4	147.7	0	**

* Perforation; ** Bulging.

5.63 [Kojima, 1991]

5.63.1 *General data on shields and projectiles*

Shields. Single and spaced double MRC slabs ($c_{l,w} = 1.2\,m$), with steel lining or without lining; $f_c' = 27.0\,MPa$, $f_{cr} = 3.3\,MPa$. Detailed structure of the meshes is not described.

Table 1. Characteristics of reinforcement.

Slab label	Characteristics of reinforcement
Slab-1	$t^{F,R} = 15\,mm$, $\rho_{l,w}^{F,R} = 0.6\%$
Slab-2	$t^{F,R} = 15\,mm$, $\rho_{l,w}^{F,R} = 0.6\%$, $T_{lin}^{R} = 3.2\,mm$

Table 2. Mechanical properties of steel rebars used in the meshes.

Diameter (mm)	σ_y (MPa)	σ_u (MPa)	Diameter (mm)	σ_y (MPa)	σ_u (MPa)
6	389	539	10	419	569
13	389	530			

Projectiles. Non-deformed hemispherical-nose projectiles (Fig. 3.4.1); $D = 60\,mm$, $L_{cyl} = 70\,mm$, $m = 2\,kg$.

Table 3. Shield damage types (SDTs).

Notation	Description	Notation	Description
(1)	Penetration	(2)	Scabbing
(3)	Perforation		

Additional notation in the tables. N_{rr}^{sh} – number of rebars ruptured.

5.63.2 *Test results*

Table 4. Test results for single Slab-1.

T (mm)	v_s (m/s)	P (mm)	SDT	N_{rr}^{sh}
240	209	60	(1)	1
180	211	78	(2)	2
120	215	–	(3)	3
120	164	100	(2)	1
120	95	44	(2)	0

Table 5. Test results for the shield consisting of two Slab-1 plates with air gap of 120 mm.

T_1 (mm)	T_2 (mm)	v_s (m/s)	P (mm)	SDT, front plate	SDT, rear plate	N_{rr}^{sh}
90	90	210	106	(3)	(2)	3
60	120	206	59	(3)	(2)	4
60	120	202	25	(3)	(2)	0

Table 6. Test results for single Slab-2 shield.

T (mm)	v_s (m/s)	P (mm)	SDT	N_{rr}^{sh}
180	206	66	(1)	1
120	212	125	(1)	2

5.64 [Lai *et al.*, 2015]

5.64.1 *General data on shields and projectiles*

Shields. Monolithic ultra-high performance concrete (UHPC) shields ($D^{sh} = T = 300\,mm$) having various mix proportions.

Fibers. Copper coated steel fibers ($d^{fib} = 0.2\,mm$, $L^{fib} = 13\,mm$) and/or basalt fiber ($d^{fib} = 0.017\,mm$, $L^{fib} = 20\,mm$).

Table 1. Characteristics of shields.

Shield	Steel fibers ρ_V (%)	Basalt fibers ρ_V (%)	f_c' (MPa)	Shield	Steel fibers ρ_V (%)	Basalt fibers ρ_V (%)	f_c' (MPa)
Shield 1	0	0	124	Shield 2	3	0	232
Shield 3	3	1	234	Shield 4	3	0	248
Shield 5	3	1	251				

Projectiles. Standard bullets; $D = 14.8\,mm$, $L = 52\,mm$, $m = 43\,g$.

5.64.2 *Test results*

Table 2. Test results for multiple impact.

Shield	First impact		Second impact		Third impact	
	v_s (m/s)	P (mm)	v_s (m/s)	P (mm)	v_s (m/s)	P (mm)
Shield 1	945	242	–	–	–	–
Shield 2	960	165	970	242	–	–
Shield 3	979	135	968	205	970	272
Shield 4	1002	183	1008	225	–	–
Shield 5	978	178	985	202	–	–

5.65 [Lin *et al.*, 2012]

5.65.1 *General data on shields and projectiles*

Shields. Single PC slabs, $f_c' = 76.6\,MPa$.
Projectiles. Heat treated 45 steel ogive-nose projectiles (Fig. 3.3.1); $D = 20\,mm$, $\psi = 3$, $L = 120\,mm$, $\sigma_y > 355\,MPa$, $m \approx 250\,g$.

5.65.2 *Test results*

Table 1. Test results on DOP.

θ_s (deg)	v_s (m/s)	P (mm)	θ_s (deg)	v_s (m/s)	P (mm)
0	224	59	0	216	53
10	231	55	10	213	52
20	242	48	20	181	42

5.66 [Liu *et al.*, 2003]

Note. Data below are taken from [Gao *et al.*, 2009].

5.66.1 *General data on shields and projectiles*

Shields. Single PC slabs; $\gamma = 2400\,kg/m^3$, $f_c' = 30\,MPa$.
Projectiles. Ogive-nose projectiles (Fig. 3.3.1) with $D = 62\,mm$ of two types: *Type 1* ($\psi = 2.85$ and $\rho = 176.8\,mm$) and *Type 2* ($\psi = 1.53, \rho = 94.5\,mm$).

5.66.2 *Test results*

Table 1. Test results for projectiles of Type 1.

m (kg)	v_s (m/s)	P (m)	m (kg)	v_s (m/s)	P (m)
3.777	763	0.83	3.747	666	0.56

Table 2. Test results for projectiles of Type 2.

m (kg)	v_s (m/s)	P (m)	m (kg)	v_s (m/s)	P (m)	m (kg)	v_s (m/s)	P (m)
3.034	577	0.34	3.022	538	0.37	3.154	630	0.46

5.67 [Lixin *et al.*, 2000]

See also [Lixin, 1997].

5.67.1 *General data on shields and projectiles*

Summary. Two series of experiments for truncated ogive-nose projectiles (Fig. 3.6.2) penetrating into PC slabs ($T = 400\,mm$).

Table 1. Series characteristics.

Series	Shield		Projectile			
	$c_{l,w}$ (mm)	f'_c (MPa)	D (mm)	d (mm)	ρ (mm)	L (mm)
Series-1	600	30	25.3	12.6	90.1	170
Series-2	3200	32	65	21.8	340	630

5.67.2 *Test results*

Table 2. Test results on DOP.

Series	m (kg)	v_s (m/s)	P (mm)	Series	m (kg)	v_s (m/s)	P (mm)
Series-1	0.284	338	93	Series-1	0.277	339	95
Series-1	0.284	415	120	Series-1	0.281	417	120
Series-1	0.279	523	155	Series-1	0.284	491	145
Series-1	0.282	600	205	Series-1	0.280	613	195
Series-2	7.85	290	316				

5.68 [Lok, 1994]

5.68.1 *General data on shields and projectiles*

Shields. Single PC (*Shield 1*), MRC (*Shield 2*), and FRC (*Shields 3-10*) slabs; $c_{l,w} = 810\,mm$. $T = 50\,mm$.

Table 1. Types of steel fibers (in Shields 3-10).

Label	Description	L^{fib} (mm)	σ_u (MPa)
F-1a	Flat wavy cold drawn carbon steel fibers.	38	1240
F-1b	Flat wavy cold drawn carbon steel fibers.	51	1240

Label	Description	L^{fib} (mm)	σ_u (MPa)
F-2	Uniform square steel fibers with slightly enlarged crimped ends.	18	800
F-3	Bundles of round steel fibers held together with water-soluble glue that disintegrates during mixing. Each strand of steel fiber had a hook at each end.	60	≥ 1100

Mesh (*in Shield 2*). Intermediate mesh, $d_{l,w} = 6\,mm$, $g_{l,w} = 200\,mm$.

Table 2. Parameters of shields.

Shield No.	Reinforcement	Fibers ρ_V (%)	f'_c cube (MPa)	f'_c cylinder (MPa)	Slump (mm)	γ (kg/m^3)
1	No	–	43.6	34.6	80	2378
2	MRC	–	40.2	31.8	90	2370
3	F-1a	0.5	40.3	31.5	25	2374
4	F-1a	1.0	41.6	31.5	10	2412
5	F-1b	0.5	44.7	36.6	15	2365
6	F-1b	1.0	45.4	36.2	51	2456
7	F-3	0.5	41.8	31.4	0	2406
8	F-3	1.0	40.4	31.5	53	2425
9	F-2	0.5	41.6	31.0	5	2355
10	F-2	1.0	42.3	32.6	10	2381

Projectiles. Steel alloy probe had a diameter of 6.35 mm with a shallow conical tip. About 15 mm from the tip, the diameter enlarges to 7.85 mm along the entire shaft. Steel firing head was screwed onto one threaded end of the probe and was guided into the barrel by an attached finned plastic sleeve. The mass of impact probe together with the steel firing head was 44.69 g.

Impact velocity. $v_s = 183\,m/s$.

Tests feature. Each shield was impacted at 9 points. The locations of these points in coordinates associated with frontal face of the shield (the origin is located in the corner and the axes are directed along the edges of the frontal plane) are given in Table 3.

Table 3. Location of impact points (mm).

Shot no.	Abscissa	Ordinate	Shot no.	Abscissa	Ordinate
1	375	375	2	625	125
3	125	125	4	125	625
5	625	625	6	375	125
7	125	375	8	375	625
9	625	375			

5.68.2 Test results

Table 4#. Depth of penetration, P (mm).

Shot no. $\rightarrow$ / Shield no. $\downarrow$	1	2	3	4	5	6	7	8	9
1	50.0	38.6	32.6	50.1	50.1	32.8	50.2	50.1	50.1
2	30.7	33.5	28.6	33.4	28.9	33.6	36.6	28.9	38.6
3	36.8	50.2	37.9	50.1	33.9	37.8	30.8	50.2	46.9
4	32.7	37.7	33.0	40.8	30.0	31.0	50.2	33.8	34.9
5	40.8	39.7	30.9	31.8	50.2	33.7	33.6	50.2	35.8
6	34.8	32.6	35.7	50.3	37.8	37.8	36.8	50.2	33.6
7	31.0	35.7	29.8	29.8	34.9	29.9	41.6	33.7	35.7
8	31.6	36.6	25.0	34.7	30.8	32.6	31.6	32.4	32.3
9	32.7	28.6	33.7	40.4	33.7	50.1	33.7	32.7	36.4
10	41.5	37.6	31.8	32.7	33.5	37.5	34.8	38.4	31.7

5.69   [Luo *et al.*, 2000]

See also [Luo *et al.*, 2001].

5.69.1   *General data on shields and projectiles*

Shields. Two types of single FRC slabs; $c_{l,w} = 400\,mm$, $T = 300\,mm$. Values of f_c' and f_r were the test results of testing cube-shaped specimens having the length of the edge 100 mm and $100 \times 100 \times 400$ mm beam specimens, correspondingly.

Table 1. Properties of shields and fibers.

Shield	Fiber		Shield properties		
	L^{fib} (mm)	d^{fib} (mm)	f_c' (MPa)	f_r (MPa)	ρ_V (%)
CZ fiber shield	31	0.35	116.1	54.6	7
YL fiber shield	21	0.60	107.1	36.5	10

Projectiles. Cone-nose projectiles (Fig. 3.2.1); $D = 37\,mm$, $L_{cyl} = 30\,mm$, $L_{nose} = 120\,mm$, $\beta = 8.8°$.

5.69.2 *Test results*

Table 2. Test results on DOP.

Shield	m (g)	v_s (m/s)	P (mm)	Final status of projectile
CZ fiber shield	901.0	368	132	Rebounded
CZ fiber shield	901.5	375	147	Rebounded
CZ fiber shield	905.6	369	132	Rebounded
CZ fiber shield	904.2	378	151	Rebounded
CZ fiber shield	909.0	375	150	Rebounded
YL fiber shield	904.2	369	161	Imbedded
YL fiber shield	906.0	365	177	Imbedded

Note. Trajectories of projectiles are curvilinear.

5.70 [Maalej *et al.* 2004]

See also [Maalej *et al.* 2005].

5.70.1 *General data on shields and projectiles*

Shields. Single hybrid (steel and PE) fiber ECC; $c_l = 300\,mm$, $c_w = 170\,mm$.

Table 1. Shields.

ρ_V (%)		300×75×15 mm specimens		50×50×50 mm cube	Ø100×200 mm cylinder	
Steel fibers	PE fibers	σ_u (MPa)	ε_u (%)	f_c' (MPa)	f_c' (MPa)	E (GPa)
0.5	1.5	5	4	70	55	18

Table 2. Chracteristics of fibers.

Material	L^{fib} (mm)	d^{fib} (mm)	E (GPa)	σ_u (MPa)
Steel	13	0.160	200	2500
PE	12	0.039	66	2610

Projectiles. Hardened steel (HRC 50) ogive-nose projectile (Fig. 3.3.1); $D = 12.60\,mm$, $L_{cyl} = 5.03\,mm$, $L_{nose} = 18.9\,mm$, $\psi = 2.5$, $m = 15\,g$.

5.70.2 *Test results*

Table 3#. Test results for shields with $T = 55$ *mm*.

v_s (m/s)	P (mm)	v_s (m/s)	P (mm)	v_s (m/s)	P (mm)
303	21	407	35	465	39
465	42	559	55		

Table 4#. Test results for shields with $T = 75$ *mm*.

v_s (m/s)	P (mm)	v_s (m/s)	P (mm)	v_s (m/s)	P (mm)
302	20	431	36	518	42
578	52	671	75		

Table 5#. Test results for shields with $T = 100$ *mm*.

v_s (m/s)	P (mm)	v_s (m/s)	P (mm)	v_s (m/s)	P (mm)	v_s (m/s)	P (mm)
301	20	403	33	411	31	474	36
525	44	586	46	664	60	749	73

Table 6#. Test results for shields with $T = 120$ *mm*.

v_s (m/s)	P (mm)	v_s (m/s)	P (mm)	v_s (m/s)	P (mm)
398	29	398	31	460	34
547	42	615	51	721	56

Table 7#. Test results for shields with $T = 150$ *mm*.

v_s (m/s)	P (mm)	v_s (m/s)	P (mm)	v_s (m/s)	P (mm)
346	25	404	31	541	39
622	46	711	57		

5.71 [Máca *et al.*, 2014]

See also [Máca and Sovják, 2012].

5.71.1 *General data on shields and projectiles*

Shields. Single ultra-high performance concrete (UHPC) and ultra-high performance fiber reinforced concrete (UHPFRC) slabs; $c_l = 300\,mm$, $c_w = 400\,mm$, $T = 50\,mm$. Ø100×200 mm cylinders were tested to measure f_c' and E. Tensile tests were carried out on dog-bone shaped specimens with the length of 330 mm, without a notch and the cross-section of the narrowed part of 30×30 mm.

Fibers. $L^{fib} = 13\,mm$, $d^{fib} = 0.15\,mm$, $\sigma_u = 2800\,MPa$.

Table 1. Average mechanical properties of the UHPC and UHPFRC mixtures.

Mixture	$\tilde{\rho}_m$	f_c' (MPa)	f_r (MPa)	σ_u (MPa)	E (GPa)
UHPC	–	132.4	9.9	6.6	41.1
UHPFRC-1	0.1	148.5	27.0	7.1	45.1
UHPFRC-2	0.2	151.7	40.1	10.4	56.3
UHPFRC-3	0.3	148.1	47.5	11.7	51.5

Table 2. Shield damage types (SDTs).

Notation	Description
(1)	Projectile passed through the slab
(2)	Projectile was retained in the slab
(3)	Panel was punched but the projectile bounced back

Projectiles. Deformable (DP) and non-deformable (NDP) projectiles.

Table 3. Ogive-nose jacketed projectiles [Sovják *et al.*, 2015].

Projectile label	D (mm)	L (mm)	m (g)	Core Material	Core Mass (g)	Jacket Material	Jacket Mass (g)
DP	7.92	23.2	8.04	Soft lead	5.26	Steel	2.78
NDP	7.92	26.6	8.04	Mild steel	3.65	Steel	4.15

Note. For mild steel, $\sigma_y = 550\,MPa$.

5.71.2 *Test results*

Table 4. Test results for DP (UHPC and UHPFRC mixtures).

Mixture	v_s (m/s)	SDT	m^{fr} (g)	P (mm)
UHPC	710	(1)	–	50
UHPC	714	(1)	–	50
UHPFRC-1	715	(3)	53	20
UHPFRC-1	728	(3)	122	21
UHPFRC-2	692	(3)	63	19
UHPFRC-2	706	(3)	44	21
UHPFRC-3	716	(3)	50	19
UHPFRC-3	718	(3)	170	19

Table 5. Test results for NDP and UHPFRC-2 mixtures.

v_s (m/s)	SDT	m^{fr} (g)	v_s (m/s)	SDT	m^{fr} (g)
712	(2)	46	706	(1)	35
706	(1)	129	718	(2)	32
721	(1)	90	718	(2)	32

5.72 [Magnusson *et al.*, 2001]

5.72.1 *General data on shields and projectiles*

Shields. Single and layered PC, steel FRC and MRC barriers.

Table 1. Concrete types.

Label	Concrete type	Label	Concrete type
NSC	Normal strength concrete	AFC	Abetong and Fredrikstad concrete
OC	Optiroc concrete	AC	Abetong concrete
TC	Trondeim concrete	DC	Densit concrete
DSC	Densit special concrete	CRC	Compact reinforced composite
B	Ballistocrete		

Table 2. Types of reinforcement.

Notation of reinforcement	Rebars		Fibers
	$d_{l,w}$ (mm)	ρ_V (%)	ρ_V (%)
(1)	25	1.55	1.0
(2)	25	6.0	–
(3)	25	7.0	–
(4)	25	7.0	1.5
(5)	25	25.0	2.0
(6)	25	25.0	4.0
(7)	32	6.5	1.0
(8)	–	–	1.0
(9)	–	–	1.5
(10)	–	–	2.0
(11)	–	–	2.6
(12)	–	–	2.7
(13)	Rock boulders		
(14)	Rails		

Table 3. **Projectile-1** (57 mm caliber discarding-sabot projectile).

Part of projectile body	Material	Mass (kg)	Diameter (mm)	Length (mm)
Whole body	–	4.5	57	233
Core	Tungsten carbide	2.56	43	129
Core support	Heavy tungsten alloy	0.89	42	66

Projectile-2. 152 mm caliber semi-armor piercing grenade; $m = 44.76\,kg$ (including mass of explosive charge of 2.7 kg), $D = 152\,mm$, $L = 435\,mm$.

Projectile-3. 75 mm caliber armor piercing projectile made of tempered steel; $m = 6.28\,kg$, $D = 75\,mm$, $L = 225\,mm$, $\psi = 1.69$ (value of ψ from [Hansson, 2003]).

Projectile-4. Steel 34CrNiMoG ogive-nose projectile (Fig. 3.3.1); $D = 152\,mm$, $\psi = 2.5$, $m = 46.328\,kg$, $\sigma_y = 886\,MPa$, HB 270-330.

5.72.2 *Test conditions and results*

Notes.

- Thickness of layer is giver in meters. Notation $10\times[(NSC)\,0.2]$ indicates that the shield consists of 10 layers and each NSC layer has the thickness of 0.2 m.
- Asterisk indicates perforation.

Table 4. Test conditions and results for Projectile-1.

$c_{l,w}$ (m)	D^{sh} (m)	Shield structure	Reinforcement	f_c' (MPa)	v_s (m/s)	P (m)	P^{cr} (m)
2.0	–	$10\times[(NSC)\,0.2]$	–	30	1394	*	–
–	1.4	$1\times[(NSC)\,3.0]$	–	39	1364	2.62	0.30
–	1.4	$1\times[(AC)\,2.5]$	–	149	1387	1.35	0.40
–	1.4	$1\times[(AC)\,2.0]$	–	158	985	0.76	0.50
2.0	–	$5\times[(DC)\,0.2]$	(12)	180	1380	*	–
2.0	–	$10\times[(DC)\,0.2]$	(12)	180	1394	1.25	–
0.64	–	$6\times[(CRC)\,0.2]$	(5)	220	1391	0.52	–

Table 5. Test conditions and results for Projectile-2.

$c_{l,w}$ (m)	D^{sh} (m)	Shield structure	Reinforcement	f_c' (MPa)	v_s (m/s)	v_r (m/s)	P (m)	P^{cr} (m)
–	1.6	$10\times[(NSC)\,0.2]$	–	38	484	–	0.68	–
–	1.6	$1\times[(NSC)\,2.0]$	–	38	483	–	0.66	–
–	1.6	$1\times[(NSC)\,2.0]$	–	38	482	–	0.66	–
–	1.6	$1\times[(NSC)\,2.0]$	–	38	647	–	0.99	–
–	1.6	$1\times[(AC)\,1.5]$	(9)	83	571	–	0.41	–
–	1.6	$1\times[(OC)\,1.5]$	(8)	108	653	–	0.56	–
–	1.6	$1\times[(TC)\,1.2]$	(8)	141	647	–	0.44	–
–	1.4	$1\times[(AC)\,0.4]$	–	153	615	276	*	0.15
–	1.4	$1\times[(AC)\,0.4]$	–	153	618	303	*	0.20

$c_{l,w}$ (m)	D^{sh} (m)	Shield structure	Reinforcement	f'_c (MPa)	v_s (m/s)	v_r (m/s)	P (m)	P^{cr} (m)
–	1.4	1×[(AC) 0.4]	–	153	612	293	*	0.20
–	1.4	1×[(AC) 0.8]	–	153	620	–	0.45	0.30
–	1.4	1×[(AC) 0.8]	–	153	612	–	0.54	0.26
–	1.4	1×[(AC) 0.8]	–	153	619	–	0.51	0.21
–	1.4	1×[(AC) 0.4]	(2)	153	617	–	*	0.06
–	1.4	1×[(AC) 0.4]	(2)	153	616	–	*	0.06
–	1.4	1×[(AC) 0.4]	(2)	153	619	260	*	0.08
–	1.6	1×[(DC) 1.0]	(10)	180	485	–	0.25	–
–	1.6	1×[(DC) 1.0]	(10)	180	489	–	0.24	–
–	1.6	1×[(DC) 0.5]	(10)	180	485	–	0.24	–
1.5	–	5×[(DC) 0.2]	(10)	200	480	–	0.20	–
1.5	–	4×[(DC) 0.2]	(10)	200	650	–	0.44	–

Table 6. Test conditions and results for Projectile-3.

$c_{l,w}$ (m)	D^{sh} (m)	Shield structure	Reinforcement	f'_c (MPa)	v_s (m/s)	v_r (m/s)	P (m)
1.5	–	2×[(NSC) 1.0]	(3)	34	481	–	0.67
1.5	–	5×[(NSC) 0.2]	–	36	478	160	*
–	1.6	1×[(NSC) 3.0]	–	36	576	–	1.70
–	1.6	1×[(NSC) 3.0]	–	41	468	–	1.03
–	1.6	1×[(NSC) 3.0]	–	41	481	–	1.02
–	1.6	1×[(AC) 1.5]	(9)	83	488	–	0.56
–	1.6	1×[(AFC) 1.5]	–	90	479	–	0.63
–	1.6	1×[(AFC) 1.5]	–	90	583	–	0.98
–	1.6	1×[(OC) 1.5]	(8)	108	486	–	0.59
–	1.6	1×[(AC) 1.5]	–	134	581	–	0.83
–	1.6	1×[(TC) 1.5]	(8)	141	584	–	0.81
–	1.6	1×[(TC) 1.2]	(8)	141	480	–	0.60
1.5	–	1×[(TC) 1.2]	(7)	141	480	–	0.42
1.5	–	5×[(DC) 0.2]	(11)	180	473	–	0.60
–	1.6	1×[(DC) 1.5]	(10)	183	580	–	0.53
–	1.6	1×[(DSC) 1.0]	(9)	199	480	–	0.45
1.5	–	1×[(CRC) 1.0]	(4)	200	480	–	0.30
–	1.6	1×[(DC) 1.2]	–	203	480	–	0.45
1.5	–	5×[(CRC) 0.2]	(6)	220	478	–	0.35

Table 7. Test conditions and results for Projectile-4.

D^{sh} (m)	Shield structure	Reinforcement	f'_c (MPa)	v_s (m/s)	v_r (m/s)	P (m)	P^{cr} (m)
1.6	1×[(NSC) 2.0]	–	39	466	–	1.09	0.58
2.4	1×[(NSC) 2.0]	–	39	468	–	1.01	0.50

D^{sh} (m)	Shield structure	Reinforcement	f'_c (MPa)	v_s (m/s)	v_r (m/s)	P (m)	P^{cr} (m)
2.4	1×[(OC) 0.75]	–	103	460	183	*	–
2.4	1×[(OC) 0.75]	–	103	455	204	*	–
2.4	1×[(OC) 0.75]	–	103	459	181	*	–
1.6	1×[(TC) 1.2]	(8)	120	464	–	0.57	0.35
1.6	1×[(TC) 1.2]	(1)	120	467	–	0.50	0.20
2.4	1×[(AC) 1.2]	–	149	468	–	0.38	0.34
3.0	1×[(AC) 1.2]	–	149	465	–	0.44	0.40
1.0	1×[(AC) 1.2]	–	153	466	–	0.75	0.60
1.6	1×[(AC) 1.2]	–	153	467	–	0.66	0.47
1.6	1×[(AC) 1.5]	(13)	158	576	–	0.79	–
1.6	1×[(AC) 1.5]	(14)	158	573	–	0.44	–

5.73 [Malone and Tom, 1998]

Note. In experiments with multiple shots we present below only the results of the first shot.

5.73.1 *General data on shields and projectiles*

Shields. Single foamed fiber-reinforced concrete, or shock-absorbing concrete (SACON) blocks were cast from foamed concrete (or mortar) reinforced with steel or polypropylene fiber; details on preparation of SACON are presented in [Fabian *et al.*, 1996].

Projectiles. M16A2 rifle firing the 5.56-mm NATO ball M855 rounds was used. The M855 bullet is designed to contain 1.30 g of copper alloy, 0.65g of steel and 2.07g of lead alloy; $m = 4.02g$, $L = 23mm$, $v_s = 990 - 1000 \, m/s$.

5.73.2 *Test results*

Table 1#. Test results on DOP.

Fiber material	γ (kg/m^3)	f'_c (MPa)	P (mm)
Steel 70	1236	2.8-3.5	101.8
Steel 60	1080	2.7-2.9	133.3
Steel 50	915	2.1-2.4	101.8
Polypropylene	800	2.1-2.4	196.2

5.74   [Mayseless *et al.*, 2008]

5.74.1   *General data on shields and projectiles*

Shields.   Single   B-600   PC   slabs; $c_{l,w} = 350\,mm$, $T = 100\,mm$, $\gamma = 2411\,kg/m^3$, $f_c' = 60\,MPa$.

Projectiles. B32-API projectile, $D = 14.5\,mm$.

5.74.2   *Test results*

Table 1. Test results.

v_s (m/s)	θ_s (deg)	Result	v_r (m/s)	θ_r (deg)
973	70	R	–	124
976	70	R	–	140
975	70	R	315	155
970	70	R	–	132
971	65	R	155	155
972	60	R	–	150
979	55	R	–	145
973	50	P	–	–

R – Ricochet; P – Perforation.

v_r is measured in the direction of the line of fire.

5.75   [Miwa *et al.*, 2006]

5.75.1   *General data on shields and projectiles*

Shields. Single PC slabs: $c_{l,w} = 500\,mm$, $f_c' = 25\,MPa$.

Projectiles. Hard mushroom-shaped projectiles (Fig. 3.4.2), $D = 25\,mm$, $m = 50\,g$.

Shield damage types (SDTs). (1) – Spalling, (2) – Scabbing, (3) – Perforation.

5.75.2   *Test results*

Table 1#. Test results on DOP.

T (mm)	v_s (m/s)	$P\#$ (mm)	SDT	T (mm)	v_s (m/s)	$P\#$ (mm)	SDT
30	180	–	(3)	30	210	–	(3)
60	200	15.0	(2)	70	310	–	(3)

T (mm)	v_s (m/s)	$P\#$ (mm)	SDT	T (mm)	v_s (m/s)	$P\#$ (mm)	SDT
70	310	28.0	(2)	80	210	16.7	(1)
80	310	24.0	(2)	80	415	–	(3)
80	415	25.0	(2)	90	420	30.0	(2)
90	420	31.0	(2)	100	310	20.7	(1)
100	490	–	(3)	120	415	25.5	(1)
130	490	36.7	(1)				

5.76 [Mohamed *et al.*, 2009]

5.76.1 *General data on shields and projectiles*

Shields. Layered shields ($\tilde{c}_{l,w} = 500\,mm$) composed of two or three blocks in contact. Every block ($T_i = 200\,mm$) was made of plain concrete (block B-0), or of the same concrete with one (block B-1), two (block B-2) or three (block B-3) reinforcement mesh/meshes located at the frontal and rear faces.

Concrete mix. Dolomite aggregate ($q_{max}^{agg} = 19\,mm$, $\gamma = 1400$ kg/m^3); $\gamma = 2350\,kg/m^3$, $f_c' = 35\,MPa$, $\sigma_u = 3.1\,MPa$, $E = 29\,GPa$.

Mesh. Galvanized woven wire mesh, $d_{l,w} = 2\,mm$, $g_{l,w} = 50\,mm$. For packages of meshes, $t^{F,R} = 20\,mm$. Material: 1006 steel, $\gamma = 7850\,kg/m^3$, $\sigma_y = 250\,MPa$, $\sigma_u = 460\,MPa$, $E = 210\,GPa$.

Table 1. Spherical-segment nose projectiles (Fig. 3.8.1).

D (mm)	L (mm)	ρ (mm)	Material	HB	σ_y (MPa)	σ_u (MPa)	ε_f (%)	m (g)
23	64	17	Steel	475	1726	1900	7	175

5.76.2 *Test results*

Table 2. Test results on DOP.

Layer of the shield			v_s (m/s)	P (mm)
First	Second	Third		
B-0	B-0	B-0	974	400
B-0	B-0	–	976	400
B-3	B-0	–	976	400
B-3	B-3	–	978	280
B-2	B-2	–	996	287
B-1	B-1	–	994	290
B-2	B-1	–	979	285

5.77 [Mohan *et al.*, 2014]

5.77.1 *General data on shields and projectiles*

Shields. Single micro reinforced concrete slab ($c_{l,w} = 300\,mm$, $T = 100\,mm$): plain concrete matrix with 30 layers of mesh; $f'_c = 55\,MPa$ (cube-shaped specimens having the length of the edge 150 mm were tested).

Mesh. Mild steel wire, $d_{l,w} = 0.3\,mm$, $g_{l,w} = 3\,mm$, $\sigma_y = 334\,MPa$.

Projectiles. Ogive-nose projectile (Fig. 3.3.1) – 5.56×45 mm NATO bullet; $D = 5.7\,mm$, $L_{cyl} = 7.57\,mm$, $L_{nose} = 10.28\,mm$, $\psi = 3.5$, $m = 4\,g$.

5.77.2 *Test results*

$v_s = 960\,m/s$, $P = 68\,mm$.

5.78 [Mu and Zhang, 2011]

5.78.1 *General data on shields and projectiles*

Shields. Single PC slabs ($c_{l,w} = 360\,mm$, $T = 400\,mm$), $f'_c = 51\,MPa$ (cube-shaped specimens having the length of the edge 100 mm were tested), $\gamma = 2440\,kg/m^3$.

Projectiles. Heat treated 35CrMnSiA steel ogive-nose projectile (Fig. 3.3.1); $D = 12\,mm$, $L_{cyl} = 34\,mm$, $\psi = 3$, HRC 42.

5.78.2 *Test results*

Table 1. Test results.

m (g)	v_s (m/s)	P (mm)	m_r (g)	m (g)	v_s (m/s)	P (mm)	m_r (g)
26.6	621	75	26.4	26.5	715	92	26.0
26.7	859	120	25.7	26.7	913	132	25.7
26.6	1112	140	24.6	26.7	1191	144	24.1
26.4	1291	140	23.5	26.6	1449	162	19.4

5.79 [Nia *et al.*, 2014]

5.79.1 *General data on shields and projectiles*

Shields. Single PC and FRC slabs; natural sand aggregate ($q_{\max}^{agg} = 4.75\,mm$).

Fibers. Some slabs have hooked end steel fibers, $L^{fib} = 50\,mm$, $d^{fib} = 0.75\,mm$, $\gamma = 7800\,kg/m^3$.

Table 1. Properties of shield material.

Label	f'_c (MPa)	σ_u (MPa)	Peak stress (MPa)	Peak Strain (%)	Rupture Strain (%)	γ (kg/m^3)
Shield-1	39.0	2.3	39.3	0.08	0.12	2409
Shield-2	51.9	2.7	51.7	0.02	0.16	2535
Shield-3	44.1	4.4	44.8	0.06	0.51	2437
Shield-4	65.6	4.0	64.3	0.08	0.10	2517

Table 2. Phosphor bronze cone-nose projectiles (Fig. 3.2.1).

D (mm)	L_{nose} (mm)	L (mm)	γ (kg/m^3)	σ_y (MPa)	σ_u (MPa)	δ (%)
5.0	4.3	50.25	8800	380	491	11

5.79.2 *Test results*

Note. Parameter $\Delta L/L$ was taken from the plot.

Table 3. Test results for Shield-1.

v_s (m/s)	P (mm)	$\Delta L/L$ (%)	v_s (m/s)	P (mm)	$\Delta L/L$ (%)	v_s (m/s)	P (mm)	$\Delta L/L$ (%)
1134	92	100	956	82	91.9	825	81	83.9
950	80	95.7	920	82	93.8	710	60	91.0

Table 4. Test results for Shield-2.

v_s (m/s)	P (mm)	$\Delta L/L$ (%)	v_s (m/s)	P (mm)	$\Delta L/L$ (%)	v_s (m/s)	P (mm)	$\Delta L/L$ (%)
1142	88	99.8	933	68	91.9	806	70	83.9
980	76	89.7	900	72	93.9	750	62	83.9

Table 5. Test results for Shield-3.

v_s (m/s)	P (mm)	$\Delta L/L$ (%)	v_s (m/s)	P (mm)	$\Delta L/L$ (%)	v_s (m/s)	P (mm)	$\Delta L/L$ (%)
879	83	96.0	783	71	90.1	672	62	85.9

Table 6. Test results for Shield-4.

v_s (m/s)	P (mm)	$\Delta L/L$ (%)	v_s (m/s)	P (mm)	$\Delta L/L$ (%)	v_s (m/s)	P (mm)	$\Delta L/L$ (%)
871	49	90.0	737	57	83.9	666	52	80.0

5.80 [O'Neil, 1999]

5.80.1 *General data on shields and projectiles*

Shields. Single fiber reinforced very high strength concrete slabs ($D^{sh} = 762\,mm$, $T = 914\,mm$); Ø102×203 mm cylinders were tested to determine f_c' .

Fibers. Hooked-ended, steel fibers; $L^{fib} = 30\,mm$, $d^{fib} = 0.5\,mm$; $\rho_V = 3\%$.

Projectiles. Hardened 4340 Steel (HRC 43-45) ogive-nose projectile (Fig. 3.3.1); $D = 26.9\,mm$, $L_{cyl} = 203.8\,mm$, $L_{nose} = 35.6\,mm$, $\rho = 53.63\,mm$.

5.80.2 *Test results*

Table 1. Test results on DOP.

m	f_c'	v_s	P	m	f_c'	v_s	P
(g)	(MPa)	(m/s)	(mm)	(g)	(MPa)	(m/s)	(mm)
905	157	229	80.7	904	162	287	131.8
907	163	397	161.4	902	159	406	180.2
904	160	573	293.2	902	157	587	295.9
903	160	747	443.9	904	155	754	460.0

5.81 [Ohno *et al.*, 1994]

See also [Tsubota *et al.*, 1993], [Shirai *et al.*, 1997], [Shirai *et al.*, 1993].

5.81.1 *General data on shields and projectiles*

Shields. Single slabs without liners and with liners on the rear face; double-layer slabs without liners (layers in-contact and slabs with air gaps or an absorber between layers); $c_{l,w} = 600\,mm$, $f_c' = 25.4\,MPa$; steel rebars and liners (see Table 1).

Table 1. Reinforcement.

Shield	T_{lin}^R	$t^{F,R}$	$d_{l,w}^{F,R}$	$g_{l,w}^{F,R}$
	(mm)	(mm)	(mm)	(mm)
Slab-1	–	–	6	100
Slab-2	Variable	–	6	100
Slab-3	–	10	7	70

Projectiles. Mild steel cylindrical projectiles (Fig. 3.1.1), $D = 35\,mm$.
Impact velocity. $v_s = 170\,m/s$.

Table 2. Shield damage types (SDTs).

Notation	Description	Notation	Description
(1)	Perforation	(2)	Scabbing
(3)	Penetration	(4)	Splitting
(5)	Bulging		

5.81.2 *Test results for single slab*

Table 3. Test results for Slab-1.

T (mm)	SDT	T (mm)	SDT	T (mm)	SDT	T (mm)	SDT
50	(1)	60	(1)	70	(1)	80	(1)
85	(1)	90	(2)	100	(2)	110	(2)
120	(3)	140	(3)	160	(3)		

Table 4. Test results for Slab-2, $T^{lin} = 0.8$ *mm.*

T (mm)	SDT	T (mm)	SDT	T (mm)	SDT
50	(1)	60	(4)	70	(5)
80	(5)	90	(5)	100	(5)
105	(3)	120	(3)	140	(3)

Table 5. Test results for Slab-2, $T^{lin} = 1.2$ *mm.*

T (mm)	SDT	T (mm)	SDT	T (mm)	SDT
50	(1)	60	(4)	65	(5)
70	(5)	80	(5)	90	(5)
95	(5)	100	(3)	120	(3)

Table 6. Test results for Slab-2, $T^{lin} = 1.6$ *mm.*

T (mm)	SDT	T (mm)	SDT	T (mm)	SDT
50	(4)	55	(5)	60	(5)
70	(5)	80	(5)	90	(5)
95	(3)	100	(3)	120	(3)

Table 7. Test results for Slab-2, $T^{lin} = 2.0$ *mm.*

T (mm)	SDT	T (mm)	SDT
50	(5)	90	(3)

5.81.3 *Test results for double-layer slabs without liner*

Table 8. Test results for layered shield consisting of two Slabs-3.

T_1 (mm)	T_2 (mm)	Interlayer (C/A/G)	SDT (front slab)	SDT (rear slab)
45	45	C	(2)	(2)
30	60	C	(1)	(2)
60	30	C	–	(2)
45	45	G	(1)	(1)
30	60	G	(1)	(2)
60	30	G	(1)	(1)
45	45	A	(1) (2)	(2)
30	60	A	(1)	–
60	30	A	(1) (2)	(2)

A – absorber, C – in-contact, G – 15 mm air gap.

5.82 [Quek *et al.*, 2010]

5.82.1 *General data on shields and projectiles*

Shields. Single PC and functionally graded cementitious panels consisting of PE-fibrous ferrocement, calcined bauxite aggregates and conventional mortar; $c_{l,w} = 200\,mm$, $T = 100\,mm$, Ø100×200 mm cylinders and 300×75×20 mm specimens were used for compressive and tensile tests, correspondingly.

Table 1. Material properties.

Notation	Material	f_c' (MPa)	E (GPa)	σ_u (MPa)
Mat-1	PE-fibrous ferrocement material	83.68	19.83	6.45
Mat-2	Bauxite aggregate concrete	110.13	40.42	3.19
Mat-3	Granite aggregate concrete	106.11	28.51	2.64
Mat-4	Mortar	110.77	32.72	2.46

Table 2. Shields structure.

Shields	Layer 1	Layer 2	Layer 3	Layer 4
Shield 1		Monolithic, Mat-4		
Shields 2-5	Mat-1	Mat-2	Mat-4	Mat-1
Shield 6	Mat-1	Mat-3	Mat-4	Mat-1

Table 3. Shields details.

Shields	T_1 (mm)	T_2 (mm)	T_3 (mm)	T_4 (mm)	ρ_V (%)
Shield 1	Monolithic, $T = 100\,mm$				–
Shield 2	10	10	70	10	0.26
Shield 3	20	10	60	10	0.39
Shield 4	10	10	60	20	0.39
Shields 5-6	10	20	60	10	0.26

Projectiles. Ogive-nose projectiles (Fig. 3.3.1) manufactured of hardened ASSAB grade 8407 supreme tool steel (HRC 50); $D = 13.35\,mm$, $\psi = 2.5$, $m \approx 17\,g$.

5.82.2 *Test results*

Table 1. Test results for different shields.

Shield	v_s (m/s)	P (mm)	v_s (m/s)	P (mm)	v_s (m/s)	P (mm)	v_s (m/s)	P (mm)
Shield 1	302.7	16.7	401.6	19.7	507.1	36.9	581.4	41.8
Shield 2	305.2	19.8	405.4	22.9	514.6	28.4	587.2	37.0
Shield 3	305.6	20.1	399.5	24.5	493.4	30.6	589.5	37.6
Shield 4	305.5	17.4	395.8	23.6	508.7	28.8	573.9	35.8
Shield 5	299.4	16.6	397.9	21.1	506.8	27.1	593.1	32.5
Shield 6	308.7	16.7	407.6	24.4	494.4	29.1	584.9	33.6

5.83 [Romander and Sliter, 1984]

5.83.1 *General data on shields and projectiles*

Shields. Single RC barriers.

Projectiles. Turbine disks (considered as a projectile) were modeled as a simple equivalent T-section (for details see the original paper) made from 4340 steel heat treated to HRC 30. In impact tests below (except for last two tests), missiles impacted the panels in a piercing orientation where the sharp leading edge of the missile strikes the panel first. Last two test correspond to blunt turbine missiles orientation when the projected area of the missile onto the panel is maximized. Equivalent missile diameter, $\widetilde{D}$, is a diameter of a circle with the same area as the projected area of the turbine disc missile.

Notes.

- Penetration depth can exceed the wall thickness because of bulging of the rear surface.
- Data on ρ_V is taken from [Walter and Wolde-Tinsae, 1984].

5.83.2 *Test results*

Table 1. Test parameters and results.

$\tilde{D}$ (mm)	m (kg)	v_s (m/s)	T (mm)	f'_c (MPa)	ρ_V (%)	P (mm)	v_r (m/s)
70.4	2.84	200.6	124.7	28.8	4.9	149.9	–
70.4	2.84	128.6	124.7	34.0	4.9	63.5	–
70.4	2.84	201.2	97.0	26.0	4.4	–	94.5
70.4	2.84	156.1	97.0	33.1	4.4	–	42.7
70.4	2.84	129.2	97.0	20.8	4.4	79.5	–
70.4	2.84	171.0	83.1	24.9	1.0	–	111.3
70.4	2.84	130.1	83.1	18.8	1.0	–	64.3
70.4	2.84	51.8	55.4	18.8	1.0	–	17.4
70.4	2.84	35.7	55.4	22.9	1.0	31.8	–
47.8	1.37	203.3	124.7	24.0	4.9	105.4	–
47.8	1.37	198.4	97.0	27.7	4.4	–	26.5
47.8	1.37	164.9	97.0	25.1	4.4	94.2	–
47.8	1.37	204.8	83.1	22.9	1.0	–	119.8
47.8	1.37	164.0	83.1	31.7	1.0	–	67.1
47.8	1.37	82.3	55.4	31.7	1.0	–	24.4
47.8	1.37	51.8	55.4	24.1	1.0	35.6	–
49.5	1.10	196.6	97.0	20.4	4.4	100.3	–
49.5	1.10	213.4	83.1	23.6	1.0	–	94.5
49.5	1.10	83.8	55.4	24.9	1.0	96.5	–
105.4	2.84	210.0	97.0	23.5	4.4	–	99.1
72.1	1.37	146.9	83.1	21.4	1.0	127.0	–
72.1	1.37	75.6	55.4	23.6	1.0	78.7	–

5.84 [Schulz and Heimdahl, 1981]

5.84.1 *General data on shields and projectiles*

Shields. Single slab made of Thorite (® Standard Dry Wall Products), fast-setting concrete, $f'_c = 27.2\,MPa$.

Projectiles. Projectiles (Figs. 3.1.2b and 3.1.2c) were hollow cylinders ($D = 12.70\,mm$, $L = 50.80\,mm$, $d = 11.68\,mm$, $L_1 = 6.35\,mm$, $m \approx 20\,g$) machined from heat-treated 4340 steel (HRC 38-40); usually, there was a groove at the internal surface of a striker (Fig. 3.1.2c).

Table 1. Additional parameters of projectiles $\left[\chi = 0.5(d_1 - d)\right]$.

Projectile label	Shape	L_2 (mm)	χ (mm)
Projectile-1	Fig. 3.1.2b	–	–
Projectile-2	Fig. 3.1.2c	11.68	0.10
Projectile-3	Fig. 3.1.2c	11.68	0.20
Projectile-4	Fig. 3.1.2c	18.03	0.10
Projectile-5	Fig. 3.1.2c	18.03	0.20

5.84.2 *Test results*

Additional notation. PFC – projectile final condition (* – bulged, ** – broke up).

Table 2. Test results for Projectile-1.

v_s (m/s)	P (mm)	PFC	v_s (m/s)	P (mm)	PFC	v_s (m/s)	P (mm)	PFC
664	83.3	*	728	76.0	*	744	88.2	*
756	83.3	*	759	80.9	*	767	71.1	**
768	68.6	**	768	73.5	**	770	71.1	**
779	68.6	**						

Table 3. Test results for Projectile-2.

v_s (m/s)	P (mm)	PFC	v_s (m/s)	P (mm)	PFC	v_s (m/s)	P (mm)	PFC
657	63.7	*	674	76.0	*	707	80.9	*
734	61.3	**	756	73.5	**			

Table 4. Test results for Projectile-3.

v_s (m/s)	P (mm)	PFC	v_s (m/s)	P (mm)	PFC	v_s (m/s)	P (mm)	PFC
634	66.2	*	649	73.5	*	652	51.5	**
668	51.5	**	674	58.8	**			

Table 5. Test results for Projectile-4.

v_s (m/s)	P (mm)	PFC	v_s (m/s)	P (mm)	PFC	v_s (m/s)	P (mm)	PFC
750	78.4	*	754	58.8	**	759	71.1	**
773	71.1	**	774	78.4	**			

Table 6. Test results for Projectile-5.

v_s (m/s)	P (mm)	PFC	v_s (m/s)	P (mm)	PFC	v_s (m/s)	P (mm)	PFC
716	76.0	*	742	76.0	*	746	76.0	*
759	78.4	**	770	73.5	**			

5.85   [Schulz *et al.*, 1982]

5.85.1   *General data on shields and projectiles*

Shields. Single slab, made of Thorite, fast-setting concrete, $f_c' = 27.2\,MPa$.

Projectiles. Machined from ZK6OA magnesium cylindrical homogeneous projectiles and projectiles with internal cavity filled with plasticine; $D = 12.7\,mm$, $L = 63.5\,mm$, $m \approx 14\,g$.

Table 1. Geometric parameters of projectiles.

Projectile label	Shape	d (mm)	L_1 (mm)
Projectile-1	Fig. 3.1.1	–	–
Projectile-2	Fig. 3.1.2b	6.35	6.35
Projectile-3	Fig. 3.1.2b	7.94	6.35
Projectile-4	Fig. 3.1.2b	9.25	6.35

5.85.2   *Test results*

Table 2. Test results for different projectiles.

Projectile	v_s (m/s)	P (mm)	d_{bul} (mm)	v_s (m/s)	P (mm)	d_{bul} (mm)	v_s (m/s)	P (mm)	d_{bul} (mm)
Projectile-1	247	9.1	13.2	347	23.1	13.4	376	23.4	–
	399	23.1	13.4	412	21.1	–	430	21.3	–
	511	32.3	–						

Projectile	v_s (m/s)	P (mm)	d_{bul} (mm)	v_s (m/s)	P (mm)	d_{bul} (mm)	v_s (m/s)	P (mm)	d_{bul} (mm)
Projectile-2	430	27.4	13.4	442	28.2	13.6	448	14.2	–
	454	29.7	–	463	15.0	–			
Projectile-3	281	11.4	13.4	295	15.0	–	299	14.2	13.6
	322	18.5	13.6	355	16.8	–			
Projectile-4	129	2.8	13.6	191	4.3	–	215	6.4	–
	241	5.1	–	273	5.6	–			

d_{bul} is residual cavity bulge diameter of projectile.

5.86 [Schulz *et al.*, 1984]

5.86.1 *General data on shields and projectiles*

Shields. Single reinforced slab ($c_{l,w} = 254\,mm$) made of Thorite, fast-setting concrete, $f_c' = 27.2\,MPa$. Reinforcement can be considered as a set of 12.7 mm layers whereby each layer is made of concrete with soft steel wire mesh (hardware cloth, 1.27 mm in diameter) in the middle.

Projectiles. Machined from heat-treated 4340 steel (HRC 38-40) hollow cylindrical projectiles having internal cavity filled with plasticine or unfilled internal cavity (Fig. 3.1.2c, $D = 12.7\,mm$, $d = 10.7\,mm$, $L = 50.8\,mm$, $L_1 = 6.35\,mm$).

Table 1. Notations in tables below.

Symbol	Explanation
=	Parameter could not be measured, projectile broke up or damaged too badly
*	Projectile did not perforate, parameter given is penetration depth, mm
#	Projectile did not perforate
-	Exit velocity could not be determined
~	Estimated value, impact velocity could not be determined
^	Projectile damaged, measurement results are uncertain

5.86.2 *Test results*

Additional notation. d_{bul}^{cav} is bulge diameter of residual cavity of projectile.

Table 2. Test results for projectile having unfilled cavity and $\theta_s = 0°$.

T	m	v_s	v_r	D_r	d_{bul}^{cav}	L_r
(mm)	(g)	(m/s)	(m/s)	(mm)	(mm)	(mm)
12.7	19.94	547	458	12.80	13.08	50.55
12.7	20.14	549	441	12.88	13.03	50.52
12.7	19.87	643	501	13.06	13.26	50.17
12.7	20.17	651	544	12.93	13.36	50.14
12.7	19.97	722	590	13.21	13.51	49.78
12.7	20.03	732	579	13.18	14.05	49.53
25.4	20.06	544	427	12.93	13.08	50.50
25.4	20.06	561	408	12.90	13.18	50.39
25.4	19.94	643	462	12.90	13.61	49.89
25.4	19.95	658	494	13.11	13.64	49.86
25.4	20.11	713	530	13.08	14.30	49.23
25.4	19.97	733	555	13.41	14.73	48.87
63.5	20.29	538	222	12.78	13.08	50.52
63.5	20.24	549	232	12.85	13.23	50.34
63.5	19.96	626	293	12.93	13.61	50.19
63.5	19.82	643	293	12.90	13.69	49.89
63.5	20.09	721	373	13.23	14.38	49.17
127.0	19.87	550	*1.0	12.80	13.21	50.50
127.0	20.20	552	*0.9	12.75	13.23	50.34
127.0	19.92	649	*0.8	12.88	13.56	49.86
127.0	19.73	658	72.5	12.90	13.56	50.01
127.0	19.96	735	–	13.41	14.30	49.15
127.0	20.01	739	0	13.16	9.86	48.59

Table 3. Test results for projectile with filled cavity and $\theta_s = 0°$.

T	m	v_s	v_r	D_r	d_{bul}^{cav}	L_r
(mm)	(g)	(m/s)	(m/s)	(mm)	(mm)	(mm)
12.7	26.34	539	490	12.88	13.23	50.47
12.7	26.51	544	486	12.80	13.21	50.44
12.7	26.36	631	543	13.13	13.46	50.14
12.7	26.45	651	545	13.08	13.56	50.06
12.7	26.47	725	646	13.16	13.92	49.73
12.7	26.34	727	654	13.16	13.97	49.73
25.4	26.39	556	443	12.75	13.16	50.47
25.4	26.42	561	437	12.78	13.21	50.52
25.4	26.18	631	480	13.03	13.59	49.96
25.4	26.41	642	510	13.06	13.72	49.89
25.4	26.33	722	586	13.16	14.94	49.07
25.4	26.42	725	610	13.18	14.99	49.05
63.5	26.47	535	328	12.78	13.26	50.39
63.5	26.13	550	313	12.93	13.21	49.89

T	m	v_s	v_r	D_r	d_{bul}^{cav}	L_r
(mm)	(g)	(m/s)	(m/s)	(mm)	(mm)	(mm)
63.5	26.39	643	373	12.95	13.72	49.86
63.5	26.42	646	384	13.08	14.15	49.68
63.5	26.49	719	448	13.26	15.01	48.90
63.5	26.42	735	453	13.18	15.85	48.36
127.0	26.38	539	*1.6	12.75	13.28	50.37
127.0	26.41	541	38	12.78	13.34	50.34
127.0	26.40	623	133	12.93	14.07	49.76
127.0	26.43	649	210	13.00	14.07	49.66
127.0	26.17	725	244	13.21	15.06	48.84
127.0	26.51	732	232	13.31	15.75	48.36

Table 4. Test results for projectile with unfilled cavity and $\theta_s = 45°$.

T	m	v_s	v_r	D_r	d_{bul}^{cav}
(mm)	(g)	(m/s)	(m/s)	(mm)	(mm)
12.7	19.20	551	431	12.75	12.95
12.7	19.63	561	415	12.78	13.13
12.7	19.49	~655	479	12.90	13.89
12.7	19.50	660	523	12.95	13.82
12.7	19.66	744	602	13.08	14.30
12.7	19.88	~752	589	12.95	14.35
25.4	19.77	535	309	12.90	12.98
25.4	19.68	549	321	12.78	13.03
25.4	19.71	648	405	12.85	14.22
25.4	19.60	652	407	12.93	14.00
25.4	19.42	744	494	12.95	=
25.4	19.35	760	475	12.93	=
63.5	19.68	538	#	12.80	13.13
63.5	19.58	549	#	12.73	13.18
63.5	19.67	654	#	12.83	14.50
63.5	19.83	661	#	12.85	14.35
63.5	19.95	704	=	12.80	=
63.5	19.54	722	=	12.88	=

Table 5. Test results for projectile with filled cavity and $\theta_s = 45°$.

T	m	v_s	v_r	D_r	d_{bul}^{cav}
(mm)	(g)	(m/s)	(m/s)	(mm)	(mm)
12.7	26.26	540	440	12.78	38.46
12.7	26.04	551	452	12.80	13.18
12.7	26.25	636	564	12.85	13.28
12.7	26.21	646	545	12.90	13.92
12.7	26.07	725	618	12.98	14.63

T	m	v_s	v_r	D_r	d_{bul}^{cav}
(mm)	(g)	(m/s)	(m/s)	(mm)	(mm)
12.7	26.07	745	634	12.95	15.01
25.4	26.10	562	384	12.78	13.39
25.4	26.16	568	391	12.78	13.49
25.4	26.23	639	465	12.98	14.71
25.4	26.00	661	486	12.90	15.27
25.4	26.14	721	507	12.93	17.42
25.4	26.08	733	522	13.00	17.04
63.5	26.23	553	122	12.80	13.59
63.5	26.17	554	108	12.78	13.67
63.5	26.06	648	#	12.80	^17.02
63.5	26.09	649	#	12.78	^18.57
63.5	26.23	664	=	12.83	=
63.5	26.13	681	=	12.90	=

5.87 [Shan *et al.*, 2014]

5.87.1 *General data on shields and projectiles*

Shields. Layered concrete shields consisting of PC slabs in-contact; $D^{sh} = 550\,mm$, $T_i = 630\,mm$, $q^{agg} = 10 - 15\,mm$.

Projectiles. Monolithic and hollow ogive-nose projectiles (Fig. 3.3.1); $D = 15\,mm$, $\psi = 3$.

Table 1. Material properties of projectiles.

Material	σ_y (MPa)	HRC	Material	σ_y (MPa)	HRC	Material	σ_y (MPa)	HRC
AerMet 100	1900	53	AD 300	1650	49	DT 300	1603	48.5

Table 2. Notations (numbers in parentheses in column "Note") in the table with test results.

Notation	Description
(1)	Penetrated through 2 slabs
(2)	Projectile exited from the side of the shield at depth of 1 m
(3)	Ricochet; the depth of crater is 60 mm
(4)	Shield was used in 14 experiments; the total penetration depth is 1.37 m
(5)	Projectile exited from the side of the shield at depth of 0.85 m
(6)	Projectile exited from the side of the shield at depth of 1.4 m

5.87.2 *Test results*

Table 3. Test results.

Projectile material	Projectile type	N_{sl}	f_c' (MPa)	m (g)	v_s (m/s)	P (mm)	Δm (g)	Note
AerMet 100	Monolithic	2	40	131.75	1089	1055	3.78	–
AerMet 100	Monolithic	2	36	137.70	1136	–	–	(1)
AerMet 100	Hollow	2	38	103.70	1066	885	3.16	–
AerMet 100	Hollow	2	35	103.23	1193	935	3.50	–
AerMet 100	Hollow	2	39	103.07	1230	1185	3.76	–
DT 300	Hollow	2	34	100.14	1240	1195	3.10	–
DT 300	Hollow	2	36	99.18	1121	860	2.60	–
DT 300	Hollow	2	33	99.29	1083	885	4.44	–
AD 300	Hollow	2	39	103.58	1242	1075	3.61	–
AD 300	Hollow	2	32	103.53	1172	1020	3.18	–
AD 300	Hollow	2	36	103.50	1361	–	–	(2)
AD 300	Hollow	2	32	103.55	1258	1115	3.59	–
AD 300	Hollow	2	33	103.33	1093	935	3.01	–
AerMet 100	Monolithic	3	35	131.90	1430	–	–	(3)
DT 300	Monolithic	3	35	128.07	1220	1307	4.41	(4)
DT 300	Monolithic	3	35	128.13	1330	–	–	(5)
AerMet 100	Monolithic	3	32	131.80	1271	–	–	(6)

N_{sl} – number of slabs (layers) in the shield.

5.88 [Shiqiao *et al.*, 2006]

5.88.1 *General data on shields and projectiles*

Shields. Three-layer PC shields with air gaps of $1m$; $D^{sh} = 1500\,mm$.
Projectiles. Ogive-nose projectile (Fig. 3.3.1); $D = 62\,mm$, $\psi = 1.21$, $\rho = 75\,mm$, $m = 3\,kg$.

5.88.2 *Test results*

Table 1. Test results ($T_2 = T_3 = 200\,mm$).

v_s (m/s)	First plate			Second plate		Third plate	
	T_1 (mm)	f_c' (MPa)	P^{cr} (mm)	f_c' (MPa)	P^{cr} (mm)	f_c' (MPa)	P^{cr} (mm)
590	300	45	150	35	110	30	–
670	200	35	100	35	80	30	90
669	200	35	110	35	90	30	90
653	200	35	100	30	90	30	85
650	200	45	150	30	–	30	–

5.89   [Sjøl and Teland, 2000]

5.89.1   *General data on shields and projectiles*

Note. Taken from compendium of experimental results on penetration into concrete by [Sjøl and Teland, 2000].

Table 1. Initial data on shields and projectiles for FFI and Bofors experiments.

Set label	Reference	Projectile type	Shape	D (mm)	ψ	m (kg)
FFI experiments	[Sjøl *et al.*, 1998]	Cylindrical	Fig. 3.1.1	12	–	0.0205
Bofors experiments (series 1)	–	Ogive-nose	Fig. 3.3.1	152	0.7	44.76
Bofors experiments (series 2)	–	Ogive-nose	Fig. 3.3.1	75	1.5	6.28

Table 2. Initial data on shields and projectiles for Bergman experiments [Bergman, 1950] with ogive-nose projectiles (Fig. 3.3.1) with $\psi = 1.5$.

Series No.	D (mm)	m (kg)	f_c' (MPa)	Series No.	D (mm)	m (kg)	f_c' (MPa)
1	75	5.45	33.9	2	37	0.867	39.2
3	155	40.2	33.0	4	76	6.81	35.0
5	307	454	38.6	6	11	0.02	30.2
7	12.7	0.0291	46.3				

5.89.2   *Test results*

Table 3. Test results for FFI experiments, $f_c' = 35\,MPa$.

v_s (m/s)	P (mm)	v_s (m/s)	P (mm)	v_s (m/s)	P (mm)
414	25	445	30	567	37
572	54	749	65	754	53

Table 4. Test results for Bofors data (series 1).

f_c' (MPa)	v_s (m/s)	P (mm)	f_c' (MPa)	v_s (m/s)	P (mm)
36	576	1700	30	481	1020
41	468	1030	34	481	670
90	583	980	90	479	630
90	486	590	90	488	560
132	581	830	140	584	810
140	480	595	140	480	420

f'_c (MPa)	v_s (m/s)	P (mm)	f'_c (MPa)	v_s (m/s)	P (mm)
180	473	600	200	580	530
250	481	460	200	480	300
203	480	450	200	480	450
220	478	350			

Table 5. Test results for Bofors data (series 2).

f'_c (MPa)	v_s (m/s)	P (mm)	f'_c (MPa)	v_s (m/s)	P (mm)
35	647	990	38	484	680
38	483	655	38	482	660
90	653	560	90	571	410
140	647	440	180	485	250
180	489	235	180	485	240
200	480	200	200	650	440

Table 6. Test results for Bergman data.

Series	v_s (m/s)	P (mm)	v_s (m/s)	P (mm)	v_s (m/s)	P (mm)	v_s (m/s)	P (mm)
Series 1	160	107	168	103	318	222	329	220
	439	311	493	332	493	348	596	434
	661	506	685	498	700	549		
Series 2	233	71	246	79	313	114	317	105
	408	142	508	188	517	160	638	250
	658	265	700	296	788	364	857	463
	857	438	858	429				
Series 3	260	330	320	400	634	980	672	1100
	730	1300						
Series 4	254	177	296	215	471	437	638	627
	704	754						
Series 5	295	1140	305	1060	305	1150	310	1150
Series 6	280	19	575	56	575	64	595	56
	600	59						
Series 7	260	24	388	36	446	47	575	61
	692	85	788	106	908	127		

5.90 [Sjøl *et al.*, 2002]

5.90.1 *General data on shields and projectiles*

Shields. Single PC slabs, $f'_c = 35\,MPa$.

Projectiles. Ogive-nose (Fig. 3.3.1), $L_{nose} \approx D$ (estimated from the drawing) and cylindrical (Fig. 3.1.1) projectiles; $D = 12\,mm$.

5.90.2 *Test results for cylindrical projectiles*

Table 1. Steel projectiles.

m	v_s	P	P^{cr}	m	v_s	P	P^{cr}
(g)	(m/s)	(mm)	(mm)	(g)	(m/s)	(mm)	(mm)
20.5	368	27	–	20.5	498	43	–
20.5	654	48	–	65.83	312	71	33
65.14	362	91	26	65.94	434	123	28
65.24	505	126	32	65.82	614	171	30
65.94	680	206	30				

Table 2. Tungsten projectiles.

m	v_s	P	P^{cr}	m	v_s	P	P^{cr}
(g)	(m/s)	(mm)	(mm)	(g)	(m/s)	(mm)	(mm)
64.79	356	75	22	122.56	219	60	25
122.96	288	93	21	123.22	329	125	21
122.28	361	147	22	122.28	405	147	26

5.90.3 *Test results for ogive-nose projectiles*

Table 3. Steel projectiles.

m	v_s	P	P^{cr}	m	v_s	P	P^{cr}
(g)	(m/s)	(mm)	(mm)	(g)	(m/s)	(mm)	(mm)
25	359	20	–	24.21	414	23	–
24.01	436	48	26	24.3	500	85	26
24.22	506	67	–	25.0	506	30	–
23.87	564	108	22	25.0	567	71	–
24.22	643	116	20	24.25	810	158	27

Table 4. Tungsten projectiles.

m	v_s	P	P^{cr}	m	v_s	P	P^{cr}
(g)	(m/s)	(mm)	(mm)	(g)	(m/s)	(mm)	(mm)
51.08	245	64	25	50.84	305	68	25
51.17	358	82	27	51.11	398	107	25
50.22	512	158	25	50.36	664	224	22
50.62	742	279	22	65.30	275	72	22
66.28	373	108	24	65.72	482	178	26
65.97	614	251	19	66.10	699	320	24
66.28	702	212	30	65.00	734	217	35
65.84	754	299	24	63.23	799	162	23
123.00	198	77	22	123.44	264	109	22
123.46	353	181	20	122.80	381	186	20

m	v_s	P	P^{cr}	m	v_s	P	P^{cr}
(g)	(m/s)	(mm)	(mm)	(g)	(m/s)	(mm)	(mm)
124.58	458	252	17	123.53	526	345	24
124.68	576	322	39	123.05	616	378	27

5.91 [Sliter, 1980]

Notes.

- Study by [Sliter, 1980] contains compendium of experimental results obtained by many researchers. In the following we present some of these results which were not included in the corresponding sections of the handbook devoted to the original investigations where these results were obtained.
- Many of the experimental results presented in this study can be also found in [Haldar and Miller, 1982].

5.91.1 *General data on shields and projectiles*

Shields. Single PC or lightly RC.
Projectiles. Cylindrical projectiles (Fig.3.1.1).
Notations for Shield Damage Types (SDTs). N – No scabbing (and no perforation), S – Scabbing, P – Perforation.

5.91.2 *Test results*

Table 1. Data and test results from [Vassallo, 1975], $D = 203$ *mm*.

m	T	f'_c	v_s	P	SDT
(kg)	(mm)	(MPa)	(m/s)	(mm)	
97	305	39.8	37.2	19	S
97	305	31.4	65.2	–	S
96	305	31.4	103.6	–	P
96	457	31.0	49.1	51	N
96	457	33.8	102.7	229	S
96	610	32.9	89.9	76	N
97	610	31.0	114.9	114	S

Table 2. Data and test results from [Jankov *et al.*, 1976], $D = 76$ *mm*.

m	T	f'_c	v_s	P	SDT
(kg)	(mm)	(MPa)	(m/s)	(mm)	
5.66	114	22.1	27.1	5	N

m (kg)	T (mm)	f_c' (MPa)	v_s (m/s)	P (mm)	SDT
5.75	152	26.3	30.8	4	N
5.75	152	26.3	41.1	3	N
5.71	152	30.3	55.8	9	N
5.75	152	26.3	62.5	11	S
10.65	152	26.3	29.0	5	N
10.65	152	26.3	39.0	5	S

Table 3. Data and test results from [Barber, 1973], $D = 25$ *mm*, $m = 3.64$ kg.

T (mm)	f_c' (MPa)	v_s (m/s)	P (mm)	SDT
76	40.7	66.4	–	P
152	33.2	45.7	30	N
152	40.7	64.9	41	N
152	41.0	67.1	51	N
229	39.7	61.9	43	N

Table 4. Data and test results from [Stephenson, 1977], $D = 25$ *mm*, $m = 3.62$ *kg*.

T (mm)	f_c' (MPa)	v_s (m/s)	P (mm)	SDT
457	25.2	92.4	91	N
305	24.4	132.6	147	N

Table 5. Data and test results from [Langheim, 1977], $D = 45$ *mm*, $T = 231$ *mm*, $m = 1.11$ *kg*.

f_c' (MPa)	v_s (m/s)	P (mm)	SDT	f_c' (MPa)	v_s (m/s)	P (mm)	SDT
32.6	165.8	43	N	34.3	164.0	46	N
35.7	150.9	39	N	40.4	168.9	41	N
32.6	209.7	55	N	34.3	205.7	58	N
35.7	216.7	59	N	40.4	205.7	58	N

Table 6. Data and test results from [Langheim, 1977], $D = 45$ *mm*, $T = 231$ *mm*, $m = 2.25$ *kg*.

f_c' (MPa)	v_s (m/s)	P (mm)	SDT	f_c' (MPa)	v_s (m/s)	P (mm)	SDT
32.6	154.8	68	N	34.3	154.8	78	N
35.7	157.0	71	N	40.4	157.0	71	N
32.6	212.8	106	S	34.3	212.8	130	S
35.7	203.0	102	S	40.4	203.0	106	S

f_c' (MPa)	v_s (m/s)	P (mm)	SDT	f_c' (MPa)	v_s (m/s)	P (mm)	SDT
40.7	299.9	–	S	40.7	311.8	–	P
45.7	311.8	–	S	49.2	307.8	–	P

Table 7. Data and test results from [Langheim, 1976], SDT=N.

m (kg)	D (mm)	T (mm)	f_c' (MPa)	v_s (m/s)	P (mm)
0.892	40	351	42.0	278.9	58
0.892	40	399	38.2	225.9	59
0.892	40	399	35.4	146.0	40
0.892	40	399	42.1	234.1	60
0.458	40	351	38.2	146.0	26
0.458	40	399	39.5	164.0	31
0.458	40	399	39.5	291.1	44
0.109	20	249	42.3	253.9	25
0.118	20	300	42.3	189.9	24
0.118	20	300	39.4	277.1	33

5.92 [Soe *et al.*, 2013]

5.92.1 *General data on shields and projectiles*

Shields. ECC shields ($f_c' \approx 67\,MPa$) made of the same basic PC mix with fiber additives: *ECC-1* shields [Polyvinyl alcohol (PVA) fibers with $\rho_V = 1.5\%$ and steel fibers with $\rho_V = 0.5\%$] and *ECC-2* [PVA fibers with $\rho_V = 1.75\%$ and steel fibers with $\rho_V = 0.58\%$], as well as *C45* and *C90* standard concrete shields; $c_l = 300\,mm$, $c_w = 170\,mm$, $T = 55\,mm$.

Table 1. Characteristics of straight fibers.

Material	L^{fib} (mm)	d^{fib} (mm)	γ (kg/m^3)	E (GPa)	σ_u (MPa)
Steel	12	0.200	7800	200	2600
PVA	8	0.039	1300	66	1600

Table 2. Ogive-nose projectiles (Fig. 3.3.1).

D (mm)	L_{cyl} (mm)	L_{nose} (mm)	ψ	m (g)	Material	HRC
12.60	5.03	18.9	2.5	15	Hardened steel	50

Note. Double impact was also considered in the study.

5.92.2 *Test results*

Table 3. Test results on DOP.

Shield	v_s (m/s)	P (mm)	SDT	Shield	v_s (m/s)	P (mm)	SDT
ECC-1	306	12	N	ECC-1	463	25	N
ECC-1	528	27	N	ECC-1	658	–	P
ECC-1	657	–	P	ECC-2	327	13	N
ECC-2	460	19	N	ECC-2	524	25	N
ECC-2	652	–	P	ECC-2	658	–	P
C45	313	–	B	C45	441	–	B
C45	524	–	B	C90	322	9	N
C90	463	–	B	C90	517	–	B

SDT versions. N – non perforation, P – perforation, B – broken into pieces.

5.93 [Sovják *et al.*, 2015]

See also [Sovják *et al.*, 2014], [Sovják *et al.*, 2013].

5.93.1 *General data on shields and projectiles*

Shields. Single high strength concrete (HSC), FRC, ultra high performance concrete (UHPC) and ultra-high performance fiber reinforced concrete (UHPFRC) slabs; $q_{\max}^{agg} = 8\,mm$, $c_l = 300\,mm$, $c_w = 400\,mm$. Ø100×200 mm cylinders were tested to measure f_c' and E. Tensile tests were carried out on dog-bone shaped specimens without a notch with the length of 330 mm and the cross-section of the narrow part of 30×30 mm.

Table 1. Characteristics of HSC, FRC and UHPC mixtures.

Mixture	f_c' (MPa)	σ_u (MPa)	Fibers			
			L^{fib} (mm)	d^{fib} (mm)	$\widetilde{\rho}_m$	μ (kg/m^3)
HSC	69	4.5	–	–	–	–
FRC*	38	3.1	30	0.375	0.14	50
UHPC	132	6.6	13	0.15	–	–

* $\sigma_u = 2000\,MPa$ for fibers.

Table 2. Characteristics of UHPFRC mixtures ($L^{fib} = 13\,mm$, $d^{fib} = 0.15\,mm$).

Mixture	$\widetilde{\rho}_m$	μ (kg/m^3)	f_c' (MPa)	σ_u (MPa)
UHPFRC-0.5	0.05	40	154	7.5
UHPFRC-1.0	0.10	80	149	7.8
UHPFRC-1.5	0.15	120	149	9.9
UHPFRC-2.0	0.20	160	164	9.9
UHPFRC-2.5	0.25	200	159	8.9
UHPFRC-3.0	0.30	240	148	10.9

Table 3. Shield damage types (SDTs).

Notation	Description
(1)	Projectile passed through the shield
(2)	Projectile was retained in the shield
(3)	Shield was punched but projectile bounced back

Projectiles. "Deformable" (DP) and "Non-deformable" (NDP) projectiles.

Table 4. Ogive-like nose jacketed projectiles.

Projectile designation	D (mm)	L (mm)	m (g)	Core Material	Core Mass (g)	Jacket Material	Jacket Mass (g)
DP	7.92	23.2	8.04	Soft lead	5.26	Steel	2.78
NDP	7.92	26.6	8.04	Mild steel	3.65	Steel	4.15

Note. $\sigma_y = 550\,MPa$ for mild steel.

5.93.2 *Test results*

Table 5. Test results for DP and shields with $T = 50$ *mm*.

Mixture	SDT	m^{fr} (g)	P (mm)	Mixture	SDT	m^{fr} (g)	P (mm)
UHPC	(1)	–	–	UHPFRC-1.0	(3)	88	20
UHPFRC-2.0	(3)	54	20	UHPFRC-3.0	(3)	110	19
HSC	(1)	569	–	FRC	(1)	202	–

Table 6. Test results for NDP and shields with $T = 50$ *mm*.

Mixture	SDT	m^{fr} (g)	Mixture	SDT	m^{fr} (g)
UHPFRC-2.0	(2)	60	HSC	(1)	283
FRC	(1)	130			

Table 7. Test results for NDP and shields with $T = 45\,mm$.

Mixture	v_s (m/s)	v_r (m/s)	m^{fr} (g)	Mixture	v_s (m/s)	v_r (m/s)	m^{fr} (g)
UHPC	691	260	560	UHPFRC-0.5	705	226	295
UHPFRC-1.0	703	228	99	UHPFRC-1.5	712	161	109
UHPFRC-2.0	720	71	98	UHPFRC-2.5	711	47	126
FRC	695	349	102	HSC	693	124	295

5.94 [Stronge and Schulz, 1981]

5.94.1 *General data on shields and projectiles*

Shield. Single PC slab made of Thorite, a fast-setting high-strength wet and dry concrete.

Projectiles. Machined from 4340 steel (HRC 38-40) cylindrical projectiles with internal cavity; $D = 12.7\,mm$, $L = 50.8\,mm$.

Table 1. Characteristics of projectiles.

Projectile label	Shape	d (mm)	L_1 (mm)	m (g)
Projectile-1	Fig. 3.1.2a	10.16	6.35	21.3
Projectile-2	Fig. 3.1.2a	8.89	6.35	28.2
Projectile-3	Fig. 3.1.2b	10.16	6.35	21.9

5.94.2 *Test results*

Additional notation. d_{bul} is residual bulge diameter of projectile.

Table 2. Test results for Projectile-1.

Shield cure	v_s (m/s)	P (mm)	d_{bul} (mm)	Shield cure	v_s (m/s)	P (mm)	d_{bul} (mm)
Wet	532	67.6	13.0	Dry	533	81.3	13.2
Dry	579	86.4	13.2	Wet	582	81.0	13.1
Dry	623	73.7	–	Dry	623	99.1	13.4
Wet	637	93.0	13.3	Wet	642	92.7	13.2
Dry	652	86.4	13.5	Dry	654	95.3	13.5
Wet	655	60.5	–	Dry	661	86.4	–
Wet	722	55.6	–	Dry	722	66.0	–

Table 3. Test results for Projectile-2.

Shield cure	v_s (m/s)	P (mm)	d_{bul} (mm)	Shield cure	v_s (m/s)	P (mm)	d_{bul} (mm)
Wet	558	100.1	12.8	Wet	607	119.9	12.8
Wet	649	119.1	13.0	Dry	668	129.5	13.0
Wet	728	173.7	13.0	Dry	802	171.5	13.1

Table 4. Test results for Projectile-3.

Shield cure	v_s (m/s)	P (mm)	d_{bul} (mm)	Shield cure	v_s (m/s)	P (mm)	d_{bul} (mm)
Wet	588	95.3	13.1	Wet	652	112.8	13.1
Wet	692	121.4	13.4	Wet	742	106.4	14.0
Dry	765	74.7	–	Dry	783	108.7	14.7
Wet	786	122.9	14.4	Dry	792	86.6	–
Dry	829	89.7	–	Dry	901	78.5	–

5.95 [Sugano *et al.*, 1993]

5.95.1 *General data on shields and projectiles*

Shields. Single MRC slabs.

Table 1. Characteristics of shields.

Type of test	$c_{l,w}$ (mm)	Steel rebars			σ_y (MPa)	σ_u (MPa)
		$t^{F,R}$ (mm)	$d_{l,w}^{F,R}$ (mm)	$\rho_{l,w}^{F,R}$ (%)		
SST	1500	15	6	0.4	447	585
IST	2500	30	16	0.4	361	560

SST – Small-Scale Test, IST – Intermediate-Scale Test.

Projectiles. Cylindrical rigid projectiles (Fig. 3.1.1); *Projectile-1* ($D = 101 mm$, $m = 3.6 \, \text{kg}$) and *Projectile-2* ($D = 300 mm$, $m = 100 \, \text{kg}$).

Table 2. Shield damage types (SDTs).

Notation	Description	Notation	Description
(1)	Penetration	(2)	Just scabbing
(3)	Scabbing	(4)	Perforation
*	Boring		

5.95.2 Test results

Table 3. Results for SSTs (Projectile-1, $f_c' = 23.5\,MPa$).

T (mm)	v_s (m/s)	SDT	P^{cr} (mm)	T (mm)	v_s (m/s)	SDT	P^{cr} (mm)
350	198	(1)	42	300	199	(1)	39
210	83	(1)	11	210	128	(3)	24
210	214	(3)	37	180	102	(3)	10
180	124	(3)	19	180	217	(3)	*
150	97	(3)	10	150	141	(3)	23
150	198	(4)	–	120	94	(3)	14

Table 4. Results for ISTs (Projectile-2, $f_c' = 23.5\,MPa$).

T (mm)	v_s (m/s)	SDT	P^{cr} (mm)	T (mm)	v_s (m/s)	SDT	P^{cr} (mm)
600	105	(1)	14	600	155	(2)	40
600	215	(3)	120	600	248	(3)	140
550	214	(3)	125	450	215	(4)	*

Table 5. Results for ISTs (Projectile-2, $f_c' = 35.3\,MPa$).

T (mm)	v_s (m/s)	SDT	P^{cr} (mm)	T (mm)	v_s (m/s)	SDT	P^{cr} (mm)
600	209	(3)	33	550	99	(1)	5
550	152	(3)	25	550	216	(3)	95
550	247	(3)	140	450	223	(3)	215
350	210	(4)	–				

5.96 [Svinsås *et al.*, 2011]

5.96.1 General data on shields and projectiles

Shields. PC slabs; $D^{sh} = 240\,cm$, $T = 750\,mm$; $f_c' = 103\,MPa$ (cube-shaped specimens with the length of the edge of 150 mm were tested); $\sigma_{tsp} = 6.5\,MPa$ and $E = 44.0\,GPa$ (Ø100×200 mm cylinders were tested).

Projectiles. Ogive-nose projectiles (Fig. 3.3.1); $D = 152\,mm$, $\rho = 380\,mm$, $m = 46.2\,g$.

5.96.2 *Test results*

Table 1. Test results on v_r.

v_s (m/s)	v_r (m/s)	v_s (m/s)	v_r (m/s)	v_s (m/s)	v_r (m/s)
460.0	183	455.5	204	458.8	181

5.97 [Szczepanski and Collins, 1983]

5.97.1 *General data on shields and projectiles*

Shields. Single PC slabs; $c_{l,w} = 2.44\,m$; $f_c' = 34.5\,MPa$; limestone aggregate ($f_c' \geq 117\,MPa$, $q_{\max}^{agg} = 19\,mm$).
Projectiles. Two types of projectiles having $D = 85.1\,mm$ were used. *Projectile-1* had conical nose with blunt cylindrical tip of 25.4 mm in diameter and $L = 686\,mm$. *Projectile-2* had ogive nose with blunt cylindrical tip of 25.4 mm in diameter and $L = 673\,mm$. For both projectiles, $m = 16.3\,kg$; they had cavities filled with a material that had a density of about 1600 kg/m^3; jacket material was heat treated E4340 steel (HRC 42-46).

5.97.2 *Test results*

Additional notations. In the Outcome column: R – ricochet; P – penetration or rebound.

Table 1. Test results for Projectile-1.

T (mm)	v_s (m/s)	θ_s (deg)	Outcome	T (mm)	v_s (m/s)	θ_s (deg)	Outcome
762	356	23	P	762	314	28.9	P
762	304	35.7	R	762	239	24.5	R
762	226	20.6	P	762	226	29.3	R
305	353	34.3	P	305	306	45	P
305	277	52.9	R	305	220	39.4	R
305	234	34.4	P	102	224	50.9	P
102	228	59.9	P	102	222	68.9	R
102	329	76.5	R	102	341	66.6	R
102	332	60.3	P	762	296	33.6	R
305	339	50.1	R	762	334	33.8	R

Table 2. Test results for Projectile-2.

T (mm)	v_s (m/s)	θ_s (deg)	Outcome	T (mm)	v_s (m/s)	θ_s (deg)	Outcome
305	361	50.2	R	305	378	44.9	P
305	234	40.8	P	305	235	45.6	R
102	230	63.8	P	102	234	69.3	R
102	361	70.8	R	102	351	65.6	P
762	223	23.6	R	762	227	18.3	P
762	333	35.6	P	762	337	39.6	R

5.98 [Tai, 2009]

See also [Tai and Tang, 2006].

5.98.1 *General data on shields and projectiles*

Shields. Single reactive powder concrete (RPC) and normal strength concrete (NSC) slabs with fibers and without fibers; $\widetilde{c}_{l,w} = 320\,mm$, $T = 50\,mm$.

Fibers. Cold drawn low-carbon steel; $L^{fib} = 12\,mm$, $d^{fib} = 0.175\,mm$.

Projectiles. Hardened SKH-51 tool steel (HRC 60) cylindrical projectiles (Fig. 3.1.1); $D = 25\,mm$, $L = 75\,mm$, $m = 297\,g$.

Table 1. Shield damage types (SDTs).

Notation	Description	Notation	Description
(1)	Fully perforated	(2)	Projectile caught
(3)	Projectile rebounded	(4)	Data on DOP is unavailable (shield ruptured or perforated)
(5)	Slightly damaged	(6)	Heavily damaged

5.98.2 *Test results*

Table 2. Test results for RPC shields.

ρ_m (%)	f_c' (MPa)	σ_{tsp} (MPa)	f_r (MPa)	v_s (m/s)	P (mm)	SDT
–	25.0	2.6	10.6	27.0	0.6	(5), (3)
–	25.0	2.6	10.6	35.7	NA	(4), (6), (1)
–	25.0	2.6	10.6	56.8	NA	(4), (6), (1)
2	25.2	3.1	13.3	41.7	5.3	(5), (3)
2	25.2	3.1	13.3	56.8	NA	(4), (6), (1)
2	25.2	3.1	13.3	64.1	NA	(4), (6), (1)

Table 3. Test results for NSC shields.

ρ_m (%)	f'_c (MPa)	σ_{tsp} (MPa)	f_r (MPa)	v_s (m/s)	P (mm)	SDT
–	161.9	7.3	22.0	34.7	–	(4), (6), (2)
–	161.9	7.3	22.0	58.5	–	(4), (6), (1)
–	161.9	7.3	22.0	76.0	–	(4), (6), (1)
1	175.3	13.8	27.2	58.2	1.8	(5), (3)
1	175.3	13.8	27.2	76.0	11.9	(5), (2)
1	175.3	13.8	27.2	104.0	–	(4), (5), (2)
2	178.3	21.9	35.8	76.0	3.4	(5), (3)
2	178.3	21.9	35.8	85.0	4.7	(5), (3)
2	178.3	21.9	35.8	104.0	5.5	(5), (3)
5	192.8	31.6	72.8	58.5	1.4	(5), (3)
5	192.8	31.6	72.8	78.1	2.9	(5), (3)
5	192.8	31.6	72.8	104.1	4.7	(5), (3)

5.99 [Teland and Sjøl, 2000]

5.99.1 *General data on shields and projectiles*

Shields. PC and RC monolithic and layered (plates in-contact) shields.

Table 1. Undeformed during penetration projectiles.

Projectile	Shape	D (mm)	ρ (mm)	ψ	m (g)
Ogive-nose projectiles	Fig. 3.3.1	75	140	1.87	6.28
Truncated double cone-nosed projectiles	*	152	–	–	44.76

* Photograph of the projectile is presented in original study.

5.99.2 *Test results*

Notations.
- Asterisk in column D^{sh} indicates shield having square cross-section. Most likely that in this case D^{sh} is the length of a side of a square.
- In column Remarks F indicates fiber reinforcement with the percentage of fibers in parenthesis (if any) while R indicates non-fiber reinforcement.

Table 2. Test results for ogive-nose projectiles.

f_c' (MPa)	D^{sh} (m)	Shield structure	v_s (m/s)	P (mm)	v_r (m/s)	Remarks
38	1.6	10×[200]	484	680	0	–
200	1.5*	5×[200]	480	200	0	F(2.0%)
38	1.6	1×[2000]	483	655	0	–
38	1.6	1×[2000]	482	660	0	–
180	1.6	1×[1000]	485	250	0	F(2.0%)
180	1.6	1×[1000]	489	235	0	F(2.0%)
180	1.6	1×[500]	485	240	0	F(2.0%)
90	1.6	–	571	410	0	F
90	1.6	1×[1500]	653	560	0	F(1.0%)
140	1.6	1×[1200]	647	440	0	F(1.0%)
200	1.5*	4×[200]	650	440	0	F(2.0%)
30	1.6	1×[2000]	647	990	0	–
153	1.4	1×[400]	617	–	276	–
153	1.4	1×[400]	618	–	303	–
153	1.4	1×[400]	618	–	293	–
153	1.4	1×[800]	620	450	0	–
153	1.4	1×[800]	612	540	0	–
153	1.4	1×[800]	619	510	0	–
153	1.4	1×[400]	619	–	260	R

Table 3. Test results for truncated double cone-nose projectiles.

f_c' (MPa)	D^{sh} (m)	Shield structure	v_s (m/s)	P (mm)	v_r (m/s)	Remarks
250	1.5	4×[200]	476	–	~0	F
36	1.5	5×[200]	478	–	160	–
220	1.5	5×[200]	478	350	0	F(4.0%), R
180	1.5	5×[200]	473	600	0	F, R
41	1.6	1×[3000]	468	1030	0	–
250	2.2	5×[200]	481	460	0	F
200	1.5*	1×[1000]	480	300	0	F(1.50%), R
34	1.5	2×[1000]	481	670	0	R
203	1.6	1×[1200]	480	450	0	–
200	1.6	1×[1000]	480	450	0	F
90	1.6	1×[1500]	479	630	0	–
108	1.6	1×[1500]	486	590	0	F(1.0%)
83	1.6	1×[1500]	488	560	0	F(1.5%)
36	1.6	1×[3000]	576	1700	0	–
30	1.6	1×[3000]	481	1020	0	–
140	1.6	1×[1200]	480	595	0	F(1.0%)

f_c' (MPa)	D^{sh} (m)	Shield structure	v_s (m/s)	P (mm)	v_r (m/s)	Remarks
140	1.5*	1×[1200]	480	420	0	F(1.0%), R
140	1.6	1×[1500]	584	810	0	F(1.0%)
90	1.6	1×[1500]	583	980	0	–
200	1.6	1×[1500]	580	530	0	F(2.0%)
132	1.6	1×[1500]	581	830	0	–
149	3.0	1×[1200]	465	440	0	–
39	2.4	1×[2000]	468	1010	0	–
149	2.4	1×[1200]	468	380	0	–
153	1.6	1×[1200]	467	660	0	–
153	1.0	1×[1200]	466	750	0	–
39	1.6	1×[2000]	466	1090	0	–
120.5	1.6	1×[1200]	464	570	0	F(1.0%)
120.5	1.6	1×[1200]	467	500	0	F(1.0%), R
158	1.6	1×[1500]	573	440	0	Rail

5.100 [Teland and Sjøl, 2004]

Note. Data from [Sjøl *et al.*, 1998], [Sjøl *et al.*, 2000].

5.100.1 *General data on shields and projectiles*

Shields. Single PC slabs, $f_c' = 35\,MPa$.

Table 1. Cylindrical projectiles (Fig. 3.1.1), $D = 12\,mm$.

Notation	L (mm)	Material	m (g)
Projectile-1	24	Steel	20.5
Projectile-2	42	Tungsten	65.8
Projectile-3	72	Tungsten	122.8

5.100.2 *Test results*

Table 2#. Test results for different projectiles.

Projectile	v_s (m/s)	P (mm)	v_s (m/s)	P (mm)	v_s (m/s)	P (mm)	v_s (m/s)	P (mm)
Projectile-1	368	27	414	25	444	30	497	43
	567	37	572	54	654	48	748	65
	753	53						

Projectile	v_s (m/s)	P (mm)	v_s (m/s)	P (mm)	v_s (m/s)	P (mm)	v_s (m/s)	P (mm)
Projectile-2	312	71	356	75	362	91	434	123
	505	126	614	171	679	206		
Projectile-3	219	60	288	93	329	125	361	147
	405	147						

5.101 [Teland, 2001]

Note. Author indicated that data were provided by F.J. Mostert (private communication).

5.101.1 *General data on shields and projectiles*

Shields. Predrilled RC shields (shields with an inline cylindrical hole); $D^{sh} = T = 300\,mm$, $\gamma = 2000\,kg/m^3$, $f'_c = 20\,MPa$.

Projectiles. Ogive-nose projectiles (Fig. 3.3.1); $D = 20\,mm$, $\psi = 2.11$ $m = 141.6\,g$.

Initial projectile velocity. $v_s \approx 350\,m/s$.

5.101.2 *Test results*

Table 1#. Test results (through-hole in the shield).

ξ^{hole}	P (mm)	ξ^{hole}	P (mm)	ξ^{hole}	P (mm)	ξ^{hole}	P (mm)
0.00	98	0.25	128	0.45	234	0.65	299

Table 2#. Test results (depth of the hole is 150 mm).

ξ^{hole}	P (mm)	ξ^{hole}	P (mm)	ξ^{hole}	P (mm)
0.00	146	0.25	138	0.35	213
0.65	269	0.75	241		

5.102 [Tsubota *et al.*, 1993]

5.102.1 *General data on shields and projectiles*

Shields. Single RC barriers with rebars ($c_{l,w} = 0.6\,m$; $f'_c = 24.5\,MPa$) of different thickness with steel liners and without steel liners.

Rebars. $d_{l,w}^{F,R} = 10\,mm$, $g_{l,w}^{F,R} = 100\,mm$.

Impact velocity. $v_s = 170\,m/s$.

Projectiles. "It consisted of three parts: a head, a body and a tail. The head was solid mild steel with a diameter of 35 mm and it had a flat nose shape. The tail was made of plastic and was used to locate the projectile within a launching tube with press ring. Mild steel pipe with an outer diameter of 39.5 mm and a wall thickness of 0.42 mm was employed for the body. Since the head of the projectile was made of mild steel and the axial compressive strength of the body was very high, the projectile was regarded as a rigid missile. The total mass of the projectile was 0.43 kg and its total length was 125 mm."

Table 1. SDTs for shields without liners.

Notation	Description
(1)	Projectile penetrated through the shield
(2)	Considerable concrete debris spalled off the rear face of the shield
(3)	Crater was formed at the frontal face of the shield but no scabbing occurred on the rear face

Table 2. SDTs for shields with liners.

Notation	Description
(4)	Projectile penetrated through the shield as well as through the steel liner
(5)	Steel liner was splitt; projectile did not perforate the shield
(6)	Residual deformation of the liner; steel liner bulged
(7)	Crater was formed at the frontal face of the shield but no damage or deformation to the steel liner

5.102.2 *Test results*

Note. P is the DOP into concrete part of slab.

Table 3. Test results for shields without liners.

T (mm)	P (mm)	SDT	T (mm)	P (mm)	SDT	T (mm)	P (mm)	SDT
50	–	(1)	60	–	(1)	70	–	(1)
80	–	(1)	85	–	(1)	90	18	(2)
100	27	(2)	110	17	(2)	120	33	(3)
140	29	(3)	160	29	(3)			

Table 4. Test results for shields with liner of $T_{lin}^R = 0.8\,mm$.

T (mm)	P (mm)	h^{def} (mm)	SDT	T (mm)	P (mm)	h^{def} (mm)	SDT
50	–	–	(4)	60	60	43	(5)
70	70	26	(6)	80	46	18	(6)
90	23	902	(6)	100	25	3.5	(6)
105	17	~0	(7)	120	27	0	(7)
140	34	0	(7)				

Table 5. Test results for shields with liner of $T_{lin}^R = 1.2\,mm$.

T (mm)	P (mm)	h^{def} (mm)	SDT	T (mm)	P (mm)	h^{def} (mm)	SDT
50	–	–	(4)	60	60	46	(5)
65	65	29	(6)	70	70	22	(6)
80	42	10	(6)	90	22	6	(6)
95	20	~0	(7)	100	33	0	(7)
120	20	0	(7)				

Table 6. Test results for shields with liner of $T_{lin}^R = 1.6\,mm$.

T (mm)	P (mm)	h^{def} (mm)	SDT	T (mm)	P (mm)	h^{def} (mm)	SDT
50	50	–	(5)	55	55	36	(6)
60	60	28	(6)	70	49	20	(6)
80	34	13	(6)	90	27	7	(6)
95	16	0	(7)	100	19	0	(7)
120	30	0	(7)				

Table 7. Test results for other shields with liners.

T (mm)	T_{lin}^F	T_{lin}^R	P (mm)	h^{def} (mm)	SDT
50	–	2.0	50	34	(6)
90	–	2.0	20	0	(7)
50	–	0.8	–	–	(4)
70	–	0.8	35	18	(6)
80	–	0.8	42	11	(6)
60	–	1.2	60	24	(6)
70	–	1.2	56	14.3	(6)
70	–	1.6	63	18	(6)
90	–	1.6	22	5.2	(6)
100	–	1.6	24	3	(6)
70	0.8	–	–	–	(4)
70	0.8	0.8	20	22	(6)

5.103　[Unosson and Nilsson, 2006]

See also [Unosson, 2000].

5.103.1　*General data on shields and projectiles*

Shield 1. PC slabs, $D^{sh} = 1400\,mm$, $T = 800\,mm$.
Shield 2. PC slabs, $D^{sh} = 1400\,mm$, $T = 400\,mm$.
Shield 3. RC with rebars and 3 identical meshes; $D^{sh} = 1400\,mm$, $T = 400\,mm$; $t^{F,R} = 67.5\,mm$; $t^{I} = 200\,mm$ (for intermediate mesh), and $d_{l,w} = 25\,mm$, $g_{l,w} = 120\,mm$ (for all 3 meshes).
Concrete. Cube-shaped specimens having the length of the edge 150 mm were tested to measure $f_c' = 153\,MPa$; $E = 58\,GPa$ and $\sigma_u = 8\,MPa$ were determined by testing $\varnothing 100 \times 200$ mm cylinders.
Projectiles. Steel ogive-nose projectile (Fig. 3.3.1); $D = 75\,mm$, $L = 225\,mm$, $\rho = 127\,mm$, $m = 6.3\,g$.

5.103.2　*Test results*

Table 1. Test results.

Shield	v_s (m/s)	P (m)	v_r (m/s)	Shield	v_s (m/s)	P (m)	v_r (m/s)
Shield 1	623	0.45	–	Shield 1	618	0.54	–
Shield 1	622	0.51	–	Shield 2	613	–	276
Shield 2	616	–	303	Shield 2	621	–	293
Shield 3	624	–	260				

5.104　[Vossoughi *et al.*, 2007]

5.104.1　*General data on shields and projectiles*

Shields. PC and FRC specimens having fabric on rear face or on both faces; $c_{l,w} = 305\,mm$; $q_{\max}^{agg} = 6.35\,mm$.
Fabrics. Polypropylene (P) and Zylon (Z).
Note. "For the samples with fabric, the fabric was placed at the bottom of the mold and concrete was placed on top of the fabric. Another sheet

of fabric was then positioned on top of the concrete. No glue was used in the process. The fabric placed on the top did not bond as well with the concrete compared to the fabric at the bottom of the mold."

Projectiles. Cone-nose projectiles (Fig. 3.2.1); $D = 12.7\,mm$, $L = 25.4\,mm$, $\beta = 22.5°$, $m = 34\,g$.

Table 1. Shield damage types (SDTs).

Notation	Description	Notation	Description
(1)	Complete penetration of the projectile.	(2)	Projectile stuck
(3)	Shield was perforated and then projectile bounced		

5.104.2 *Test results*

Table 2. Test results for shields with $T = 38.1\,mm$.

f'_c (MPa)	Fabrics on both sides	v_s (m/s)	v_r (m/s)	SDT
41	–	167	–	(1)
41	–	192	22	(1)
40	–	200	0	(2)
40	–	189	0	(2)
40	–	188	9	(1)
43	Z	187	0	(3)
43	Z	184	0	(3)
43	Z	196	0	(2)
30	–	187	13	(1)
30	–	186	24	(1)
30	–	171	9	(1)

Table 3. Test results for shields with $T = 23 - 28\,mm$.

T (mm)	f'_c (MPa)	Fabrics at frontal face	Fabrics at rear face	v_s (m/s)	v_r (m/s)	SDT
25.4	30	–	–	99	21	(1)
25.4	30	–	–	92	11	(1)
26.7	30	–	–	82	0	(3)
25.4	35	–	–	84	0	(3)
25.4	35	–	–	72	0	(3)
27.9	35	–	–	90	0	(3)
22.9	30	–	P	142	72	(1)
21.6	30	–	P	103	34	(1)
24.1	30	–	P	89	14	(1)

T (mm)	f_c' (MPa)	Fabrics at frontal face	Fabrics at rear face	v_s (m/s)	v_r (m/s)	SDT
25.4	37	P	P	106	0	(3)
21.6	37	P	P	95	12	(1)
25.4	37	P	P	87	0	(2)
22.9	37	–	Z	106	0	(3)
22.9	37	–	Z	113	0	(3)
22.9	37	–	Z	123	0	(3)
24.1	39	Z	Z	131	0	(2)
25.4	39	Z	Z	143	0	(2)
24.1	39	Z	Z	142	17	(1)

5.105 [Wang, 2011]

5.105.1 *General data on shields and projectiles*

Shields. Plain high strength concrete (HSC), steel (ST) fiber reinforced high strength concrete (SFHSC), polyethylene (PE) fibers high strength concrete (PEHSC), hybrid fiber reinforced high strength concrete (HYHSC) slabs ($c_l = 300\,mm$, $c_w = 170\,mm$, $T = 150\,mm$); granite aggregate, $q_{max}^{agg} = 10\,mm$.

Table 1. Characteristics of concrete mixtures (Series 1 of experiments).

Shield material	γ (kg/m^3)	f_c' (MPa)	f_r (MPa)	Fiber	μ (kg/m^3)	ρ_V (%)
HSC	2397	83	6	–	–	–
SFHSC	2393	85	7	ST	39	0.5

Note. Cube-shaped specimens having the length of the edge 100 mm were tested to measure f_c'.

Table 2. Characteristics of concrete mixtures (Series 2 of experiments).

Shield material	γ (kg/m^3)	f_c' (MPa)	f_r (MPa)	Fiber	μ (kg/m^3)	ρ_V (%)	γ (kg/m^3)
HSC	2398	85.8/86.9	5.8	36.2/38.1	–	–	–
SFHSC	2393	85.4/90.7	7.2	38.3/38.1	ST	39	0.5
PEHSC	2381	84.9/84.5	8.3	35.0/37.0	PE	4.85	0.5
HYHSC	2384	85.2/81.8	8.1	34.6/36.3	ST+PE	39+4.85	1.0

Note. Values of f_r were determined by testing 400×100×100 mm beams. Values of f_c' and E were determined by testing 400×100×100 mm beams (nominator) and Ø77×154 mm cylinders (denominator).

Table 3. Material and parameters of fibers.

Material	L^{fib} (mm)	d^{fib} (mm)	σ_u (MPa)	E (GPa)
ST	13	0.16	2500	200
PE	12	0.039	2610	66

Projectiles. Ogive-nose projectile (Fig. 3.3.1); $D = 13.35\,mm$, $L_{cyl} = 5.33\,mm$, $L_{nose} = 20.03\,mm$.

5.105.2   *Test results*

Table 4. Test results (Series 1).

Shield material	m (g)	v_s (m/s)	P (mm)	Shield material	m (g)	v_s (m/s)	P (mm)
HSC	18.8	641	43.8	HSC	18.5	649	36.4
HSC	18.5	641	40.2	SFHSC	18.8	650	41.8
SFHSC	18.8	641	36.6	SFHSC	18.8	633	36.8

Table 5. Test results (Series 2).

Shield material	m (g)	v_s (m/s)	P (mm)	Shield material	m (g)	v_s (m/s)	P (mm)
HSC	18.8	642	43.8	HSC	18.5	649	36.4
HSC	18.5	641	40.2	SFHSC	18.8	650	41.8
SFHSC	18.8	641	36.6	SFHSC	18.8	633	36.8
PEHSC	18.5	65	40.3	PEHSC	18.9	641	38.9
PEHSC	18.9	650	41.2	HYHSC	18.5	650	36.8
HYHSC	18.8	650	44.0	HYHSC	18.9	641	38.9

5.106   [Weerheijm *et al.*, 2015]

5.106.1   *General data on shields and projectiles*

Shields. Various monolithic shields made of porous concrete, $c_{l,w} = 500\,mm$, $T = 100\,mm$.

Table 1. Properties of shield materials.

Notation	Porosity (volume %)	Compressive strength (MPa)	
		Static	Dynamic*
Shield 1	20.3	41.9	76.8
Shield 2	18.8	50.5	86.0
Shield 3	20.1	44.8	79.7
Shield 4	18.6	48.8	84.4
Shield 5	24.8	15.9	26.3

* In drop weight tests at velocity of 4.5 m/s.

Projectiles. AP8 FFV projectiles.

5.106.2 *Test results*

Table 2. Test results (perforation).

Shield	v_s (m/s)	v_r (m/s)	Shield	v_s (m/s)	v_r (m/s)	Shield	v_s (m/s)	v_r (m/s)
Shield 3	889	480	Shield 4	712	128	Shield 4	706	148
Shield 4	717	183	Shield 4	882	520	Shield 4	707	170
Shield 5	707	169						

Table 3. Test results (penetration).

Shield	v_s (m/s)	P (mm)	Shield	v_s (m/s)	P (mm)	Shield	v_s (m/s)	P (mm)
Shield 1	708	78	Shield 1	710	79	Shield 3	804	64
Shield 3	661	60	Shield 3	665	62			

5.107 [Wu *et al.*, 2012]

5.107.1 *General data on shields and projectiles*

Shields. Single PC C35 concrete slabs with quartz aggregates, $f_c' = 27.3\,MPa$.

Projectiles. 30CrMnSi steel ($\sigma_y = 1130\,MPa$) ogive-nose projectile (Fig. 3.3.1), $D = 157\,mm$, $L_{cyl} = 82\,mm$, $L = 105\,mm$, $\psi = 3$.

5.107.2 *Test results*

Table 1. Test results on DOP.

m (g)	v_s (m/s)	P (mm)	m_r (g)	m (g)	v_s (m/s)	P (mm)	m_r (g)
127.34	664	495	124.24	127.61	822	730	123.27
127.69	1010	827	122.08	127.50	1142	790	121.26
127.62	1269	897	12.75	127.50	1354	–	98.46

5.108   [Wu *et al.*, 2015a]

5.108.1   *General data on shields and projectiles*

Shields. Single UHPCC slabs ($D^{sh} = 750\,mm$) with additions of steel fibers (equivalent diameter of 0.175 mm, $L^{fib} = 13\,mm$ $\sigma_u = 1800\,MPa$) and basalt coarse aggregates ($q_{max}^{agg} = 10\,mm$, $f_c' = 120\,MPa$).

Projectiles. Ogive-nose   projectiles (Fig. 3.3.1) of the same shape ($D = 25.3\,mm$, $\psi = 3$, $L = 151.9\,mm$, $L_{cyl} = 109.9\,mm$) manufactured of two different materials: DT300 steel with $\sigma_y = 1500\,MPa$ (*Projectile-1*) and D6A steel with $\sigma_y = 1420\,MPa$ (*Projectile-2*). Projectiles of types 1 and 2 had slightly different shape of cavities filled with polymer inert materials ($\gamma = 1500\,kg/m^3$).

5.108.2   *Test results*

Table 1. Test results for Projectile-1 and $v_s = 510\,m/s$.

T (m)	f_c' (MPa)	Fibers ρ_V (%)	Basalt μ (kg/m^3)	m (g)	P (mm)	P^{cr} (mm)
0.65	67.5	1.5	978	333.1	148	50
0.65	67.5	1.5	978	344.8	168	51
0.55	87.3	–	1042	336.7	115	50
0.55	87.3	–	1042	341.9	132	76
0.55	99.3	1.5	870	345.6	133	65
0.55	99.3	1.5	870	341.7	116	71
0.50	125.2	2.0	800	341.0	124	42
0.50	125.2	2.0	800	341.0	120	56
0.50	114.0	3.0	–	341.1	134	52
0.50	114.0	3.0	–	341.6	133	37

Table 2. Test results for Projectile-1 and $v_s = 850\,m/s$.

T (m)	f_c' (MPa)	Fibers ρ_V (%)	Basalt μ (kg/m^3)	m (g)	P (mm)	P^{cr} (mm)
1.00	67.5	1.5	978	350.1	345	62
1.00	67.5	1.5	978	341.3	346	69
0.65	87.3	–	1042	340.6	260	110
0.65	87.3	–	1042	341.1	278	91
0.65	99.3	1.5	870	342.3	255	58
0.65	99.3	1.5	870	346.4	260	56

T (m)	f_c' (MPa)	Fibers ρ_V (%)	Basalt $\mu\ (kg/m^3)$	m (g)	P (mm)	P^{cr} (mm)
0.60	125.2	2.0	800	342.7	252	59
0.60	125.2	2.0	800	341.5	261	60
0.60	114.0	3.0	0	345.6	288	48
0.60	114.0	3.0	0	344.1	268	43

Table 3. Test results for Projectile-2 and $v_s = 1150\,m/s$.

T (m)	f_c' (MPa)	Fibers ρ_V (%)	Basalt $\mu\ (kg/m^3)$	m (g)	P (mm)	P^{cr} (mm)
1.50	35.0	–	750	392.3	–	120
1.00	88.0	2.0	870	395	564	110
1.00	128.0	3.0	800	394.5	402	115
1.00	142.0	4.0	–	393.8	420	70

Table 4. Test results for Projectile-2 and $v_s = 1250\,m/s$.

T (m)	f_c' (MPa)	Fibers ρ_V (%)	Basalt $\mu\ (kg/m^3)$	m (g)	P (mm)	P^{cr} (mm)
1.50	35.0	–	750	392.3	673	150
1.00	88.0	2.0	870	393.3	540	–
1.00	128.0	3.0	800	394.3	503	119
1.00	142.0	4.0	–	394.3	473	75

Table 5. Test results for Projectile-2 and $v_s = 1320\,m/s$.

T (m)	f_c' (MPa)	Fibers ρ_V (%)	Basalt $\mu\ (kg/m^3)$	m (g)	P (mm)	P^{cr} (mm)
1.50	35.0	–	750	392.3	748	95
1.00	88.0	2.0	870	395	681	102
1.00	128.0	3.0	800	393.6	130	130
1.00	142.0	4.0	–	394.1	490	88

5.109 [Wu *et al.*, 2015b]

5.109.1 *General data on shields and projectiles*

Shields. Layered shields consisting of several RC spaced and in-contact slabs with rebars (air gaps of 200-300 mm) and with 1 mm thickness liner welded onto the rear face of the last slab. Each slab contained *several meshes* with the same reinforcement at both sides; $c_{l,w} = 675\,mm$, $f_c' = 41\,MPa$, $q_{max}^{agg} = 10\,mm$.

Meshes. $t^{F,R} = 25\,mm$ and $d_{l,w} = 6\,mm$, $\rho_{l,w} = 0.5\%$ (for all meshes).

Table 1. Parameters of slabs.

T (mm)	N_{mesh}	$g_{l,w}$ (mm)	l_{mesh} (mm)	T (mm)	N_{mesh}	$g_{l,w}$ (mm)	l_{mesh} (mm)
100	2	98.5	64	150	3	98.5	57
200	3	75	82	300	4	67.5	88

N_{mesh} – number of meshes; l_{mesh} – distance between adjacent meshes.

Projectiles. Heat treated 45CrNiMoV steel (HRC 45, $\sigma_y = 1420\,MPa$) ogive-nose projectile (Fig. 3.3.1); $D = 25.3\,mm$, $L_{cyl} = 152\,mm$, $L_{nose} = 42\,mm$, $\psi = 3$.

5.109.2 *Test results*

Notation. Asterisk in P^{cr} column indicates layers in-contact, otherwise test is conducted with spaced shield.

Table 2. Test results for one-layer ($T = 300$ *mm*) shields.

m (g)	m_r (g)	L_r (mm)	v_s (m/s)	v_r (m/s)	P^{cr} (mm)
429.0	419.8	147	641.5	272	55
428.3	420.9	147	540	0	55
427.1	419.4	147	601	180	60
428.4	417.7	147	679	329	60
426.9	418.1	148	737	422	57

Table 3. Test results for two-layer ($T_1 = 100$ *mm*, $T_2 = 200$ *mm*) shields.

m (g)	m_r (g)	L_r (mm)	v_s (m/s)	1st layer v_r (m/s)	1st layer P^{cr} (mm)	2nd layer v_r (m/s)	2nd layer P^{cr} (mm)
428.5	423.0	148	539	436	35	137	50
428.1	420.8	148	601	482	45	207	45
428.4	420.8	148	634	544	50	304	50
429.4	420.7	147	643	–	65	269	*
428.1	420.7	147	729	651	45	451	55

Table 4. Test results for two-layer ($T_1 = 200\,mm$, $T_2 = 100\,mm$) shields.

m (g)	m_r (g)	L_r (mm)	v_s (m/s)	1st layer		2nd layer	
				v_r (m/s)	P^{cr} (mm)	v_r (m/s)	P^{cr} (mm)
427.7	421.9	147	540	292	50	159	45
429.9	423.1	148	597	404	55	270	50
427.6	421.5	147	641	438	60	295	50
428.7	420.9	147	638	–	50	286	*
427.0	418.5	147	731	527	60	484	50

Table 5. Test results for two-layer ($T_1 = T_2 = 150\,mm$) shields.

m (g)	m_r (g)	L_r (mm)	v_s (m/s)	1st layer		2nd layer	
				v_r (m/s)	P^{cr} (mm)	v_r (m/s)	P^{cr} (mm)
429.7	423.4	148.0	536	379	50	161	45
430.7	424.1	148.0	603	448	50	246	55
426.9	423.9	148.0	643	511	50	333	50
428.8	424.0	148.2	644	–	50	317	*
428.9	424.0	148.3	721	620	50	444	50

Table 6. Test results for three-layer ($T_1 = T_2 = T_3 = 100\,mm$) shields.

m (g)	m_r (g)	L_r (mm)	v_s (m/s)	1st layer		2nd layer		3rd layer	
				v_r (m/s)	P^{cr} (mm)	v_r (m/s)	P^{cr} (mm)	v_r (m/s)	P^{cr} (mm)
428.9	423.6	148.0	536	423	50	307	50	182	50
428.8	422.9	147.5	607	520	50	465	50	309	50
427.9	420.9	147.0	640	549	50	486	50	354	50
427.3	421.2	147.4	639	–	40	–	*	311	*
428.3	421.3	147.0	729	665	50	567	50	482	50

5.110 [Wu *et al.*, 2015c]

5.110.1 *General data on shields and projectiles*

Shields. Monolithic ultra-high performance steel fiber reinforced concrete shields with the corundum coarse aggregate (UHP-CASFRC) and high performance concrete (HSC) shields.

Projectiles. Ogive-nose projectiles (Fig. 3.3.1) of the same shape ($D = 25.3\,mm$, $\psi = 3$, $L \approx 151.9\,mm$, $L_{cyl} = 109.9\,mm$) made of DT300 steel with $\sigma_y = 1500\,MPa$; $m \approx 341\,g$. Projectiles had cavities filled with polymer inert materials ($\gamma = 1500\,kg/m^3$).

5.110.2   *Test results for UHP-CASFRC shields*

Table 1. Perforation data (for fibers, $\rho_V = 2\%$).

Shot no.	v_s (m/s)	f_c' (MPa)	T (cm)	q^{agg} (mm)	Corundum μ (kg/m^3)	P (mm)	P^{cr} (mm)
1	508	110.7	50	5-20	1126	98	54
2	508	110.7	50	5-20	1126	101	55
3	503	125.6	50	35-45	1126	21	21
4	515	125.6	50	35-45	1126	68	68
5	510	129.2	50	65-75	1126	59	59
6	517	129.2	50	65-75	1126	37	37
7	505	128.8	50	5-20	1700	145	54
8	507	128.8	50	5-20	1700	106	75
9	524	102.5	50	5-20	1126	106	56
10	511	102.5	50	5-20	1126	88	51
11	708	110.7	60	5-20	1126	171	84
12	722	110.7	60	5-20	1126	176	86
13	832	110.7	60	5-20	1126	185	82
14	846	110.7	60	5-20	1126	191	84

Table 2. Change of mass and length of penetrators.

Shot no.	m (g)	L (mm)	m_r (g)	L_r (mm)	Shot no.	m (g)	L (mm)	m_r (g)	L_r (mm)
1	341.2	150.7	329.2	140.5	2	341.2	150.7	329.5	141.6
3	341.2	150.9	Broken	Broken	4	341.2	151.9	Broken	Broken
5	341.3	150.8	Broken	Broken	6	341.2	151.1	Broken	Broken
7	341.2	150.9	332.0	143.6	8	341.2	151.9	328.4	139.2
9	341.2	151.2	327.7	143.1	10	340.7	150.8	329.3	142
11	341.3	151.1	318.6	140.2	12	341.2	150.4	323.5	139.7
13	341.2	151.0	312.7	139.2	14	341.2	150.9	315.2	136.4

5.110.3 *Test results for HSC shields*

Table 3. Perforation data.

Shot no.	v_s (m/s)	f_c' (MPa)	T (cm)	q^{agg} (mm)	Corundum μ (kg/m^3)	P (mm)	P^{cr} (mm)
15	510	61.8	60	5-20	1090	151	93
16	516	61.8	60	5-20	1090	143	85

Table 4. Change of mass and length of penetrators.

Shot no.	m (g)	L (mm)	m_r (g)	L_r (mm)	Shot no.	m (g)	L (mm)	m_r (g)	L_r (mm)
15	341.2	151.2	336.1	149.2	16	341.2	150.8	336.3	150.3

5.111   [Wu *et al.*, 2015d]

5.111.1 *General data on shields and projectiles*

Shields. PC and steel fiber reinforced high strength concrete (SFRHSC) shields, $D^{sh} = 1m$. Reinforcing cage includes two layers of two-way steel mesh ($d_{l,w}^{F,R} = 6mm$, $g_{l,w}^{F,R} = 100mm$) as well as stirrups (5 mm in diameter) welded at the cross nodes of steel mesh; $t^{F,R} = 25mm$. Steel liner (if any) was welded by stirrups to the rear face of the SFRHSC slab, $T_{lin}^R = 3mm$. Values of f_c', σ_{tsp} and bending strength σ_{bend} were tested using cylindrical specimens, cube-shaped specimens having the length of the edge 150 mm and $100 \times 100 \times 400$ mm beams, correspondingly. Some shields had the rear sandy soil layer.

Table 1. Characteristics of shields.

Shot no.	T (mm)	f_c' (MPa)	σ_{tsp} (MPa)	σ_{bend} (MPa)	ρ_V* (%)	Liner	Soil	Rebars
1	200	83.7	12	11	3	+	+	+
2	200	89	11	–	3	+	–	+
3	200	85.5	14	15	3	+	+	+
4	200	100.8	10	–	3	+	–	+
5	200	98	12	14	3	+	–	+
6	200	88	13	17	3	–	+	+
7	200	92.7	12	15	3	–	+	+
8	200	81	8	11	3	–	+	+

Shot no.	T (mm)	f_c' (MPa)	σ_{tsp} (MPa)	σ_{bend} (MPa)	ρ_V * (%)	Liner	Soil	Rebars
9	250	56.7	8	–	3	+	+	+
10	250	53	9	–	3	+	+	+
11	250	54	9	11	3	+	+	–
12	250	54.3	8	–	3	+	+	–
13	250	49.5	8	–	3	+	+	–
14	250	59.4	8	12	3	+	+	+
15	250	57.6	9	–	3	+	+	+
16	350	46	4	–	0	+	+	+
17	350	47	4	7	0	+	+	+
18	350	47	5	8	0	+	+	+

* ρ_V is steel fiber volumetric ratio.

Projectiles. 35GrMnSi steel ($\sigma_y = 1620\,MPa$, HB 241) projectiles (Fig. 3.6.4b), $m = 693\,g$.

5.111.2   *Test results*

Table 2. Test results.

Shot no.	v_s (m/s)	P (mm)	Shot no.	v_s (m/s)	P (mm)	Shot no.	v_s (m/s)	P (mm)
1	568	**	2	488	192	3	514	130
4	576	*	5	502	170	6	524	155
7	528	168	8	558	166	9	678	*
10	680	*	11	661	*	12	580	210
13	593	220	14	669	*	15	595	215
16	583	200	17	705	*	18	679	265

* shield was totally perforated by projectile.

** rear steel liner was pierced out and projectile stuck.

5.112   **[Zhang *et al.*, 2005]**

5.112.1   *General data on shields and projectiles*

Shields. Single PC and FRC barriers with different fiber reinforcement; $c_l = 300\,mm$, $c_w = 170\,mm$, $T = 150\,mm$. Values of f_c' for mixes with granite and with quartz aggregate were measured using cube-shaped specimens having the length of the edges 100 mm and 50 mm, correspondingly.

Fibers. Straight steel fibers, $L^{fib} = 13\,mm$ and $d^{fib} = 0.2\,mm$.

Table 1. Characteristics of concrete mixtures.

Label	f_c' (MPa)	f_r (MPa)	μ (kg/m^3)	ρ_V (%)	Aggregate material	q_{max}^{agg} (mm)
Mix-1	45.5	–	–	–	Granite	20
Mix-2	58.3	5.1	–	–	Granite	20
Mix-3	87.8	–	–	–	Granite	10
Mix-4	93.5	–	118	1.5	Granite	10
Mix-5	112.5	–	–	–	Granite	10
Mix-6	115.0	–	118	1.5	Granite	10
Mix-7	150.9	13.0	–	–	Quartz	1.18
Mix-8	187.2	31.5	119	1.5	Quartz	1.18
Mix-9	183.6	32.6	119	1.5	Quartz	1.18
Mix-10	203.5	32.8	119	1.5	Quartz	1.18
Mix-11	237.0	33.0	119	1.5	Quartz	1.18

Projectiles. Hardened steel (HRC 50) ogive-nose projectiles (Fig. 3.3.1); $D = 12.60\,mm$, $L_{cyl} = 5.03\,mm$, $L_{nose} = 18.9\,mm$, $\psi = 2.5$, $m = 15\,g$.

5.112.2 *Test results*

Table 2. Test results for shields made of different concrete mixtures.

Label	v_s (m/s)	P (mm)	v_s (m/s)	P (mm)	v_s (m/s)	P (mm)	v_s (m/s)	P (mm)
Mix-1	668.5	48.0	675.6	48.5	667.7	46.5	–	–
Mix-2	694.4	46.0	684.9	45.0	657.9	39.0	657.9	38.0
Mix-3	675.5	41.0	670.7	38.5	679.3	57.4	–	–
Mix-4	665.0	38.0	640.5	36.5	–	–	–	–
Mix-5	677.5	31.0	670.6	28.5	678.2	30.5	–	–
Mix-6	678.0	33.5	650.0	28.0	–	–	–	–
Mix-7	646.6	31.0	634.2	36.5	675.7	31.5	684.9	33.5
Mix-8	644.3	39.0	694.4	30.5	704.2	39.5	–	–
Mix-9	637.6	35.5	621.3	35.5	694.4	30.0	–	–
Mix-10	653.0	35.0	644.5	31.0	625.0	51.2	–	–
Mix-11	620.0	28.5	647.5	30.5	636.0	28.5	–	–

5.113 [Zhang *et al.*, 2007]

See also [Sharif, 2005], [Astarlioglu *et al.*, 2013].

5.113.1 *General data on shields and projectiles*

Shields. Single PC and FRC barriers with different fiber reinforcement; $c_i = 300\,mm$, $c_w = 170\,mm$, $T = 150\,mm$. Cube-shaped specimens with the length of the edge 100 mm were tested to measure f'_c; $100 \times 100 \times 400$ mm beam specimens were tested to determine E and f_r.

Table 1. Parameters of fibers ($L^{fib} = 13mm$).

Material	E (GPa)	γ (kg/m^3)	σ_u (MPa)	d^{fib} (mm)
Steel (ST)	200	7800	2600	0.200
Polypropylene (PP)	4	910	30	0.015
Polyethylene (PE)	66	970	2400	0.039

Table 2. Properties of concrete mixtures.

Label	Type	Fiber type	ρ_V (%)	q^{agg}_{max} (mm)
Mix-1	PC	–	0	20
Mix-2	PC	–	0	20
Mix-3	PC	–	0	20
Mix-4	PC	–	0	10
Mix-5	FRC	ST	1.0	20
Mix-6	FRC	ST	1.0	20
Mix-7	FRC	PE	1.0	20
Mix-8	FRC	PP	1.0	20
Mix-9	FRC	ST	0.5	20
Mix-10	FRC	ST	1.0	20
Mix-11	FRC	ST	1.5	20
Mix-12	FRC	ST+PP	0.75+0.25	20
Mix-13	FRC	ST	1.0	10

Projectiles. Ogive-nose projectiles (Fig. 3.3.1); $D = 12.60\,mm$, $L_{cyl} = 5.03\,mm$, $L_{nose} = 18.9\,mm$, $m = 15\,g$.

5.113.2 *Test results*

Table 3. Test results for shields made of different concrete mixtures.

Label	γ (kg/m^3)	f_c' (MPa)	f_r (MPa)	E (GPa)	v_s (m/s)	P (mm)
Mix-1	2372	46.3	5.6	41.5	684	45.0
Mix-1	2372	46.3	5.6	41.5	667	43.0
Mix-1	2372	46.3	5.6	41.5	656	41.0
Mix-2	2377	63.3	6.3	42.6	634	36.0
Mix-2	2377	63.3	6.3	42.6	634	37.0
Mix-2	2377	63.3	6.3	42.6	627	35.0
Mix-3	2410	96.9	6.0	45.9	690	28.3
Mix-3	2410	96.9	6.0	45.9	676	25.0
Mix-3	2410	96.9	6.0	45.9	675	25.0
Mix-4	2422	111.6	8.0	48.3	675	27.0
Mix-4	2422	111.6	8.0	48.3	676	25.0
Mix-4	2422	111.6	8.0	48.3	670	24.0
Mix-5	2418	69.9	7.0	41.9	699	39.0
Mix-5	2418	69.9	7.0	41.9	680	39.0
Mix-5	2418	69.9	7.0	41.9	690	37.0
Mix-6	2426	75.9	8.5	44.3	680	35.0
Mix-6	2426	75.9	8.5	44.3	676	32.0
Mix-6	2426	75.9	8.5	44.3	680	29.0
Mix-7	2388	93.0	7.4	46.5	676	28.0
Mix-7	2388	93.0	7.4	46.5	675	27.0
Mix-7	2388	93.0	7.4	46.5	662	31.0
Mix-8	2341	85.3	6.3	42.0	610	26.0
Mix-8	2341	85.3	6.3	42.0	690	33.0
Mix-8	2341	85.3	6.3	42.0	699	31.0
Mix-9	2438	112.5	7.5	49.2	709	22.0
Mix-9	2438	112.5	7.5	49.2	709	26.2
Mix-9	2438	112.5	7.5	49.2	685	23.3
Mix-10	2467	126.7	9.8	49.2	676	24.0
Mix-10	2467	126.7	9.8	49.2	676	23.0
Mix-10	2467	126.7	9.8	49.2	690	26.9
Mix-11	2494	123.1	9.5	50.5	685	25.3
Mix-11	2494	123.1	9.5	50.5	662	21.2
Mix-11	2494	123.1	9.5	50.5	641	21.8
Mix-12	2432	105.9	9.2	48.0	615	22.0
Mix-12	2432	105.9	9.2	48.0	638	25.0
Mix-12	2432	105.9	9.2	48.0	638	24.0
Mix-13	2483	129.7	9.9	51.2	704	26.0
Mix-13	2483	129.7	9.9	51.2	699	25.0
Mix-13	2483	129.7	9.9	51.2	690	25.0

5.114 [Zhang *et al.*, 2012a]

5.114.1 *General data on shields and projectiles*

Shields. Single PC and steel fiber reinforced concrete (SFRC) slabs with different sizes of aggregate; $c_{l,w} = 360\,mm$, $T = 400\,mm$; 100 mm edge cube were tested to measure f_c'.

Fibers. For SFRC slabs, steel fibers with sizes $0.5 \times 2.0 \times 35.0$ mm; $\sigma_f = 600\,MPa$; $\rho_V = 1\%$.

Table 1. Values of f_c' (MPa) for different shields.

Shield label	Type of concrete	q^{agg} (mm)	f_c' (MPa)
PC_5-10	PC	5-10	49
PC_10-15	PC	10-15	60
PC_15-20	PC	15-20	54
SFRC_5-10	SFRC	5-10	68
SFRC_10-15	SFRC	10-15	72
SFRC_15-20	SFRC	15-20	65

Projectiles. 38CrSi hardened steel fragment (HRC 42); $D = 12.6\,mm$.

5.114.2 *Test results*

Table 2. Test results for PC shields.

Shield	m (g)	$\Delta m/m$ (%)	v_s (m/s)	P (mm)
PC_5-10	23.60	77.2	1370	76
PC_5-10	23.12	65.2	1320	86
PC_5-10	22.90	80.1	1321	64
PC_10-15	23.19	75.6	1400	69
PC_10-15	23.27	74.3	1403	68
PC_10-15	23.11	74.2	1404	61
PC_15-20	23.20	70.2	1384	65
PC_15-20	23.23	74.9	1410	70
PC_15-20	23.17	69.4	1352	68
PC_15-20	23.14	68.2	1250	59

Table 3. Test results for SFRC shields.

Shield	m (g)	$\Delta m/m$ (%)	v_s (m/s)	P (mm)
SFRC_5-10	23.07	70.8	1371	62
SFRC_5-10	22.90	70.3	1365	70

Shield	m (g)	$\Delta m/m$ (%)	v_s (m/s)	P (mm)
SFRC_5-10	23.00	69.2	1400	65
SFRC_10-15	23.30	71.2	1390	63
SFRC_10-15	22.96	67.1	1385	62
SFRC_10-15	23.05	76.6	1373	62
SFRC_15-20	23.13	74.8	1365	65
SFRC_15-20	23.05	68.0	1400	65
SFRC_15-20	23.00	17.5	1412	71

5.115 [Zhang *et al.*, 2013a]

See also [Zhang *et al.*, 2013b].

5.115.1 *General data on shields and projectiles*

Shields. Single PC barriers; $f'_c = 30 - 35\,MPa$, $\gamma = 2300\ kg/m^3$; $D^{sh} = 1m$, $T = 2m$ for *Projectile-1* and $c_{l,w} = 3.5m$, $T = 8m$ for *Projectile-2*.

Table 1. Hard steel 4340 ($\sigma_y = 220$ MPa) projectiles.

Notation	Projectile	Shape	D (mm)	L (mm)	ψ
Projectile-1	Blunted ogive–cylindrical projectile	Fig. 3.3.2	Sizes are given in Fig. 3.3.2		
Projectile-2	Ogive-nose projectile	Fig. 3.3.1	144	680	3

5.115.2 *Test results*

Table 2. Test results for Projectile-1 ($L = 195\,mm$).

m (kg)	v_s (m/s)	P (mm)	m_r (kg)	L_r (mm)
2.00	1119	267	1.86	192
1.99	1249	333	1.80	188
1.83	1327	370	1.69	187.5
1.83	1443	343	1.63	171

Table 3. Test results for Projectile-2 ($m = 47\,kg$).

v_s (m/s)	m_r (kg)	L_r (mm)	v_s (m/s)	m_r (kg)	L_r (mm)
1100	46.52	672	1300	46.14	675

Chapter 6

Penetration into geological media

6.1 Cumulative indexes of experiments

Classification of experiments according to types of projectiles is presented in в Tables 1–6.

Table 1. Experiments on normal penetration of cylindrical projectiles (Fig. 3.1.1).

Section	D (mm)	Material of shield
6.8	14.9	Eglin sand
6.11	7.8	Cook's Bayou sand
6.20	6.35	Rock (diorite)
6.26	7.87	Hanover silt soil
6.27	6.35	Dry and fluid-filled sandstone; High and low porosity green shale
6.33	15	Quartz sands
6.36	112; 229	Different types of one- and multilayered geological materials

Table 2. Experiments on normal penetration of cone-nose projectiles (Fig. 3.2.1).

Section	D (mm)	Material of shield
6.3	13	Dry quartz sand
6.7	5	Ottawa coarse silica sand
6.20	6.35; 12.7	Rock (diorite)
6.27	6.35	Dry and fluid-filled sandstone; High and low porosity green shale
6.36	216	Different types of one- and multilayered geological materials

Table 3. Experiments on penetration of ogive-nose projectiles (Fig. 3.3.1).

Section	D (mm)	Material of shield	θ_s
6.5	203	Madera limestone; Welded tuff; Layered earth shields	0
6.6	165–258	Welded tuff; Sandstone; Welded agglomerate; Granite	0
6.7	5	Ottawa coarse silica sand	0
6.12	22	Granite; Limestone	0
6.13	7.11	Limestone	0
6.14	95.2	Soil target at the Antelope Lake site	0
6.15	7.1-25.4	Limestone	0
6.16	4.17	Coarse foundry sand	0
6.18	165	Welded Tuff, Sandstone	0
6.24	40-165	Thin alternating layers of sand; Silt and clay sized particles	0
6.25	203; 259	Sandstone, welded tuff; Welded agglomerate; Weathered granite; Madera limestone	0
6.28	76.2	Moist, sandy clay	0
6.29	20	Rock	0
6.32	7.11	Salem limestone	0
6.36	79-229; 457	Different types of one- and multilayered geological materials	0
6.37	152; 165	Soil shield at the Antelope Lake site	>0
6.38	156	Soil shield in Pedro dry lake site	0

Table 4. Experiments on penetration of hemispherical-nose projectiles (Fig. 3.4.1).

Section	D (mm)	Material of shield	θ_s
6.7	5	Ottawa coarse silica sand	0
6.8	5; 20	Eglin sand; Ottawa sand	0
6.20	6.35; 12.7	Rock (diorite)	0
6.21	12.7	Rock (diorite and spessartite)	0
6.22	39	Clay; Clay and Sand; Soil,	≥ 0
6.26	7.87	Hanover silt soil	0
6.27	6.35	Dry and fluid-filled sandstone; High and low porosity green shale	0

Table 5. Experiments on penetration of spherical projectiles (Fig. 3.5.1).

Section	D (mm)	Material of shield	θ_s
6.9	4	Ottawa sand	0
6.17	6.35	Rock (green shale and diorite)	0
6.23	4.76; 7.14	Granite; Sandstone	0
6.30	9.52; 25.4	Sand	≥ 0
6.34	50	Sand	≥ 0
6.35	9.53	Eglin sand; Glass beads	0

Table 6. Experiments on normal penetration of shields by projectiles having other shapes and bullets.

Section	Shape of projectile	Material of shield
6.2	5.56-mm steel cubes	Hanover silt
	5.56-mm aluminum cubes	Snow
6.4	Truncated cone-nose, Fig. 3.6.1; $D = 13$ mm	Dry quartz sand
6.9	Three-cylindrical; $D = 3.1$ mm	Ottawa sand
6.10	7.62 mm AP, Fig. 3.11.1	Sand from Årdal in Norway
6.19	Truncated ogive-nose projectile, Fig. 3.6.2, $D = 7.62$ mm	River sands; Clay
6.31	Cal. 0.50 AP M2 20 mm AP M75 37 mm AP M74	Granite; Diabase; Limestone (Indiana); Sandstone
	Cal. 0.50 AP M2 20 mm AP M75	Quartzite
	75 mm AP M61 75 mm AP M72 76 mm AP M62 90 mm AP M77 90 mm AP M82	Limestone (Warsaw); Limestone (St. Louis)

6.2 [Aitken *et al.*, 1976]

6.2.1 *General data on shields and projectiles*

Shields. Hanover silt (frozen and unfrozen) and snow were investigated. Preparation of specimens for investigation of shields and determining mechanical properties of shield materials were described in details in the original report.

Table 1. Mechanical properties of Hanover silt ($\gamma = 1906\ kg/m^3$).

T^{sh} ($^\circ C$)	E (GPa)	σ_y (MPa)	T^{sh} ($^\circ C$)	E (GPa)	σ_y (MPa)	T^{sh} ($^\circ C$)	E (GPa)	σ_y (MPa)
−3	0.84	14.3	−10	1.48	22.2	−25	2.76	41.4

Table 2. Mechanical properties of snow.

γ (kg/m^3)	E (GPa)	σ_y (kPa)	γ (kg/m^3)	E (GPa)	σ_y (kPa)	γ (kg/m^3)	E (GPa)	σ_y (kPa)
300	0.20	67	400	6.32	105	500	9.91	1034

Projectiles. 5.56-mm steel and aluminum cubes.
Additional notations. S_s and S_r are areas of frontal faces of projectile before and after penetration event, correspondingly.

6.2.2 *Test results for 5.56-mm steel cubes penetrating into Hanover silt*

Table 3#. DOP versus v_s , unfrozen silt.

v_s (m/s)	P (mm)	v_s (m/s)	P (mm)	v_s (m/s)	P (mm)	v_s (m/s)	P (mm)
258	43.9	339	51.0	354	57.9	551	68.8
566	70.8	731	92.9	746	91.0	851	97.2

Table 4#. DOP versus v_s , $T^{sh} = -3^\circ C$.

v_s (m/s)	P (mm)	v_s (m/s)	P (mm)	v_s (m/s)	P (mm)	v_s (m/s)	P (mm)
351	19.1	521	34.1	622	45.8	721	50.9
904	62.7	1026	65.9	1096	61.2	1141	56.0

Table 5#. DOP versus v_s, $T^{sh} = -10^\circ C$.

v_s (m/s)	P (mm)	v_s (m/s)	P (mm)	v_s (m/s)	P (mm)	v_s (m/s)	P (mm)
379	20.8	412	20.1	409	24.8	546	36.2
570	36.8	578	36.9	713	45.9	762	51.9
774	51.9	809	56.1	896	61.1	904	55.1
923	55.2	920	59.1	984	54.8	991	57.0
1046	52.7	1077	53.3	1126	65.0	1131	61.3
1138	60.3	1154	52.3				

Table 6#. DOP versus v_s, $T^{sh} = -25^\circ C$.

v_s (m/s)	P (mm)	v_s (m/s)	P (mm)	v_s (m/s)	P (mm)	v_s (m/s)	P (mm)
333	15.2	443	17.4	606	24.1	627	24.3
628	29.6	702	31.2	746	41.2	845	43.9
888	42.5	919	48.4	1047	54.7	1069	54.7

Table 7#. Deformation coefficient versus v_s, unfrozen silt.

v_s (m/s)	S_r/S_s	v_s (m/s)	S_r/S_s	v_s (m/s)	S_r/S_s
260	0.07	344	0.08	740	0.16
751	0.07	854	0.20	1022	0.40

Table 8#. Deformation coefficient versus v_s, $T^{sh} = -3^\circ C$.

v_s (m/s)	S_r/S_s	v_s (m/s)	S_r/S_s	v_s (m/s)	S_r/S_s	v_s (m/s)	S_r/S_s
356	0.08	525	0.09	628	0.09	731	0.18
905	0.29	1026	0.40	1097	0.88	1142	1.15

Table 9#. Deformation coefficient versus v_s, $T^{sh} = -10^\circ C$.

v_s (m/s)	S_r/S_s	v_s (m/s)	S_r/S_s	v_s (m/s)	S_r/S_s	v_s (m/s)	S_r/S_s
390	0.08	412	0.08	427	0.08	555	0.08
573	0.08	585	0.08	717	0.20	765	0.32
770	0.30	772	0.29	814	0.48	898	0.33
908	0.07	924	0.48	929	0.67	984	0.40
993	0.53	1049	0.94	1079	0.99	1127	1.51
1133	1.66	1138	1.55	1127	0.83		

Table 10#. Deformation coefficient versus v_s , $T^{sh} = -25°C$.

v_s (m/s)	S_r/S_s	v_s (m/s)	S_r/S_s	v_s (m/s)	S_r/S_s
614	0.09	651	0.09	1079	0.39

6.2.3 *Test results for 5.56-mm aluminum cubes penetrating into snow*

Table 11. Properties of shields.

Shield label	γ (kg/m^3)	T^{sh} $(°C)$	Shield label	γ (kg/m^3)	T^{sh} $(°C)$	Shield label	γ (kg/m^3)	T^{sh} $(°C)$
Shield-1	3160	−15.6	Shield-2	3280	−12.9	Shield-3	2840	−6.4
Shield-4	2820	−3.1	Shield-5	3340	−0.6			

Table 12#. DOP versus v_s .

Shield	v_s (m/s)	P (mm)	Shield	v_s (m/s)	P (mm)
Shield-1	341	122.6	Shield-1	438	129.4
Shield-1	812	172.8	Shield-1	1080	145.3*
Shield-2	365	123.5	Shield-2	493	119.5
Shield-2	594	146.6	Shield-3	270	118.3
Shield-3	467	157.4	Shield-3	686	170.0
Shield-3	1069	171.1*	Shield-4	170	123.6
Shield-4	372	168.2	Shield-4	823	184.6
Shield-4	1071	191.8*	Shield-5	415	122.8
Shield-5	555	165.2	Shield-5	902	179.8*
Shield-5	1092	167.0*			

* Cube deformed.

6.3 [Allen *et al.*, 1957b]

6.3.1 *General data on shields and projectiles*

Shields. $c_{l,w} = 305\,mm$, $T = 1830\,mm$; shield material was dry quartz sand that passed U. S. Sieve No. 14 and was retained on U. S. Sieve No. 40; $q^{grain} \approx 1\,mm$. This material is also known as "Crystal Amber Sand," by the Monterey Sand Company, Monterey, CA. Static loading properties of the sand were determined by testing of 350 g dry material sample placed in a steel 58.7 mm in diameter cylinder, and the applied force reached the maximum value of 1644 kN. As a result, the sand was compressed from the initial density of $1640\,kg/m^3$ to the maximum density of $2380\,kg/m^3$.

Projectiles. Hardened 4130 Steel (HRC 45) cone-nose (Fig. 3.2.1) and cylindrical (Fig. 3.1.1) projectiles; $D = 13.0\,mm$, $L = 130\,mm$.

6.3.2 *Test results*

Table 1. Test results on DOP.

β	m	v_s	P	β	m	v_s	P
(deg)	(g)	(m/s)	(cm)	(deg)	(g)	(m/s)	(cm)
90^	80.30	674	137	90^	81.20	716	128
85	80.60	721	133	80	80.45	714	135
75	80.35	687	140	70	80.15	707	142
65	79.90	703	144	60	79.70	712	145
60	79.40	673	140	55	79.35	712	146
35	77.25	608	102	30	76.80	733	74
50	79.05	730	153	50	80.05	695	156
45	78.60	718	154	40	78.00	713	148
15	71.30	760	55*				

* Projectile hit side of box and broke into two pieces.

^ Cylinder.

6.4   [Allen *et al.*, 1957a]

6.4.1 *General data on shields and projectiles*

Shields. $c_{l,w} = 305\,mm$, $T = 3660\,mm$; shield material was dry quartz sand that passed U. S. Sieve No. 14 and was retained on U. S. Sieve No. 40; $q^{grain} \approx 1\,mm$ (for details see [Allen *et al.*, 1957b]).

Projectiles. Hardened 4130 Steel (HRC 45) truncated cone-nose projectiles (Fig. 3.6.1); $D = 13.0\,mm$, $L = 130\,mm$, $L_{nose} = 25.4\,mm$.

6.4.2 *Test results*

Table 1. Test results on DOP.

d (mm)	m (g)	v_s (m/s)	P (cm)	d (mm)	m (g)	v_s (m/s)	P (cm)
9.14	80.00	638	200	9.14	80.15	644	236
9.91	82.05	651	202	9.91	81.45	689	208
10.67	81.85	649	172	10.67	82.60	652	214
10.67	81.93	667	191	11.43	83.20	686	146
11.43	83.15	752	211	12.19	85.10	631	170
12.19	84.35	645	168	12.19	84.53	655	154
12.95	84.81	667	138	12.95	85.00	683	160
12.95	85.30	713	156	12.98	80.86	640	127
12.98	81.15	641	128	12.98	81.11	645	126
12.98	81.05	649	126	12.98	80.42	661	130
12.98	81.00	664	124	12.98	81.10	666	147
12.98	80.96	673	133	12.98	80.30	678	137
12.98	80.75	682	129	12.98	81.25	692	136
12.98	81.20	716	128				

6.5 [Bernard, 1976]

6.5.1 *Shallow penetration into rock*

Note. The author referred to [Patterson, 1972] and [Thigpen, 1974] presented data on experiments with *Madera limestone* and *welded tuff*; properties of these materials are presented in Table 1.

Projectiles. Ogive-nose projectiles (Fig. 3.3.1).

Table 1. Tests description.

Material	γ (kg/m^3)	f_c' (MPa)	m (kg)	D (mm)	ψ	v_s (m/s)	P (m)
Limestone	2690	94.4	306	203	9.25	174	0.87
Tuff	1850	38.0	455	229	6.0	212	1.51

6.5.2 *Deep penetration into multilayer Earth shield*

Results of projectile penetration tests (see Table 2) conducted at the Watching Hill Blast Range near Ralston, Alberta, Canada were reported by the author with reference to [Hadala, 1975].

Table 2. Characteristics of projectiles and test results on DOP.

m (kg)	D (mm)	L (mm)	v_s (m/s)	P (m)
127.01	152.4	1219	178/174	22.31
181.44	165.1	1524	131/125	14.72
127.01	152.4	1219	125/123	12.13
127.01	152.4	1219	158/156	20.03
181.44	165.1	1524	150/146	20.70
181.44	165.1	1524	93/93	9.08
90.72	104.8	1524	–/198	30.18
90.72	104.8	1524	189/186	26.24

Note. Values of v_s are measured using different techniques.

6.5.3 *Deep penetration into idealized four-layer Earth shield*

Note. Data were presented by the author with reference to [Hadala, 1975].

Table 3. Properties of layers.

Depth of layer (m)	γ (kg/m^3)	f_c' (MPa)	Depth of layer (m)	γ (kg/m^3)	f_c' (MPa)
0–2.44	1490	31.0	2.44–4.88	1419	20.7
4.88–7.32	1858	20.7	>7.32	1954	20.7

Projectiles. Ogive-nose projectiles (Fig. 3.3.1).

Table 4#. Characteristics of projectiles and test results on DOP.

m (kg)	D (mm)	ψ	v_s (m/s)	P (m)
91	104.8	6.0	185	26.13
91	104.8	6.0	197	30.11
127	152.4	6.0	124	12.10
127	152.4	6.0	157	20.06
127	152.4	6.0	176	22.31
182	165.1	9.25	92	9.15
182	165.1	9.25	124	14.89
182	165.1	9.25	147	20.79

6.5.4 *Very deep penetration into idealized four-layer Earth shield*

Note. Data were obtained on the basis of tests conducted at the Tonopah Test Range near Tonopah, Nevada (Antelope Dry Lake Site); reported by [Patterson, 1969], [Peterson and Hadala, 1976].

Table 5. Properties of layers.

Depth of layer (m)	γ (kg/m^3)	Lower-Bound f'_c (MPa)	Upper-Bound f'_c (MPa)
0-7.62	1650	68.9	344.8
7.62-11.28	1586	0.1	344.8
11.28-33.22	1778	517.2	1034.5
>33.22	1650	344.8	1034.5

Projectiles. Ogive-nose projectiles (Fig. 3.3.1).

Table 6#. Tests results on DOP ($D = 229 mm$, $\psi = 6$, $m = 295 kg$).

v_s (m/s)	P (m)	v_s (m/s)	P (m)	v_s (m/s)	P (m)
453	26.10	587	32.73	788	59.88

6.6 [Bernard, 1977]

Note. The author referred to experiments conducted in Sandia Research Laboratories under Energy Research and Development Agency (ERDA) or Defense Nuclear Agency (DNA) sponsorship.

6.6.1 *General data on shields and projectiles*

Table 1. Ogive-nose projectiles (Fig. 3.3.1).

Projectile label	D (mm)	ψ	m (kg)
Projectile-1	165.1	6.00	181*
Projectile-2	228.6	9.25	390
Projectile-3	258.8	9.25	1166
Projectile-4	203.2	9.25	613

* Without sabot.

Table 2. Rock shields.

Shield label	Material	γ (kg/m^3)	f'_c (MPa)	RQD (%)
Shield-1	Welded tuff	1950	60±15%	100
Shield-2	Sandstone	2080	23.4±30%	82
Shield-3	Welded agglomerate	1920	27.5±20%	60
Shield-4	Sandstone	2120	48.9±10%	37
Shield-5	Sandstone	2140	40.8±5%	32
Shield-6	Granite	2620	46.2±15%	32

6.6.2 *Test results*

Table 3. Test results on DOP.

Shield	Projectile	v_s (m/s)	P (cm)
Shield-1	Projectile-1	372	222
Shield-1	Projectile-1	411	259
Shield-1	Projectile-1	475	360
Shield-1	Projectile-1	501	335
Shield-1	Projectile-1	503	335
Shield-2	Projectile-1	444	357
Shield-2	Projectile-1	459	372
Shield-3	Projectile-2	325	396
Shield-4	Projectile-3	251	311
Shield-5	Projectile-4	268	305*
Shield-6	Projectile-4	262	381

* Sandstone was covered by 76 cm of soil.

6.7   [Bless *et al.*, 2009]

6.7.1 *General data on shields and projectiles*

Shields. Ottawa coarse silica sand (92% of particles with $q^{grain} = 0.4-0.6\,mm$); $c_{l,w} = 20\,cm$, $T = 244\,cm$.

Projectiles. Ogive-nose (Fig. 3.3.1), hemispherical-nose (Fig. 3.4.1) and cone-nose (Fig 3.2.1, $L_{nose} = 2.5\,mm$) projectiles with $D = 5\,mm$, $L = 50\,mm$ (including the stabilizer length of 13.7 mm) and $m = 14\,g$, fabricated from 93%W-Ni-Co. Projectiles were flight stabilized with either a fin set or flare made from aluminum and press-fit onto the rear. Shapes of stabilizers are given in the original paper and they are not described here.

6.7.2 *Test results*

Table 1. Test results on DOP.

Projectile nose shape	θ_s (deg)	v_s (m/s)	P (mm)	Stabilizer
Cone	–	606	1500	Fin
Cone	1.1	617	1560	Fin
Hemisphere	2.2	613	1945	Fin
Hemisphere	–	619	965	Fin
Hemisphere	0.4	620	685	Flare
Hemisphere	3.2	656	1755	Fin
Hemisphere	3.0	778	675	Flare
Hemisphere	0.3	1439	385	Flare
Ogive	6.4	503	1050	Fin
Ogive	3.3	609	1525	Fin
Ogive	–	613	370	Flare
Ogive	2.9	618	1650	Fin
Ogive	8.2	623	355	Flare
Ogive	0.4	633	1605	Fin
Ogive	0.0	707	1830 ± 200	Fin
Ogive	4.6	739	575	Flare
Ogive	8.6	744	255	Flare
Ogive	8.9	774	2125 ± 225	Flare
Ogive	1.9	915	815	Fin
Ogive	8.2	930	555	Flare
Ogive	8.2	1020	370	Flare
Ogive	3.8	1430	420	Flare
Ogive	6.3	1448	360	Flare
Ogive	1.3	1697	320	Flare
Ogive	1.3	1867	380	Flare
Ogive	5.4	2124	360	Flare

6.8 [Bless *et al.*, 2011]

Note. Some data from [Bless *et al.*, 2012]; see also [Bless *et al.*, 2010].

6.8.1 *General data on shields and projectiles*

Table 1. Characteristics of shields ($q^{grain} \approx 0.5\,mm$).

Test label	Investigator	Shield material
OU	K. Watanabe at Osaka University (OU)	Eglin sand
AFRL	W. Cooper at the Air Force Research Laboratory (AFRL)	Eglin sand
IAT-St	S. Bless and B. Peden at the Institute for Advanced Technology (IAT) at The University of Texas at Austin	Ottawa sand
IAT-TA	Ibid	Ibid

Table 2. Characteristics of projectiles.

Test label	Material	γ (kg/m^3)	Shape	D (mm)	L (mm)	HS/VS
OU	Steel	7210	Cylinder (Fig. 3.1.1)	14.9	26	VS
AFRL	Steel	7800	Hemispherical–nose projectile (Fig. 3.4.1)	20	127	HS
IAT-St	Steel	7800	Ibid	5	50	HS
IAT-TA	Tungsten alloy	17500	Ibid	Ibid	Ibid	Ibid

6.8.2 *Test results*

Table 3. Test results on DOP.

Test label	Sand γ (kg/m^3)	L (mm)	m (g)	v_s (m/s)	P (mm)
IAT-St	1560	50	16.3	810	914
IAT-TA	1560	50	15.1	656	1753
AFRL	1490	127	208	566	1470
AFRL	1490	127	208	831	1470
AFRL	1490	127	208	849	1470
AFRL	1580	127	205	631	1060
AFRL	1580	127	205	863	1060
AFRL	1580	127	199	925	1060
OU	1370	26.0	32.6	455	245
OU	1410	26.1	32.4	480	210
OU	1370	26.0	32.7	282	237

6.9 [Borg *et al.*, 2013]

6.9.1 *General data on shields and projectiles*

Shields. Ottawa sand ($q^{grain} = 0.45 - 0.55\,mm$, $\gamma = 1560\,kg/m^3$) consisting of 99% pure Quartz; $c_l = 25\,cm$, $c_w = 18\,cm$, $T = 35\,cm$; VS. **Projectile-1** consisted of three cylindrical parts ($L = 76\,mm$, $m = 4.67$ g) made from machine brass. Base diameter, $D = 3.1\,mm$, was reduced by 1.6 mm starting 15 mm from the nose and extended approximately 15 mm along the cylinder.
Projectile-2 was a full bore copper ball (Fig. 3.5.1), $D = 4\,mm$, $m = 0.235\,g$.

6.9.2 *Test result*

Table 1. Test results on DOP.

Projectile label	v_s (m/s)	P (mm)	Projectile label	v_s (m/s)	P (mm)
Projectile-1	35	201	Projectile-1	35	209
Projectile-1	35	210	Projectile-1	35	199
Projectile-1	35	205	Projectile-2	127	55
Projectile-2	128	55	Projectile-2	141	60
Projectile-2	141	55	Projectile-2	159	60
Projectile-2	159	59	Projectile-2	212	65
Projectile-2	212	61			

6.10 [Børvik *et al.*, 2013]

6.10.1 *General data on shields and projectiles*

Shields. Sand from Årdal in Norway; $\eta_{FM} = 2.25$, $\gamma = 1856\,kg/m^3$ for *wet sand* and $\gamma = 1726\,kg/m^3$ for *dry sand*; $E = 31.8\,GPa$ and $\sigma_f = 233\,MPa$ for bulk material.
Projectiles. 7.62 mm AP bullets (Fig. 3.11.1, Table 3.11.1).

6.10.2 *Test results*

Table 1#. DOP versus v_s for wet sand.

v_s (m/s)	P (mm)	v_s (m/s)	P (mm)	v_s (m/s)	P (mm)
906	270	916	260	918	250
920	240	926	290	936	270

Table 2#. DOP versus v_s for dry sand.

v_s (m/s)	P (mm)	v_s (m/s)	P (mm)	v_s (m/s)	P (mm)
901	190	907	170	916	220
919	225	939	215		

6.11 [Butler, 1975c]

See also [Butler, 1976].

6.11.1 *General data on shields and projectiles*

Shields. Cook's Bayou sand, $\gamma = 1660$–$1760\,kg/m^3$; cube-shaped container having 30.48 cm long edges.

Table 1. Cylindrical projectiles (Fig. 3.1.1).

Projectile label	Material	D (mm)	L (mm)	m (g)	γ (kg/m^3)	σ_y (MPa)	HB
Projectile-1	Steel (SAE 1020)	7.87	8.0	3.00	7710	350	160
Projectile-2	Steel (AISI C-1141)	7.8	7.8	2.87	7700	680	252
Projectile-3	Brass (ASTM B-16)	7.8	7.8	3.00	8050	260	110

6.11.2 *Test results*

Additional notations. S_s and S_r are areas of the frontal faces of projectile before and after penetration event, correspondingly.

Table 2. Test results for Projectile-1 (Series 1).

Shield γ (kg/m^3)	v_s (m/s)	P (mm)	S_r/S_s	m_r/m_s
1720	729	85.7	1.20	0.96
1670	954	88.8	1.29	0.97
1740	962	83.8	(hit on side)	0.87
1700	976	84.8	1.95	0.97
1710	980	76.8	2.15	0.99
1650	1134	92.2	2.31	1.00
1690	1167	84.0	2.35	0.99

Table 3. Test results for Projectile-1 (Series 2).

Shield γ (kg/m^3)	v_s (m/s)	P (mm)	S_r/S_s	m_r/m_s
1750	1072	81.4	2.22	1.00
1690	1121	92.4	2.26	1.00
1780	1339	97.8	2.69	0.91
1760	1449	76.8	1.44	0.73
1760	1505	94.0	3.29	0.91
1780	1624	88.5	1.50	0.60
1750	1707	74.5	2.84	0.98
1750	1786	83.3	2.78	0.61
1750	1836	85.3	2.00	0.50
1750	1958	88.1	3.57	0.84

Table 4. Test results for Projectile-2.

Shield γ (kg/m^3)	v_s (m/s)	P (mm)	S_r/S_s	m_r/m_s
1750	878	96.3	1.34	0.98
1760	1194	95.8	1.82	0.96
1750	1348	105.0	–	–
1720	1481	86.7	2.37	0.98
1740	1628	100.1	2.23	0.69
1720	1648	105.7	1.36	0.75
1750	1679	113.0	1.43	0.58
1740	1946	104.4	1.80	0.55

Table 5. Test results for Projectile-3.

Shield γ (kg/m^3)	v_s (m/s)	P (mm)	S_r/S_s	m_r/m_s
1750	1169	106.7	1.52	0.63
1760	1802	93.7	–	0.16

6.12 [Folsom, 1987]

6.12.1 *General data on shields and projectiles*

Shields. Predrilled (shields with an inline cylindrical hole) granite and limestone shields.

Table 1. Characteristics of shields ($D^{sh} = 102\,mm$, including aluminum case).

Shield material	γ (kg/m^3)	E (GPa)	f'_c (MPa)
Granite	2650	45.5	150
Limestone	2370	21.4	38

Projectiles. Hardened 4340 Steel (HRC 32-36) ogive-nose projectiles (Fig. 3.3.1); $D = 22.0\,mm$, $L_{cyl} = 44.0\,mm$, $L_{nose} = 22.0\,mm$, $\rho = 27.6\,mm$, $m = 88.7\,g$.

6.12.2 *Test results*

Table 2. Test results on DOP.

Shield material	ξ	v_s (m/s)	P (mm)
Granite	0.885	311	90
Granite	0.886	394	146
Granite	0.956	317	148
Granite	0.964	224	96
Limestone	0.888	311	175
Limestone	0.959	267	257

6.13 [Forrestal and Hanchak, 2002]

6.13.1 *General data on shields and projectiles*

Shields. Properties of nominal limestone shield material: $\gamma = 2310\,kg/m^3$, $\eta_{wat}^{sh} = 0.15\%$, $\eta_{por} = 15\%$, $f'_c = 60\,MPa$.

Projectiles. 4340 Steel (HRC 45) ogive-nose projectiles (Fig. 3.3.1); $D = 7.11\,mm$, $L = 71.1\,mm$, $\psi = 3$, $m = 20.5\,g$.

6.13.2 *Test results*

Table 1. Test results on DOP.

v_s (m/s)	P (mm)	v_s (m/s)	P (mm)	v_s (m/s)	P (mm)
242	30.9	271	35.0	289	36.2
308	36.8	320	40.0	331	44.4

6.14 [Forrestal and Luk, 1992]

6.14.1 *General data on shields and projectiles*

Shields. Sandia Tonopah Test Range, NV, a soil shield at the Antelope Lake site. Mechanical properties of the shield material are presented in the original publication.

Projectiles.   Ogive-nose   projectiles   (Fig. 3.3.1); $D = 95.2\,mm$, $L_{cyl} = 516\,mm$, $L_{nose} = 158\,mm$, $\psi = 3$, $m = 23.1\,kg$.

6.14.2 *Test results*

Table 1. Test results on P^{cr}.

v_s (m/s)	P^{cr} (mm)	v_s (m/s)	P^{cr} (mm)	v_s (m/s)	P^{cr} (mm)
280	6.48	280	4.98	278*	5.18
280*	5.18	280	5.02	280	4.82

* Oblique impact.

6.15 [Frew *et al.*, 2000]

6.15.1 *General data on shields and projectiles*

Shields. Limestone shields; $c_{l,w} = 500\,mm$; Ø521×108 mm cylinders were tested to determine f'_c.

Table 1. Properties of nominal shield material.

γ (kg/m^3)	η^{sh}_{wat} (%)	η_{por} (%)	f'_c (MPa)
2300-2320	0.11-0.16	14.4-15.1	58-61

Table 2. Ogive-nose projectiles (Fig. 3.3.1); $L/D = 10$, $\psi = 3$.

Series label	D (mm)	Material	HRC	L_{cyl} (mm)	L_{nose} (mm)	m (g)
Series-1	12.7	4340 Steel	44-45	106	21	117
Series-2a	7.1	4340 Steel	44-46	59.3	11.8	20.5
Series-2b	7.1	Aer Met 100	53	59.3	11.8	20.5
Series-3	25.4	4340 Steel	45-46	212	42	931

6.15.2 *Test results*

Table 3. Results for Series-1 tests (average $f_c' = 58\,MPa$).

T (m)	v_s (m/s)	P (m)	$\Delta m/m$ (%)	T (m)	v_s (m/s)	P (m)	$\Delta m/m$ (%)
0.61	459	0.141	0.16	0.61	608	0.232	0.24
0.91	853	0.362	0.96	0.91	956	0.523	1.29
0.91	1134	0.562	2.59	1.22	1269	0.812	3.87
1.22	1404	0.924	5.25	1.22	1502	1.017	5.90

Table 4. Results for Series-2a tests.

T (m)	v_s (m/s)	P (m)	$\Delta m/m$ (%)	f_c' (MPa)
0.3	497	0.067	0.34	–
0.61	597	0.105	0.34	64.7
0.61	787	0.165	0.78	–
0.61	1037	0.271	2.11	–
1.02	1365	0.430	4.90	58.2
1.02	1516	0.516	7.34	–
0.46	795	0.178	0.83	60.1
0.61	1060	0.294	2.49	60.3
0.61	1230	0.392	3.76	55.7
1.02	1340	0.437	4.84	57.5

Table 5. Results for Series-2b tests.

T (m)	v_s (m/s)	P (m)	$\Delta m/m$ (%)	f_c' (MPa)
1.02	1266	0.379	4.59	63.8
1.02	1674	0.581	10.85	63.9

Table 6. Results for Series-3 tests.

T	v_s	P	$\Delta m/m$	f_c'
(m)	(m/s)	(m)	(%)	(MPa)
0.61	407	0.26	0	56.2
0.76	566	0.39	0.1	61.9
1.07	800	0.79	0.3	66.7
1.52	917	1.02	1.1	59.8
1.98	1177	1.50	2.8	61.2

6.16   [Hankins *et al.*, 2006]

6.16.1   *General data on shields and projectiles*

Shields. Coarse foundry sand shields.
Projectiles. 7075-T6 Aluminum ogive-nose projectiles (Fig. 3.3.1); $L = 41.66\,mm$, $D = 4.17\,mm$, $\psi = 0.5$.

6.16.2   *Test results*

Table 1#. Test results on DOP.

v_s	P	v_s	P	v_s	P	v_s	P
(m/s)	(mm)	(m/s)	(mm)	(m/s)	(mm)	(m/s)	(mm)
23	7.1	78	41.9	83	55.9	88	59.2
143	68.2	170	68.8	227	89.3	255	88.5
344	112.8	366	114.4	404	133.6	446	123.8
476	120.8						

6.17   [Kabo *et al.*, 1977]

6.17.1   *General data on shields and projectiles*

Shields. Rock (green shale and diorite) shields; $\widetilde{D}^{sh} = 140\,mm$.

Table 1. Properties of shield material.

Material	γ	ν	E	f_c'	η_{por}	q^{grain}
	(kg/m^3)		(GPa)	(MPa)	(%)	(mm)
Green shale	1690	0.25	1-3.5	1-10	10-30	–
Diorite	2670	0.25	58.0	221	–	1.5–2.0

Projectiles. Steel spherical projectiles (Fig. 3.5.1); $D = 6.35\,mm$.

6.17.2 *Test results*

Table 2. Test results for diorite shields.

v_s (m/s)	P (mm)	v_s (m/s)	P (mm)	v_s (m/s)	P (mm)	v_s (m/s)	P (mm)
167.1	2.12	181.0	2.00	195.0	1.65	206.0	2.13
213.4	2.15	224.0	2.34	236.0	2.46	246.0	2.24

Table 3. Test results for green shale shields (small velocities).

mv_s (kg·m/s)	P (mm)	mv_s (kg·m/s)	P (mm)	mv_s (kg·m/s)	P (mm)	mv_s (kg·m/s)	P (mm)
0.054	1.53	0.105	3.53	0.137	4.54	0.155	5.10
0.172	5.82	0.192	5.61	0.205	6.45	0.228	7.06
0.236	7.37	0.246	7.81	0.258	8.40		

Table 4. Test results for green shale shields (high velocities).

mv_s (kg·m/s)	P (mm)	mv_s (kg·m/s)	P (mm)	mv_s (kg·m/s)	P (mm)	mv_s (kg·m/s)	P (mm)
0.350	12.11	0.453	14.35	0.491	15.81	0.510	22.27
0.520	20.20	0.525	29.22	0.536	16.14	0.554	15.12
0.556	25.11	0.570	19.94	0.577	29.17	0.577	31.94
0.583	29.13	0.597	26.50	0.601	32.97	0.616	25.25
0.663	29.83	0.707	25.17	0.864	37.37	0.877	31.77
0.905	31.36	0.912	38.19	0.917	40.00	0.933	32.10
0.972	40.35	0.972	44.32	1.007	42.15	1.556	60.27
1.655	63.63	1.665	65.28	1.685	63.81	1.776	89.00

6.18 [Kar, 1977]

Note. Data obtained by D.K. Butler.

6.18.1 *General data on shields and projectiles*

Shields. Welded tuff, $f_c' = 38.0\,MPa$ (Series 1); sandstone, $f_c' = 23.4\,MPa$ (Series 2).

Projectiles. Ogive-nose projectiles (Fig. 3.3.1) with $D = 165\,mm$, $\psi = 9.25$, $m = 181.4\,kg$.

6.18.2 *Test results*

Table 1. Test results on DOP

Series	v_s (m/s)	P (mm)	Series	v_s (m/s)	P (mm)
Series 1	372	2.23	Series 1	411	2.59
Series 1	475	3.60	Series 1	501	3.35
Series 1	503	3.35	Series 2*	444	3.56
Series 2*	459	3.72			

* Data from [Butler, 1975b]

6.19 [Khromov, 2004]

6.19.1 *General data on shields and projectiles*

Projectiles. Projectile shape is shown in Fig. 3.6.2; $D = 7.62\,mm$, $L = 19.7\,mm$ and $L_{nose} \ll L$, $d \approx D$.
Shields. River sands, clay.

6.19.2 *Test results*

Table 1. DOP for river sand ($\eta_{wat}^{sh} = 5\%$), $P\,(mm)$.

T^{sh}	Impact velocity, $v_s\,(m/s)$			
(°C)	207	315	397	482
−10	53	105	110	147
0	78	126	143	182
+10	83	130	147	191
+20	85	131	152	199
+30	87	133	155	187

Table 2. DOP for river sand ($\eta_{wat}^{sh} = 10\%$), $P\,(mm)$.

T^{sh}	Impact velocity, $v_s\,(m/s)$			
(°C)	207	315	397	482
−10	57	107	111	144
0	59	118	129	153

T^{sh}	Impact velocity, $\quad v_s\,(m/s)$			
(°C)	207	315	397	482
+10	72	120	132	157
+20	80	124	135	158
+30	75	119	135	160

Table 3. DOP for clay, P (mm).

T^{sh}	Impact velocity, $\quad v_s\,(m/s)$			
(°C)	207	315	397	482
−10	67	80	120	132
0	94	111	140	166
+10	98	120	147	169
+20	100	125	128	173
+30	95	90	111	170

6.20   [Kumano and Goldsmith, 1982a]

6.20.1   *General data on shields and projectiles*

Shields. Rock (diorite) shields; $\widetilde{D}^{sh} = 140\,mm$.

Table 1. Properties of shield material.

γ (kg/m^3)	E (GPa)	G (GPa)	f_c' (MPa)	σ_u (MPa)	q^{grain} (mm)
2810	57.7	23.6	142	13.2	~1.0

Table 2. Characteristics of projectiles.

Projectile label	Projectile shape	D (mm)	L (mm)	β (deg)	m (g)
Projectile-1	Hemispherical-nose (Fig. 3.4.1)	12.7	21	–	18.5
Projectile-2	Cone-nose (Fig. 3.2.1)	12.7	23	45	18.5
Projectile-3	Cone-nose (Fig. 3.2.1)	12.7	26	30	18.5
Projectile-4	Hemispherical-nose (Fig. 3.4.1)	6.35	10.6	–	2.3
Projectile-5	Cone-nose (Fig. 3.2.1)	6.35	11.6	45	2.3
Projectile-6	Cone-nose (Fig. 3.2.1)	6.35	13.0	30	2.3
Projectile-7	Cylindrical (Fig. 3.1.1)	6.35	9.6	–	2.3

6.20.2 *Test results*

Table 3. Test results on DOP.

Projectile label	v_s (m/s)	P (mm)	Projectile label	v_s (m/s)	P (mm)
Projectile-1	34.5	1.47	Projectile-1	43.5	1.67
Projectile-1	57.2	2.34	Projectile-2	22.0	1.59
Projectile-2	32.3	2.09	Projectile-2	43.3	2.54
Projectile-2	53.0	2.35	Projectile-3	23.2	1.91
Projectile-3	34.4	2.35	Projectile-3	43.2	3.07
Projectile-3	54.1	3.55	Projectile-4	70.8	1.25
Projectile-4	124.2	1.92	Projectile-5	72.2	1.24
Projectile-5	124.4	2.35	Projectile-6	69.7	1.65
Projectile-6	125.8	2.70	Projectile-7	120.0	1.92

6.21 [Kumano and Goldsmith, 1982b]

6.21.1 *General data on shields and projectiles*

Shields. Rock (diorite and spessartite) shields; $\widetilde{D}^{sh} = 140\,mm$.

Table 1. Properties of shield material.

Material	γ (kg/m^3)	E (GPa)	G (GPa)	f_c' (MPa)	σ_u (MPa)	q^{grain} (mm)
Diorite	2810	57.7	23.6	142	13.2	~1.0
Spessartite	2860	73.7	27.8	194	35.1	~0.2

Projectiles. Cone-nose projectiles (Fig. 3.2.1; $D = 12.7\,mm$, $\beta = 30°$) and hemispherical-nose projectiles (Fig. 3.4.1; $D = 12.7\,mm$).

6.21.2 *Test results*

Table 2. Test results for diorite shields.

Projectile shape	m (g)	v_s (m/s)	P (mm)
Cone-nose	18.59	23.2	1.9
Cone-nose	18.71	34.4	2.4
Cone-nose	18.69	43.2	3.1

Projectile shape	m (g)	v_s (m/s)	P (mm)
Cone-nose	18.70	54.1	3.6
Hemispherical-nose	18.54	34.5	1.5
Hemispherical-nose	18.54	43.5	1.7
Hemispherical-nose	18.61	57.2	2.3

Table 3. Test results for spessartite shields.

Projectile type	m (g)	v_s (m/s)	P (mm)
Cone-nose	18.59	27.9	2.1
Cone-nose	18.69	60.1	3.4
Hemispherical-nose	18.54	58.8	3.4

6.22   [Lorenz *et al.*, 2000]

6.22.1   *General data on shields and projectiles*

Shields. Shields made of different materials.

Projectiles. Steel hemispherical-nose projectiles (Fig. 3.4.1); $D = 39\,mm$, $L = 100\,mm$, $m = 670\,g$.

Additional notation. α is the angle of attack.

Note. Penetrator is an element of complex structure designed for exploration of Mars. Test results presented below are obtained on the Earth.

6.22.2   *Test results*

Table 1. Test results (presented only results of real experiments without restored and unreliable data).

Shield material	k_s	m (g)	θ_s (deg)	α (deg)	v_s (m/s)	P (cm)
Clay	5 - 7	655	15	0	168	26
Clay	5 - 7	655	15	7	175	17
Clay	5 - 7	655	15	14	195	14
Clay	5 - 7	650	15	14	175	32

Shield material	k_S	m (g)	θ_s (deg)	α (deg)	v_s (m/s)	P (cm)
Clay + Sand	20 - 40	616	15	0	180	69
Clay + Sand	5 - 8	657	15	0	175	33
Clay	5 - 8	667	0	0	170	37
Clay	5 - 9	577	15	14	190	29
Clay	3 - 4	657	15	12	180	25
Clay	3 - 5	515	15	14	198	16
Clay	4 - 6	657	25	12	214	41
Clay	4 - 7	505	30	0	191	23
Soil	3 - 3.5	623	20	6	194	41
Soil	3 - 3.5	644	20	0	205	33
Soil	3 - 3.5	581	25	-3	198	35
Soil	1 - 3	576	25	15	188	20
Clay	3.5 - 6.5	620	20	0	186	20
Clay	4.5 - 6.5	640	25	1	185	28
Clay	3 - 9	545	25	2	188	25
Clay	3 - 4.5	642	25	−15	187	35
Clay	3 - 4	660	25	10	190	50
Clay	4 - 6	615	25	−1	191	30

6.23   [Maurer and Rinehart, 1960]

6.23.1   *General data on shields and projectiles*

Table 1#. Properties of shield material.

Material	γ (kg/m^3)	σ_u (MPa)	f_c' (MPa)	σ_{ss} (MPa)
Granite	2723	7.65	114.72	9.11
Sandstone	2385	0.50	18.87	2.11

Note. γ was taken from the table.

Projectiles. Steel spherical projectiles (Fig. 3.5.1) of two types: *Projectile-1* with $D = 4.76mm$, $m = 0.442\,g$ and *Projectile-2* with $D = 7.14mm$, $m = 1.503\,g$.

Impact velocities. $v_s = 90 - 1830\,m/s$.

Additional notations. $\chi_s = mv_s/(0.25\pi D^2)$, $1Mg = 10^6 g$.

6.23.2 *Test results*

Table 2#. Test results for penetration of Projectile-1 into granite.

χ_s $(Mg \cdot s/m^2)$	P (mm)	χ_s $(Mg \cdot s/m^2)$	P (mm)	χ_s $(Mg \cdot s/m^2)$	P (mm)	χ_s $(Mg \cdot s/m^2)$	P (mm)
1.18	2.0	1.74	2.6	2.20	3.1	3.22	4.9
3.88	5.6	4.18	6.4	4.55	6.9	4.69	6.9
4.85	7.4	4.98	7.1				

Table 3#. Test results for penetration of Projectile-2 into granite.

χ_s $(Mg \cdot s/m^2)$	P (mm)	χ_s $(Mg \cdot s/m^2)$	P (mm)	χ_s $(Mg \cdot s/m^2)$	P (mm)	χ_s $(Mg \cdot s/m^2)$	P (mm)
0.56	0.4	4.24	5.8	6.05	8.1	8.54	12.4

Table 4#. Test results for penetration of Projectile-1 into sandstone.

χ_s $(Mg \cdot s/m^2)$	P (mm)	χ_s $(Mg \cdot s/m^2)$	P (mm)	χ_s $(Mg \cdot s/m^2)$	P (mm)
0.64	3.2	1.62	8.1	3.21	16.2
4.21	22.6	4.36	22.0		

Table 5#. Test results for penetration of Projectile-2 into sandstone.

χ_s $(Mg \cdot s/m^2)$	P (mm)	χ_s $(Mg \cdot s/m^2)$	P (mm)	χ_s $(Mg \cdot s/m^2)$	P (mm)	χ_s $(Mg \cdot s/m^2)$	P (mm)
0.87	4.7	2.58	14.4	5.34	30.1	7.45	40.9

6.24 [Patterson *et al.*, 1975]

6.24.1 *General data on shields and projectiles*

Shields. Penetration tests were conducted in the vicinity of the Prairie Flat High Explosive Simulation Test (HEST) facility at the Defense Research Establishment Suffield (DRES), Ralston, Alberta, Canada. The site is located on an old glacial lake bed and the near surface deposits consist of thin alternating layers of sand, silt, and clay sized particles.

Projectiles. Ogive-nose projectiles (Fig. 3.3.1) fabricated from heat treated DOAC steel (HRC 42, $\sigma_y = 1240 - 1310\,MPa$).

Table 1. Characteristics of projectiles.

Projectile label	D (mm)	L (mm)	m (kg)	ψ
Projectile-1	152.4	1219.2	128.4	6.0
Projectile-2	165.1	1524.0	181.4	9.25
Projectile-3	104.9	1524.0	90.7	9.25
Projectile-4	77.7	934.7	13.9	6.0
Projectile-5	39.6	482.6	4.2	6.0
Projectile-6	76.2	947.4	26.0	6.0

6.24.2 *Test results*

Table 2. Test results on DOP.

Projectile label	Average k_S	v_s (m/s)	P (mm)
Projectile-1	15.5	178/174/178	22.3
Projectile-1	13.0	125/123/123	12.1
Projectile-1	15.5	161/156/159	20.0
Projectile-2	13.5	131/125/123	14.7
Projectile-2	14.6	150/146/151	20.7
Projectile-2	12.2	93/93/89	9.1
Projectile-3	13.6	– /198/202	30.2
Projectile-3	12.6	186/189/ –	26.2
Projectile-4	5.0	– /127/ –	3.1
Projectile-4	4.7	– /95/ –	2.0
Projectile-5	4.0	– /139/ –	3.1
Projectile-5	3.8	– /104/ –	2.0
Projectile-6	5.6	– /94/ –	3.0

Note. Values of v_s were obtained using different methods of measurement.

6.25 [Patterson, 1973]

Note. Data from [Rohani, 1975], [Butler, 1975a].

6.25.1 *General data on shields and projectiles*

Shields. Rock shields.
Projectiles. Ogive-nose projectiles (Fig. 3.3.1).

6.25.2 *Test results*

Table 1. Test results on DOP.

Parameters of projectile			Characteristics of shield			v_s	P
m (kg)	D (mm)	ψ	Name	Type *	RQD* (%)	(m/s)	(m)
520.7	259	9.25	Sandstone	DH	37	251.5	3.10
461.8	229	6	Tonopah Test Range welded tuff	DL	80	198.1	2.79
389.6	229	9.25	Tonopah Test Range welded agglomerate	D	60	324.6	3.96
278.1	203	9.25	Weathered granite	HD	32	262.1	3.81
305.7	203	9.25	Madera limestone	CH	–	289.6	3.15

* See [TM 5-818-1/AFM 88-3, 1983 (Ch. 7)] for notations in classification.

6.26 [Richmond, 1980]

6.26.1 *General data on shields and projectiles*

Shields. Hanover silt soil in cube-shaped boxes with the length of edge 305 mm; $\gamma = 1700 \, kg/m^3$.

Additional notations. T_{st}^{sh} , T_{fin}^{sh} – temperature at the beginning and at the end of test series, correspondingly; nominator and denominator in the column for v_s denotes impact velocity measured at 1.5 m from muzzle and at 0.5 m from shield, correspondingly.

Table 1. Steel (AIAI C-1215) projectiles.

Projectile label	Projectile	Shape	D (mm)	L (mm)	m (g)
Projectile-1	Hemispherical-nose projectiles	Fig. 3.4.1	7.87	17	6.03
Projectile-2	Hemispherical-nose projectiles	Fig. 3.4.1	7.87	33	12.05
Projectile-3	Cylindrical projectiles	Fig. 3.1.1	7.87	16	6.03
Projectile-4	Cylindrical projectiles	Fig. 3.1.1	7.87	32	12.05

Table 2. Description of notations in column Note in Tables 3–7.

Notation	Description	Notation	Description
(1)	Yaw at impact	(2)	Tumbled in flight
(3)	Impact velocity lost	(4)	Slight deformation near the nose of projectile

6.26.2 *Test results*

Table 3. Test results for Series 1 ($\eta_{wat}^{sh} = 19.4\%$, $T_{st}^{sh} = -3.5°C$, $T_{fin}^{sh} = -9.1°C$).

Projectile label	v_s (m/s)	P (mm)	Note	Projectile label	v_s (m/s)	P (mm)	Note
Projectile-1	409/399	54	(1)	Projectile-1	464/456	124	–
Projectile-1	632/623	159	–	Projectile-2	491/484	89	(1)
Projectile-2	468/548	79	(1)	Projectile-3	420/410	64	–
Projectile-3	588/577	119	–	Projectile-4	524/511	84	–
Projectile-4	447/493	79	(2)	Projectile-4	668/661	224	–

Table 4. Test results for Series 2 ($\eta_{wat}^{sh} = 18.4\%$, $T_{st}^{sh} = -11.2°C$, $T_{fin}^{sh} = -10.7°C$).

Projectile label	v_s (m/s)	P (mm)	Note	Projectile label	v_s (m/s)	P (mm)	Note
Projectile-1	347/318	54	–	Projectile-1	477/470	109	–
Projectile-2	347/270	–	(2)	Projectile-2	595/590	84	(1)
Projectile-2	347/234	–	(2)	Projectile-2	368/360	–	(2)
Projectile-2	347/246	–	(2)	Projectile-3	346/277	49	–
Projectile-3	651/638	94	–	Projectile-4	346/282	44	(1)
Projectile-4	651/645	204	–				

Table 5. Results for Series 3 tests ($\eta_{wat}^{sh} = 19.0\%$, $T_{st}^{sh} = -11.7°C$, $T_{fin}^{sh} = -11.1°C$).

Projectile label	v_s (m/s)	P (mm)	Note	Projectile label	v_s (m/s)	P (mm)	Note
Projectile-1	–/241	24	–	Projectile-1	348/326	59	–
Projectile-2	–/294	–	(2)	Projectile-2	362/361	104	–
Projectile-2	503/498	144	(1)	Projectile-2	349/319	64	(1)
Projectile-3	–/268	24	–	Projectile-3	497/484	69	–
Projectile-4	349/196	39	–	Projectile-4	355/354	104	–
Projectile-4	349/292	74	–				

Table 6. Results for Series 4 tests ($\eta_{wat}^{sh} = 19.7\%$, $T_{st}^{sh} = -12.4°C$, $T_{fin}^{sh} = -11.9°C$).

Projectile label	v_s (m/s)	P (mm)	Note	Projectile label	v_s (m/s)	P (mm)	Note
Projectile-1	174/38	15	–	Projectile-1	187/38	10	–
Projectile-1	542/–	99	(3)	Projectile-1	487/480	94	–
Projectile-1	466/459	104	–	Projectile-3	597/583	89	–

Projectile label	v_s (m/s)	P (mm)	Note	Projectile label	v_s (m/s)	P (mm)	Note
Projectile-3	526/515	84	–	Projectile-3	399/388	54	–
Projectile-4	426/406	–	(2)	Projectile-4	555/547	164	(1)
Projectile-4	478/470	109	(1)				

Table 7. Results for Series 5 tests ($\eta_{wat}^{sh} = 19.1\%$, $T_{st}^{sh} = -24\,^{\circ}C$, $T_{fin}^{sh} = -23.8\,^{\circ}C$) for Projectile-3.

v_s (m/s)	P (mm)	Note	v_s (m/s)	P (mm)	Note
228/–	–	(1)	105/–	–	(1)
347/296	34	–	349/325	44	–
361/354	54	–	–/364	54	–
531/519	69	–	527/515	69	–
651/635	89	–	672/661	94	(4)
354/–	64	(3)	540/527	84	–

6.27 [Rogers *et al.*, 1986]

6.27.1 *General data on shields and projectiles*

Shields. Dry and fluid-filled sandstone, low porosity green shale, high porosity green shale shields; 76×51×38 mm blocks. The fluid-filled samples were prepared by submerging the dry specimens in a kerosene bath for at least 12 hours prior to use.

Table 1. Properties of shield material (in dry conditions).

Shield label	Shield material	γ (kg/m^3)	E (GPa)	f_c' (MPa)	σ_u (MPa)	η_{por} (%)
Shield-1	Sandstone	2220	46.0/26.2*	52.2	5.84	8
Shield-2	Low porosity green shale	2110	13.3	15.7	1.52	7
Shield-3	High porosity green shale	1670	3.77	7.15	0.37	28

* Values in nominator and in denominator refer to directions along and perpendicular to the fibers, respectively.

Projectiles. Cone-nose projectiles (Fig. 3.2.1; $\beta = 30\,^{\circ}$, $m = 22.4\,g$), hemispherical-nose projectiles (Fig. 3.4.1; $D = 6.35\,mm$, $m = 31.5\,g$), cylindrical projectiles (Fig. 3.1.1; $m = 32.2\,g$). For all steel (HRC 60) projectiles $D = 6.35mm$.

6.27.2 *Test results*

Notation in Tables below. FF − fluid-filled material.

Table 2. Test results for cone-nose projectiles.

Shield label	Material condition	v_s (m/s)	P (mm)	Shield label	Material condition	v_s (m/s)	P (mm)
Shield-1	Dry	16.3	2.55	Shield-1	Dry	18.2	2.58
Shield-1	Dry	22.0	3.05	Shield-1	FF	17.8	2.07
Shield-1	FF	21.5	2.50	Shield-2	Dry	12.6	2.12
Shield-2	Dry	19.1	3.22	Shield-2	FF	15.1	1.83
Shield-2	FF	18.6	2.74	Shield-2	FF	22.3	3.40
Shield-3	Dry	14.7	3.71	Shield-3	Dry	18.0	4.23
Shield-3	Dry	22.4	3.78	Shield-3	FF	11.9	3.62
Shield-3	FF	18.3	5.88	Shield-3	FF	22.4	6.41

Table 3. Test results for hemispherical-nose projectiles.

Shield label	Material condition	v_s (m/s)	P (mm)	Shield label	Material condition	v_s (m/s)	P (mm)
Shield-1	Dry	13.3	0.76	Shield-1	Dry	15.5	0.92
Shield-1	Dry	18.7	1.01	Shield-1	FF	10.2	0.62
Shield-1	FF	16.1	1.01	Shield-1	FF	18.6	1.22
Shield-2	Dry	13.9	1.06	Shield-2	Dry	16.3	1.05
Shield-2	Dry	18.1	1.24	Shield-2	FF	13.5	1.19
Shield-2	FF	16.0	1.36	Shield-3	Dry	12.4	1.96
Shield-3	Dry	15.8	2.41	Shield-3	Dry	18.8	3.30
Shield-3	FF	11.4	2.05	Shield-3	FF	15.8	2.79

Table 4. Test results for cylindrical projectiles.

Shield label	Material condition	v_s (m/s)	P (mm)	Shield label	Material condition	v_s (m/s)	P (mm)
Shield-1	Dry	11.9	0.20	Shield-1	Dry	15.5	0.18
Shield-1	Dry	17.9	0.26	Shield-1	FF	12.4	0.32
Shield-1	FF	16.1	0.35	Shield-1	FF	18.5	0.31
Shield-2	Dry	13.0	0.41	Shield-2	Dry	16.3	0.33
Shield-2	Dry	18.3	0.46	Shield-2	FF	12.0	0.69
Shield-2	FF	16.0	0.77	Shield-3	Dry	11.9	1.37
Shield-3	Dry	15.0	1.68	Shield-3	Dry	17.6	2.07
Shield-3	FF	11.9	1.27	Shield-3	FF	15.9	2.31

6.28 [Rohani, 1973]

Note. Data from [Bernard and Hanagud, 1975].

6.28.1 *General data on shields and projectiles*

Shields. Moist sandy clay at one of the range areas at Jefferson Proving Ground, IN, $\sigma_y = 0.345\,MPa$, $E = 20.7\,MPa$, $\gamma = 2000\,kg/m^3$.

Projectiles. Terminal delivery vehicle (TDV). Ogive-nose (Fig. 3.3.1), $D = 76.2\,mm$, $L = 450\,mm$, $\psi = 2.35$. Total mass $m = 7.85\,kg$. However, when the full length of the projectile was embedded in the shield, the projectile base separated and remained at the shield surface. The projectile mass was reduced in this case to $m = 6.17\,kg$.

6.28.2 *Test results*

Table 1#. DOP versus v_s.

v_s (m/s)	P (m)	v_s (m/s)	P (m)	v_s (m/s)	P (m)	v_s (m/s)	P (m)
106.8	1.36	107.3	1.70	107.4	1.45	107.8	1.55
113.6	1.76	131.9	2.31	136.5	2.57	136.7	2.10
136.7	2.25	136.9	2.55	137.0	1.90	137.1	2.35
137.1	2.74	137.2	1.79	137.2	1.93	137.3	1.83
139.3	2.28	139.9	2.50	170.5	2.54	171.1	2.29
185.8	2.05						

6.29 [Seah *et al.*, 2011]

6.29.1 *General data on shields and projectiles*

Table 1. Characteristics of rock shield.

$c_{l,w}$ (mm)	T (mm)	E (GPa)	f_c' (MPa)	σ_u (MPa)	v	γ (kg/m^3)
600	100	54	163	7.1	0.27	2626

Projectiles. Oil-hardened Arne tool steel (HRC 51-53, $\sigma_y = 1850 - 1900\,MPa$) ogive-nose projectiles (Fig. 3.3.1); $D = 20\,mm$, $L_{cyl} = 61.74\,mm$, $L_{nose} = 33.16\,mm$, $\rho = 60\,mm$, $\psi = 3$, $m = 197\,g$.

6.29.2 *Test results*

$$v_{bl} = 406\,m/s, \quad v_{pl} = 288\,m/s.$$

Note. Protection limit velocity, v_{pl}, was defined as the average of two striking velocities, one of which was the highest velocity such that daylight could not be seen through the plate at the point of impact and the other of which was the lowest velocity such that the projectile created a through hole but did not perforate the shield.

6.30 [Soliman *et al.*, 1976]

6.30.1 *General data on shields and projectiles*

Shields. Wooden tank filled with sand ($c_{l,w}=1.22\,m$, $T = 2.44\,m$); $\gamma = 2700\,kg/m^3$, $q^{grain} \approx 1.0\,mm$.
Projectiles. Steel ($\gamma = 7800\,kg/m^3$) spherical projectiles (Fig. 3.5.1).

6.30.2 *Test results*

Table 1#. Ricochet conditions ($D = 9.52\,mm$).

θ_s (deg)	v_s (m/s)	Ricochet (Y/N)	θ_s (deg)	v_s (m/s)	Ricochet (Y/N)	θ_s (deg)	v_s (m/s)	Ricochet (Y/N)
70	164	Y	70	168	Y	70	171	Y
70	175	Y	70	181	Y	70	183	Y
69	155	Y	69	160	Y	69	164	Y
69	167	Y	69	167	N	69	169	N
69	171	N	69	173	N	68	139	Y
68	142	Y	68	145	Y	68	146	Y
68	149	Y	68	151	Y	68	152	Y
68	155	N	68	157	N	68	160	N
68	163	N	67	129	Y	67	132	Y
67	135	Y	67	139	Y	67	140	Y
67	142	Y	67	143	Y	67	145	Y
67	148	N	67	153	N	67	156	N
66	125	Y	66	132	Y	66	135	Y
66	138	Y	66	139	Y	66	141	N
66	145	N	66	147	N	65	123	Y
65	125	Y	65	128	Y	65	132	Y
65	133	Y	65	134	N	65	136	N
65	139	N	65	141	N	64	118	Y
64	121	Y	64	123	Y	64	126	Y
64	127	Y	64	128	Y	64	130	N

θ_s (deg)	v_s (m/s)	Ricochet (Y/N)	θ_s (deg)	v_s (m/s)	Ricochet (Y/N)	θ_s (deg)	v_s (m/s)	Ricochet (Y/N)
64	132	N	64	134	N	64	136	N
63	114	Y	63	117	Y	63	118	Y
63	120	Y	63	123	Y	63	124	N
63	127	N	63	130	N	63	133	N
62	113	Y	62	114	Y	62	118	Y
62	119	Y	62	121	N	62	122	N
62	128	N	62	140	Y	61	119	N
61	122	N	61	124	N	61	129	N
61	6-117	Y	60	6-129	N			

Y/N=Yes\No

Table 2#. Plough length, l_p, vs. θ_s for $v_s = 47\,m/s$, $D = 25.4\,mm$.

θ_s (deg)	l_p (mm)	θ_s (deg)	l_p (mm)	θ_s (deg)	l_p (mm)	θ_s (deg)	l_p (mm)
88	306	86	281	84	267	82	257
80	256	77	276	75	294	73	307
70	331	67	356	65	370	63	396
61	434						

Table 3#. θ_r vs. θ_s for $v_s = 47\,m/s$, $D = 25.4\,mm$.

θ_s (deg)	θ_r (deg)	θ_s (deg)	θ_r (deg)	θ_s (deg)	θ_r (deg)	θ_s (deg)	θ_r (deg)
88	92	86	94	84	95	82	97
80	98	77	100	75	102	73	104
70	106	67	108	65	110	63	111
61	114						

6.31 [Tolch and Bushkovitch, 1947]

6.31.1 *General data on shields and projectiles*

Shields. $102\times102\times76$ blocks made of different types of rocks.

Table 1. Parameters of shields.

Shield material	f'_c (MPa)	Shield material	f'_c (MPa)
Diabase	212.90	Granite	155.71
Quartzite	176.38	Limestone (Warsaw)	137.80
Limestone (St. Louis)	134.36	Limestone (Indiana)	
Sandstone	41.34		

Table 2. Characteristics of projectiles.

Projectile label	Projectile description	Projectile label	Projectile description
Projectile-1	Cal. 0.50 AP M2	Projectile-2	20 mm AP M75
Projectile-3	37 mm AP M74	Projectile-4	75 mm AP M61
Projectile-5	75 mm AP M72	Projectile-6	76 mm AP M62
Projectile-7	90 mm AP M77	Projectile-8	90 mm AP M82

Note. Generally, values of v_s and P in tables are averaged over results of several experiments.

6.31.2 *Test results for single rounds (one shot at the same location at the shield)*

Table 3. Test results for granite shields.

Projectile label	v_s (m/s)	P (mm)	Projectile label	v_s (m/s)	P (mm)
Projectile-1	278	27.9	Projectile-1	375	27.9
Projectile-1	555	35.6	Projectile-1	831	50.8
Projectile-2	352	43.2	Projectile-2	610	81.3
Projectile-2	833	104.1	Projectile-3	297	68.6
Projectile-3	303	63.5	Projectile-3	601	129.5
Projectile-3	878	200.7			

Table 4. Test results for diabase shields.

Projectile label	v_s (m/s)	P (mm)	Projectile label	v_s (m/s)	P (mm)
Projectile-1	363	21.3	Projectile-1	548	35.6
Projectile-1	814	50.8	Projectile-2	350	43.2
Projectile-2	601	71.1	Projectile-2	798	94.0
Projectile-3	351	76.2	Projectile-3	575	114.3
Projectile-3	805	203.2			

Table 5. Test results for quartzite shields.

Projectile label	v_s (m/s)	P (mm)	Projectile label	v_s (m/s)	P (mm)
Projectile-1	356	27.9	Projectile-1	543	40.6
Projectile-1	800	63.5	Projectile-2	346	50.8
Projectile-2	596	83.8	Projectile-2	804	114.3

Table 6. Test results for limestone (Indiana) shields.

Projectile label	v_s (m/s)	P (mm)	Projectile label	v_s (m/s)	P (mm)
Projectile-1	370	38.1	Projectile-1	541	63.5
Projectile-1	812	109.2	Projectile-2	348	63.5
Projectile-2	610	139.7	Projectile-3	353	106.7
Projectile-3	587	193.0			

Table 7. Test results for sandstone shields.

Projectile label	v_s (m/s)	P (mm)	Projectile label	v_s (m/s)	P (mm)
Projectile-1	386	40.6	Projectile-1	542	68.6
Projectile-1	808	96.5	Projectile-2	350	66.0
Projectile-2	603	124.5	Projectile-3	363	114.3

Table 8. Test results for limestone (Warsaw).

Projectile label	v_s (m/s)	P (mm)	Projectile label	v_s (m/s)	P (mm)
Projectile-4	622	241.3	Projectile-5	619	304.8
Projectile-6	792	279.4	Projectile-7	823	355.6
Projectile-8	823	482.6			

Table 9. Test results for limestone (St. Louis) shields.

Projectile label	v_s (m/s)	P (mm)	Projectile label	v_s (m/s)	P (mm)
Projectile-4	619	381.0	Projectile-5	619	482.6
Projectile-6	792	711.2	Projectile-7	823	914.4
Projectile-8	808	609.6			

6.31.3 *Test results for repeated impacts in the same crater*

Additional notation. N_{shot} – number of shots.

Table 10. DOP (mm) versus v_s and N_{shot} for granite shields.

Projectile label	$v_s\ (m/s)$	N_{shot}					
		1	2	3	5	6	9
Projectile-1	350	28	–	41	48	–	61
Projectile-1	507	41	–	48	–	89	–
Projectile-1	792	56	61	107	–	–	–

Projectile label	$v_s\ (m/s)$	N_{shot}					
		1	2	3	5	6	9
Projectile-2		48	–	94	124	–	–
Projectile-2		74	124	–	–	–	–
Projectile-2		86	127	157	–	–	–

Table 11. DOP (mm) versus v_s and N_{shot} for diabase shields.

Projectile label	$v_s\ (m/s)$	N_{shot}					
		1	2	3	6	7	9
Projectile-1	371	25	–	36	–	48	56
Projectile-1	489	36	–	48	58	–	–
Projectile-1	548	19	28		–	–	–
Projectile-1	803	43	–	81	119	–	–
Projectile-2	348	43	–	79	–	112	–
Projectile-2	599	74	–	145	216	–	–

Table 12. DOP (mm) versus v_s and N_{shot} for quartzite shields and Projectile-2.

$v_s\ (m/s)$	N_{shot}		
	1	2	3
352	56	86	–
593	137	137	165
810	112	196	–

Table 13. DOP (mm) versus v_s and N_{shot} for limestone (Indiana) shields and Projectile-2.

$v_s\ (m/s)$	N_{shot}			
	1	2	3	4
354	66	86	94	124
613	127	203	221	–

Table 14. DOP (mm) versus v_s and N_{shot} for sandstone shields.

Projectile label	$v_s\ (m/s)$	N_{shot}		
		1	2	3
Projectile-2	358	61	94	–
Projectile-2	611	127	234	–
Projectile-2	827	203	284	–
Projectile-3	358	102	178	246

Table 15. DOP (mm) versus v_s and N_{shot} for limestone (Warsaw) shields.

Projectile label	v_s (m/s)	N_{shot}			
		1	5	6	16
Projectile-4	622	241	864	–	1067
Projectile-5	619	305	–	1219	–
Projectile-6	792	279	1295	–	–

6.32   [Warren *et al.*, 2004]

6.32.1   *General data on shields and projectiles*

Shields. Salem limestone from Elliot Stone Co. located in Bedford, IN; $\gamma = 2310\,kg/m^3$, $\eta_{wat}^{sh} < 0.2\%$, $\eta_{por} = 15\%$, $f_c' = 60\,MPa$.

Projectiles. VAR 4340 steel ($\sigma_y = 1430\,MPa$, $FT = 104\,MPa \cdot m^{0.5}$) ogive-nose projectiles (Fig. 3.3.1); $D = 7.11\,mm$, $L = 71.1\,mm$, $\psi = 3$.

6.32.2   *Test results*

Table 1. Test results.

θ_s (deg)	T (mm)	m (g)	Projectile HRC	v_s (m/s)	$\widetilde{P}$ (mm)	m_r (g)
15	610	20.513	44.1	411	59	–
15	406	20.503	44.0	510	86	20.458
15	610	20.47	44.6	686	130	20.387
15	610	20.42	43.7	773	154	20.277
15	610	20.521	44.2	940	215	20.270
15	610	20.46	44.4	1030	241	20.046
15	610	20.46	44.8	1255	378	19.504
30	406	20.514	43.9	413	77/32*	20.507
30	406	20.521	44.3	10	91/46*	20.503
30	406	20.521	43.9	610	115	20.476
30	610	20.48	44.7	701	42*	20.422
30	610	20.522	43.9	748	159	20.423
30	610	20.516	43.9	804	170	20.373
30	610	20.43	43.8	863	182	20.253
30	610	20.46	44.7	1023	263	20.102
30	610	20.49	44.6	1249	351	19.557

* Rod exited from the frontal face of the shield. Depth of penetration measured at the deepest part of crater from the frontal face of the shield.

6.33 [Watanabe *et al.*, 2011]

6.33.1 *General data on shields and projectiles*

Shields. Shields were made from quartz sands with the following properties of grains: $q^{grain} = 0.1 - 1.0\,mm$, $\gamma = 2650\,kg/m^3$. In the experiments grains with $q^{grain} = 0.3 - 0.5\,mm$ and sand with, $\gamma = 1490 - 1560\,kg/m^3$, $\eta_{por} = 40 - 43\%$ were used. Sand was placed in two kinds of cylindrical containers of height 200 mm fabricated from polymethyl methacrylate (PMMA). **Container B** had the inner diameter 80 mm and wall thickness 5 *mm* while the corresponding parameters for the **Container C** were 190 mm and 10 mm.

Projectiles. Cylindrical projectile ($D = 15\,mm$) consisting of cylindrical impactor with the length $L = 5\,mm$ fabricated from stainless steel or brass (mass of the projectile was 12 g or 12.5 g, respectively) and polycarbonate sabot with length of 20 *mm* mounted at the rear part of the impactor.

6.33.2 *Test results*

Table 1#. Test results for brass impactor penetrating into Container B.

v_s (m/s)	P (mm)	v_s (m/s)	P (mm)	v_s (m/s)	P (mm)
325	71.6	352	59.9	356	44.0
384	40.9	515	62.4	521	58.7

Table 2#. Test results for brass impactor penetrating into Container C.

v_s (m/s)	P (mm)	v_s (m/s)	P (mm)	v_s (m/s)	P (mm)	v_s (m/s)	P (mm)
154	98.8	252	131.9	258	131.5	335	147.9
343	139.4	353	142.8	394	144.9	486	125.0
519	101.2	583	108.7	677	110.2	945	61.8
1044	78.0	1161	34.8				

Table 3#. Test results for steel impactor penetrating into Container C.

v_s (m/s)	P (mm)	v_s (m/s)	P (mm)	v_s (m/s)	P (mm)
375	131.1	378	111.4	713	116.2
983	115.9	1148	93.3	1392	89.1

6.34 [Xu *et al.*, 2014]

6.34.1 *General data on shields and projectiles*

Shields. Sand was packed in pool box ($c_l = 1.2\,m$, $c_w = 0.5\,m$, $T = 0.4\,m$). Sand grains size distribution: for 60%, 30%, 10% of grains, $q^{grain} < 1.25\,mm$, $q^{grain} < 0.55\,mm$, $q^{grain} < 0.20\,mm$, correspondingly; minimum and maximum sand densities $1550\,kg/m^3$ and $1860\,kg/m^3$; specific gravity 2.61; $\eta_{wat}^{sh} = 4.34\%$. [ASTM D4254-91] and the [ASTM D4254-93] standards were used in testing material properties.

Projectiles. Three shapes of concrete projectiles were employed: spherical projectiles (Fig. 3.5.1; $D = 50\,mm$); cube-shaped projectiles with the length of edges 60 *mm* and 100 *mm* and with 1/3 edge length chamfer at their eight corners.

Additional notations. In Fig. 6.35.1 α_s is the angle between the vector of impact velocity, $\vec{v}_s$, and the frontal surface of the shield (FSS). Generally, after impact a striker travels inside the shield and emerges with arbitrarily directed velocity, $\vec{v}_r$. In order to determine direction of $\vec{v}_r$ one has to construct a parallelepiped with the faces $OAA'O'$ and $CBB'C'$ that are perpendicular to FSS and parallel to the plane containing

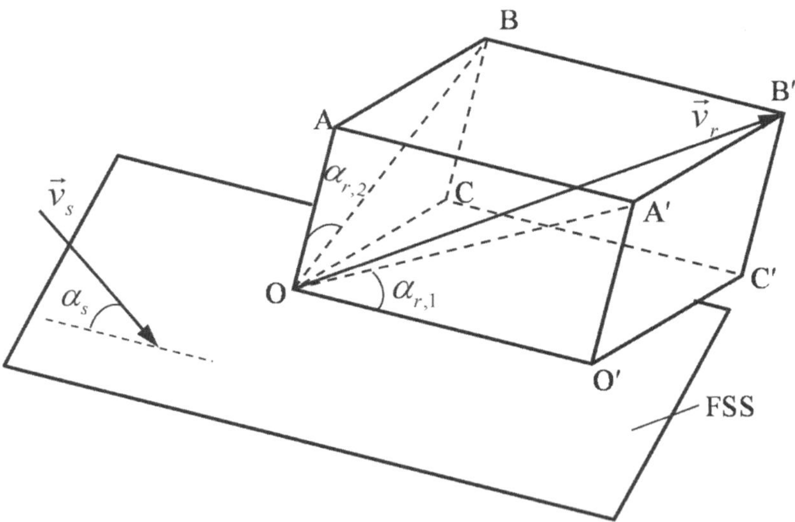

Fig. 6.35.1. Additional notations.

$\vec{v}_s$ and normal to FSS. $\vec{v}_r$ is directed along the diagonal OB' of the parallelepiped, OA' and OB' − projections of the diagonal onto faces $OAA'O'$ and $OABC$, respectfully. Direction of $\vec{v}_r$ is determined by two angles, $\alpha_{r,1}$ and $\alpha_{r,2}$.

Test results

Table 1. Test results for spherical projectiles.

α_s (deg)	v_s (m/s)	$\alpha_{r,1}$ (deg)	$\alpha_{r,2}$ (deg)	v_r (m/s)
10	17.14	6.84	1.60	13.33
10	25.5	9.39	2.44	24.00
10	40.1	11.65	2.15	30.01
10	51.8	11.65	2.15	40.01
10	61.7	11.96	1.93	52.19
10	70.8	12.27	1.79	60.02
10	80	12.60	0.76	63.16
10	92.31	13.08	1.35	70.61
15	25.5	9.05	2.12	21.33
15	40.2	10.83	2.12	30.01
15	51.8	12.19	1.63	40.01
15	61.8	12.24	2.26	45.73
15	70.9	12.75	0.72	60.00
15	80	10.66	2.19	66.71
15	92.31	11.81	1.27	75.02
20	17.14	7.91	1.43	15.00
20	30	11.16	1.49	23.08
20	40.1	10.92	3.97	24.01
20	48	12.74	0.75	30.00
20	62	12.49	0.73	33.71
20	66.67	13.21	1.16	40.00
20	80	13.37	1.49	48.01
20	92.31	13.68	1.83	60.02
30	17.14	11.54	0.88	13.33
30	30	11.77	1.11	13.79
30	39.63	18.23	1.59	13.91
30	48	17.48	0.34	24.00
30	62.23	18.72	1.34	22.51
30	66.67	18.33	0.09	26.67
30	80	19.18	0.17	30.00
30	92.31	20.39	0.21	34.29
40	52	17.16	6.38	17.09
40	72.2	19.85	7.00	19.42
50	51.9	18.40	1.11	8.82
50	72.03	21.92	5.56	12.25

Table 2. Test results for 60 *mm* chamfered projectiles.

α_s (deg)	v_s (m/s)	$\alpha_{r,1}$ (deg)	$\alpha_{r,2}$ (deg)	v_r (m/s)
10	23	9.35	1.97	17.14
10	34.3	11.94	0.25	21.82
10	41.2	12.56	1.23	24.00
10	48	12.95	1.74	30.00
10	59.8	13.45	2.24	34.30
10	70.6	13.13	4.13	38.71
10	82.8	11.44	2.04	60.00
10	92.3	12.56	4.42	63.17
15	23	10.66	0.81	10.91
15	34.3	14.12	4.21	20.00
15	41.2	13.29	7.20	24.01
15	48	12.79	3.86	30.00
15	59.8	12.45	3.51	36.36
15	70.6	12.32	3.34	38.40
15	82.8	14.98	3.58	40.01
15	92.3	17.57	5.65	42.88
20	23	13.38	3.14	12.00
20	34.3	15.56	0.79	16.67
20	41.2	13.41	1.24	18.35
20	48	17.11	1.81	20.69
20	59.8	12.75	1.51	23.61
20	70.6	14.43	6.58	27.92
20	82.8	13.77	4.06	29.27
20	92.3	15.95	10.05	31.62
30	23	15.28	2.50	11.43
30	34.3	15.47	1.14	13.33
30	41.2	16.40	2.65	10.75
30	48	17.08	0.23	13.04
30	59.8	15.16	2.91	13.66
30	70.6	16.46	2.73	18.46
30	82.8	16.80	4.99	21.83
30	92.3	17.45	0.82	23.53
40	48	15.65	1.66	11.45
40	70.6	16.58	3.55	12.34
50	48	17.39	4.34	9.68
50	70.6	18.21	2.16	10.18

Table 3. Test results for chamfered projectiles with length of edge 100 *mm*.

α_s (deg)	v_s (m/s)	$\alpha_{r,1}$ (deg)	$\alpha_{r,2}$ (deg)	v_r (m/s)
10	19.67	14.09	0.52	8.00
10	30	19.23	3.99	12.00
10	40	16.22	4.39	24.01

α_s (deg)	v_s (m/s)	$\alpha_{r,1}$ (deg)	$\alpha_{r,2}$ (deg)	v_r (m/s)
10	52.17	12.47	9.28	30.02
10	60	11.65	9.37	40.02
10	70.59	17.55	10.68	48.08
10	80	17.98	10.79	60.10
10	92.31	18.99	11.09	70.73
15	19.67	14.64	0.78	8.57
15	30	13.62	0.34	11.43
15	40	17.103	3.30	17.14
15	52.17	18.72	2.55	21.82
15	60	17.29	3.79	24.00
15	70.59	14.22	6.02	26.68
15	80	16.35	3.21	30.00
15	92.31	16.43	2.57	34.29
20	19.67	9.46	4.40	7.06
20	30	14.22	1.13	8.00
20	40	16.20	1.33	10.05
20	52.17	16.37	1.43	13.33
20	60	16.85	4.93	16.69
20	70.59	18.07	3.79	24.01
20	80	18.65	4.00	26.68
20	92.31	19.23	4.22	29.28
30	17.14	19.71	0.09	6.86
30	27.27	20.57	0.13	9.23
30	40	21.42	0.18	11.76
30	52.17	22.57	0.37	14.12
30	60	22.24	0.57	16.11
30	70.59	20.89	0.86	20.00
30	80	20.304	1.05	23.08
30	92.31	20.05	1.15	25.53
40	52.17	12.83	0.16	16.05
40	70.59	19.68	1.66	20.01
50	52.17	14.30	2.72	9.16
50	70.59	15.25	3.79	12.56

6.35 [Yamamato *et al.*, 2011]

6.35.1 *General data on shields and projectiles*

Shields. Florida coastal sand called Eglin Sand (Quikrete Commercial Grade Fine Sand No. 1961) and glass beads (Ballotini Glass Beads, size6, Batch SR 091208). Penetrator entered vertically into the open top cylinder container ($D^{sh} = 100\,mm$, $T = 150 - 300\,mm$) or open top box ($c_l = 100\,mm$, $c_w = 30\,mm$, $T = 200\,mm$).

Projectiles. Stainless steel spherical projectiles (Fig. 3.5.1); $D = 9.53mm$, $m = 3.6g$.

6.35.2   *Test results*

Table 1. Test results for Eglin sand ($q^{grain} = 355 - 500\,mm$).

γ (kg/m^3)	v_s (m/s)	P (mm)	γ (kg/m^3)	v_s (m/s)	P (mm)	γ (kg/m^3)	v_s (m/s)	P (mm)
1490	940	163	1560	1000	99	1500	1030	138
1550	1130	97	1550	1140	80			

Table 2. Test results for Eglin sand ($q^{grain} = 500 - 710\,mm$).

γ (kg/m^3)	v_s (m/s)	P (mm)	γ (kg/m^3)	v_s (m/s)	P (mm)
1540	970	209	1580	1130	97
1570	1250	75	1570	1260	75

Table 3. Test results for glass beads.

γ (kg/m^3)	v_s (m/s)	P (mm)	γ (kg/m^3)	v_s (m/s)	P (mm)
1550	1210	149	1550	1210	211
1540	1230	171	1540	1270	166

6.36   [Young, 1967]

See also [Young, 1969].

6.36.1   *General data on shields and projectiles*

Table 1. Shield description.

Shield label	Shield location	Layer number	Layer material	Depth range (m)
Shield 1	Tonopah Test Range, NV, Main Dry Lake	1 of 3	Clayey silt, silty clay, dense, hard	0–2.44
Shield 1	Ibid	2 of 3	Sand, silty, very dense, dry, cemented	2.44–7.62
Shield 1	Ibid	3 of 3	Sand, silty, clayey, very dense	>7.62
Shield 2	Salt target, White Sands Missile Range, NM	1 of 2	Clay, silty, soft, wet, brown	0–0.49

Shield label	Shield location	Layer number	Layer material	Depth range (m)
Shield 2	Ibid	2 of 2	Ibid	>0.49
Shield 3	Great Salt Lake Desert, UT	1 of 1	Clay, soft, wet, gray, medium varved to high plasticity	0–3.05
Shield 4	Northrup Strip, Salt target White Sands Missile Range, NM	1 of 1	Gypsite, selenite, hard, moist, very homogeneous	0–3.05
Shield 5	Gulkana Glacter, AK	1 of 1	Ice, glacier	0–6.1
Shield 6	Eielson AFB, AK	1 of 1	Silt, clayey, frozen (permafrost)	0–9.14
Shield 7	Eglin AFB, FL	1 of 1	Sand, loose, moist	0–21.34
Shield 8	Nevada Test Site, Tonopah Test Range, NV	1 of 1	Rock, highly welded, fine-grained agglomerate	0–3.05
Shield 9	Grants, NM	1 of 1	Sandstone	0–9.14
Shield 10	Skaggs Island, CA	1 of 1	Clay, silty, wet (bay mud)	0–15.24
Shield 11	Tonopah Test Range, NV	1 of 1	Sand, silty, clayey, dense (desert alluvium)	0–3.05

Table 2. Properties of layer material.

Shield label	Layer number	f_c' (kPa)	γ (kg/m^3)	η_{wat}^{sh} (%)	k_S
Shield 1	1	766	1714	15	5.2
Shield 1	2	766	1858	10	2.5
Shield 1	3	479	1874	13	5.2
Shield 2	1	24	1762	37	40
Shield 2	2	192	2083	25	6.5
Shield 3	1	29	1682	71	50
Shield 4	1	527	1954	37	2.5
Shield 5	1	–	–	–	4.15
Shield 6	1	–	–	–	3.8
Shield 7	1	–	1826	7	6.5
Shield 8	1	–	–	–	1.07*
Shield 9	1	–	–	–	1.3*
Shield 10	1	57	–	75	40*
Shield 11	1	–	–	–	4.4*

* Data from [Young, 1969].

Projectiles. Ogive-nose (Fig. 3.3.1), cone-nose (Fig. 3.2.1) and cylindrical (Fig. 3.1.1) projectiles.

Note. Original source contains some additional characteristics of soils.

6.36.2 *Test results*

Table 3. Test results for Shield 1 and ogive-nose projectiles.

k_S	D (mm)	ψ	v_s (m/s)	P (m)	k_S	D (mm)	ψ	v_s (m/s)	P (m)
5.4	79	9.25	34.1	1.128	5.2	79	9.25	42.7	1.554
5.3	97	9.25	32.9	0.884	5.3	97	9.25	42.4	1.402
6.1	112	9.25	33.5	0.945	5.9	112	6.0	33.8	0.823
5.3	112	2.2	34.4	0.610	5.2	112	6.0	49.7	1.402
5.9	112	2.2	50.6	1.311	5.2	112	2.2	51.2	1.219
5.1	112	9.25	51.2	1.615	5.3	112	2.2	51.8	1.219
5.4	112	2.2	51.8	1.250	5.4	112	2.2	51.8	1.250
5.2	112	2.2	52.4	1.219	5.0	112	2.2	52.4	1.189
4.9	112	9.25	52.4	1.554	5.4	112	2.2	53.3	1.311
5.6	112	2.2	60.4	1.737	5.3	112	6.0	61.3	1.981
5.5	112	2.2	64.9	1.859	6.9	137	2.2	32.9	0.610
5.9	137	9.25	42.1	1.311	5.7	137	2.2	43.6	1.006
5.5	137	2.2	50.3	1.219	5.9	137	9.25	51.2	1.829
5.4	137	9.25	52.4	1.433	5.3	137	9.25	53.0	1.433
4.6	137	2.2	53.3	0.884	4.9	137	9.25	54.3	1.372
4.7	137	9.25	54.3	1.311	5.6	137	9.25	54.9	1.615
4.9	137	2.2	63.7	1.280	5.7	137	9.25	64.0	2.103
5.1	137	2.2	78.9	1.981	5.2	203	2.2	52.1	0.732
4.8	203	6.0	78.0	1.646	–	203	9.25	100.6	3.505
–	203	6.0	106.4	2.591	–	203	9.25	115.8	3.901
–	203	9.25	117.3	3.871	–	203	9.25	118.0	3.962
–	203	6.0	140.2	2.896	4.3	203	6.0	153.0	3.810
–	203	9.25	156.7	5.243	–	203	9.25	158.5	4.145
–	203	9.25	170.1	5.029	–	203	9.25	171.3	5.547
–	203	9.25	172.2	5.700	–	203	9.25	173.7	5.395
–	203	9.25	203.9	5.639	–	203	9.25	228.6	6.096
4.2	216	9.25	81.7	1.798	–	229	9.25	74.1	3.109
–	229	2.2	78.0	2.804	–	229	6.0	93.0	3.901
–	229	6.0	120.1	5.060	–	229	2.2	136.2	4.298
–	229	6.0	139.6	4.328	–	229	6.0	142.3	4.206
–	229	6.0	150.0	5.182	–	229	9.25	207.6	7.437
–	229	6.0	217.3	11.582	–	229	9.25	276.1	17.374
–	229	6.0	292.3	16.307	–	229	9.25	299.9	16.002
–	229	6.0	316.1	16.307	–	229	2.2	320.0	16.886
–	229	2.2	323.1	16.002	–	229	9.25	328.3	22.250
–	457	12.5	76.8	3.962	–	457	12.5	274.6	19.812

Table 4. Test results for Shield 2 and ogive-nose projectiles.

D (mm)	ψ	v_s (m/s)	P (m)	D (mm)	ψ	v_s (m/s)	P (m)
112	2.2	32.0	0.792	112	2.2	33.5	1.158
112	2.2	52.1	1.676	137	2.2	34.1	0.975
137	9.25	35.4	1.372	137	2.2	52.7	1.524
137	9.25	52.7	1.951	137	9.25	65.2	2.286
203	2.2	34.4	0.823	203	2.2	41.5	1.006
203	9.25	48.2	1.494	203	2.2	51.5	1.402
203	2.2	62.2	1.768				

Table 5. Test results for Shields 3-11 and ogive-nose projectiles.

Shield label	k_S	D (mm)	ψ	v_s (m/s)	P (m)
Shield 3	50	203	2.2	38.4	4.206
Shield 4	3.0	112	9.25	51.8	0.945
Shield 4	2.6	137	9.25	52.7	0.701
Shield 4	3.3	137	9.25	52.7	0.884
Shield 4	2.5	137	9.25	77.7	1.280
Shield 4	2.3	137	9.25	83.5	1.341
Shield 5	4.1	229	6.0	146.3	5.486
Shield 5	4.2	229	6.0	146.3	5.639
Shield 6	3.8	229	6.0	121.9	4.724
Shield 6	3.7	229	6.0	121.9	4.572
Shield 6	3.9	229	6.0	176.8	7.772
Shield 6	3.8	229	6.0	237.7	10.516
Shield 7	8.2	229	6.0	122.2	10.516
Shield 7	5.7	229	6.0	130.1	7.864
Shield 7	5.9	229	6.0	195.4	15.392
Shield 7	6.3	229	6.0	208.5	15.392
Shield 7	6.3	229	9.25	267.3	21.641
Shield 7	5.9	229	6.0	291.1	19.294
Shield 7	7.4	229	2.2	320.6	21.336
Shield 8	1.07	229	9.25	324.6	3.962
Shield 9	1.3	203	9.25	268.2	3.048
Shield 9	1.3	259	9.25	249.9	3.109
Shield 10	40	229	9.25	23.8	4.877
Shield 11	5.0	229	2.2	291.1	13.716
Shield 11	4.1	229	2.2	297.5	11.034

Table 6. Test results for Shield 1 and cylindrical projectiles.

k_S	D (mm)	v_s (m/s)	P (m)	k_S	D (mm)	v_s (m/s)	P (m)
5.6	112	32.0	0.396	5.7	112	32.0	0.396
5	112	40.2	0.518	4.8	112	42.7	0.549
4.8	112	50.9	0.732	5.4	112	52.4	0.884
4.2	112	62.5	0.884	–	229	90.5	2.362
–	229	167.0	4.145	5	229	179.8	5.182
4.8	229	182.9	3.993	4.1	229	198.1	4.663
–	229	220.4	5.791				

Table 7. Test results for cone-nose projectiles.

Shield label	k_S	D (mm)	L_{nose}/D	v_s (m/s)	P (m)
Shield 1	6	216	2	49.7	1.067
Shield 1	5.5	216	3	50.9	1.250
Shield 1	5.3	216	3	59.4	1.554
Shield 1	5.3	216	3	59.4	1.920
Shield 1	5.7	216	3	67.1	2.377
Shield 1	4.2	216	2	80.8	1.707
Shield 1	4.5	216	3	80.8	2.195
Shield 1	4.3	216	3	82.0	2.164
Shield 1	4.2	216	3	96.0	2.682
Shield 4	2.5	216	3	78.3	1.158
Shield 4	2.2	216	3	80.8	1.097
Shield 4	1.9	216	3	106.7	1.433

6.37   [Young, 1976]

Note. Data from [Bernard and Creighton, 1979].

6.37.1   *General data on shields and projectiles*

Shields. Soil penetration tests were conducted in Antelope dry lake site on Tonopah Test Range. Values of k_S are given in Table 1.

Table 1. Values of k_S.

Depth (m)	k_S	Depth (m)	k_S	Depth (m)	k_S
0–8.2	5	8.2–10.7	2.5	10.7–15.2	5.5
15.2–24.4	7	24.4–25.9	3	25.9–33.2	8
33.2–51.8	11.5	> 51.8	13.5		

Projectiles. Ogive-nose projectiles (Fig. 3.3.1).

6.37.2 *Test results*

Table 2. Parameters of projectiles and test results.

m (kg)	D (mm)	L (m)	ψ	v_s (m/s)	θ_s (deg)	$\widetilde{P}$ (m)
145	152	1.219	9.25	711	30	39.62*
181	165	1.651	9.25	493	75	38.25**

* Rotated 80°and stopped 41.16 m below surface at angle of 20° from horizontal.

** Rotated 2° towards surface.

6.38 [Young, 1978]

Note. Data from [Bernard and Creighton, 1979].

6.38.1 *General data on shields and projectiles*

Shields. Soil penetration tests were conducted in the Pedro dry lake site on Tonopah Test Range, $k_S = 3.6$.

Projectiles. Ogive-nose projectiles (Fig. 3.3.1); $D = 156\,mm$, $L = 1.556m$, $\psi = 6$, $m = 158\,kg$.

Note. Slightly oblique impact.

6.38.2 *Test results*

Table 1. Test results on $\widetilde{P}$.

v_s (m/s)	$\widetilde{P}$ (m)	v_s (m/s)	$\widetilde{P}$ (m)
427	14.97	506	17.68

Bibliography

[Abadel *et al.*, 2014] Abadel, A., Almusallam, T., Abbas, H., Al-Salloum, Y., Siddiqui, N., Iqbal, R. (2014). Impact behavior of hybrid-fiber RC slabs. In: *Proc. of the 7th International Conference on FRP Composites in Civil Engineering* (August 20-22, 2014, Vancouver, Canada), paper 230.

[Abdel-Kader and Fouda, 2014] Abdel-Kader, M., Fouda, A. (2014). Effect of reinforcement on the response of concrete panels to impact of hard projectiles. *Int. J. of Impact Engineering*, **63**: 1-17.

[Abdel-Wahed *et al.*, 2010] Abdel-Wahed, M.A., Salem, A.M., Zidan, A.S., Riad, A.M. (2010). Penetration of a small caliber projectile into single and multi-layered targets. In: *Proc. of 4th European Conf. on Computational Mechanics (ECCM 2010)*, (May 16-21, 2010, Paris, France).

[Aitken *et al.*, 1976] Aitken, G.W., Swinzow, G.K., Farrell, D.R. (1976). Projectile and fragment penetration in snow and frozen soil. U.S. Army Cold Regions Research and Engineering Laboratory, Hanover, NH.

[Allen *et al.*, 1957a] Allen, W.A., Mayfield, E.B., Morrison, H.L. (1957). Dynamics of a projectile penetrating sand. Part II. *J. of Applied Physics*, **28**(11): 1331-1335.

[Allen *et al.*, 1957b] Allen, W.A., Mayfield, E.B., Morrison, H.L. (1957). Dynamics of a projectile penetrating sand. *J. of Applied Physics*, **28**(3): 370-376.

[Allison, 1992] Allison, W.H. (1992). Armor Steel Ballistic Test. 20-mm API-T M602 Projectiles. TECOM Project No. 1-EG-965-000-773.

[Almansa and Cánovas, 1996] Almansa, E.M., Cánovas, M.F. (1996). Design of SFRC barriers against impact of small arms projectiles. In: Jones, N., Bulson, P.S., eds., *Structures Under Shock and Impact IV* [*Proc. of the 4th Int. Conf. on Structures under Shock and Impact* (July 1996, Udine, Italy)], Computational Mechanics Publ., Southampton, pp. 79-88.

[Almansa and Cánovas, 1999] Almansa, E.M., Cánovas, M.F. (1999). Behaviour of normal and steel fiber reinforced concrete under impact of small projectiles. *Cement Concrete Research*, **29**(11): 1807-1814.

[Almohandes *et al.*, 1996] Almohandes, A.A., Abdel-Kader, M.S., Eleiche, A.M. (1996). Experimental investigation of the ballistic resistance of steel-fiberglass reinforced polyester laminated plates. *Composites. Part B*, **27**(5): 447-458.

[Almusallam *et al.*, 2013] Almusallam, T.H., Siddiqui, N.A., Iqbal, R.A., Abbas, H. (2013). Response of hybrid-fiber reinforced concrete slabs to hard projectile impact. *Int. J. of Impact Engineering*, **58**: 17-30.

[Almusallam *et al.*, 2015b] Almusallam, T., Al-Salloum, Y., Alsayed, S., Iqbal, R., Abbas, H. (2015). Effect of CFRP strengthening on the response of RC slabs to hard projectile impact. *Nuclear Engineering and Design*, **286**: 211–226.

[Almusallam *et al.*, 2015a] Almusallam, T.H., Abadel, A.A., Al-Salloum, Y.A., Siddiqui, N.A. Abbas, H. (2015). Effectiveness of hybrid-fibers in improving the impact resistance of RC slabs. *Int. J. of Impact Engineering*, **81**: 61-73.

[Anderson *et al.*, 1984] Anderson, W. F., Watson, A.J., Armstrong, P.J. (1984). Fibre reinforced concretes for the protection of structures against high velocity impact. In: Morton, J., ed., *Structural Impact and Crashworthiness, vol. 2 [Proc. of the Int. Conf. on Structural Impact and Crashworthiness* (July 16-20, 1984, Imperial College, London, U.K.)]. London, Elsevier Applied Science Publishers, pp. 687-695.

[Anderson *et al.*, 1992] Anderson, C.E., Jr., Morris, B.L., Littlefield, D.L. (1992). A penetration mechanics database. Rep. No. 3593/001, Southwest Research Institute.

[Anderson *et al.*, 1999] Anderson, C.E., Jr., Hohler, V., Walker, J.D., Stilp, A.J. (1999). The influence of projectile hardness on ballistic performance. *Int. J. of Impact Engineering*, **22**(6): 619-632.

[Ansari *et al.*, 2010] Ansari, R., Khan, S.H., Khan, A.H. (2010). Oblique impact of cylindro-conical projectile on thin aluminium plates. In: *Proc. of Int. Conf. on Theoretical, Applied, Computational and Experimental Mechanics (ICTACEM 2010)*, (December 27-29, 2010, Kharagpur, India).

[Ansari, 1998] Ansari, R. (1998). Normal and oblique impact of projectiles on single and layered thin plates. Ph.D. thesis, Indian Institute of Technology, Delhi.

[Astarlioglu *et al.*, 2013.] Astarlioglu, S, Krauthammer, T., Bui, L. (2013). Penetration Resistance of Normal-strength and Ultra-high-performance Concrete. Rep. to DTRA, CIPPS-TR-007-2013, Tasks 501 and 5.2. Center for Infrastructure Protection and Physical Security, University of Florida.

[ASTM D4254-91] Standard test method for minimum index density and unit weight of soils and calculation of relative density [ASTM D4254-91] (1992). American Society of Testing and Materials, West Conshohocken, PA.

[ASTM D4254-93] Standard classification of soils for engineering purposes (Unified Soil Classification System) [ASTM D4254-93] (1993). American Society of Testing and Materials, West Conshohocken, PA.

[Awerbuch and Bodner, 1974] Awerbuch, J., Bodner, S.R. (1974). Experimental investigation of normal perforation of projectiles in metallic plates. *Int. J. of Solids and Structures,* **10**(6): 685-699.

[Awerbuch and Bodner, 1977] Awerbuch, J., Bodner, S.R. (1977). An investigation of oblique perforation of metallic plates by projectiles. *Experimental Mechanics,* **17**(4): 147-153.

[Awerbuch, 1969] Awerbuch, J. (1969). Projectile penetration in metals and composite materials. Master Thesis, Israel Institute of Technology, Haifa, Israel (in Hebrew).

[Awerbuch, 1970] Awerbuch, J. (1970). A mechanics approach to projectile penetration. *Israel J. of Technology*, **8**(4): 375-383.

[Babaei *et al.*, 2011] Babaei, B., Shokrieh, M.M., Daneshjou, M. (2011). The ballistic resistance of multi-layered targets impacted by rigid projectiles. *Materials Science and Engineering A*, **530**: 208–217.

[Backman and Finnegan, 1976] Backman, M.E., Finnegan, S.A. (1976). Dynamics of the oblique impact and ricochet of non-deforming spheres against thin plates. Rep. NWC TP 5844. Naval Weapons Center, China Lake, CA.

[Backman and Goldsmith, 1978] Backman, M., Goldsmith, W. (1978). The mechanics of penetration of projectiles into targets. *Int. J. of Engineering Science*, **16**(1): 1-99.

[Baranov *et al.*, 1991] Baranov, A.V., Konanykhin, Y.P., Pasechnik, L.P., Sugak, S.G. (1991). *Investigation of Dynamic Effect of Impactor on Multilayer Shields.* Universitet Druzhby Narodov, Chernogolovka (in Russian).

[Barber, 1973] Barber, R.B. (1973). *Steel Rod/Concrete Slab Impact Test: Experimental simulation.* Bechtel Corporation.

[Barr *et al.*, 1983] Barr, P., Carter, P.G., Howe, W.D., Neilson, A.J., Richards, A.E. (1983). Experimental studies of the impact resistance of steel faced concrete composition. In: *Trans. of the 7th Int. Conf. on Structural Mechanics in Reactor Technology (SMiRT-7)*, (Aug 22-26, 1983, Chicago, IL), paper j8/4, pp. 395-402.

[Barr, 1990] Barr, P., compiler. (1990) *Guidelines for the Design and Assessment of Concrete Structures Subjected to Impact.* UK Atomic Energy Authority, Safety and Reliability Directorate, UK.

[Bartus, 2009] Bartus, S.D. (2009). Evaluation of Titanium-5Al-5Mo-5V-3Cr (Ti-5553) alloy against fragment and armor-piercing projectiles. Tech. Rep. ARL-TR-4996. U.S. Army Research Laboratory, Weapons and Materials Research Directorate, Aberdeen Proving Ground, MD.

[Baty *et al.*, 2003] Baty, R.S., Lundgren, R.G., Patterson, W.J. (2003). On pilot-hole assisted penetration. In: *Proc. of 11th Int. Symp. on Interaction of the Effects of Munitions with Structures* (May 5-9, 2003, Mannheim, Germany).

[Ben-Dor *et al.*, 2001] Ben-Dor, G., Dubinsky, A., Elperin, T. (2001). A class of models implying the Lambert-Jonas relation. *Int. J. of Solids and Structures*, **38**(40-41): 7113-7119.

[Ben-Dor *et al.*, 2002] Ben-Dor, G., Dubinsky, A., Elperin, T. (2002). On the Lambert-Jonas approximation for ballistic impact. *Mechanics Research Communications*, **29**(2-3): 137-139.

[Ben-Dor *et al.*, 2006] Ben-Dor, G., Dubinsky, A., Elperin, T. (2006). *Applied High-Speed Plate Penetration Dynamics.* Springer, Dordrecht, 2006.

[Ben-Dor *et al.*, 2012] Ben-Dor, G., Dubinsky, A., Elperin, T. (2012). Investigation and optimization of protective properties of metal multi-layered shields: A Review. *Int. J. of Protective Structures*, **3**(3): 275 -291.

[Ben-Dor *et al.*, 2013] Ben-Dor, G., Dubinsky, A., Elperin, T. (2013). *High-Speed Penetration Dynamics: Engineering Models and Methods*. World Scientific Publishing, 2013.

[Beppu*et al.*, 2008] Beppu, M., Miwa, K., Itoh, M., Katayama, M., Ohno, T. (2008). Damage evaluation of concrete plates by high-velocity impact. *Int. J. of Impact Engineering*, **35**(12):1419–1426.

[Bergman, 1950] Bergman, S.G.A. (1950). Inträngning av panserbrytande projectiler och bomber i armerad betong FortF. Rapp. B2, Stockholm, Sweden (in Swedish).

[Bernard and Creighton, 1979] Bernard, R.S., Creighton, D.C. (1979). Projectile penetration in soil and rock: analysis for non-normal impact. Rep. WES/TR/SL-79-15. U.S. Army Engineer Waterways Experiment Station, Vicksburg, MS.

[Bernard and Hanagud, 1975] Bernard, R.S., Hanagud, S.V. (1975). Development of a projectile penetration theory. Rep. S-75-9. Penetration theory for shallow to moderate depths (Rep. 1 of Series). U.S. Army Engineer Waterways Experiment Station, Vicksburg, MS.

[Bernard, 1976] Bernard, R.S. (1976). Development of a projectile penetration theory: Rep. S-75-9. Deep penetration theory of homogeneous and layered targets. (Rep. 2 of Series). Soils and Pavements Laboratory, U.S. Army Engineer Waterways Experiment Station, Vicksburg, MS.

[Bernard, 1977] Bernard, R.S. (1977). Empirical analysis of projectile penetration in rock. Miscellaneous paper S-77-16. U.S. Army Engineer Waterways Experiment Station, Vicksburg, MS.

[Berriaud *et al.*, 1977] Berriaud, C., Sokolovsky, A., Gueraud, R., Dulac, J., Labrot, R. (1977). Local behaviour of reinforced concrete walls under missile impact. In: *Proc. of 4th Int. Conf. on Structural Mechanics in Reactor Technology (SMiRT-4)*, (August 15-19, 1977, San Francisco, California), paper J7/9.

[Berriaud *et al.*, 1978] Berriaud, C., Sokolovsky, A., Gueraud, R., Dulac, J., Labrot, R. (1978). Local behaviour of reinforced concrete walls under missile impact. *Nuclear Engineering and Design*, **45**(2): 457-469 (in French).

[Berriaud *et al.*, 1979] Berriaud, C., Verpeaux, P., Hoffman, A., Jamet, P., Avet-Flancard, R. (1979). Tests and calculations of the local behaviour of concrete structures under missile impact. In: *Trans. of 5th Int. Conf. on Structural Mechanics in Reactor Technology (SMiRT-5)*, (August 13-17, 1979, Berlin (West), Germany), paper J7/1.

[Berriaud *et al.*, 1982] Berriaud, C., Verpeaux, P., Jamet, P. (1982). Concrete wall perforation by rigid missile. In: *RILEM-CEB-IABSE-IASS Interassociation Symp. on Concrete Structures under Impact and Impulsive Loading* (June 2-4, 1982, Berlin, Germany, F.R), pp. 358–367.

[Beth, 1945] Beth, R.A. (1945). Concrete penetration, Division 2. NDRC Rep. A-319, OSRD Rep. 4856. National Defense Research Committee (NDRC) of the Office of Scientific Research and Development (OSRD). Washington, DC.

[Bless *et al.*, 2009] Bless, S.J., Berry, D.T., Pedersen, B., Lawhorn, W. (2009). Sand penetration by high-speed projectiles. *AIP Conference Proc.*, 1195: 1361-1364.

[Bless *et al.*, 2010] Bless, S.J., Berry, D.T., Pedersen, B., Satapathy, S., Warren, T., Lawhorn, W. (2010). High-Speed Projectile Penetration in Sand. In: *Proc. of the 25th Int. Symp. on Ballistics* (May 17-21, 2010, Beijing, China), pp. 1339-1346.

[Bless *et al.*, 2011] Bless, S.J., Cooper, W., Watanabe, K. (2011). Penetration of Rigid Rods into Sand. In: *Proc. of the 26th Int. Symp.on Ballistics* (September 12-16, 2011, Miami, FL), pp. 1252-1257.

[Bless *et al.*, 2012] Bless, S.J., Cooper, W., Watanabe, K., Peden, R. (2012). Deceleration of projectiles in sand. *AIP Conference Proc.*, 1426: 45-47.

[Borg *et al.*, 2013] Borg, J.P., Morrissey, M.P., Perich, C.A., Vogler, T.J., Chhabildas, L.C. (2013). In situ velocity and stress characterization of a projectile penetrating a sand target: Experimental measurements and continuum simulations. *Int. J. of Impact Engineering* , **51**: 23-35.

[Børvik *et al.*, 1998] Børvik, T., Langseth, M., Hopperstad, O.S., Malo, K.A. (1998). Empirical equations for ballistic penetration of metal plates. Fortifikatorisk Notat No. 260/98. The Norwegian Defence Construction Service, Central Staff – Tech. Division, Oslo, Norway.

[Børvik *et al.*, 1999] Børvik, T., Langseth, M., Hopperstad, O.S., Malo, K.A. (1999). Ballistic penetration of steel plates. *Int. J. of Impact Engineering*, **22**(9-10): 855-886.

[Børvik *et al.*, 2002a] Børvik, T., Langseth, M., Hopperstad, O.S., Malo, K.A. (2002). Perforation of 12mm thick steel plates by 20mm diameter projectiles with flat, hemispherical and conical noses. Part I: Experimental study. *Int. J. of Impact Engineering*, **27**(1): 19–35.

[Borvik *et al.*, 2002b] Borvik, T., Langseth, M., Hopperstad, O.S., Polanco-Loria, M.A. (2002). Ballistic perforation resistance of high performance concrete slabs with different unconfined compressive strengths. In: Brebbia, C.A., de Wilde, W.P., eds., *High Performance Structures and Composites* [*Proc. of 1st Int. Conf. on high performance structures and composites* (Sevilla, Spain)]. WIT Press, Southampton, pp. 273-282.

[Børvik *et al.*, 2003] Børvik, T., Hopperstad, O.S., Langseth, M., Malo, K.A. (2003). Effect of target thickness in blunt projectile penetration of Weldox 460 E steel plates. *Int. J. of Impact Engineering*, **28**(4): 413-464.

[Børvik *et al.*, 2004] Børvik, T., Clausen, A. H., Hopperstad, O.S., Langseth, M. (2004). Perforation of AA5083-H116 aluminium plates with conical-nose steel projectiles – experimental study. *Int. J. of Impact Engineering*, **30**(4): 367-384.

[Børvik *et al.*, 2007] Børvik, T., Gjørv, O.E., Langseth, M. (2007). Ballistic perforation resistance of high-strength concrete slabs. *Concrete Int.*, **29**(6): 45-50.

[Børvik *et al.*, 2009] Børvik, T., Dey, S., Clausen, A.H. (2009). Perforation resistance of five different high-strength steel plates subjected to small-arms projectiles. *Int. J. of Impact Engineering*, **36**(7): 948–964.

[Børvik *et al.*, 2010a] Børvik, T., Hopperstad, O.S., Pedersen, K.O. (2010). Quasi-brittle fracture during structural impact of AA7075-T651 aluminium plates. *Int. J. of Impact Engineering*, **37**(5): 537-551.

[Børvik *et al.*, 2010b] Børvik, T., Forrestal, M.J., Warren, T.L. (2010). Perforation of 5083-H116 aluminum armor plates with ogive-nose rods and 7.62 mm APM2 bullets. *Experimental Mechanics*, **50**(7): 969-978.

[Børvik *et al.*, 2011] Børvik, T., Dey, S., Olovsson, L., Langseth, M. (2011). Normal and oblique impact of small arms bullets on AA6082-T4 aluminium protective plates. *Int. J. of Impact Engineering*, **38**(7):577-589.

[Børvik *et al.*, 2013] Børvik, T., Olovsson, L., Dey, S. (2013). Penetration of bullets into granular targets. In: *Proc. of the 27th Int. Symp. on Ballistics* (April 22-26, 2013, Freiburg, Germany), paper 14156549. DEStech Publications, Inc, pp. 1856-1865.

[Brooks and Erickson, 1971] Brooks, P.N., Erickson, W.H. (1971). Ballistic evaluation of materials for armor penetrators. Rep. No. DREV R-643/71. Defence Research Establishment Valcartier, Quebec, Canada.

[Brooks, 1973] Brooks, P.N. (1973). On the prediction of crater profiles produced in ductile targets by the impact of rigid penetrators at ballistic velocities. Rep. DREV R-686/73. Defence Research Establishment Valcartier, Quebec, Canada.

[Buchar *et al.*, 2001] Buchar, J., Krejčí, J., Forejt, M. (2001). Response of concrete to the impact of fragment simulating projectiles. *Engineering Mechanics*, **8**(1): 63-69.

[Buchar *et al.*, 2002a] Buchar, J., Voldřich, J., Rolc, S., Lazar, M. (2002). Ballistic performance of the dual hardness armor. In: *Proc. of the 20th Int. Symp. on Ballistics* (23-27 September, 2002, Orlando, FL).

[Buchar *et al.*, 2002b] Buchar, J., Voldřich, J., Rolc, S., Lazar, M. (2002). Response of concrete to the impact of fragments simulating projectiles. In: *Proc. of the 20th Int. Symp. on Ballistics* (23-27 September, 2002, Orlando, FL).

[Butler 1975c] Butler, D.K. (1975). Development of a high-velocity powder gun and analysis of fragment penetration tests into sand. Project 4AI61102B52E. Final Rep. U.S. Army Engineer Waterways Experiment Station, Soils and Pavements Laboratory, Vicksburg, MS.

[Butler, 1976] Butler, D.K. (1976). High-velocity fragment penetration into sand: a comparison of experimental results with theoretical predictions. Project 4AI61102B52E. U.S. Army Engineer Waterways Experiment Station, Vicksburg, MS.

[Butler, 1975a] Butler, D.K. (1975). An analytical study of projectile penetration into rock. Rep. S-75-7. Soils and Pavements Laboratory, U.S. Army Engineer Waterways Experiment Stations, Vicksburg, MS.

[Butler, 1975b] Butler, D.K. (1975). Pretest penetration for DNA rock penetration experiments at a sandstone site near San Ysidro, New Mexico. Soils and Pavements Laboratory, U.S. Army Engineer Waterways Experiment Stations, Vicksburg, MS.

[Buyuk *et al.*, 2009] Buyuk, M., Kan, S., Loikkanen, M.J. (2009). Explicit finite-element analysis of 2024-T3/T351 aluminum material under impact loading for airplane engine containment and fragment shielding. *J. of Aerospace Engineering*, **22**(3): 287-295.

[Buzaud *et al.*, 1999a] Buzaud, E., Laurensou, R., Darrigade, A., Belouet, P., Lissayou, C. (1999). Hard target defeat: An analysis of reinforced concrete perforation process. In: *Proc. of the 9th Int. Symp. on Interaction of the Effects of Munitions with Structures* (May 3-7, 1999, Berlin-Strausberg, Germany), pp. 283-290.

[Buzaud *et al.*, 1999b] Buzaud, E., Don, D., Chapelle, S., Gary, G., Bailly, P. (1999). Perforation studies into MB50 concrete slabs. In: *Proc. of the 9th Int. Symp. on Interaction of the Effects of Munitions with Structures* (May 3-7, 1999, Berlin-Strausberg, Germany), pp. 407–414.

[Calder and Goldsmith, 1971] Calder, C.A., Goldsmith, W. (1971). Plastic deformation and perforation of thin plates resulting from projectile impact. *Int. J. of Solids and Structures*, **7**(7): 863-881.

[Canfield and Clator, 1966] Canfield, J.A., Clator, I.G. (1966). Development of a scaling law and techniques to investigate penetration in concrete. Tech. Rep.2057. U.S. Naval Weapons Laboratory, Dahlgren, Va.

[Cargile *et al.*, 1993] Cargile, J.D., Giltrud, M.E., Luk, V.K. (1993). Perforation of thin unreinforced concrete slabs. Rep. SAND-93-0150C. In: *Proc. of the Special Session during the 6th Int. Symp. On Interaction of Nonnuclear Munitions with Structures* (May 3-7, 1993, Panama City Beach, FL), Sandia National Laboratories, Albuquerque, NM,

[Cargile *et al.*, 2002] Cargile, J.D., O'Neil, F., Neeley, B.D. (2002). Very high strength concretes for use in blast and penetration resistant structures. *AMPTIAC Quarterly* **6**(4): 61-66.

[Cargile, 1999] Cargile, J.D. (1999). Development of a constitutive model for numerical simulations of projectile penetration into brittle geomaterials. Tech. Rep. SL-99-11. Army Engineer Research and Development Center, Vicksburg, MS.

[Cheeseman *et al.* 2008] Cheeseman, B.A., Gouch, W.A., Burkins, M.S. 2008. Ballistic evaluation of aluminum 2139-T8. In: *Proc. of the 24th Int. Symp. on Ballistics* (September 22-26, 2008, New Orleans, LA), vol. 1, pp. 651-659.

[Chen *et al.*, 2010] Chen, X.W., Li, Q.M., Zhang, F.J., He, L.L. (2010). Investigation of the structural failure of penetration projectiles. *Int. J. of Protective Structures*, **1**(1): 41-65.

[Chen *et al.*, 2011] Chen, J.S., Chi, S.W., Lee, C.H., Lin, S.P., Marodon, C., Roth, M.J., Slawson, T.R. (2011). A multiscale meshfree approach for modeling fragment penetration into ultra high-strength concrete. Rep. ERDC/GSL TR-11-35. Geotechnical and Structures Laboratory, Department of Civil and Environmental Engineering, University of California, Los Angeles, CA.

[Chen *et al.*, 2014] Chen, C., Zhu, X., Hou, H., Zhang, L., Shen, X., Tang, T. (2014). An experimental study on the ballistic performance of FRP-steel plates completely penetrated by a hemispherical-nosed projectile. *Steel Composite Structures*, **16**(3): 269-288.

[Chocron *et al.*, 1999] Chocron, S., Grosch, D.J., Anderson, C.E., Jr. (1999). DOP and V_{50} predictions for the 0.30-caliber APM2 projectile. In: *Proc. of the 18th Int. Symp. on Ballistics* (November 15-19, 1999, San Antonio, Texas), vol. 2, pp. 769-776.

[Cimpoeru, 2002] Cimpoeru, S.J. (2002). Analytical modelling of the perforation of multi-layer metallic Targets by fragment simulating projectiles. In: *Proc. of the 20th Int. Symp. on Ballistics* (September 23-27, 2002, Orlando, FL), pp. 778-785.

[Contiliano *et al.*, 1977] Contiliano, R., Donaldson, C. (1977). The Development of a theory for the design of lightweight armor. Rep. No. AFDL-TR-77-114. Aeronautical Research Associates of Princeton, Inc., NJ.

[Contiliano *et al.*, 1978] Contiliano, R.M., McDonough, T.B., Swanson, C.V. (1978). Application of the integral theory of impact to the qualification of materials and the development of a simplified rod penetrator model. Rep. No. ARAP-368. Aeronautical Research Associates of Princeton, Inc., NJ.

[Copland and Scheffler, 2003] Copland, A., Scheffler, D. (2003). Influence of air gaps on long rod penetrators attacking multi-plate target arrays. Rep. No. ARL-TR-2906. U.S. Army Research Laboratory, Aberdeen Proving Ground MD.

[Copland *et al.*, 2005] Copland, A., Bjerke, T.W., Weeks, D. (2005). Semi-infinite target design effects on terminal ballistics performance - influence of plate spacing on penetrator energy partitioning. Rep. No. ARL-TR-3688, Army Research Laboratory, Aberdeen Proving Ground, MD.

[Corbett and Reid, 1993] Corbett, G.G., Reid, S.R. (1993). Local loading of simply-supported steel-grout sandwich plates. *Int. J. of Impact Engineering*, **13**(3): 443-461.

[Corran *et al.*, 1983a] Corran, R.S.J., Shadbolt, P.J., Ruiz, C. (1983). Impact loading of plates-an experimental investigation. *Int. J. of Impact Engineering*, **1**(1): 3-22.

[Corran *et al.*, 1983b] Corran, R.S.J., Ruiz, C., Shadbolt, P.J. (1983). On the design of containment shield. *Computers and Structures*, **16**(1-4): 563-572.

[Crouch *et al.*, 1990] Crouch, I.G., Baxter, B.J., Woodward, R.L. (1990). Empirical tests of a model for thin plate perforation. *Int. J. of Impact Engineering*, **9**(1): 19-33.

[Ćwik *et al.*, 2012] Ćwik, T.K., Lannucci, L., Curtis, P., Pope, D. (2012). Project Rep.: Ballistic performance of Armox 370T and Armox 440T steels subjected to 20mm FSPs impact. Department of Aeronautics, Imperial College London.

[Dancygier and Yankelevsky, 1994] Dancygier, A.N., Yankelevsky D.Z. (1994). Comparative penetration tests of fiber reinforced concrete plates. In: *Proc. of the 3rd Int. Conf. on Structures Under Shock and Impact (SUSI)*, (June 1994, Madrid, Spain), pp. 273-280.

[Dancygier and Yankelevsky, 1996] Dancygier, A.N., Yankelevsky, D.Z. (1996). High strength concrete response to hard projectile impact. *Int. J. of Impact Engineering*, **18**(6): 583-599.

[Dancygier and Yankelevsky, 1999] Dancygier, A.N., Yankelevsky, D.Z. (1999). Effect of reinforced concrete properties on resistance to hard projectile impact. *ACI Structural J.*, **96**(2): 259-267.

[Dancygier *et al.*, 1995] Dancygier, A.N., Yankelevsky, D.Z., Ben-Menashe, Y. (1995). Resistance of high strength concrete plates to hard projectiles impact. In: *Trans. of 13th Int. Conf. on Structural Mechanics in Reactor Technology (SMiRT-13)*, (August 13-18, 1995, Porto Alegre, Brazil), paper H033/1, pp. 213-218.

[Dancygier *et al.*, 1999] Dancygier, A.N., Yankelevsky, D.Z., Baum, H. (1999). Behavior of reinforced concrete walls with interior plaster coating under internal hard projectile impact. *ACI Materials J.*, **96**(1): 116-125.

[Dancygier *et al.*, 2007] Dancygier, A.N., Yankelevsky, D.Z., Jaegermann, C. (2007). Response of high performance concrete plates to impact of non-deforming projectiles. *Int. J. of Impact Engineering*, **34**(11): 1768-1779.

[Dancygier *et al.*, 2014] Dancygier, A.N., Katz, A., Benamou, D., Yankelevsky, D.Z. (2014). Resistance of double-layer reinforced HPC barriers to projectile impact. *Int. J. of Impact Engineering*, **67**: 39-51.

[Dancygier, 1997] Dancygier, A.N. (1997). Effect of reinforcement ratio on the resistance of reinforced concrete to hard projectile impact. *Nuclear Engineering and Design*, **172**(1-2): 233-245.

[Dancygier, 1998] Dancygier, A.N. (1998). Rear face damage of normal and high-strength concrete elements caused by hard projectile impact. *ACI Structural J.*, **95**(3): 291-304.

[Darrigade and Buzaud, 1999] Darrigade, A., Buzaud, E. (1999). High performance concrete: A Numerical and experimental study. In: *Proc. of the 18th Int. Symp. on Ballistics* (November 15-19, 1999, San Antonio, TX), Lancaster, Penn.: Technomic Pub. Co., pp. 845-852.

[Davis *et al.*, 2004] Davis, R.N., Neely, A.M., Jones, S.E. (2004). Mass loss and blunting during high-speed penetration. *J. of Mechanical Engineering Science (Proc. of the Institution of Mechanical Engineers. Part C)*, **218**(9): 1053-1062.

[Davis, 2003] Davis, R.N. (2003). Modeling of high-speed friction using multi-step incrementation of the coefficient of sliding friction. In: *Proc. of the AIAA 54th Annual Southeastern Regional Student Conf.* (27-28 March, 2003, Kill Devil Hills, NC), AIAA-RSC2-2003-U-004, pp. 1-10.

[Dean *et al.*, 2009] Dean, J., Dunleavy, C.S., Brown, P.M., Clyne, T.W. (2009). Energy absorption during projectile perforation of thin steel plates and the kinetic energy of ejected fragments. *Int. J. of Impact Engineering*, **36**(10-11): 1250-1258.

[Delsasso *et al.*, 1943] Delsasso, L.A., Emrich, R.J., Kramer, R.L., Robertson, H.P., Slutz R.J. (1943). Ballistic tests of STS armor plate, using 37-mm projectiles. Rep. No. A-156. DIV. 2. National Defense Research Committee, Armor and Ordnance.

[Deng *et al.*, 2012] Deng, Y., Zhang, W., Cao, Z. (2012). Experimental investigation on the ballistic resistance of monolithic and multi-layered plates against hemispherical-nosed projectiles impact. *Materials and Design*, **41**: 266–281.

[Deng *et al.*, 2013] Deng, Y., Zhang, W., Cao, Z. (2013). Experimental investigation on the ballistic resistance of monolithic and multi-layered plates against ogival-nosed rigid projectiles impact. *Materials and Design*, **44**: 228–239.

[Deng *et al.*, 2014a] Deng, Y., Zhang, W., Yang, Y., Shi, L., Wei, G. (2014). Experimental investigation on the ballistic performance of double-layered plates subjected to impact by projectile of high strength. *Int. J. of Impact Engineering*, **70**: 38-49.

[Deng *et al.*, 2014b] Deng, Y., Zhang, W., Yang, Y., Wei, G. (2014). The ballistic performance of metal plates subjected to impact by projectiles of different strength. *Materials and Design*, **58**: 305-315.

[Deng *et al.*, 2014c] Deng, Y., Zhang, W., Qing, G., Wei, G., Yang, Y., Hao, P. (2014). The ballistic performance of metal plates subjected to impact by blunt-nosed projectiles of different strength. *Materials and Design*, **54**: 1056–1067.

[Deniz and Yildirim, 2013] Deniz, T., Yildirim, R.O. (2013). Ballistic penetration of hardened steel plates. In: *Proc. of the 27th Int. Symp. on Ballistics* (April 22-26, 2013, Freiburg, Germany), DEStech Publications, Inc, pp. 1390-1401.

[Deniz, 2010] Deniz, T. (2010). Ballistic penetration of hardened steel plates. Master Thesis. Middle East Technical University, Ankara, Turkey.

[Deshpande and Gupta, 2000] Deshpande, P.U., Gupta, N.K. (2000). Normal impact of spherical balls on metallic plates. In: Zhao, X.L., Grzebieta, R.H., eds. *Proc. of the 7th Int. Symp. Structural Failure and Plasticity [IMPLAST 2000]*, (October 4-6, 2000, Melbourne, Australia), pp. 73-78.

[Dey *et al.*, 2004] Dey, S., Børvik, T., Hopperstad, O.S., Leinum, J.R., Langseth, M. (2004). The effect of target strength on the perforation of steel plates using three different projectile nose shapes. *Int. J. of Impact Engineering*, **30**(8–9): 1005–1038.

[Dey *et al.*, 2007] Dey, S., Børvik, T., Teng, X, Wierzbicki, T., Hopperstad, O.S. (2007). On the ballistic resistance of double-layered steel plates: an experimental and numerical investigation. *Int. J. of Solids and Structures*, **44**(20): 6701–6723.

[Dey *et al.*, 2011] Dey, S., Børvik, T, Hopperstad, O.S. (2011). Computer-aided design of protective structures: Numerical simulations and experimental validation. *Applied Mechanics and Materials*, **82**: 686-691.

[Dikshit and Sundararajan, 1992a] Dikshit, S.N., Sundararajan, G. (1992). Effect of clamping rigidity of the armour on ballistic performance. *Defence Science J.*, **42**(2): 117-120.

[Dikshit and Sundararajan, 1992b] Dikshit, S.N., Sundararajan, G. (1992). The penetration of thick steel plates by ogive shaped projectile - experiment and analysis. *Int. J. of Impact Engineering*, **12**(3): 373–408.

[Dikshit *et al.*, 1995] Dikshit, S.N., Rao, K., Sundararajan, G. (1995). The influence of plate hardness on the ballistic penetration of thick steel plate. *Int. J. of Impact Engineering*, **16**(2): 293-320.

[Dikshit, 1998a] Dikshit, S.N. (1998). Ballistic behaviour of tempered steel armour plates under plane strain condition. *Defence Science J.*, **48**(2): 167-172.

[Dikshit, 1998b] Dikshit, S.N. (1998). Oblique Impact study in thin steel armour plate. *Defence Science J.*, **48**(2): 185-195.

[Dikshit, 1998c] Dikshit, S.N. (1998). Ballistic behaviour of thick steel armour plate under oblique Impact: Experimental investigation. *Defence Science J.*, **48**(3): 271-276.

[Dikshit, 1999] Dikshit, S.N. (1999). Ballistic behaviour of thick steel armour plate under oblique Impact: Experimental investigation II. *Defence Science J.*, **49**(3): 257-262.

[Dikshit, 2000] Dikshit, S.N. (2000). Influence of hardness on perforation velocity in steel armour plates. *Defence Science J.*, **50**(1): 95-99.

[Dolinski and Rittel, 2015] Dolinski, M., Rittel, D. (2015). Experiments and modeling of ballistic penetration using an energy failure criterion. *J. of the Mechanics and Physics of Solids*, **83**: 1–18.

[Durmuş *et al.*, 2011] Durmuş, A., Güden, M., Gülçimen, B., Ülkü, S., Musa, E. (2011). Experimental investigations on the ballistic impact performances of cold rolled sheet metals. *Materials and Design*, **32**(3): 1356–1366.

[Edwards and Man, 2000] Edwards, M.R., Man, J.H. (2000). Estimation of ballistic limits - A comparative study of. prediction methods. In: *Proc. of European Forum on Ballistics of Projectiles* (April, 11-14, 2000, Saint-Louis, France), pp.453-460.

[Edwards and Mathewson, 1997] Edwards, M.R., Mathewson, A. (1997). The ballistic properties of tool steel as a potential improvised armour plate. *Int. J. of Impact Engineering*, **19**(4): 297-309.

[Eleiche *et al.*, 1996] Eleiche, A.M., Abdel-Kader, M.S., Almohandes, A. (1996). Penetration resistance of laminated plates from steel and flbreglass-reinforced polyester. In: Jones, N., Bulson, P.S., eds., *Structures under shock and impact IV* [*Proc. of the 4th Int. Conf. on Structures under Shock and Impact* (July 1996, Udine, Italy)], Computational Mechanics Publ., Southampton, pp. 117-126.

[Elek *et al.*, 2005] Elek, P., Jaramaz, S., Mickovic, D. (2005). Modeling of perforation of plates and multi-layered metallic targets. *Int. J. of Solids and Structures,* **42**(3-4): 1209–1224.

[Erice *et al.*, 2010] Erice, B., Gálvez, F., Cendón, D., Sánchez-Gálvez, V. (2010). Mechanical behavior of FV535 steel against ballistic impact at high temperatures. In: *Proc. of the IBM POWER Systems and System Storage Sym.* 2010 (May 17 – 21, 2010, Beijing, China).

[Fabian *et al.*, 1996] Fabian, G.L., O'Donnell, R.H., Tom, J.G., Malone, P.G. (1996). Use of shock-absorbing concrete (SACON) as an environmentally compatible bullet-trapping medium on small-arms training ranges. In: *Proc. of the Tri-Service Environmental Technology Workshop, Enhancing Readiness Through Environmental Quality Technology* (20-22 May 1996, Hershey, PA). No. SFIM-AEC-ET-CR-96187. U. S. Army Environmental Center, Aberdeen Proving Ground, MD, pp. 187-196.

[Fan *et al.*, 2013] Fan, P., Wang, M., Song, C. (2013). Anti-strike capability of steel-fiber reactive powder concrete. *Defence Science J.*, **63**(4): 363-368.

[Fang *et al.*, 2006] Fang, Q., Hou, X.F., Zhang, Y.N., Zhang, Y.D. (2006). Experimental and numerical investigations on projectile penetration into polypropylene fiber reinforced concrete targets. In: Jones, N., Brebbia, C.A., eds. *Structures Under Shock and Impact IX* [*Proc. of the 9th Int. Conf. on Structures under Shock and Impact* (July 3-5, 2006, New Forest, UK)]. WIT Press, Southampton, pp. 215-223.

[Favorsky *et al.*, 2011] Favorsky, V., Assaf, Z., Aizik, F., Ran, E., Borenstein, A., Vizel, A., Ravid, M., Shapira, N. (2011). Experimental-numerical study of inclined impact in Al7075-T7351 targets by 0.3" AP projectiles. In: *Proc. of the 26th Int. Symp. on Ballistics* (September 12-16, 2011, Miami, FL), pp. 1644-1654.

[Fiquet and Dacquet, 1977] Fiquet, G., Dacquet, S. (1977). Study of the perforation of reinforced concrete slabs by rigid missiles — Experimental study, part II. *Nuclear Engineering and Design*, **41**(1): 103–120.

[Folsom, 1987] Folsom, E.N., Jr., (1987). Projectile penetration into concrete with an inline hole. Master's Thesis. Rep. UCRL-53786. Lawrence Livermore National Laboratory, University of California, Livermore, CA.

[Forquin *et al.*, 2015] Forquin, P., Sallier, L., Pontiroli, C. (2015). A numerical study on the influence of free water content on the ballistic performances of plain concrete targets. *Mechanics of Materials*, **89**: 176-189.

[Forrestal and Hanchak, 1999] Forrestal, M.J., Hanchak, S.J. (1999). Perforation experiments on HY-100 steel plates with 4340 R_c 38 and maraging T-250 steel rod projectiles. *Int. J. of Impact Engineering*, **22**(9-10): 923-933.

[Forrestal and Hanchak, 2002] Forrestal, M.J., Hanchak, S.J. (2002). Penetration limit velocity for ogive-nose projectiles and limestone targets. *J. of Applied Mechanics*, **69**(6): 853-854.

[Forrestal and Luk, 1992] Forrestal, M.J., Luk, V.K. (1992). Penetration into soil targets. *Int. J. of Impact Engineering*, **12**(3): 427-444.

[Forrestal and Piekutowski, 2000] Forrestal, M.J., Piekutowski, A.J. (2000). Penetration experiments with 6061-T6511 aluminum targets and spherical-nose steel projectiles at striking velocities between 0.5 and 3.0 km/s. *Int. J. of Impact Engineering*, **24**(1): 57-67.

[Forrestal *et al.*, 1987] Forrestal, M.J., Rosenberg, Z., Luk, V.K., Bless, S.J. (1987). Perforation of aluminum plates with conical-nosed rods. *J. of Applied Mechanics*, **54**(1): 230-232.

[Forrestal *et al.*, 1988] Forrestal, M.J., Okajima, K., Luk, V.K. (1988). Penetration of 6061-T651 aluminum target with rigid long rods. *J. of Applied Mechanics*, **55**(4): 755-760.

[Forrestal *et al.*, 1990] Forrestal, M.J., Luk, V.K., Brar, N.S. (1990). Perforation of aluminum armor plates with conical-nose projectiles. *Mechanics of Materials*, **10**(1-2): 97-105.

[Forrestal *et al.*, 1991] Forrestal, M.J., Brar, N.S., Luk, V.K. (1991). Perforation of strain-hardening targets with rigid spherical-nose rods. *J. of Applied Mechanics,* **58**(1): 7-10.

[Forrestal *et al.*, 1992] Forrestal, M.J., Luk, V.K., Rosenberg, Z., Brar N.S. (1992). Penetration of 7075-T651 aluminum targets with ogival-nose rods. *Int. J. of Solids and Structures,* **29**(14-15): 1729-1736.

[Forrestal *et al.*, 1994] Forrestal, M.J., Altman, B.S., Cargile, J.D., Hanchak, S.J. (1994). An empirical equation for penetration depth of ogive-nose projectiles into concrete targets. *Int. J. of Impact Engineering,* **15**(4): 395-405.

[Forrestal *et al.*, 1996] Forrestal, M.J., Frew, D.J., Hanchak, S.J., Brar, N.S. (1996). Penetration of grout and concrete targets with ogive-nose steel projectiles. *Int. J. of Impact Engineering,* **18**(5): 465-476.

[Forrestal *et al.*, 2003] Forrestal, M.J., Frew, D.J., Hickerson, J.P., Rohwer, T.A. (2003). Penetration of concrete targets with deceleration-time measurements. *Int. J. of Impact Engineering,* **28**(5): 479-497.

[Forrestal *et al.*, 2010] Forrestal, M.J., Børvik, T., Warren, T.L. (2010). Perforation of 7075-T651 aluminum armor plates with 7.62 mm APM2 bullets. *Experimental Mechanics,* **50**(8): 1245-1251.

[Fras *et al.*, 2015a] Fras, T., Colard, L., Reck, B. (2015). Modeling of ballistic impact of fragment simulating projectiles against aluminum plates. In: *Proc. of the 10th European LS-DYNA Conf.* (June 15-17, 2015, Würzburg, Germany).

[Fras *et al.*, 2015b] Fras, T., Colard, L., Lach, E., Rusinek, A., Reck, B. (2015). Thick AA7020-T651 plates under ballistic impact of fragment-simulating projectiles. *Int. J. of Impact Engineering,* **86**: 6-353.

[Frew *et al.*, 1998] Frew, D.J., Hanchak, S.J., Green, M.L., Forrestal, M.J. (1998). Penetration of concrete targets with ogive-nose steel rods. *Int. J. of Impact Engineering,* **21**(6): 489-497.

[Frew *et al.*, 2000] Frew, D.J., Forrestal, M.J., Hanchak, S.J. (2000). Penetration experiments with limestone targets and ogive-nose steel projectiles. *J. of Applied Mechanics,* **67**(4): 841-845.

[Frew *et al.*, 2006] Frew, D.J., Forrestal, M.J., Cargile, J.D. (2006). The effect of concrete target diameter on projectile deceleration and penetration depth. *Int. J. of Impact Engineering,* **32**(10): 1584–1594.

[Fugelso and Bloedow, 1966] Fugelso, L.E., Bloedow, F. H. (1966). Studies in the perforation of thin metallic plates by projectile impact: I. Normal impact of circular cylinders. Rep. MR-1250. American Transportation Corp., Niles, IL.

[Gallardy, 2011] Gallardy, D. (2011). Ballistic evaluation of aluminum 7085-T7E01 and T7E02. In: *Proc. of the 26th Int. Symp. on Ballistics* (September 12-16, 2011, Miami, FL), pp. 1257-1363.

[Gallardy, 2012] Gallardy, D. (2012). Ballistic evaluation of 7085 aluminum. Tech. Rep. ARL-TR-5952.U.S. Army Research Laboratory, Aberdeen Proving Ground, MD 21005-5066.

[Gao *et al.*, 1995] Gao, S., Liu, M., Tan, H. (1995). Dynamic analysis of a penetrator penetrating against half-infinite concrete target. *Acta Armamentarii*, **16**(4): 46–50 (in Chinese).

[Gao *et al.*, 2003] Liu, H., Gao, S., Li, K. (2003). Measurement technologies and result analysis on experiment of penetration of steel projectile into thick concrete target. In: Wen, T. (ed). *Proc. of 5th Int. Symp. of Test and Measurement*. International Academic Publishers, Beijing, 4: 3339–3342.

[Gao *et al.*, 2006] Gao, S., Liu, H., Li, K., Huang, F., Jin, L. (2006). Normal expansion theory for penetration of projectile against concrete target. *Applied Mathematics and Mechanics (English Edition)*, **27**(4): 485-492.

[Gao *et al.*, 2009] Gao, S., Liu, H., Lei, J. (2009). A fuzzy model of the penetration resistance of concrete targets. *Int. J. of Impact Engineering*, **36**(4): 644–649.

[Goel *et al.*, 1986] Goel, R.A., Chandra, S., Singh, H., Abrol, A.K., Singh, B. (1986). Effect of L/D ratio on bomb ricochet. *Defence Science J.*, **36**(4): 389-394.

[Goel *et al.*, 1988] Goel, R.A., Chandra, S., Abrol, A.K., Singh, B. (1988). Behaviour of a kinetic energy projectile on angular impact. *Defence Science J.*, **38**(3): 293-299.

[Goel *et al.*, 1990] Goel, R.A., Chandra, S., Chandola, U.C., Abrol, A.K., Kumar, R. (1990). Behaviour of concrete on high velocity impact. In: *Minutes of the 24th Explosives Safety Seminar* (August 28-30, 1990, St. Louis, Missouri, US), vol. 2, pp. 2277-2299.

[Goel *et al.*, 2000] Goel, R.A., Chandra, S., Abrol, A.K. (2000). Concrete piercing warhead – a model study. In: Brebbia, C.A., Jones, N. (eds.). *Structures Under Shock and Impact VI* [*Proc. of the 6th Int. Conf. on Structures under Shock and Impact* (July 2000, Cambridge, UK]. WIT Press, Southampton, pp. 195-204.

[Gogolewski and Morgan, 1999] Gogolewski, R. P., Morgan, B. R. (1999). Ballistic experiments with titanium and aluminum targets. Rep. UCRL-ID-136691. Lawrence Livermore National Laboratory.

[Gogolewski *et al.*, 1996] Gogolewski, R.P., Cunningham, B.J., Riddle, R.O., Lesuer, D. R., Syn, C.K. (1996). On the importance of target material interfaces during low speed impact. In: *Proc. of the 16th Int. Symp. on Ballistics* (September 23-28, 1996, San Francisco, CA), vol. 3, pp. 751–760.

[Gold *et al.*, 1993] Gold, V.M., Pearson, J.C., Turci, J.P. (1993). Resistance of concrete and reinforced concrete structures impacted by Cu and Ta projectiles. In: *Proc. of the 6th Int. Symp. On Interaction of Nonnuclear Munitions with Structures* (May 3-7, 1993, Panama City Beach, Florida), pp. 44-49.

[Gold *et al.*, 1996] Gold, V.M., Vradis, G.C., Pearson, J.C. (1996). Concrete penetration by eroding projectiles: experiments and analysis. *J. of Engineering Mechanics*, **122**(2): 145-152.

[Gold, 1996] Gold, V.M. (1996). Concrete penetration by eroding projectiles: experiments and analysis. Tech. Rep., ARAED-TR-96014. U.S. Army Armament Research, Development and Engineering Center, Picatinny Arsenal, NJ.

[Goldsmith and Finnegan, 1971] Goldsmith, W., Finnegan, S.A. (1971). Penetration and perforation processes in metal targets at and above ballistic velocities. *Int. J. of Mechanical Science,* **13**(10): 843-866.

[Goldsmith and Finnegan, 1985] Goldsmith, W., Finnegan, S.A. (1985). Normal and oblique impact of cylindro-conical and cylindrical projectiles on metallic plates. Tech. Rep. NWC TP 6479.Naval Weapons Center, China Lake, CA 93555-6001.

[Goldsmith and Finnegan, 1986] Goldsmith, W., Finnegan, S.A. (1986). Normal and oblique impact of cylindro-conical and cylindrical projectiles on metallic plates. *Int. J. of Impact Engineering,* **4**(2): 83-105.

[Goldstein *et al.,* 1977] Goldstein, S., Berriaud, C., Labrot, R. (1977). Study of the perforation of reinforced concrete slabs by rigid missiles —Experimental study, part III. *Nuclear Engineering and Design,* **41**(1): 121-128.

[Golubev and Medvedkin, 2001] Golubev, V.K., Medvedkin, V.A. (2001). Penetration of a rigid rod into a thick steel plate at elevated velocities. *Strength of Materials,* **33**(4): 400-405.

[Gomez and Shukla, 2001] Gomez, J.T., Shukla, A. (2001). Multiple impact penetration of semi-infinite concrete. *Int. J. of Impact Engineering,* **25**(10): 965-979.

[Gooch *et al.,* 2007] Gooch, W.A., Burkins, M.S., Squillacioti, R.J. (2007) Ballistic testing of commercial aluminum alloys and alternative processing techniques to increase the availability of aluminum armor. In: *Proc. of the 23rd International Symp. on Ballistics* (April 16-20, 2007, Tarragona, Spain).

[Grabarek, 1971] Grabarek, C.L. (1971). Penetration of armor by steel and high density penetrators. Memorandum Rep. BRL MR 2134. Ballistic Research Laboratory, Aberdeen Proving Ground, MD.

[Grace, 1995] Grace, F.I. (1995). Long-rod penetration into targets of finite thickness at normal impact. *Int. J. of Impact Engineering,* **16**(3): 419–433.

[Gran and Frew, 1997] Gran, J.K., Frew, D.J. (1997). In-target radial stress measurements from penetration experiments into concrete by ogive-nose steel projectiles. *Int. J. of Impact Engineering,* **19**(8): 715-726.

[Gran *et al.,* 1999] Gran, J., Moxley, R.E., Adley, M.D. (1999). Measurements of stress in concrete during deep penetration. In: *Proc. of the 9th Int. Symp. on Interaction of the Effects of Munitions with Structures* (May 3-7, 1999, Berlin-Strausberg, Germany), pp. 307-314.

[Graves and Beatty, 1994] Graves, J.H., Beatty, J.H. (1994). Ballistic performance and adiabatic shear behavior of AerMet 100 Steel. Rep. ARL-TR-454. Army Research Laboratory, Watertown, MA.

[Griffin, 1959] Griffin, T.J. (1959). Ballistic evaluation of aluminum armor. First Rep. on Project TB4-005. Development and Proof Services, Aberdeen Proving Ground, MD.

[Guo *et al.,* 2014a] Guo, L., He, Y., Zhang, X.F., Pang, C.X., Qiao, L., Guan, Z.W. (2014). Study mass loss at microscopic scale for a projectile penetration into concrete. *Int. J. of Impact Engineering,* **72**: 17-25.

[Guo *et al.*, 2014b] Guo, L., He, Y., Zhang, X., He, Y., Qiao, L. (2014). Thermal effect on mass loss of projectile during penetration into concrete: experimental and numerical study. In: Ames, R.G., Boeka, R.D., eds., *Proc. of the 28th Int. Symp. on Ballistics* (September 24-26, 2014, Atlanta, Georgia, USA), vol.**2**: 958-967.

[Gupta and Madhu, 1992] Gupta, N.K., Madhu, V. (1992). Normal and oblique impact of a kinetic energy projectile on mild steel plates. *Int. J. of Impact Engineering,* **12**(3): 333-343.

[Gupta and Madhu, 1997] Gupta, N.K., Madhu, V. (1997). An experimental study of normal and oblique impact of hard-core projectile on single and layered plates. *Int. J. of Impact Engineering,* **19**(5-6): 395-414.

[Gupta *et al.*, 2001] Gupta, N.K., Ansari, R., Gupta, S.K. (2001). Normal impact of ogive nosed projectiles on thin plates. *Int. J. of Impact Engineering,* **25**(7): 641-660.

[Gupta *et al.*, 2006] Gupta, N.K., Iqbal, M.A., Sekhon, G.S. (2006). Experimental and numerical studies on the behavior of thin aluminum plates subjected to impact by blunt- and hemispherical-nosed projectiles. *Int. J. of Impact Engineering,* **32**(12): 1921-1944.

[Gupta *et al.*, 2007] Gupta, N.K., Iqbal, M.A., Sekhon, G.S. (2007). Effect of projectile nose shape, impact velocity, and target thickness on deformation behavior of aluminum plates. *Int. J. of Solids and Structures,* **44**(10): 3411–3439.

[Gupta *et al.*, 2008] Gupta, N.K., Iqbal, M.A., Sekhon, G.S. (2008). Effect of projectile nose shape, impact velocity, and target thickness on the deformation behavior of layered plates. *Int. J. of Impact Engineering,* **35**(1): 37-60.

[Hadala, 1975] Hadala, P.F. (1975). Evaluation of empirical and analytical procedures used for predicting the rigid body motion of an earth penetrator. Miscellaneous Paper S-75-15. U.S. Army Engineer Waterways Experiment Station, Vicksburg, MS.

[Haight *et al.*, 2012] Haight, C., McNamara, K., Courtney, M. (2012). Does V50 depend on armor mass? Research Rep. U.S. Air Force Academy, 2354 Fairchild Drive, USAF Academy, CO.

[Haldar and Miller, 1982] Haldar, A., Miller, F.J. (1982). Penetration depth in concrete for nondeformable missiles. *Nuclear Engineering and Design,* **71**(1): 79-88.

[Hanchak *et al.*, 1993] Hanchak, S.J., Altman, B.S., Forrestal, M.J. (1993). Perforation of HY-100 steel plates with long rod projectiles. In: *Proc. of 13th Army symposium on solid mechanics* (17-19 August 1993, Plymouth, MA), pp. 247-256.

[Hanchak, 1992] Hanchak, S. J., Forrestal, M. J., Young, E.R., Ehrgott, J.Q. (1992). Perforation of concrete slabs with 48 MPa (7 ksi) and 140 MPa (20 ksi) unconfined compressive strengths. *Int. J. of Impact Engineering,* **12**(1): 1-7.

[Hankins *et al.*, 2006] Hankins, M.D., Stoltz, B.C., Torres, K.L., Jones, S.E. (2006). Penetration of a coarse sand target by rigid projectiles. In: Jones, N., Brebbia, C.A., eds. *Structures under shock and impact IX* [*Proc. of the 9th Int. Conf. on Structures under Shock and Impact* (July 3-5, 2006, New Forest, UK)]. WIT Press, Southampton, pp. 187–196.

[Hansson and Skoglund, 2002] Hansson, H., Skoglund, P. (2002). Simulation of concrete penetration in 2D and 3D with the RHT material model. Rep. FOI-R-0720-SE. Swedish Defence Research Agency (FOI), Tumba, Sweden.

[Hansson, 2001] Hansson, H. (2001). Numerical simulation of concrete penetration with Euler and Lagrange formulation. Rep. FOI-R-0190-SE. Swedish Defence Research Agency (FOI), Tumba, Sweden.

[Hansson, 2002] Hansson, H. (2002). Modelling of concrete perforation. In: Jones, N. *et al.*, eds., *Structures under Shock and Impact VII* [*Proc. of the 7th Int. Conf. on Structures under Shock and Impact (Montreal, Quebec)*]. Computational Mechanics Publ., Southampton, pp. 79-90.

[Hansson, 2003] Hansson, H. (2003). A note on empirical formulas for the prediction of concrete penetration. Rep. FOI-R-0968-SE. Swedish Defence Research Agency (FOI), Tumba, Sweden.

[Hansson, 2004] Hansson, H. (2004). 3D simulation of concrete penetration using SPH formulation and the RHT material model. In: Jones, N., Brebbia, C.A., eds., *Structures under Shock and Impact VIII* [*Proc. of the 8th Int. Conf. on Structures under Shock and Impact* (Crete, Greece)]. WIT Press, Southampton, UK, pp. 201-220.

[Hansson, 2005a] Hansson, H. (2005). Penetration in concrete for projectiles with L/D≈9. Rep. FOI-R-1659-SE. Swedish Defence Research Agency (FOI), Tumba, Sweden.

[Hansson, 2005b] Hansson, H. (2005). Simulation of penetration in normal strength concrete for projectiles with L/D≈9. Rep. FOI-R-1759-SE. Swedish Defence Research Agency (FOI), Tumba, Sweden.

[Hansson, 2011] Hansson, H. (2011). Warhead penetration in concrete protective structures. Licentiate Thesis. Royal Institute of Technology (KTH), Stockholm, Sweden.

[Hashimoto *et al.,* 2005] Hashimoto, J., Takiguchi, K., Nishimura, K., Matsuzawa, K., Tsutsui, M., Ohashi, Y., Kojima, I., Torita, H. (2005). Experimental study on behavior of RC panels covered with steel plates subjected to missile impact. In: *Trans. of 18th Int. Conf. on Structural Mechanics in Reactor Technology (SMiRT-18)*, (August 7-12, 2005, Beijing, China), paper J05-4, pp. 2604-2615.

[Heider and Günter, 1998] Heider, N., Günter, U. (1998). Modern geopenetrators and relevant revision of concrete penetration models. In: Jones, N. et al. (eds.). *Structures under Shock and Impact V* [*Proc. of the 5th Int. Conf. on Structures under Shock and Impact* (June 24-26, 1998, Thessaloniki, Greece)]. WIT Press, Southampton, pp. 807-815.

[Heimdahl and Schulz, 1982] Heimdahl, O.E.R., Schulz, J.C. (1982). Effect of longitudinal grooves on survivability of cylindrical steel projectiles fired against simulated concrete targets. Rep. NWC-TP-6402, Naval Weapons Center, China Lake, CA.

[Hiermaier and Thoma, 1997] Hiermaier, S., Thoma, K. (1997). Numerical simulation of penetration in concrete using different types of code. In: *Proc. of the 8th Int. Symp. on Interaction of the Effects of Munitions with Structures* (April 22-25, 1997, McLean, VA).

[Holmen *et al.*, 2013] Holmen, J.K., Johnsen, J., Jupp, S., Hopperstad, O.S., Børvik, T. (2013). Effects of heat treatment on the ballistic properties of AA6070 aluminium alloy. *Int. J. of Impact Engineering*, **57**: 119–133.

[Holt *et al.*, 1993] Holt, W.H., Mock, W., Jr., Soper, W.G., Coffey, C.S., Ramachandran, V., Armstrong, R.W. (1993). Reverse-ballistic impact study of shear plug formation and displacement in Ti$_6$Al$_4$V alloy. *J. of Applied Physics,* **73**(8): 3753-3759.

[Huang *et al.*, 2011] Huang, G.Y., Wu, G., Shun Shan Feng, S.S., Zhou, T. (2011). Theory and experiment study of normal penetration of slender nose projectile into steel target. *Applied Mechanics and Materials*, **80-81**: 137-142 [selected, peer reviewed papers from the 2011 Int. Conf. on Information Engineering for Mechanics and Materials (ICIMM 2011), Aug. 13-14, 2011, Shanghai, China].

[Hutchings *et al.*, 1981] Hutchings, I.M., Macmillan, N.H., Rickerby, D.G. (1981). Further studies of the oblique impact of a hard sphere against a ductile solid. *Int. J. of Mechanical Science*, **25**(11): 639-646.

[Ipson *et al.*, 1973] Ipson, T.W., Recht, R.F., Schmeling, W.A. (1973). The effect of projectile nose shape upon ballistic limit velocity, residual velocity, and ricochet obliquity. Tech. Rep. NWC TP 5607. Denver Research Institute of the University of Denver. Contract Number N00123-72-C-0267.

[Iqbal and Gupta, 2008] Iqbal, M.A., Gupta, N.K. (2008). Energy absorption characteristics of aluminum plates subjected to projectile impact. *Latin American J. of Solids and Structures*, **5**(4): 259-287.

[Iqbal and Gupta, 2011] Iqbal, M.A., Gupta, N.K. (2011). Ballistic limit of single and layered aluminium plates. *Strain. An Int. J. for Experimental Mechanics.* **47**(s1): e205–e219.

[Iqbal *et al.*, 2006] Iqbal, M.A., Gupta, N.K., Sekhon, G.S. (2006). Behaviour of thin aluminium plates subjected to impact by ogive-nosed projectiles. *Defence Science J.*, **56**(5): 841-852.

[Iqbal *et al.*, 2013] Iqbal, M.A., Khan, S.H., Ansari, R., Gupta, N.K. (2013). Experimental and numerical studies of double-nosed projectile impact on aluminum plates. *Int. J. of Impact Engineering,* **54**: 232-245.

[Iqbal *et al.*, 2015] Iqbal, M.A., Tiwari, G., Gupta, P.K., Bhargava, P. (2015). Ballistic performance and energy absorption characteristics of thin aluminium plates. *Int. J. of Impact Engineering*, **77**: 1-15.

[Iremonger *et al.*, 1989] Iremonger, M.J., Claber, K.J., Ho, K.Q. (1989). Small arms penetration of concrete. In: *Proc. of the 4th Int. Symp. on the Interaction of Non-Nuclear Munitions with Structures* (April 17-21, 1989, Panama City Beach, FL), v.1, pp. 74-79.

[Jagannathan, 2010] Jagannathan, M. (2010). Energy Partitioning in Ballistic Impact Fragmentation of Titanium Boride. Master thesis. University of Utah.

[Jankov *et al.,* 1976] Jankov, Z.D., Shanahan, J.A., White, M.P. (1976). Missile tests of quarter-scale reinforced concrete barriers. In: *Proc. of a Symp. on Tornadoes, Assessment of Knowledge and Implications for Man* (June 1976, Lubbock, TX).

[Jankowiak and Rusinek, 2015] Jankowiak, T., Rusinek, A. (2015). Perforation of sheet steel by rigid projectiles. In: Stewart, M., Netherton, M. (eds.). Design and analysis of protective structures. *Proc. of the 3rd Int. Conf. on Protective Structures* (3-6 February 2015, Newcastle, Australia), pp. 242-249.

[Jankowiak *et al.,* 2014] Jankowiak, T., Rusinek, A., Kpenyigba, K.M., Pesci, R. (2014). Ballistic behavior of steel sheet subjected to impact and perforation. *Steel and Composite Structures,* **16**(6): 595-609.

[Jena *et al.,* 2010a] Jena, P.K., Mishra, B., Babu, M.R., Babu, A., Singh, A.K., Kumar, K.S., Bhat, T.B. (2010). Effect of heat treatment on mechanical and ballistic properties of a high strength armour steel. *Int. J. of Impact Engineering,* **37**(3): 242-249.

[Jena *et al.,* 2010b] Jena, P.K., Jagtap, N., Kumar, K. S., Bhat, T.B. (2010). Some experimental studies on angle effect in penetration. *Int. J. of Impact Engineering,* **37**(5): 489–501.

[Jenq *et al.,* 1988] Jenq, S.T., Goldsmith, W.,. Kelly, J.M. (1988). Effect of target bending in normal impact of a flat-ended cylindrical projectile near the ballistic limit. *Int. J. of Solids and Structures,* **24**(12): 1243-1266.

[Jerome *et al.,* 2000] Jerome, D.M., Tynon, R.T., Wilson, L.L., Osborn, J.J. (2000). Experimental observations of the stability and survivability of ogive-nosed, high-strength steel alloy projectiles. In: *Cementitions Materials at Striking Velocities from 800-1800 m/sec, Proc. of the 3rd Joint Classified Ballistics Symp.* (May 1-4, 2000, San Diego, CA).

[Jinzhu *et al.,* 2013] Jinzhu, L., Zhongjie, L., Hongsong, Z., Fenglei, H. (2013). Perforation experiments of concrete targets with residual velocity measurements. *Int. J. of Impact Engineering,* **57**: 1-6.

[Johnson *et al.,* 1959] Johnson, D.K., Cannon, E.T., Palmer, E.P., Grow, R.W. (1959). Cratering produced in metals by high-velocity impact. Rep. UU-4. High Velocity Laboratory, Utah University, Salt Lake City.

[Jones and Placzankis, 2011] Jones, T.L., Placzankis, B. (2011). The ballistic and corrosion evaluation of Magnesium Elektron E675 vs. Baseline Magnesium Alloy AZ31B and Aluminum Alloy 5083 for armor applications. Tech. Rep. ARL-TR-5565. U.S. Army Research Laboratory, Aberdeen Proving Ground, MD 21005-5066.

[Jones *et al.,* 2001] Jones, S.E., Toness, O., Jerome, D.M., Rule, W.K. (2001). Normal penetration of semi-infinite targets by ogive-nose projectiles, including the effects of blunting and erosion. In: Moody, F.J., ed., *Thermal hydraulics, liquid sloshing, extreme loads, and structural response: Presented at the 2001 ASME Pressure*

Vessels and Piping Conf. (July 22-26, 2001, Atlanta, Georgia), PVP Series, vol. **421**, pp. 53-59.

[Jones *et al.,* 2003] Jones, S.E., Hughes, M.L., Toness, O.A., Davis, R.N. (2003). A one-dimensional analysis of rigid-body penetration with high-speed friction. *J. of Mechanical Engineering Science (Proc. of the Institution of Mechanical Engineers. Part C),* **217**(C4): 411-422.

[Jones *et al.,* 2007] Jones, T.L., DeLorme, R.D., Matthew S. Burkins, M.S., Gooch, W.A. (2007). Ballistic evaluation of Magnesium Alloy AZ31B. Tech. Rep. ARL-TR-4077.U.S. Army Research Laboratory, Weapons and Materials Research Directorate, Aberdeen Proving Ground, MD 21005-5067.

[Jones *et al.,* 2008] Jones, T.L., DeLorme, R.D. (2008). Development of a ballistic specification for Magnesium Alloy AZ31B. Tech. Rep. ARL-TR-4664. U.S. Army Research Laboratory, Aberdeen Proving Ground, MD 21005-5067.

[Kabo *et al.,* 1977] Kabo, M., Goldsmith, W., Sackman, J.L. (1977). Impact and comminution processes in soft and hard rock. *Rock Mechanics,* **9**(4): 213-243.

[Kamal and Eltehewy, 2012] Kamal I.M., Eltehewy, E.M. (2012). Projectile penetration of reinforced concrete blocks: Test and analysis. *Theoretical and Applied Fracture Mechanics,* **60**(1): 31–37.

[Kar, 1977] Kar, A.K. (1977). Projectile penetration of earth media. In: *Trans. of the 4th Int. Conf. on Structural Mechanics in Reactor Technology (SMiRT-4),* (August 15-19, 1977, San-Francisco, CA), paper j9/8.

[Khan and Ansari, 2003] Khan, W.U., Ansari, R. (2003). Oblique impact of projectile on thin aluminium plates. *Defence Science J.,* **53**(2): 139-146.

[Khan *et al.,* 2014] Khan, S.H., Iqbal, M.A., K, S., Ansari, R., Gupta, N.K. (2014). Ballistic response of 1100-H14 thin aluminum plates against blunt and conical nose projectiles. In: *Proc. of 4th Int. Conf. on Impact Loading of Lightweight Structures (ICILLS 2014),* (January 12-16, 2014, Cape Town, South Africa).

[Khan, 2010] Khan, S.H., Azeem, M., Iqbal, D., Ansari, R. (2010). Empirical perforation equation for thin aluminum plates at low velocity impact. In: *Proc. of the National Conf. on Advances in Mechanical Engineering (NCAME 2010).* (November 27-29, 2010, Aligarh, India).

[Khoda-rahmi *et al.,* 2006] Khoda-rahmi, H., Fallahi, A., Liaghat, G.H. (2006). Incremental deformation and penetration analysis of deformable projectile into semi-infinite target. *Int. J. of Solids and Structures,* **43**(3-4): 569–582.

[Khromov, 2004] Khromov, I.V. (2004). Dynamics of penetration of rigid rotating penetrators into soil. Ph.D. Thesis. Tula State University, Tula (in Russian).

[Kim *et al.,* 2008] Kim, H.-J., Park, L., Kim, H. (2008). Study on the effect of reinforcement of concrete targets upon the penetration performance of a projectile. In: *Proc. of the 24th International Symposium on Ballistics* (September 22-26, 2008, New Orleans, LA), pp. 738-743.

[Kim *et al.,* 2015] Kim, K., Suh, Y., Moon, I., Choi, H. (2015). A study on the local impact behavior of SC walls using actual test and simulation. In: *Trans. of the 23rd*

Int. Conf. on Structural Mechanics in Reactor Technology (SMiRT-23), (August 10-14, 2015, Manchester, U.K.), Division V, paper ID 705.

[Kojima, 1991] Kojima, I. (1991). An experimental study on local behavior of reinforced concrete slabs to missile impact. *Nuclear Engineering and Design*, **130**(2): 121–132.

[Kpenyigba *et al.*, 2013] Kpenyigba, K.M., Jankowiak, T., Rusinek, A., Pesci, R. (2013). Influence of projectile shape on dynamic behavior of steel sheet subjected to impact and perforation. *Thin-Walled Structures*, **65**: 93–104.

[Kpenyigba *et al.*, 2015] Kpenyigba, K.M., Jankowiak, T., Rusinek, A., Pesci, R., Wang, B. (2015). Effect of projectile nose shape on ballistic resistance of interstitial-free steel sheets. *Int. J. of Impact Engineering*, **79**: 83-94.

[Kumano and Goldsmith, 1982a] Kumano, A., Goldsmith, W. (1982). Behavior of diorite under impact by variously-shaped projectiles. *Rock mechanics*, **15**(1): 25-40.

[Kumano and Goldsmith, 1982b] Kumano, A., Goldsmith, W. (1982). An analytical and experimental investigation of the effect of impact on coarse granular rocks. *Rock mechanics*, **15**(2): 67-97.

[Kymer and Fatzinger, 1957] Kymer, J.R., Fatzinger, H.E. (1957). Ballistic tests of 2024-T4 and 7075-T6 Aluminum Alloys (U). Rep. R-1395. Frankford Arsenal, Pitman-Dunn Laboratories Group, Philadelphia, PA.

[Lai *et al.*, 2015] Lai, J., Guo, X., Zhu, Y. (2015). Repeated penetration and different depth explosion of ultra-high performance concrete. *Int. J. of Impact Engineering*, **84**: 1-12.

[Lambert and Jonas, 1976] Lambert, J.P., Jonas, G.H. (1976). Towards standardization of in terminal ballistic testing: velocity representation. Rep. BRL-R-1852. Army Ballistic Research Laboratory, Aberdeen Proving Ground, MD.

[Lambert, 1978] Lambert, J.P. (1978). The terminal ballistics of certain 65 gram long rod penetrators impacting steel armor plate. Rep. ARBRL-TR-0272, U. S. Army Ballistic Research Laboratory, Aberdeen Proving Ground, MD.

[Landgrov and Sarkisyan, 1984] Landgrov, I.F., Sarkisyan, O.A. (1984). Piercing plastic-material barriers with a rigid punch. *J. of Applied Mechanics and Technical Physics,* **25**(5): 771-773.

[Landkof and Goldsmith, 1985] Landkof, B., Goldsmith, W. (1985). Petalling of thin, metallic plates during penetration by cylindro-conical projectiles. *Int. J. of Solids and Structures*, **21**(3): 245-266.

[Langheim, 1976] Langheim, H. (1976). Impact untersuchungen an Armierten Betonplatten. Rep. E9/76, Part 1, Ernst Mach Institute (in German).

[Langheim, 1977] Langheim, H. (1977). Impact untersuchungen an Armierten Betonplatten Rep. E14/77, Part 3, Ernst Mach Institute. (in German).

[Lin *et al.*, 2012] Lin, M.Q., Xia, Y.Y., Xiao, Z.X., Guo, X.B. (2012). Damage experiment of projectile penetration into high-strength concrete. In: Cai, M., ed., *Rock Mechanics: Achievements and Ambitions*. CRC Press, Chapter 24, pp.133-137.

[Liss and Goldsmith, 1984] Liss, J., Goldsmith, W. (1984). Plate perforation phenomena due to normal impact by blunt cylinder. *Int. J. of Impact Engineering,* **2**(1): 37-64.

[Liss *et al.,* 1983] Liss, J., Goldsmith, W., Kelly, J.M. (1983). A phenomenological penetration model of plates. *Int. J. of Impact Engineering,* **1**(4): 321-341.

[Liu and Shu, 1999] Liu, C.K., Shu, L.C. (1999). The analysis on perforation resistance of multi-layered structure. *Proc. of Conference of the Aeronautical and Astronautical Society/Explosive and Propellants Society of the Republic of China,* pp. 265-272.

[Liu and Stronge, 2000] Liu, D., Stronge, W.J. (2000). Ballistic limit of metal plates struck by blunt deformable missiles: experiments. *Int. J. of Solids and Structures,* **37**(10): 1403-1423.

[Liu *et al.,* 2003] Liu, H., Gao, S., Li, K. (2003). Measurement technologies and result analysis on experiment of penetration of steel projectile into thick concrete target. In: Wen, T., ed., *Proc. of 5th Int. conference of measurements and tests.* Beijing: World Publishing Corporation June; 2003, p. 3339–3342.

[Lixin *et al.,* 2000] Lixin, Q., Yunbin, Y., Tong, L. (2000). A semi-analytical model for truncated-ogive-nose projectiles penetration into semi-infinite concrete targets. *Int. J. of Impact Engineering,* **24**(9): 947–955.

[Lixin, 1997] Lixin, Q. (1997). Empirical prediction model for truncated-ogive-nose projectile penetration into runway target. *Explosion and Shock Waves,* **2**: 109-119 (in Chinese).

[Loikkanen *et al.,* 2005] Loikkanen, M.J., Buyuk, M., Kan, C., Meng, N. (2005). A computational and experimental analysis of ballistic impact to sheet metal aircraft structures. In: *Proc. of 5th European LS-DYNA Users Conf.* (May 25 - 26, 2005, Birmingham, UK), article 3c-79.

[Lok, 1994] Lok, T.S. (1994). Impact tests on steel fibre reinforced concrete panels. In: Bulson, P.S., ed., *Proc. of 3rd Int. Conf. on Structures under Shock and Impact* (June 1-3, 1994, Madrid, Spain). WIT Press, Southampton, pp. 243-252.

[Lorenz *et al.,* 2000] Lorenz, R.D., Moersch, J.E., Stone, J.A., Morgan A.R. Jr., Smrekar. S.E. (2000). Penetration tests on the DS-2 Mars microprobes: penetration depth and impact accelerometry. *Planetary and Space Science,* **48**(5): 419-436.

[Lou *et al.,* 2009] Lou, D.C., Solberg, J.K., Børvik, T. (2009). Surface strengthening using a self-protective diffusion paste and its application for ballistic protection of steel plates. *Materials and Design,* **30**(9): 3525–3536.

[Lowery, 1990] Lowery, B. (1990). Optimisation of a composite armour system to defeat FFV 7.62 mm AP projectiles. MSc thesis (Military Vehicle Technology), Cranfield Institute of Technology, U.K.

[Luo *et al.,* 2000] Luo, X., Sun, W., Chan, S.Y.N. (2000). Characteristics of high-performance steel fiber-reinforced concrete subject to high velocity impact. *Cement and Concrete Research,* **30**(6): 907-914.

[Luo *et al.*, 2001] Luo, X., Sun, W., Chan, S.Y.N. (2001). Steel fiber reinforced high-performance concrete: a study on the mechanical properties and resistance against impact. *Materials and Structures/Matériaux et Constructions*, **34**(3): 144-149.

[Ma and Liew, 2013] Ma, C.Y., Liew, J.Y.R. (2013). Blast and ballistic resistance of ultra-high strength steel. *Int. J. of Protective Structures*, **4**(3): 379-414.

[Maalej *et al.* 2004] Maalej, M., Zhang, J., Quek, S.T., Lee, S.C. (2004). Resistance of hybrid-fiber engineered cementitious composites. In: *Proc. of the 5th Int. Conf. on Fracture Mechanics of Concrete and Concrete Structures (FRAMCOS-5)*, (April 12-16, 2004, Vail CO), pp. 1051-1058.

[Maalej *et al.* 2005] Maalej, M., Quek, S.T., Zhang, J. (2005). Behavior of hybrid-fiber engineered cementitious composites subjected to dynamic tensile loading and projectile impact. *J. of Materials in Civil Engineering (ASCE)*, **17**(2): 143-152.

[Máca and Sovják, 2012] Máca, P., Sovják, R. (2012). Resistance of ultra high performance fibre reinforced concrete to projectile impact. In: Schleyer, G., Brebbia, C.A., eds. *Structures Under Shock and Impact XII* [*Proc. of the 12th Int. Conf. on Structures under Shock and Impact* (September 4 - 6, 2012, Kos, Greece)]. WIT Press, Southampton, pp. 261-272.

[Máca *et al.*, 2014] Máca, P., Sovják, R., Konvalinka, P. (2014). Mix design of UHPFRC and its response to projectile impact. *Int. J. of Impact Engineering*, **63**: 158-163.

[Madhu *et al.*, 2003] Madhu, V., Bhat, T.B., Gupta, N.K. (2003). Normal and oblique impacts of hard projectiles on single and layered plates – an experimental study. *Defence Science J.*, **53**(2): 147-156.

[Magnusson *et al.*, 2001] Magnusson, J., Unosson, M., Carlberg, A. (2001). High performance concrete "HPC". Field experiments and production. Tech. Rep. FOI-R-0256-SE. Swedish Defence Research Agency (FOI), Tumba, Sweden.

[Malone and Tom, 1998] Malone, P.G., Tom, J.G. (1998). Foamed, fiber-reinforced concrete as a fragment collecting medium. In: *Proc. of the 28th DoD Explosives Safety Seminar* (August 18-20, 1998, Orlando, FL).

[Manes *et al.*, 2014] Manes, A., Serpellini, F., Pagani, M., Saponara, M., Giglio, M. (2014). Perforation and penetration of aluminium target plates by armour piercing bullets. *Int. J. of Impact Engineering*, **69**: 39-54.

[Marom and Bodner, 1979] Marom, I., Bodner, S.R. (1979). Projectile perforation of multi-layered beams. *Int. J. of Mechanical Science,* **21**(8): 489-504.

[Martineau *et al.*, 2004] Martineau, R.L., Prime, M.B., Duffey, T. (2004). Penetration of HSLA-100 steel with tungsten carbide spheres at striking velocities between 0.8 and 2.5 km/s. *Int. J. of Impact Engineering*, **30**(5): 505–520.

[Maurer and Rinehart, 1960] Maurer, W.C., Rinehart, J.S. (1960). Impact crater formation in rock. *J. of Applied Physics*, **31**(7): 1247-1252.

[Mayseless *et al.*, 2008] Mayseless, M., Reifen, Y., Tibon, G., Yaziv, D., Touati, D. (2008). Ricochet of AP projectiles from hard concrete targets. In: *Proc. of the 20th Int. Symp. on Ballistics* (September 23-27, 2002, Orlando, FL), pp. 800-807.

[MIL-A-46100D, 1988] MIL-A-46100D. Military Specification. Armor plate, steel, wrought, high-hardness, 1988.

[MIL-DTL-46593B, 2006] MIL-DTL-46593B. Detail specification: projectile, calibers .22, .30, .50, and 20 mm fragment-simulating (06-jul-2006).

[Mileiko and Sarkisyan, 1981] Mileiko, S.T., Sarkisyan, O.A. (1981). Phenomenological model of punch-through, *J. of Applied Mechanics and Technical Physics,* **22**(5): 711-713.

[Mileiko *et al.,* 1979] Mileiko, S.T., Kondakov, S.F., Golofast, E.G. (1979). One case of piercing. *Strength of Materials,* **11**(12): 1413-1416.

[Miwa *et al.,* 2006] Miwa, K., Beppu, M., Ohno, T., Katayama, M. (2006). Testing and simulation of local damage in a concrete plate by the impact of a hard projectile. In: Jones, N., Brebbia, C.A., eds. *Structures Under Shock and Impact IX [Proc. of the 9th Int. Conf. on Structures under Shock and Impact* (July 3-5, 2006, New Forest, UK)]. WIT Press, Southampton, pp. 287-296.

[Mohamed *et al.,* 2009] Mohamed, M.E., Eltehawy, E.M., Kamal, I.M., Aggour, A.A. (2009). Experimental analysis of reinforced concrete panels penetration resistance. In: *Proc. of 13th Int. Conf. on Aerospace Sciences & Aviation Technology* (May 26-28, 2009, Cairo, Egypt), paper ASAT-13-TE-14.

[Mohan *et al.,* 2014] Mohan, V., Rajasankar, J., Iyer, N.R. (2014). Response simulation of a micro reinforced concrete target under ballistic impact. *Int. J. for Computational Methods in Engineering Science and Mechanics,* **15**(3): 302-308.

[Mohotti *et al.,* 2015a] Mohotti, D., Ngo, T., Raman, S.N., Mendis, P. (2015). Analytical and numerical investigation of polyurea layered aluminium plates subjected to high velocity projectile impact. *Materials and Design,* **82**: 1–17.

[Mohotti *et al.,* 2015b] Mohotti, D., Ngo, T., Mendis, P. (2015). Use of multi layered composite plate systems in ballistic impact mitigation. In: Stewart, M., Netherton, M., eds. Design and analysis of protective structures. *Proc. of the 3rd Int. Conf. on Protective Structures* (3-6 February 2015, Newcastle, Australia), pp. 366-372.

[Mu and Zhang, 2011] Mu, Z., Zhang, W. (2011). An investigation on mass loss of ogival projectiles penetrating concrete targets. *Int. J. of Impact Engineering,* **38**(8-9): 770-778.

[Mullin *et al.,* 1991] Mullin, S.A., Riegel, J.P., Tenenbaum, D.A., Erdley, D.W. (1991). Dynamic plasticity modeling of conical and blunt nosed projectiles and dual layer armor. *TACOM Combat Vehicle Survivability Symposium* (April 15-17, 1991, Gaithersburg, MD).

[Muzychenko and Postnov, 1985] Muzychenko, V. P., Postnov, V. I. (1985). Scope for forecasting alloy piercing resistance. *J. of Applied Mechanics and Technical Physics,* **25**(5): 774-776.

[Neilson, 1985] Neilson, A.J. (1985). Empirical equations for the perforation of mild steel plates. *Int. J. of Impact Engineering,* **3**(2): 137-142.

[Nia and Hoseini, 2011] Nia, A.A., Hoseini, G.R. (2011). Experimental study of perforation of multi-layered targets by hemispherical-nosed projectiles. *Materials and Design*, **32**(2): 1057–1065.

[Nia *et al.*, 2014] Nia, A.A., Zolfaghari, M., Khodarahmi, H., Nili, M., Gorbankhani, A.H. (2014). High velocity penetration of concrete targets with eroding long- rod projectiles; an experiment and analysis. *Int. J. of Protective Structures*, **5**(1): 47-63.

[Nishiwaki, 1951] Nishiwaki, J. (1951). Resistance to the penetration of a bullet through an aluminium plate. *J. of Physics Society of Japan*, **6**(5): 374-378.

[NPG Report 12-44, 1944] Penetration of homogeneous plate by 3-inch flat-nosed projectiles. Partial Rep. 12-44. Naval Proving Ground, Dahlgren, VA, 1944.

[NPG Report 16-43, 1943] Effect of cap design on 3-inch projectile performance. Partial Rep. 16-43. Naval Proving Ground, Dahlgren, VA, 1943.

[NPG Report 7-43, 1943] Penetration of homogeneous armor by 3-inch flat-nosed projectiles. Rep. 7-43. Naval Proving Ground, Dahlgren, VA, 1943.

[Nurick and Crowther, 1994] Nurick, G.N., Crowther, B.C. (1994). The measurement of the residual velocity and deflection of a projectile passing through a thin plate. In: Bulson, P.S., ed., *Proc. of 3th Int. Conf. on Structures under Shock and Impact* (June 1-3, 1994, Madrid, Spain), WIT Press, Southampton, pp. 261-272.

[Nurick and Walters, 1990] Nurick, G.N., Walters, C.E. (1990). The ballistic penetration of multiple thin plates separated by an air gap. In: *Proc. of the SEM Spring Conf. on Experimental Mechanics* (June 3-6, 1990, Albuquerque, NM), pp. 631-637.

[Nurick, 2013] Nurick, G.N. (2013). Private Communication.

[O'Daniel *et al.*, 2011] O'Daniel, J., Danielson, K., Boone, N. (2011). Modeling fragment simulating projectile penetration into steel plates using finite elements and meshfree particles. *Shock and Vibration*, **18**(3): 425–436.

[O'Neil, 1999] O'Neil, E.F., Neeley, B.D., Cargile, J.D. (1999). Tensile properties of very-high-strength concrete for penetration-resistant structures. *Shock and Vibration*, **6**(5-6): 237–245.

[Ohno *et al.*, 1994] Ohno, T., Uchida, N., Ishikawa, Y., Kasai, H., Tsubota, M., Ueda, A., Kambayashi, A., Shirai, T. (1994). Improvement on impact resistance of reinforced concrete panels against projectile impact. In: Bulson, P.S., ed., *Proc. of the 3th Int. Conf. on Structures under Shock and Impact* (June 1-3, 1994, Madrid), WIT Press, Southampton, pp. 254–260.

[Ohte *et al.*, 1982] Ohte, S., Yoshizawa, H., Chiba, N., Shida, S. (1982). Impact strength of steel plates struck by projectiles. *Bulletin of the Japan Society of Mechanical Engineering*, **25**(206): 1226–1231.

[Özşahin and Tolun, 2010] Özşahin, E., Tolun, S. (2010). Influence of surface coating on ballistic performance of aluminum plates subjected to high velocity impact loads. *Materials and Design*, **31**(3): 1276–1283.

[Özşahin *et al.*, 2011] Özşahin, E., Diltemiz, F., Tolun, S. (2011). Characterization of 6061 T651 aluminum plates subjected to high-velocity impact loads. *Anadolu*

University J. of Science and Technology – A. Applied Sciences and Engineering. **12**(1): 65-74.

[Partom *et al.*, 2001] Partom, Y., Anderson, C.E., Jr., Yaziv, D. (2001). Penetration of AP projectiles into spaced ceramic targets. In: *Proc. of the 19th Int. Symp. on Ballistics* (May 7-11, 2001, Interlaken, Switzerland), pp. 1175–1181.

[Patterson *et al.*, 1975] Patterson, W.J., Wood, W.R., Bentley, R.D. (1975). DNA/Sandia Soil penetration experiment at DRES: results and analysis. Rep. SAND 75-0001, Sandia National Laboratories, Albuquerque, NM.

[Patterson, 1969] Patterson, W.J. (1969). Terradynamic results and structural performance of a 650-pound penetrator impacting at 2570 feet per second. Rep. SC-DR-69-782. Sandia National Laboratories, Albuquerque, NM.

[Patterson, 1972] Patterson, W.J. (1972). Penetration of in situ rock by air deliverable penetrators. In: *Proc. of the Conf. on Rapid Penetration of Terrestrial Materials* (February 1972, Texas A&M University, College Station, TX), pp. 453-473.

[Patterson, 1973] Patterson, W.J. (1973). Projectile penetration of in situ rock. Rep. SLA-73-0831. Sandia National Laboratories, Albuquerque, NM.

[Pereira and Lerch, 2001] Pereira, J.M., Lerch, B.A. (2001). Effects of heat treatment on the ballistic impact properties of Inconel 718 for jet engine fan containment applications. *Int. J. of Impact Engineering*, **25**(8): 715–733.

[Pereira *et al.*, 2012] Pereira, J., Revilock, D., Ruggeri, C., Emmerling, W., Altobelli, D. (2012). Ballistic impact testing of Aluminum 2024 and Titanium 6Al-4V for material model development. In: Zacny, K., Malla, R.B., Binienda, W. *Proc. of the 13th ASCE Aerospace Division Conf. on Engineering, Science, Construction, and Operations in Challenging Environments and the 5th NASA/ASCE Workshop on Granular Materials in Space Exploration* (April 15-18, 2012, Pasadena, CA), pp. 1244-1253.

[Pereira *et al.*, 2013] [Pereira, J.M., Revilock, D.M., Lerch, B.A., Ruggeri, C.R. (2013). Impact testing of Aluminum 2024 and Titanium 6Al-4V for material model development. Tech. Rep. NASA/TM-2013-217869. NASA Glenn Research Center, Cleveland, OH, U.S.

[Pereira *et al.*, 2014] Pereira, J., Revilock, D., Ruggeri, C., Emmerling, W., Altobelli, D. (2014). Ballistic impact testing of Aluminum 2024 and Titanium 6Al-4V for material model development. *J. of Aerospace Engineering.* **27**(3): 456–465.

[Peterson and Hadala, 1976] Peterson, R.W., Hadala, P.F. (1976). Shear strength properties for the Antelope Lake target, Tonopah Test Range, Nevada. Memorandum for Record, 13 January, 1976. Soil Dynamics Division, U.S. Army Engineer Waterways Experiment Stations, Vicksburg, MS.

[Piccione, 1960] Piccione, D.A. (1960). Comparative ballistic performance of 20mm AP M95 and 20mm AP M75 projectiles against rolled homogeneous steel armor. Tech. Rep. WAL TR 160.1/1(c). Watertown Arsenal Labs., MA.

[Piekutowski *et al.,* 1996] Piekutowski, A. J., Forrestal, M. J., Poormon K. L., Warren, T. L., 1996, Perforation of aluminum plates with ogive-nose steel rods at normal and oblique impacts. *Int. J. of Impact Engineering,* **18**(7-8): 877-887.

[Piekutowski *et al.,* 1999] Piekutowski, A. J., Forrestal, M. J., Poormon K. L., Warren, T. L. (1999). Penetration of 6061-T6511 aluminum targets by ogive-nose steel projectiles with striking velocities between 0.5 and 3.0 km/s. *Int. J. of Impact Engineering,* **23**(1, part 2): 723-744.

[Project THOR, 1961] Project THOR, 1961.The resistance of various metallic materials to perforation by steel fragments; empirical relationships for fragment residual velocity and residual weigh. Tech. Rep. 47.The Johns Hopkins University Ballistic Analysis Laboratory.

[Quek *et al.,* 2010] Quek, S.T., Lin, V.W.J., Maalej, M. (2010). Development of functionally-graded cementitious panel against high-velocity small projectile impact. *Int. J. of Impact Engineering,* **37**(8): 928-941.

[Radin and Goldsmith, 1988] Radin, J., Goldsmith, W. (1988). Normal projectile penetration and perforation of layered targets. *Int. J. of Impact Engineering,* **7**(2): 229-259.

[Raguraman *et al.,* 2009] Raguraman, M., Jagadeesh, G., Deb, A. (2009). Development of an experimental facility for impact testing of armour plates. *Int. J. of Aerospace Innovations,* **1**(1): 45-55.

[Raguraman *et al.,* 2010] Raguraman, M., Jagadeesh, G., Deb, A., Barton, D.C. (2010) Experimental and numerical investigation of the behavior of aluminium plates upon ballistic impact. *Experimental Techniques,* **34**(6): 49-60.

[Ravid *et al.,* 1995] Ravid, M., Bodner, S.R., Holcman, I. (1995). Ballistic penetration in two-layered target plates. In: *Proc. of the 15th Int. Symp. on Ballistics* (May 21-24, 1995, Jerusalem, Israel), pp. 151-158.

[Recht and Ipson, 1962] Recht, R.F., Ipson, T.W. (1962). The dynamics of terminal ballistics. Ballistic evaluation procedures for armored grille designs. Final Rep. DRI No. 2025. Denver Research Institute, University of Denver.

[Recht and Ipson, 1963] Recht, R.F., Ipson, T.W. (1963). Ballistic perforation dynamics. *J. of Applied Mechanics,* **30**(3): 384-390.

[Richmond, 1980] Richmond, P.W. (1980). Influence of nose shape and L/D ratio on projectile penetration in frozen soil. Special Rep. 80-17. U.S. Army Cold Regions Research and Engineering Laboratory, Hanover, NH.

[Riedel *et al.,* 1999] Riedel, W., Thoma, K., Hiermaier, S., Schmolinske, E. (1999). Penetration of reinforced concrete by BETA-B-500 - numerical analysis using a new macroscopic concrete model for hydrocodes. In: *Proc. of the 9th Int. Symp. on Interaction of the Effects of Munitions with Structures* (May 3-7, 1999, Berlin-Strausberg, Germany), pp. 315-322.

[Rodríguez-Martínez *et al.,* 2013] Rodríguez-Martínez, J.A., Rusinek, A., Pesci, R., Zaera, R. (2013). Experimental and numerical analysis of the martensitic transformation in AISI 304 steel sheets subjected to perforation by conical and hemispherical projectiles. *Int. J. of Solids and Structures*, **50**(2): 339–351.

[Rodríguez-Millán *et al.*, 2014] Rodríguez-Millán, M., Vaz-Romero, A., Rusinek, A., Rodríguez-Martínez, J.A., Arias, A. (2014). Experimental study on the perforation process of 5754-H111 and 6082-T6 Aluminium plates subjected to normal impact by conical, hemispherical and blunt projectiles. *Experimental Mechanics,* **54**(5): 729–742.

[Rogers *et al.*, 1986] Rogers, C.O., Pang, S.-S., Kumano, A., Goldsmith, W. (1986). Response of dry- and liquid-filled porous rocks to static and dynamic loading by variously-shaped projectiles. *Rock Mechanics and Rock Engineering*, **19**(4): 235-260.

[Rohani, 1973] Rohani, B. (1973). Penetration performance of terminal delivery vehicle T-98WA into soft and frozen ground: analyses based on both field test results and theoretical calculations. U.S. Army Engineer Waterways Experiment Station, Vicksburg, MS.

[Rohani, 1975] Rohani, B. (1975). Analysis of projectile penetration into concrete and rock targets. Miscellaneous paper S-75-25. U.S. Army Engineer Waterways Experiment Station, Soils and Pavement Laboratory, Vicksburg, MS.

[Roisman *et al.*, 1999] Roisman, I.V., Weber, K., Yarin, A.L., Hohler, V., Rubin, M.B. (1999). Oblique penetration of a rigid projectile into a thick elastic-plastic target: theory and experiment. *Int. J. of Impact Engineering*, **22**(7): 707-726.

[Romander and Sliter, 1984] Romander, C.M., Sliter, G.E. (1984). Model tests of turbine missile impact on reinforced concrete. *Nuclear Engineering and Design*, **77**(3): 331-342.

[Rosenberg and Forrestal, 1988] Rosenberg, Z., Forrestal, M.J. (1988). Perforation of aluminum plates with conical-nosed rods-additional data and discussion. *J. of Applied Mechanics,* **55**(1): 236-238.

[Ryan and Cimpoeru, 2015] Ryan, S., Cimpoeru, S.J. (2015). An evaluation of the Forrestal scaling law for predicting the performance of targets perforated in ductile hole formation. In: Stewart, M., Netherton, M., eds. *Design and analysis of protective structures. Proc. of the 3rd Int. Conf. on Protective Structures* (3-6 February 2015, Newcastle, Australia), pp. 493-500.

[Saburi *et al.*, 2008] Saburi, T., Kubota, S., Yoshida, M., Wada, Y., Ogata, Y. (2008). Experimental impact study using an explosive driven projectile accelerator and numerical simulation. *Int. J. of Impact Engineering*, **35**(12): 1764–1769.

[Scheffler and Magness, 1998] Scheffler, D.R., Magness, L.S., Jr. (1998). Target strength effects on the predicted threshold velocity for hemi- and ogival-nose penetrators perforating finite aluminum targets. In: Jones, N. *et al.* (eds). *Structures under Shock and Impact V* [*Proc. of the 5th Int. Conf. on Structures under Shock and Impact* (June 24-26, 1998, Thessaloniki, Greece)]. WIT Press: Southampton. pp. 285-297.

[Scheffler, 1996] Scheffler, D.R. (1996). CTH hydrocode predictions on the effect of rod nose-shape on the velocity at which tungsten alloy rods transition from rigid body to eroding penetrators when impacting thick aluminium targets. In: Jones, N., Bulson, P.S., eds., *Structures under shock and impact IV* [*Proc. of the 4th Int. Conf. on Structures under Shock and Impact* (July 1996, Udine, Italy)], Computational Mechanics Publ., Southampton, pp. 297-310.

[Scheffler, 1997] Scheffler, D.R. (1997). Modeling the effect of penetrator nose shape on threshold velocity for thick aluminum targets. Tech. Rep. ARL-TR-1417. U.S. Army Research Laboratory, Aberdeen Proving Ground, MD.

[Schulz and Heimdahl, 1981] Schulz, J.C., Heimdahl, O.E.R. (1981). Survivability of penetrators with circumferential shear-control grooves. Rep. NWC TP 6275. Naval Weapons Center, China Lake, CA.

[Schulz *et al.*, 1982] Schulz, J. C., Heimdahl, O.E.R., Finnegan, S. (1982). Magnesium projectile impact behavior. In: *Proc. of the Army Symp. on Solid Mechanics, 1982: Critical Mechanics Problems in Systems Design* (September 21-23, 1982, Bass River, Cape Cod, MA). Army Materials and Mechanics Research Center, Watertown, MA, pp. 28-42.

[Schulz *et al.*, 1984] Schulz, J.C., Heimdahl, O.E.R., Finnegan, S. (1984). Small steel projectile firings against simulated reinforced concrete slabs at normal incidence and obliquity. . Rep. NWC TP 6501. Naval Weapons Center, China Lake, CA.

[Schwer *et al.*, 2006] Schwer, L., Hacker, K., Roe, K. (2006). Perforation of metal plates: experimental validation of numerical simulations. In: Mindle, W.L., ed., *Proc. of 9th Int. LS-DYNA Users Conf.* (June 4-6, 2006, Dearborn, MI).

[Seah *et al.*, 2011] Seah C.C., Børvik T., Remseth S., Pan T-C. (2011). Penetration and perforation of rock targets by hard projectiles. In: Zhou, Y., Zhao, J., eds. *Advances in Rock Dynamics and Applications*. CRC Press, Boca Raton, FL, pp. 143-162.

[Segletes *et al.*, 2000] Segletes, S.B., Grote, R., Polesne, J. (2000). Improving the rod-penetration algorithm for tomorrow's armors. In: *Proc. of the 22nd Army Science Conf.: Accelerating the Pace of the Transformation to the Objective Force.* (December 11-13, 2000, Baltimore, MD), pp. 66-71.

[Segletes *et al.*, 2001] Segletes, S.B., Grote, R., Polesne, J. (2001). Improving the rod-penetration algorithm for tomorrow's armors. Rep. ARL-RP-23. U.S. Army Research Laboratory, Aberdeen Proving Ground, MD.

[Segletes, 2004] Segletes, S.B. (2004). A rod ricochet model. Rep. ARL-TR-3257. U.S. Army Research Laboratory, Aberdeen Proving Ground, MD.

[Segletes, 2006] Segletes, S.B. (2006). A model for rod ricochet. *Int. J. of Impact Engineering,* **32**(9): 1403–1439.

[Senthil *et al.*, 2015] Senthil, K., Iqbal, M.A., Bhattacharjee, D., Bhargava, P., Gupta, N.K. (2015). Ballistic resistance of ARMOX 500T steel plates against 7.62 API projectiles. In: Stewart, M., Netherton, M., eds. Design and analysis of protective structures. *Proc. of the 3rd Int. Conf. on Protective Structures* (3-6 February 2015, Newcastle, Australia), pp. 501-504.

[Shadbolt *et al.,* 1983] Shadbolt, P.J., Corran, R.S.J., Ruiz, C. (1983). A comparison of plate perforation models in the sub-ordnance impact velocity range. *Int. J. of Impact Engineering,* **1**(1): 23–49.

[Shan *et al.,* 2014] Shan, Y., Huang, F., Wu, H. (2014). The influence of projectile material on mass abrasion of high velocity penetrator. In: Ames, R.G., Boeka, R.D., eds., *Proc. of the 28th Int. Symp. on Ballistics* (September 24-26, 2014, Atlanta, Georgia, USA), vol. 2, pp. 1287-1298.

[Sharif, 2005] Sharif, M.S.H. (2005). Resistance of plain and fibre-reinforced high-strength concrete to projectile impact. Master Thesis, National University of Singapore.

[Shiqiao *et al.,* 2006] Shiqiao, G., Lei, J., Haipeng, L., Kejie, L. (2006). The principle of crater-formation of a concrete target plate penetrated by a projectile. In: Jones, N., Brebbia, C.A., eds. *Structures Under Shock and Impact IX* [*Proc. of the 9th Int. Conf. on Structures under Shock and Impact* (July 3-5, 2006, New Forest, UK)]. WIT Press, Southampton, pp. 313-322.

[Shirai *et al.,* 1993] Shirai, T., Ohno, T., Ueda, M., Taniguchi, H., Kambayashi, A., Ishikawa, N. (1993). Experiment and numerical simulation of double layered RC plates under impact loading. Part 1: Impact tests for double layered RC plates. In: *Trans. of 12th Int. Conf. on Structural Mechanics in Reactor Technology (SMiRT-12)* (August 15-20, 1993, Stuttgart, Germany), paper J07/3, pp. 181-186.

[Shirai *et al.,* 1997] Shirai, T., Kambayashi, A., Ohno, T., Taniguchi, H., Ueda, M., Ishikawa, N. (1997). Experiment and numerical simulation of double layered RC plates under impact loading. *Nuclear Engineering and Design,* **176**(3): 195-205.

[Showalter *et al.,* 2008a] Showalter, D.D., Placzankis, B.E., Burkins, M.S. (2008). Ballistic performance testing of aluminum alloy 5059-H131 and 5059-H136 for armor applications. Rep. ARL-TR-4427. US Army Research Laboratory, Adelphi, MD.

[Showalter *et al.,* 2008b] Showalter, D., Gooch, W., Burkins, M., Koch, R.S. (2008). Ballistic Testing of SSAB ultra-high-hardness steel for armor applications. Rep. ARL-TR-4632. US Army Research Laboratory, Aberdeen Proving Ground, MD.

[Showalter *et al.,* 2009] Showalter, D., Gooch, W., Burkins, M., Montgomery, J., Squillacioti, R. (2009). Development and ballistic testing of a new class of auto-tempered high-hard steels under military specification MIL-DTL-46100E. Rep. ARL-TR-4997. US Army Research Laboratory, Aberdeen Proving Ground, MD.

[Shukla and Sadd, 2000] Shukla, A., Sadd, M. (2000). Penetrator resistance and target damage due to multiple impacts upon granite and concrete. Final Rep., Rhode Island University, Kingston. Dept. of Mechanical Engineering and Applied Mechanics.

[Siriphala *et al.,* 2012] Siriphala, P., Veeraklaew, T., Kulsirikasem, W., Tanapornraweekit, G. (2012). Validation of FE models of bullet impact on high strength steel armors.

In: Schleyer, G., Brebbia, C.A., eds. *Structures Under Shock and Impact XII* [*Proc. of the 12th Int. Conf. on Structures under Shock and Impact* (September 4 - 6, 2012, Kos, Greece)]. WIT Press, Southampton, pp. 75-83.

[Sjøl and Teland, 2000] Sjøl, H., Teland, J.A. (2000). Prediction of concrete penetration using Forrestal's formula. Rep. FFI/RAPPORT-99/04415. Norwegian Defence Research Establishment (FFI), Kjeller, Norway.

[Sjøl and Teland, 2001] Sjøl, H., Teland J.A. (2001). Perforation of concrete targets. Rep. FFI/RAPPORT-2001/05786. Norwegian Defence Research Establishment (FFI), Kjeller, Norway.

[Sjøl *et al.,* 1998] Sjøl, H., Teland, J.A., Kaldheim, Ø. (1998). Penetrasjon i betong med 12 mm prosjektiler. Rep. FFI/NOTAT-98/04392. Norwegian Defence Research Establishment (FFI), Kjeller, Norway (in Norwegian).

[Sjøl *et al.,* 2000] Sjøl, H., Teland, J.A., Kaldheim, Ø. (2000). Penetration into concrete – Experiments with 12 mm projectiles. Rep. FI/RAPPORT-2000/04414. Norwegian Defence Research Establishment (FFI), Kjeller, Norway.

[Sjøl *et al.,* 2002] Sjøl, H., Teland, J.A., Kaldheim, Ø. (2002). Penetration into concrete – Analysis of small scale experiments with 12 mm projectiles. Rep. FFI/RAPPORT-2002/04867. Norwegian Defence Research Establishment (FFI), Kjeller, Norway.

[Sliter, 1980] Sliter, G.E. (1980). Assessment of empirical concrete impact formulas. *J. of the Structural Division (ASCE)*, **106** (5): 1023–1045.

[Soe *et al.,* 2013] Soe, K.T., Zhang, Y.X., Zhang, L.C. (2013). Impact resistance of hybrid-fiber engineered cementitious composite panels. *Composite Structures*, **104**: 320-330.

[Soliman *et al.,* 1976] Soliman, A.S., Reid, S.R., Johnson, W. (1976). The effect of spherical projectile speed in ricochet off water and sand. *Int. J. of Mechanical Science*, **18**(6): 279-284.

[Sovják *et al.,* 2013] Sovják, R., Vavřiník, T., Máca, P., Zatloukal, J., Konvalinka, P., Song, Y. (2013). Experimental investigation of ultra-high performance fiber reinforced concrete slabs subjected to deformable projectile impact. *Procedia Engineering*, **65**: 120-125.

[Sovják *et al.,* 2014] Sovják, R., Vavřiník, T., Frydrýn, M., Mičunek, T., Zatloukal, J., Máca, P. (2014). Residual velocity of the non-deformable projectile after perforating the ultra-high performance fibre reinforced concrete. In: Schleyer, G., Bulson, P.S. (eds.). *Proc. of 13th Int. Conf. on Structures under Shock and Impact* (June 3-5, 2014, New Forest, UK). WIT Press, pp. 257-264.

[Sovják *et al.,* 2015] Sovják, R., Vavřiník, T., Zatloukal, J., Máca, P., Mičunek, T., Frydrýn, M. (2015). Resistance of slim UHPFRC targets to projectile impact using in-service bullets. *Int. J. of Impact Engineering*, **76**: 166-177.

[Squillacioti, 1994] Squillacioti, R.J. (1994) Development of an improved ballistic acceptance test for high hard armor steel plate — MIL-A-46100. Tech. Rep. ARL-TR-509. U.S. Army Research Laboratory, Watertown, MA.

[Stargel, 2005] Stargel, D.S. (2005). Experimental and numerical investigation into the effects of panel curvature on the high velocity ballistic impact response of aluminum and composite panels. Ph.D. Thesis, University of Maryland.

[Stepanov and Zubov, 1998] Stepanov G.V., Zubov, V.I. (1998). Analysis of the aluminum allow resistance to the penetration of a steel rod at impact velocities of up to 500 m/s. *Strength of Materials*, **30**(5): 536-539.

[Stephenson, 1977] Stephenson, A.E. (1977). Full-scale tornado-missile impact tests. Final Rep. NP-440. Electric Power Research Institute, Sandia Laboratories, Tonopah, NV.

[Stock and Thompson, 1970] Stock, T.A.C., Thompson, R.L. (1970). Penetration of aluminum alloys by projectiles. *Metallurgical Transactions*, **1**(1): 219-224.

[Stronge and Schulz, 1981] Stronge, W.J., Schulz, J.C. (1981). Projectile impact damage analysis. *Computers & Structures*, **13**(1-3): 287-294.

[Sugano *et al.*, 1993] Sugano, T., Tsubota, H., Kasai, Y., Koshika, N., Ohnuma, H., von Riesemann, W.A., Bickel, D.C., Parks, M.B. (1993). Local damage to reinforced concrete structures caused by impact of aircraft engine missiles. Part 1. Test program, method and results. *Nuclear Engineering and Design*, **140**(3): 387-405.

[Sullivan, 1942] Sullivan, J. (1942). Rolled armor. Ballistic properties of rolled face hardened armor and rolled homogeneous armor of various hardnessess at normal incidence and at various obliquities. Rep. 710/456. Watertown Arsenal Labs, MA.

[Sullivan, 1945] Sullivan, J.F. (1945). Resistance of various steels to perforation by fragment simulating projectiles. Memorandum Rep. WAL 710/763. Watertown Arsenal Laboratory, MA.

[Sun and Pei, 1995] Sun, T., Pei, S. (1995).The study of oblique impact of a deformable projectile at ductile spaced layered targets. In: *Proc. of the 15th Int. Symp. on Ballistics* (May 21-24, 1995, Jerusalem, Israel), paper TB49.

[Sundararajan and Dikshit, 2009] Sundararajan, G., Dikshit, S.N. (2009). The dynamic indentation behavior of steel at large depths of penetration. *J. of Materials Research*, **24**(03): 691-703.

[Svinsås *et al.*, 2011] Svinsås, E., O'Carol, C., Wentzel, C.M., Carlberg, A. (2011). Benchmark trial designed to provide validation data for modeling. In: *Proc. of the 10th Int. Symp. on Interaction of the Effects of Munitions with Structures* (May 7-11, 2001, San Diego, CA).

[Szczepanski and Collins, 1983] Szczepanski, R.D., Collins, J.A. (1983). Concrete penetration and ricochet testing of two projectile types. In: *The interaction of non-nuclear munitions with structures: Symp. proc., part 2* (May 10-13, 1983, U.S. Air Force Academy, CO), pp. 35-42.

[Tai and Tang, 2006] Tai, Y.-S., Tang, C.-C. (2006). Reactive powder concrete plate response to a flat projectile impact. In: Jones, N., Brebbia, C.A., eds. *Structures under shock and impact IX* [*Proc. of the 9th Int. Conf. on Structures under Shock and Impact* (July 3-5, 2006, New Forest, UK)]. WIT Press, Southampton, pp. 333-342.

[Tai, 2009] Tai, Y.S. (2009). Flat ended projectile penetrating ultra-high strength concrete plate target. *Theoretical and Applied Fracture Mechanics*, **51**(2): 117–128.

[Teland and Sjøl, 2000] Teland, J.A., Sjøl, H. (2000). Boundary effects in penetration into concrete. Rep. FFI/RAPPORT-2000/05414. Norwegian Defence Research Establishment (FFI), Kjeller, Norway.

[Teland and Sjøl, 2004] Teland, J.A., Sjøl, H. (2004). Penetration into concrete by truncated projectiles. *Int. J. of Impact Engineering*, **30**(4): 447-464.

[Teland, 2001] Teland, J.A. (2001). Cavity expansion theory applied to penetration of targets with pre-drilled cavities. In: *Proc. of the 19th Int. Symp. on Ballistics* (May 7-11, 2001, Interlaken, Switzerland), paper TB36, pp. 1329-1335.

[Teng *et al.*, 2008] Teng, T.-L., Shih, T.-M., Lu, C.-C. (2008). Effect of design parameters on the anti-penetration properties of space armor. *Structural Engineering and Mechanics*, **28**(6): 715-725.

[Thigpen, 1974] Thigpen, L. (1974). Projectile penetration of elastic-plastic earth media. *J. of the Geotechnical Engineering Division (Proc. of the ASCE)*, **100** (GT 3): 279-291.

[TM 43-0001-27, 1994] Army ammunition data sheets for small caliber ammunition (FSC 1305); TM 43-0001-27 (1994). HEADQUARTERS, Department of the Army, Washington, DC.

[TM 5-818-1/AFM 88-3, 1983] Technical Manual TM 5-818-1/AFM 88-3, Chapter 7 (1983). Soils and geology procedures for foundation design of buildings and other structures (except hydraulic structures). Joint Departments of the Army and Air Force, USA.

[Tolch and Bushkovitch, 1947] Tolch, N.A., Bushkovitch, A.V. (1947). Penetration and crater volume in various kinds of rocks as dependent on caliber, mass, and striking velocity of projectile. Rep. 641. Ballistic Research Laboratory, Aberdeen Proving Ground, MD.

[Tsubota *et al.*, 1993] Tsubota, H., Kasai, Y., Koshika, N., Morikawa, H., Uchida, T., Ohno, T., Kogure, K. (1993). Quantitative studies on impact resistance of reinforced concrete panels with a steel liner under impact loading. Part I: Scaled model impact test. In: *Trans. of 12th Int. Conf. on Structural Mechanics in Reactor Technology (SMiRT-12)* (August 15-20, 1993, Stuttgart, Germany), paper J07/1, pp. 169-174.

[Unosson and Nilsson, 2006] Unosson, M., Nilsson, L. (2006). Projectile penetration and perforation of high performance concrete: experimental results and macroscopic modeling. *Int. J. of Impact Engineering*, **32**(7):1068-1085.

[Unosson, 2000] Unosson, M. (2000). Numerical simulations of penetration and perforation of high performance concrete with 75mm steel projectile. Rep. FOA-R-00-01634-311-SE. Defence Research Establishment, Weapons and Protection Division, SE-147 25 Tumba, Sweden.

[Van Valkenburg *et al.*, 1956] Van Valkenburg, M. E., Clay, W.G., Huth, J. H. (1956). Impact phenomena at high speeds. *J. of Applied Physics*, 27(10): 1123-1129.

[Vassallo, 1975] Vassallo, F.A. (1975). Missile impact testing of reinforced concrete panels. Rep. HC-5609-D-1. Calspan Corporation, Buffalo, NY. Prepared for Bechtel Corporation.

[Verolme *et al.*, 1999] Verolme, J.L., Szymczak, M., Broos, J.P.F. (1999). Metallic witness packs for behind-armour debris characterisation. *Int. J. of Impact Engineering* 22(7): 693-705.

[Vijayan *et al.*, 2013] Vijayan, V., Hegde, S., Gupta, N.K. (2013). Ballistic response of thin targets impacted by truncated and tip-deforming conical projectiles. *Proc. of 27th Int. Symp. on Ballistics*, (April 22-26, 2013, Freiburg, Germany), pp.1800-1806.

[Virostek and Goldsmith, 1987] Virostek S.P., Goldsmith, W. (1987). Direct force measurement in normal and oblique impact of plates by projectiles. *Int. J. of Impact Engineering*, 6(4): 247-269.

[Vitman and Ioffe, 1948] Vitman, F.F., Ioffe, B.S. (1948). A simple method of determining dynamic hardness of metals using a double cone. *Zavodskaja Laboratorija*, 14(6): 727-732 (in Russian).

[Vitman and Stepanov, 1959] Vitman, F.F., Stepanov, V.A. (1959). Influence of Strain Rate on Resistance to Deformation of Metals at Impact Velocities of 100 to 1000 m/sec. In: *Nekotoryje Problemy Prochnosti Tvjordogo Tela. USSR Academy of Science, Moscow-Leningrad*, pp. 207-221 (in Russian). English translation: Foreign Technology Division Wright-Patterson AFB, Ohio, 1964 (AD0605234).

[Vossoughi *et al.*, 2007] Vossoughi, F., Ostertag, C.P., Monteiro, P.J.M., Johnson, G.C. (2007). Resistance of concrete protected by fabric to projectile impact. *Cement and Concrete Research*, 37(1): 96–106.

[Walter and Wolde-Tinsae, 1984] Walter, T.A., Wolde-Tinsae, A.M. (1984). Turbine missile perforation of reinforced concrete. *J. of Structural Engineering*, 110(10): 2439–2455.

[Wang, 2011] Wang, S. (2011). Experimental and numerical studies on behavior of plain and fiber-reinforced highstrength concrete subjected to high strain rate loadings. Ph.D. Thesis, National University of Singapore.

[Warren and Poormon, 2001] Warren, T. L., Poormon, K. L. (2001). Penetration of 6061-T6511 aluminum targets by ogive-nosed VAR 4340 steel projectiles at oblique angles: experiments and simulations. *Int. J. of Impact Engineering*. 25(10): 993-1022.

[Warren *et al.*, 2004] Warren, T.L., Hanchak, S.J., Poormon, K.L. (2004). Penetration of limestone targets by ogive-nosed VAR 4340 steel projectiles at oblique angles: experiments and simulations. *Int. J. of Impact Engineering*, 30(10): 1307-1331.

[Watanabe *et al.*, 2011] Watanabe, K., Tanaka, K., Iwane, K., Fukuma, S., Takayama, K., Kobayashi, H. (2011). Sand Behavior induced by high-speed penetration of projectile. Rep. AOARD-094011. Chubu University,1200 Matsumoto-cho, Kasugai City 487-8501, Japan, NA.

[Weerheijm *et al.,* 2015] Weerheijm, J., Roebroeks, G., Krabbenborg, D., Ozbek, A.S. Agar (2015). The potentials of porous concrete for ballistic protection. In: Stewart, M., Netherton, M., eds. *Design and analysis of protective structures. Proc. of the 3rd Int. Conf. on Protective Structures* (3-6 February 2015, Newcastle, Australia), pp. 686-693.

[Weidemaier *et al.,* 1993] Weidemaier, P., Senf, H., Rothenhäusler, H., Filbey, G.L., Gooch, W.A. (1993). On the ballistic resistance of laminated steel targets: experiments and numerical calculations. In: *Proc. of the 14th Int. Symp. On Ballistics* (26-29 September, 1993, Quebec, Canada), pp. 681-690.

[Williams *et al.,* 2009] Williams, E.M., Graham, S.S., Reed, P.A., Rushing, T.S. (2009). Laboratory characterization of Cor-Tuf concrete with and without steel fibers. Tech. Rep. ERDC/GSL TR-09-22. Geotechnical and Structures Laboratory, U.S. Army Engineer Research and Development Center, Vicksburg, MS.

[Woodward and Cimpoeru, 1998] Woodward, R.L., Cimpoeru, S.J. (1998). A study of the perforation of aluminium laminate targets. *Int. J. of Impact Engineering,* **21**(3): 117-131.

[Woodward and De Morton, 1976] Woodward, R.L., De Morton, M.E. (1976). Penetration of targets by flat-ended projectiles. *Int. J. of Mechanical Science,* **18**(3): 119-127.

[Woodward, 1978a] Woodward, R.L. (1978). The penetration of metal targets by conical projectiles. *Int. J. of Mechanical Science,* **20**(6): 349-359.

[Woodward, 1978b] Woodward, R.L. (1978). The penetration of metal targets which fail by adiabatic shear plugging. *Int. J. of Mechanical Science,* **20**(9): 599-607.

[Woodward, 1982] Woodward, R.L. (1982). Penetration of semi-infinite metal targets by deforming projectiles. *Int. J. of Mechanical Sciences,* **24**(2): 73-87.

[Wu *et al.,* 1994] Wu, E., Sheen, H.-J., Chen, Y.-C., Chang, L.-C. (1994). Penetration force measurement of thin plates by laser Doppler anemometry. *Experimental Mechanics,* **34**(2): 93-99.

[Wu *et al.,* 2012] Wu, H., Qian, F., Huang, F., Wang, Y. (2012). Projectile nose mass abrasion of high-speed penetration into concrete. *Advances in Mechanical Engineering,* **4**: 1-9.

[Wu *et al.,* 2015a] Wu, H., Fang, Q., Chen, X.W., Gong, Z.M., Liu, J.Z. (2015). Projectile penetration of ultra-high performance cement based composites at 510–1320 m/s. *Construction and Building Materials,* **74**: 188-200.

[Wu *et al.,* 2015b] Wu, H., Fang, Q., Peng, Y., Gong, Z.M., Kong, X.Z. (2015). Hard projectile perforation on the monolithic and segmented RC panels with a rear steel liner. *Int. J. of Impact Engineering,* **76**: 232-250.

[Wu *et al.,* 2015c] Wu, H., Fang, Q., Gong, J., Liu, J.Z., Zhang, J.H., Gong, Z.M. (2015). Projectile impact resistance of corundum aggregated UHP-SFRC. *Int. J. of Impact Engineering,* **84**: 38-53.

[Wu *et al.*, 2015d] Wu, H., Fang, Q., Gong, Z.M., Peng, Y. (2015). Hard projectile impact on layered SFRHSC composite target. *Int. J. of Impact Engineering*, **84**: 88-95.

[Xu *et al.*, 2014] Xu, J., Lee, C.K., Fan, S.C., Kang, K.W. (2014). A study on the ricochet of concrete debris on sand. *Int. J. of Impact Engineering*, **65**: 56-68.

[Xue *et al.*, 2010] Xue, L., Mock, W., Jr., Belytschko, T. (2010). Penetration of DH-36 steel plates with and without polyurea coating. *Mechanics of Materials*, **42**(11): 981–1003.

[Yamamato *et al.*, 2011] Yamamato, H., Milton, B., Hayasaka, S., Ogawa, T., Kikuchi, T., Ohtani, K., Takayama, K., Cooper, W., Tanaka, K., Watanabe, K. (2011). An experimental study of penetration of a sphere into a sand layer. In: *Phenomenological studies of the response of granular and geological media to high-speed (Mach 1-5) projectiles II.* Grant FA2386-10-1-4115, AOARD-104115 Final Rep., Chubu University, Japan.

[Yong *et al.*, 2010] Yong, M., Iannucci, L., Falzon, B.G. (2010). Efficient modelling and optimisation of hybrid multilayered plates subject to ballistic impact. *Int. J. of Impact Engineering*, **37**(6): 605-624.

[Young, 1967] Young, C.W. (1967). The development of empirical equations for predicting depth of an earth–penetrating projectile. Rep. SC-DR-67-70. Sandia National Laboratories, Albuquerque, NM.

[Young, 1969] Young, C.W. (1969). Depth predictions for penetrating projectiles. *J. of the Soil Mechanics and Foundations Division (Proc. of the ASCE)*, SM 3, May 1969: 803-817.

[Young, 1972] Young, C.W. (1972). Empirical equations for predicting penetration perforation in layered earth materials for complex penetrator configurations Rep. SC-DR-72-0523. Sandia National Laboratories, Albuquerque, NM.

[Young, 1976] Young, C.W. (1976). Status report on high velocity penetration program. Rep. SAND 76-0291. Sandia Laboratories, Albuquerque, NM.

[Young, 1978] Young, C.W. (1978). Results of December 1977 Davis gun tests at TTR. Memorandum dated 13 January 1978. Sandia Laboratories, Albuquerque, N.M.

[Young, 1997] Young, C.W. (1997). Penetration equations. Rep. SAND97-2426. Sandia National Laboratories, Albuquerque, NM.

[Zaid and Travis, 1974] Zaid, A. I. O., Travis, F. W. (1974). A comparison of single and multi-plate shields subjected to impact by a high-speed projectile. In: Harding, J., ed., *Proc. of the Conf. on the Mechanical Properties of High Rates of Strain*, (1974, Institute of Physics, London). Conf. Series No. 21, pp. 417-428.

[Zener and Peterson, 1943] Zener, C., Peterson, R.E. (1943). Mechanism of armor penetration. Watertown Arsenal, Second Partial Rep. 710/492, Watertown, MA.

[Zhang *et al.*, 2005] Zhang, M.H., Shim, V.P.W., Chewa, C.W. (2005). Resistance of high-strength concrete to projectile impact. *Int. J. of Impact Engineering*, **31**(7): 825-841.

[Zhang *et al.*, 2007] Zhang, M.H., Sharif, M.S.H., Lu, G. (2007). Impact resistance of high-strength fibre-reinforced concrete. *Magazine of Concrete Research*, **59**(3): 199–210.

[Zhang *et al.*, 2011] Zhang, Z., Qiang, H., Gao, W. (2011). Coupling of smoothed particle hydrodynamics and finite element method for impact dynamics simulation. *Engineering Structures*, **33**(1): 255–264.

[Zhang *et al.*, 2012a] Zhang, W., Mu, Z., Xiao, X. (2012). Experimental study on effect of aggregate size to anti-penetration ability of concrete targets subjected to high-velocity fragments. *Acta Armamentarii*, **33**(8): 1009-1015 (in Chinese).

[Zhang *et al.*, 2012b] Zhang, W., Deng, Y., Cao, Z., Wei, G. (2012). Experimental investigation on the ballistic performance of monolithic and layered metal plates subjected to impact by blunt projectiles. *Int. J. of Impact Engineering*, **49**: 115-129.

[Zhang *et al.*, 2013a] Zhang, X., Cao, R., Tan, D., Wang, B. (2013). Different scale experiments of high velocity penetration with concrete targets. *J. of Applied Mechanics (ASME)*, **80**(3), 031802 (6 pp.).

[Zhang *et al.*, 2013b] Zhang, X., Cao, R., Tan, D . (2013). Projectile's properties of high velocity penetration with concrete targets. In: *Proc. of the 27th Int. Symp. on Ballistics* (April 22-26, 2013, Freiburg, Germany), pp. 1098-1106.

[Zhou and Stronge, 2008] Zhou, D.W., Stronge, W.J. (2008). Ballistic limit for oblique impact of thin sandwich panels and spaced plates. *Int. J. of Impact Engineering*, **35**(11): 1339–1354.

[Zhou *et al.*, 2012] Zhou, N., Wang, J.X., Yang, R., Dong, G. (2012). Damage mechanism and anti-penetration performance of multi-layered explosively welded plates impacted by spherical projectile. *Theoretical and Applied Fracture Mechanics*, **60**(1): 23–30.

[Zook and Frank, 1985] Zook, J.A., Frank, K. (1985). Comparative penetration performance of tungsten alloy l/d= 10 long-rods with different nose shapes fired at rolled homogeneous armor. Rep. BRL-MR-3480. U.S. Army Ballistic Research Laboratory, Aberdeen Proving Ground, MD.

[Zook *et al.*, 1983] Zook, J., Slack, W., Izdebski, B. (1983). Ricochet and penetration of steel spheres impacting aluminum targets. Rep. ARBRL-MR-03243. U.S. Army Armament Research and Development Command, Ballistic Research Laboratory, Aberdeen Proving Ground, MD.

[Zook, 1977] Zook, J. (1977). An analytical model of kinetic energy projectile/fragment penetration. Rep. BRL-MR-2797. U.S. Army Ballistic Research Laboratory, Aberdeen Proving Ground, MD.

Index